The Sodium Pump: Structure, Mechanism, and Regulation

Society of General Physiologists Series • Volume 46

The Sodium Pump: Structure, Mechanism, and Regulation

Society of General Physiologists • 44th Annual Symposium

Edited by
Jack H. Kaplan
and
Paul De Weer
University of Pennsylvania

Marine Biological Laboratory
Woods Hole, Massachusetts

5–9 September 1990

© **The Rockefeller University Press**
New York

Library of Congress Catalog Card Number 91-052921
ISBN 0-87470-048-5
Printed in the United States of America

Contents

Reaction Kinetics and Transport Modes

Preface

Every few years, researchers from many nations gather to present and discuss the latest data and ideas on the workings and structure of the sodium pump. Successive meetings have been held under the auspices of the New York Academy of Sciences (1973), Aarhus University in Denmark (1978), Yale University (1981), Cambridge University in England (1984), and Aarhus University again (1987). The 6th International Sodium Pump Meeting, organized by the editors of this volume, was held in September 1990 at the Marine Biological Laboratory in Woods Hole, Massachusetts, as the 44th Annual Meeting of the Society of General Physiologists. There were nearly 400 participants.

The present volume contains the plenary presentations made at the Woods Hole symposium. The companion volume, entitled *The Sodium Pump: Recent Developments,* contains over one hundred condensed papers presented in poster form during the five-day gathering. Both volumes are divided into thematic sections that follow the organizational format of the meeting itself. We trust that the contents of these volumes convey the sense of excitement and expectation that we as participants enjoyed.

We thank all the participants in the meeting, the director and staff of the Marine Biological Laboratory, Woods Hole, Massachusetts, and the officers of the Society of General Physiologists. The conference would not have been possible without financial support from the Lucille P. Markey Charitable Trust, the National Institutes of Health, the National Science Foundation, and the U.S. Army Medical Research and Development Command. We were particularly pleased to be able to offer two dozen accommodation awards to young investigators from around the world. We are grateful to Jane Leighton and Sue Lahr for their tireless efforts to ensure the smooth running of the meeting, and to Jeanne Hess and her colleagues at the Rockefeller University Press for their patient efficiency in enabling us to bring these volumes to timely publication.

A few days after the meeting we were saddened to learn of the tragic death of Professor Peter Läuger of Konstanz. Professor Läuger had been an active contributor to our field, and his lucid presentations will be greatly missed by his friends and colleagues. This volume is dedicated to the memory of Peter Läuger.

Paul De Weer
Jack H. Kaplan

Molecular Biology and Expression

Chapter 1

Regulation of the α-Subunit Genes of the Na,K-ATPase and Determinants of Cardiac Glycoside Sensitivity

Jerry B. Lingrel, John Orlowski, Elmer M. Price, and Bhavani G. Pathak

Department of Molecular Genetics, Biochemistry and Microbiology, University of Cincinnati College of Medicine, Cincinnati, Ohio 45267-0524

The Na,K-ATPase enzyme is a plasma membrane–bound, oligomeric protein consisting of an α- and β-subunit in a one-to-one stoichiometry (Jorgensen, 1982). The α-subunit is well recognized as mediating the catalytic processes of the enzyme, whereas a clear understanding of the function of the β-subunit remains more elusive. Nonetheless, the accumulating evidence appears to support the hypothesis that the β-subunit facilitates the correct assembly and transport of the α-subunit into the plasma membrane (Geering, 1990; Noguchi et al., 1990). Na,K-ATPase is now known to belong to a multigene family. Three distinct isoforms of the α-subunit (α1, α2, and α3) have been identified by molecular genetic and immunological techniques (Sweadner, 1989; Lingrel et al., 1990). Similarly, two isoforms for the β-subunit (β1 and β2) have been identified (Vasallo et al., 1989; Lingrel et al., 1990) and, more recently, the existence of a third β-subunit isoform (β3) has been reported (Jaunin et al., 1990). The α and β isoform genes are expressed in a tissue- and cell-specific manner, and are subject to developmental and hormonal regulatory influences (for reviews see Gick et al., 1988; Sweadner, 1989; Lingrel et al., 1990). However, the precise molecular signals responsible for mediating alterations in expression of the individual α and β isoform genes in different tissues are not well defined. Furthermore, a detailed understanding of the physiological role played by each subunit isoform remains to be elucidated. The potential for the formation of distinct complexes of each α and β isoform subunit is an area of great interest and may account for the diverse functional properties of the enzyme (De Weer, 1985). To provide further insight into these questions, we have undertaken studies designed to identify the molecular stimuli that regulate Na,K-ATPase gene expression with particular focus on rodent cardiac tissue as well as an analysis of the human α isoform genes and their expression. In addition, we are examining the structure–function interactions of cardiac glycosides with Na,K-ATPase.

Regulation of Na,K-ATPase Gene Expression in Cardiac Tissue

In rat cardiac tissue, the Na,K-ATPase α isoform genes exhibit a complex pattern of expression during development (Orlowski and Lingrel, 1988). The α1 mRNA is the major α isoform transcript (~70–75% of total α mRNA abundance) expressed at all developmental stages (i.e., fetal to adult). In contrast, the α2 and α3 isoform gene transcripts are present in minor abundance (~25–30%, α2 + α3, of total α-subunit mRNA abundance) and exhibit a developmentally regulated transition in expression. The α3 isoform transcript is expressed primarily in fetal and neonatal heart. However, at 2 wk after birth the abundance of α3 isoform mRNA declines to negligible levels and is replaced quantitatively by the α2 isoform. This change persists during subsequent development of the young adult heart up to 55 d of age, the last time point measured (for summary, see Table I). The physiological significance of the reciprocal switch in α2 and α3 isoform gene expression in relation to rat cardiac function is not understood at this time. Interestingly, Herrera and co-workers (Herrera et al., 1988) have reported that the α3 isoform mRNA reappears in heart tissue of older rats, although the cause of this change is unknown.

To understand further the developmental switch of α2 and α3 isoform mRNAs in the heart, we used primary cultures of neonatal rat cardiac myocytes that are maintained in serum-free, chemically defined medium. This in vitro system allows for a direct assessment of the effects of various hormonal modulators on cardiac expression of Na,K-ATPase. The initial studies focused on thyroid hormone (triiodothyronine, T_3) since this agent has been shown to regulate Na,K-ATPase activity in heart (Curfman et al., 1977; Philipson and Edelman, 1977), as well as numerous other tissues (Lo et al., 1976; Lin and Akera, 1978; Haber and Loeb, 1988; Rayson and Edelman, 1988; Schmitt and McDonough, 1988). In addition, the influence of glucocorticoid and an α1-adrenergic agonist were also examined as these substances are known to alter Na,K-ATPase activity in organ systems (Sinha et al., 1981; Rayson and Edelman, 1982; Mernissi and Doucet, 1983; Swann and Steketee, 1989) other than heart.

Control cultures of myocardiocytes that have been maintained for 3 d in the absence of hormones other than insulin express only the α1 isoform mRNA (Orlowski and Lingrel, 1990). The α2 and α3 isoform mRNAs that are present in the freshly excised neonatal hearts (2 d of age) used for preparing the myocardial cell cultures are apparently downregulated during the initial culture period. The presence of both α2 and α3 isoform mRNAs is expected in neonatal hearts as the developmentally regulated transition in their expression is occurring during this time

TABLE I
Summary of the Relative Abundances of the Na,K-ATPase α Isoform mRNAs in Cardiac Tissue during Development and in Cultured Cardiac Myocytes in Response to Hormones

	Cardiac tissue			Cardiac myocytes			
	Fetal	Neonatal	Adult	Control	T_3	DEX	T_3 + DEX
α1	++++++	++++++	++++	++++++	++++++	++++++	++++++
α2	–	+	++	–	++	+	++
α3	+	++	–	–	++	–	–

period in vivo. The addition of T_3 to the culture media results in a several-fold induction in the abundance of the α2 and α3 isoform mRNAs, whereas α1 isoform mRNA levels remain unchanged. While the kinetics of appearance of the α2 and α3 isoforms vary somewhat, both are increased by this hormone. Therefore, T_3 cannot be solely responsible for the switch in α isoform expression. This situation is unlike that observed for the cardiac α- and β-myosin heavy chain isoform genes whereby T_3 functions as the primary modulator for the inverse regulation of these genes in vivo and in vitro (Hoh et al., 1977; Lompre et al., 1984; Gustafson et al., 1987; Orlowski and Lingrel, 1990). The results, therefore, suggest that an additional agent or agents may be responsible for the downregulation of α3 isoform mRNA that occurs during cardiac muscle development in vivo at a time when α2 isoform mRNA is increasing in abundance.

To assess further the requirements for control of myocardial Na,K-ATPase gene expression, the synthetic glucocorticoid dexamethasone (DEX) was added to the cultures alone or in combination with T_3. The addition of DEX to the cultures modestly stimulated α2 isoform mRNA accumulation without influencing the abundance of α1 and α3 isoform mRNAs. Interestingly, treatment of the cultures with both T_3 and DEX caused an increase in α2 isoform transcript levels similar to that observed with T_3 alone. However, the T_3 stimulation of α3 isoform mRNA was not observed, suggesting that DEX specifically represses the induction of α3 isoform mRNA in the presence of T_3 (for summary see Table I). Thus, the pattern of α2 and α3 isoform mRNA expression in response to thyroid plus glucocorticoid hormones in vitro appears to mimic the isoform switch observed in the intact heart in vivo. Treatment of the myocardiocytes with the α1-adrenergic agonist, norepinephrine, which is known to influence a switch in the transcription of the cardiac sarcomeric actin isogenes (Long et al., 1989), did not appear to greatly influence the expression of the Na,K-ATPase α isoform mRNAs. While additional factors may also be involved in the developmental switch, it is tempting to speculate that thyroid and glucocorticoids play an important primary role in this transition of Na,K-ATPase α isoform gene expression. Complementary evidence supporting this view can be derived from studies which show that the plasma concentrations of thyroid (Walker et al., 1980) and glucocorticoid (Henning, 1978) hormones increase during the early neonatal period in a synchronous pattern with the developmental switch in α2 and α3 isoform mRNA levels. However, to confirm an important regulatory role for these two hormones in the control of Na,K-ATPase α2 and α3 isoform gene expression, further experimentation in vivo using thyroidectomized and/or adrenalectomized animals is required. Based on the results from the myocardial cell cultures, one can predict that in hypothyroid adult animals the α1 isoform mRNA would be the major α isoform transcript present, whereas α2 and α3 isoform mRNA abundances would be at negligible levels. Administration of T_3 should augment α2 isoform mRNA levels without stimulating α3 isoform mRNA abundance, assuming that sufficient plasma glucocorticoid levels are still present to suppress α3 isoform mRNA expression. However, in adrenalectomized animals, both α2 and α3 mRNAs would be expressed in the adult heart due to the stimulatory influence of endogenous plasma T_3. Two recent publications (Gick et al., 1990; Horowitz et al., 1990) using hypothyroid animals provide supporting evidence for this view. The results show that in the hypothyroid state α1 isoform mRNA is the predominant α isoform. Administration of T_3 induces primarily the α2 isoform mRNA, and to a lesser extent the α1 isoform, without any effect on α3 isoform mRNA abundance, which remains undetectable. The stimulatory effect of T_3 on α1 isoform mRNA in vivo was not observed in the myocardial cell cultures. This difference may be accounted for by secondary or synergistic interactions of thyroid hormone with other endocrine systems that influence cardiac Na,K-ATPase α1 isoform mRNA abundance. The effect of adrenal ablation on cardiac Na,K-ATPase gene expression still remains to be assessed.

Another issue that remains to be addressed is the differential time course of induction of α2 and α3 isoform mRNA in response to T_3 in cultured myocardiocytes, with α3 isoform mRNA levels increasing more rapidly than α2 isoform mRNA. This in vitro pattern mimics the slower induction of α2 isoform mRNA observed in fetal and neonatal heart relative to that for α3 mRNA. One possible explanation for this is that the α2 and α3 isoform genes may be differentially sensitive to the T_3 receptor. Alternatively, other transcriptional or posttranscriptional regulatory factors in the fetal heart may be regulating α isoform expression. At the present time it is unknown whether the alterations in the abundances of the α isoform mRNAs in vivo and in vitro are controlled at the transcriptional or posttranscriptional levels. This is an area of continuing investigation.

Cloning and Expression of the α-Subunit Isoform Genes

A detailed understanding of the differential expression and regulation of the α-subunit genes will ultimately require analysis of nucleotide sequences involved in expression. To this end, we have isolated the human genes corresponding to the α1, α2, and α3 isoforms and the rat gene encoding that α3 isoform (Shull et al., 1989, 1990; Pathak et al., 1990). The entire human α2 isoform gene has been sequenced, including 1,500 nucleotides of the 5′-flanking region. The 5′-flanking sequences of the human α1 and α3 isoform genes and the rat α3 isoform gene have been determined as well.

An examination of the 5′ end of the human α isoform genes for potential *trans*-acting factor and hormone binding sites has revealed the presence of a number of such sites in each of the genes. A comparison of the 5′-flanking region of the human α1, α2, and α3 isoform genes is shown in Fig. 1. The human α2 and α3 isoform genes each contain more than one transcription initiation site, while the α1 isoform gene contains only one such site. Approximately 20–30 bp upstream of the transcription start sites each of the genes contains a potential TATA box (marked on Fig. 1). This sequence is not identical in the three genes. The human α3 isoform gene also contains a CCAAT box, in reverse orientation, located ~30 bp upstream of the potential TATA box. This sequence is not found in either the α1 or α2 isoform genes. Upstream of the basal promoter elements each of the human α isoform genes contains a number of potential *trans*-acting factor and hormone binding sites (indicated in Fig. 1). These sites do not appear to be conserved in the three genes. This indicates that each of the α isoform genes has its own distinct set of regulatory elements that contribute to its differential regulation. A comparison of the CpG dinucleotides in the 5′-flanking regions of the α isoform genes indicates that the α1 isoform gene has a high number of CpG dinucleotides in this region compared with the α2 and α3 isoform genes. It is known that CpG dinucleotides occur more frequently in the 5′-flanking region of housekeeping genes as compared with genes expressed in a more tissue-restrictive manner. Thus, the number of CpG dinucleotides in the 5′ ends of each of the human α isoform genes correlates well with their pattern of expression in different tissues.

A comparison of the 5′ ends of the human and rat α3 isoform genes is shown in Fig. 2. Analysis of the transcription initiation sites of both the human and rat α3 isoform genes indicates that both genes have multiple initiation sites that are located in the same region of the gene. In addition, both genes initiate transcription predominantly at an adenine, which is in close agreement with that observed for other eukaryotic genes (Breathnach and Chambon, 1981). Since this base is also the 5′-most predominant start site in both genes, we have designated it as +1 and have numbered all 5′ elements relative to this site. A comparison of the nucleotide sequences of the 5′ ends of the human and rat α3 isoform genes shows regions of highly conserved sequences. Upstream of the transcription initiation sites there is a sequence ATAT (located at −27 in rat and −26 in human) which is a potential TATA box. Approximately 30 bases upstream of the potential TATA box there is a CCAAT sequence in the reverse orientation in both genes. Both the human and rat α3 isoform genes contain a number of potential binding sites for *trans*-acting factors and hormone receptors. These are listed in Table II. Some of these binding sites are conserved in both genes and are indicated in Fig. 3. These conserved sites consist of

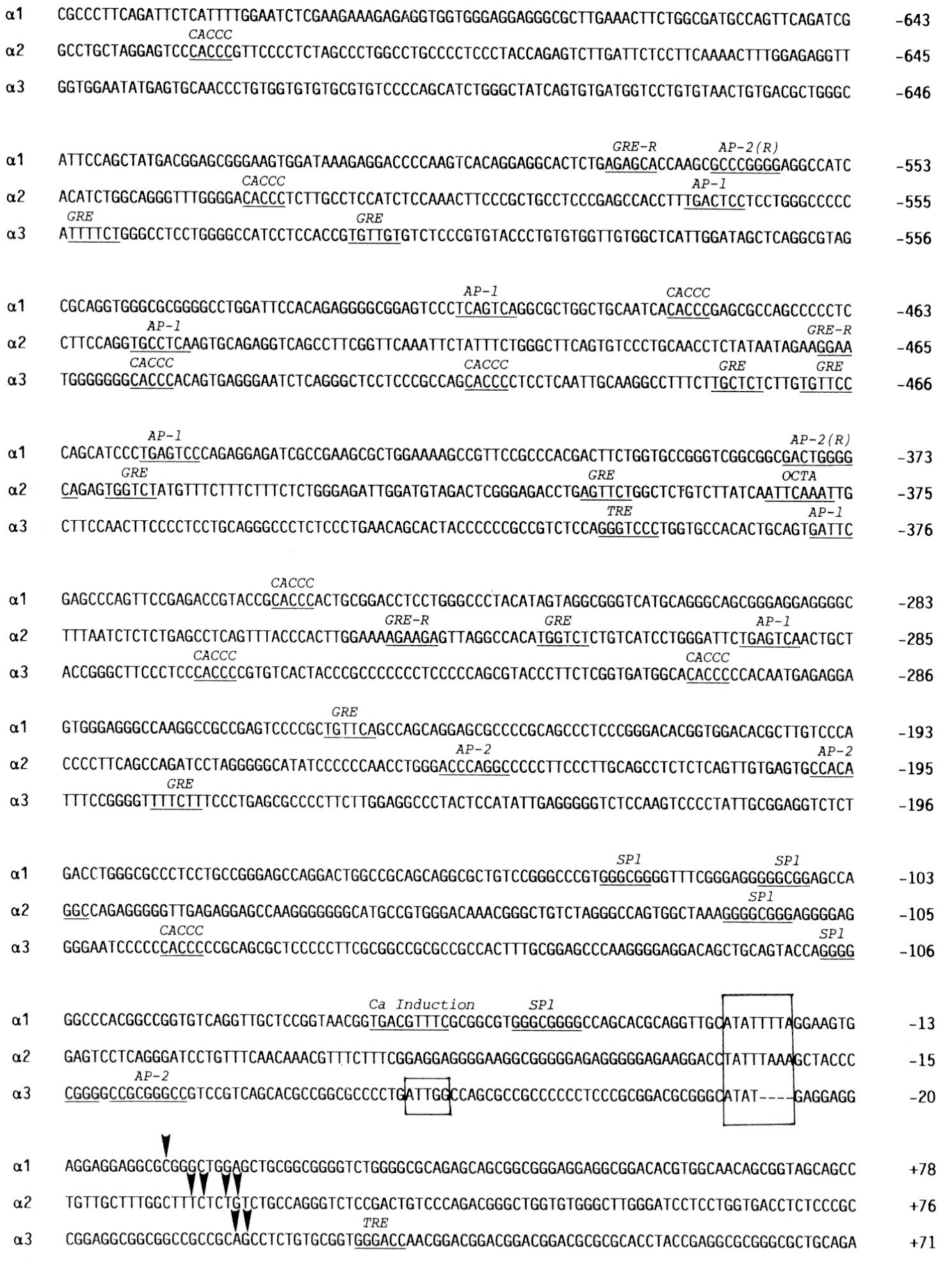

Figure 1. Comparative analysis of the nucleotide sequence of the 5′ ends of the human α1, α2, and α3 isoform genes. The transcription initiation site of each gene is marked by arrows. The numbers on the right refer to nucleotide position relative to the 5′-most transcription initiation site of each of the genes. Potential TATA and CCAAT sequences are boxed. Potential transcription factor and hormone binding sites are underlined and marked.

TABLE II
Potential Transcription Factor and Hormone Binding Sites in the 5′-flanking Sequence of the Human and Rat α3 Isoform Genes

Consensus binding element	Rat α3 gene		Human α3 gene	
CCAAT box (R)				
ATTGG	ATTGG – 681	ATTGG – 60	ATTGG – 571	ATTGG – 62
OCTA (octamer)				
ATTTGCAT	ATTTGCAg – 159		—	
OCTA (heptamer)				
CTCATGA	CTCATGc – 1105	CTCcTGA – 64	—	
AP-1				
TGANT (C/A)A	TcAGTCA – 530		TGATTCA – 375	
AP-2				
CC(C/G)C(A/G)GGC	CCGCGGGC – 36		CCGCGGGC – 93	
(T/C)C(C/G)CC(A/C)N(G/C)(G/C)(G/C)	TCCCCACCCC – 198	CCCCCCGCGG – 36		
	TCCCCCCGCG – 39		—	
SP 1				
(G/T)GGGCGG(G/A)(G/A)(T/C)	TGaGCGGGGT – 239 GGGGCcGGGC – 98		GGGCGGGGC – 100	
GGGCGG	A number of sites with 100% identity		A number of sites with 100% identity	
CCGCCC (R)	A number of sites with 100% identity		CCGCCC – 339	CCGCCC – 50
CACCC box				
GCCACACCC	GtagCACCC – 477	aatcCACCC – 322	GgggCACCC – 543	GgCACACCC – 300
			cCagCACCC – 505	cCCcCACCC – 180
	tCCcCACCC – 223	tCCaCACCC – 199	ctCcCACCC – 356	
GRE				
GGTACANNNTGTTCT	ctctggCTTTGTTgt – 576		cGctggGCATtTTCT – 639	
	tGTctcCTTTGTgCT – 562		cctccACCGTGTTgt – 607	
	tGTcttAGGTGTTCa – 534		GGccttTCTTGcTCT – 476	
	CGcgtACCATcTTCT – 442		GcTctcTTGTGTTCc – 466	
			ttTcCgGGGTtTTCT – 271	
GRE(R)				
AGAACANNNTGTACC	tGAACAGTATcaAaC– 443		—	
	AGtACAGCTgcataC – 120			
TRE				
GGG(A/T)C(C/G)	GGGTCC – 406	GGGACC + 15	GGGTCC – 399	GGGACC + 15
TRE (R)				
(C/G)G(A/T)CCC	GGTCCC – 405	GGTCCC – 227	GGTCCC – 398	GGTCCG + 109
	CGTCCC – 340	GGACCC + 16		
	GGTCCC – 257			

```
H  GTGATGGTCCTGTGTAACTGTGACGCTGGGCATTTTCTGGGCCTCCTGGGGCCATCCTCCACCG-TGTTGTGTCTCCCGTGTACCCTGTG  -588
R  GTGATGGTGGTACCTAACTGTAGTGG-GGGCATCTTCAGAGCCTCCTGGTGTGAGCTTCTGGCTTTGTTGTGTCTCCTTTGTGCTG----  -561
                                                                    GRE

H  TGGTTGTGGCTCATTGGATAGCTCAGGCGTAGTGGGGGGGCACCCACAGTGAGGGAATCTCAGGGCTCCTCCCGCC-----AGCACCCCT  -503
R  ----TGTGGGGTGGTATGTCTTAGGTGTTCAGTCATGGGGACACAGTGCAGAGGATCTTAGGAGCCCCCTCCCATCTCTGTAGCACCCCT  -475
                                                                                          CACCC

H  CCTCAATTGCAAGGCCTTTCTTGCTCTCTTGTGTTCCCTTCCAACTTCCCCTCCTGCAGGGCCCTCTCCCTGAACAGCACTACCCCCCGC  -413
R  CCACCAGTCTCCCCATCACGCGTACCATCTTCTTCCTGAACAGTATCAAACACAGATTCCTGGGGTCCCCTCTACATACTGCAAGATGGT  -385

H  CGTCTCCAGGGTCCCTGGTGCCACACTGCAGTGATTCACCGGGCTTCCCTCCCACCCCGTGTCACTACCCGCCCCCCCTCCCCCAGCGTA  -323
R  GGTACTCTGGGGCCCTGCATT---ACTGCAGTTCCTTGAGGCG--TCCCTCCTACCCTGAATCCACCCCACT------------------  -318

H  CCCTTCTCGGTGATGGCACACCCCCACAATGAGAGGATTTCCGGGGTTTTCTTTCCCTGAGCGCCCCTTCTTGGAGG--CCCTACTCCAT  -235
R  ---------------------CCCAGAAGGAGGTAGATTCTGGGGTGCTTCCTCCCTCAAAGCCTCAGTCCCTAAGGTCCCTCCCTACA  -250

H  ATTGAGGGGGTCTCCAAGTCCCCTATTGCGGAGGTCTCTGGGAATCCCCCCACCCCCGCAGCGCTCCCCCTTCGCGGCCGCGCCGCCACT  -145
R  TTGAGCGGGGTCTCTAGGTCCCCACCCTGGGGGTGTCCGGG----ATCCCCACCCCCCACTGCGCTTAGCTTCTCGGTGGCGCCGCCA-T  -165
                                                  AP-2
                                                   CACCC

H  TTGCGG-AGCCCAA---------------------GGGGAGGACAGCTGCAGTACCAGGGGCGGGGCCGCGGGCCGTCCGTCAGCACGCC  -77
R  TTGCAGAAACAAGGTTGGAGCGGTGAAGGGGGAAGGGGGAGTACAGCTGCAGTACCGGGGGCCGGGCCTCTAGCTGTCCGTCTGCTCAGG  -75
                                          GRE (R)              SP-1

H  GGCGCCCCTGATTGGCCAGCGCCGCCCCCCTCCCGCGGACGCGGGCATATGAGGAGGCGGAGGCGGCGGCCGCCGCAGCCTCTGTGCGGT  +14
R  GCCGCTCCTGATTGGCCAGAGCCGCCTCCC-CCCGCGGGCGCGGGCATATGAGGAGGCGGAGGCC-CGGCCGCCGCAGCCTCTGTGCCGT  +14
                          SP-1 (R)

H  GGGACCAACGGACGGACGGACGGACGCGCGCACCTACCGAGGCGCGGGCGCTGCAGAGGCTCCCAGCCCAAGCCTGAGCCTGAGCCCGCC  +104
R  GGGACC--C--ACGGACCGACAGACGCACGCTCCCAACGCGGCGCGGGCGCTGCAGAGGCCCCCAGCCCGAGCCCGCGCCTGAGCCCGTC  +100
    TRE

H  CCGAGGTCCCCGC-C-CC-GCCCGCCTGGCTCTCTCGCCGCGGAGCCGCCAAGATGGGG/gtaggt...Intron 1...    +160
R  CTGCGGCCACCGCTCACCAGCCTGTCCGCCGCTCTTCCCGCGGAGCCGCCAAGATGGGG/gtaggt...Intron 1...    +159
                                                    MetGly
```

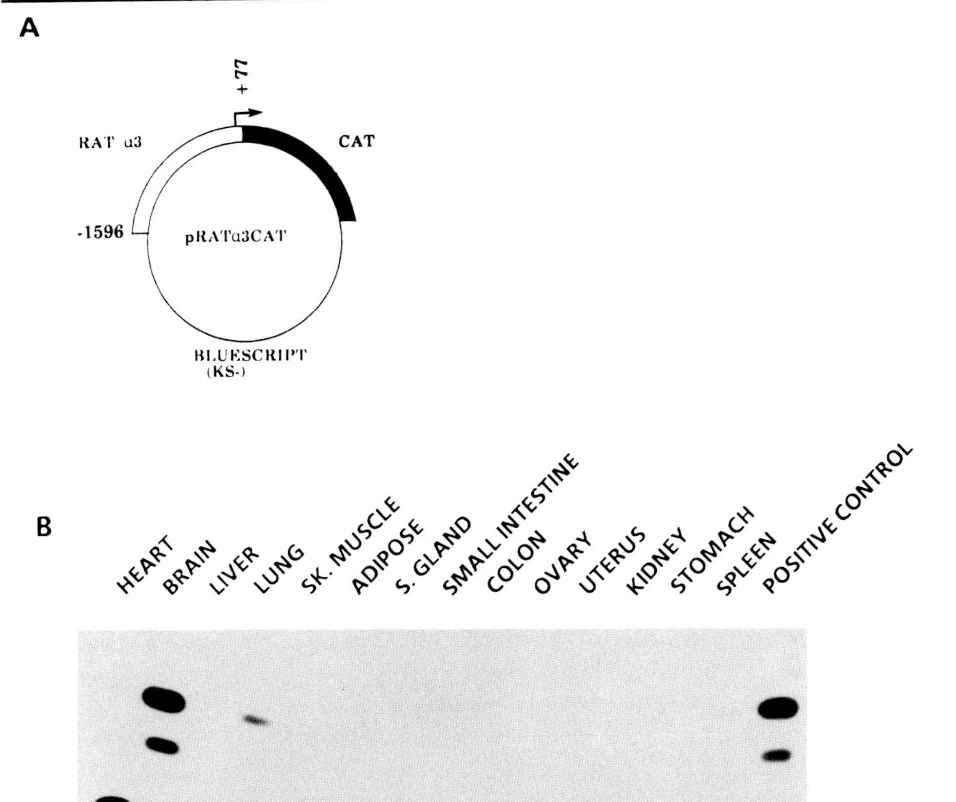

Figure 3. (*A*) Plasmid construct pRATα3 CAT. Approximately 1.6 kb of the 5′-flanking sequence of the rat α isoform gene was fused to a reporter gene, the CAT gene CAT in the plasmid vector bluescript (KS−). (*B*) CAT assay in transgenic mouse tissues. Various tissues from a transgenic mouse were homogenized in a polytron homogenizer and heat inactivated for 5 min at 65°C. Tissue extract containing 50 μg of total protein after heat inactivation was incubated with ^{14}C-labeled chloramphenicol and 4 mM acetyl CoA at 37°C for 3 h. The reaction mixture was extracted with ethyl acetate and the products were resolved by thin layer chromatography.

binding elements for transcription factors SP1 and AP-2 and also for the glucocorticoid (GRE) and thyroid hormone (TRE) receptors. There is a conserved CACCC box in both genes (located at −199 in rat and −180 in human). CACCC regions can bind a protein factor that interacts with the glucocorticoid receptor (Dierks et al., 1983; Shüle et al., 1988). This CACCC box in the α3 isoform gene is located ~70 bp

Figure 2. Nucleotide sequence of the 5′ end of human and rat α3 isoform genes. The upper sequence is that of the human gene (*H*) and the lower sequence is the rat gene (*R*). The colons indicate nucleotides that are identical in both sequences. Transcription initiation sites are indicated by arrows. Potential TATA and CCAAT sequences are boxed. Potential transcription factor and hormone binding sites that are conserved in both genes are underlined.

from a potential conserved GRE and could interact with it. Therefore, while the human α1, α2, and α3 isoform genes have diverged extensively, the α3 isoform genes, at least in rat and human, are reasonably conserved. This type of comparison helps identify sequences that may be important in expression and regulation of the gene by virtue of their conservation among species that have diverged over a reasonable time.

The availability of 5′ sequences makes it possible to test whether this region contains information required for regulated expression of the various isoform genes. Regulatory sequences often occur in this region but sometimes are present in other regions of genes (e.g., introns). Several approaches are available for studying regulated expression. One is to transfect the entire gene into appropriate cell lines and follow expression. Another is to prepare transgenic mice carrying the construct gene. Since the α isoform genes are fairly large in length it is difficult to design and manipulate constructs containing the entire gene. Another approach to studying gene regulation is to fuse the 5′ end of the gene, which in many cases is found to contain most or all of the regulatory signals, to a reporter gene such as the bacterial chloramphenicol acetyltransferase (CAT) gene. This chimeric gene construct can then be introduced either into cells or into transgenic mice. The presence of CAT is an index of transcription. We are using both approaches but have had reasonable success with transgenic mice, at least with respect to the α3 isoform gene.

The 5′-flanking sequence of the rat α3 isoform gene, encompassing nucleotides −1,596 to +77, has been attached to the CAT gene and the chimeric transgene has been injected into the fertilized eggs of mice (Fig. 3 *A*). The injected eggs were reimplanted into foster mothers and the resulting offspring were assayed for the presence of injected DNA in the genome. Five founder lines were identified. Each of these lines contains the transgene integrated at different locations in the genome. These founders were then crossed with nontransgenic mice and the resultant heterozygotes were used for assaying the expression of the transgene. CAT activity was determined in various tissues of the transgenic mice. CAT expression in tissue extracts of a mouse from one of these lines is seen in Fig. 3 *B*. The transgene is expressed at high levels in brain, as expected for the α3 isoform gene, but not in other tissues with the exception of lung. We have studied expression of the transgene in mice from a total of three different lines. In all cases it appears that the transgene is expressed primarily in the brain. Different lines have minor levels of CAT expression in a few other tissues as well. This is probably due to some chromosomal effect depending on the site of insertion of the transgene. It therefore appears that 1.6 kb of the 5′-flanking sequence of the rat α3 isoform is sufficient for tissue-specific expression. It is now important to determine whether the sequences for developmental expression are also present, i.e., whether the α3 isoform is expressed in fetal brain and fetal and neonatal heart as is the endogenous gene. With this system it should be possible to delete sequences from the 5′ end and determine which regions are responsible for the tissue-specific expression of the α3 isoform gene.

Determinants of Ouabain Sensitivity

Cardiac glycosides are known to interact with the Na,K-ATPase and inhibit its activity. This is the basis for the therapeutic use of the digitalis class of drugs to treat congestive heart failure. Ouabain, a cardiac glycoside often used in the laboratory, has long been used as a marker for selecting cells in culture. For example, the α-subunit of the murine Na,K-ATPase confers resistance to sensitive cells such as HeLa, which have a sensitive enzyme. There is a 1,000-fold difference in sensitivity between the rodent and human enzymes, with the human being more sensitive. Ouabain can thus be used in selecting somatic cell hybrids resulting from the fusion of murine and human cells.

It has previously been shown that the naturally occurring resistance of mouse and rat cells to ouabain resides in the α-subunit (Fallows et al., 1987). Advantage has been taken of this observation to confer resistance to sensitive cells via transformation. The mouse α1 isoform cDNA carried in an expression vector has been used to

TABLE III
Effect of H1–H2 Border Residues on Ouabain Sensitivity

Mutations in the H1–H2 extracellular domain	Able to confer resistance to ouabain	I_{50}
		M
Wild-type sequence		
Gln-Ala-Ala-Thr-Glu-Glu-Glu-Pro-Gln-Asn-Asp-Asn	No	8×10^{-8}
Mutant		
Arg-Ala-Ala-Thr-Glu-Glu-Glu-Pro-Gln-Asn-Asp-Asp	Yes	1×10^{-4}
Arg-Ala-Ala-Thy-Glu-Glu-Glu-Pro-Gln-Asn-Asp-Asn	Yes	1×10^{-6}
Gln-Ala-Ala-Thr-Glu-Glu-Flu-Pro-Gln-Asn-Asp-Asp	Yes	1×10^{-6}
Asp-Ala-Ala-Thr-Glu-Glu-Glu-Pro-Gln-Asn-Asp-Arg	Yes	$\sim 4 \times 10^{-3}$
Lys-Ala-Ala-Thr-Glu-Glu-Glu-Pro-Gln-Asn-Asp-Glu	Yes	5×10^{-6}
Glu-Ala-Ala-Thr-Glu-Glu-Glu-Pro-Gln-Asn-Asp-Glu	Yes	8×10^{-6}
Lys-Ala-Ala-Thr-Glu-Glu-Glu-Pro-Gln-Asn-Asp-Lys	Yes	4×10^{-4}
Ala-Ala-Ala-Thr-Glu-Glu-Glu-Pro-Gln-Asn-Asp-Ala	No	—
Gln-Ala-Ala-Thr-Glu-Glu-Glu-Pro-Gln-Asn-Asp-Ala	No	—

transfer resistance to sensitive monkey CV-1 cells (Kent et al., 1987; Emanuel et al., 1988), and the rat α1 cDNA has been used to confer resistance on human HeLa cells (Price and Lingrel, 1988). We have previously taken advantage of this system to determine the basis for the difference in sensitivity between the rat and sheep α-subunits (Price and Lingrel, 1988). Two amino acids at the border of the first extracellular region are responsible for this difference in sensitivity. The rat α1-subunit contains an arginine and an aspartate, respectively, at the borders of the first extracellular region, while the sensitive sheep enzyme contains glutamine and asparagine at these two positions, respectively. When the two border amino acids in sheep are converted to those found in rat, a 1,000-fold increase in resistance to ouabain is observed. No other combination of amino acids present in rat within this region provides this level of resistance.

Based on these findings, multiple changes in amino acids at the border residues have been constructed and assayed. The results of these studies are shown in Table

TABLE IV
Effect of the Conserved Aspartic Acid Residue in the First Extracellular Region of the Sheep Na,K-ATPase α-Subunit on Ouabain Sensitivity

Mutations in the H1–H2 extracellular domain	Able to confer resistance to ouabain	I_{50}
		M
Wild-type sequence		
Gln-Ala-Ala-Thr-Glu-Glu-Glu-Pro-Gln-Asn-Asp-Asn	No	8×10^{-8}
Mutant		
Gln-Ala-Ala-Thr-Glu-Glu-Glu-Pro-Gln-Asn-Asn-Asn	Yes	5×10^{-6}
Gln-Ala-Ala-Thr-Glu-Glu-Glu-Pro-Gln-Asn-Glu-Asn	Yes	5×10^{-6}
Gln-Ala-Ala-Thr-Glu-Glu-Glu-Pro-Gln-Asn-Lys-Asn	No?	—
Gln-Ala-Ala-Thr-Glu-Glu-Glu-Pro-Gln-Asn-Ala-Asn	Yes	3×10^{-6}
Gln-Ala-Ala-Thr-Glu-Glu-Glu-Pro-Gln-Asn-Ser-Asn	Yes	5×10^{-6}

III. The sequence of amino acids of the first extracellular domain of the sensitive sheep α1-subunit is indicated as the wild-type sequence. This α1 isoform of sheep Na,K-ATPase cannot confer resistance to the sensitive HeLa cells and has an I_{50} of 8×10^{-8} M. The various mutations that have been introduced into the border residues are shown below the wild-type sequence. As described above, expression of the sheep α1-subunit with the Gln → Arg and Asn → Asp substitutions, which mimic the rat α1 isoform, confers ouabain resistance to HeLa cells. The enzyme with the mutated sheep α1-subunit has an I_{50} of 1×10^{-4} M for ouabain, similar to the rat enzyme. The positions of the amino acid replacements are thought to reside at the membrane borders of the extracellular region. Several additional amino acid substitutions at these positions have been made and analyzed (Price et al., 1990). Replacement of uncharged amino acids as represented by the inclusion of alanine at the border residues in either one or both positions do not produce a resistant enzyme. Thus, resistance appears to depend on charged residues at these border

TABLE V
Effect of Amino Acid Substitutions of the Second Extracellular Region on Ouabain Sensitivity

Mutations in the H1–H2 extracellular domain
Wild-type sequence
Glu-Tyr-Thr-Trp-Leu-Glu
Mutant
Glu-Tyr-Thr-___-Leu-Glu
Glu-Tyr-Thr-Phe-Leu-Glu
Gly-Tyr-Thr-Trp-Leu-Glu
Gly-Tyr-Thr-Phe-Leu-Glu
Glu-Tyr-Thr-Pro-Leu-Glu
Glu-Tyr-Cys-Trp-Leu-Glu
Glu-Tyr-Thr-Trp-Leu-Gln

regions. This is confirmed by the finding that all combinations of charged amino acids confer resistance to HeLa cells, although to varying degrees. If only the Gln is substituted with Arg or only the Asn with Asp, the resistance is in between that of the rat and sheep enzymes, an I_{50} of 1×10^{-6} M ouabain. However, other double substitutions with charged amino acids confer resistance similar to the rat. With the Gln → Asp and Asn → Asp substitutions, a highly resistant enzyme is observed, an $I_{50} > 4 \times 10^{-3}$ M ouabain. This is even more resistant than the naturally occurring rat enzyme. Other combinations of charges give somewhat less resistance but nevertheless have a pronounced effect. It does not appear that the combination of charges is particularly important. For example, a negative charge on the NH_2-terminal border and a positive charge on the COOH-terminal border of this extracellular domain or substitutions with two positive or two negative amino acids still confer resistance.

While it is not known how charges at the border residues contribute to ouabain sensitivity, it is hypothesized that the charged amino acids prevent the interaction of this extracellular region with the membrane. Charges may exclude its movement or close association with the membrane, therefore weakening the interaction between ouabain and the enzyme.

Other amino acid substitutions within this region also affect ouabain sensitivity (Price et al., 1989). For example, if the Asp next to the Asn on the COOH-terminal end of this extracellular domain is altered, resistance is also observed (Table IV). In this case a substitution of the Asp for Glu, Ala, or Ser results in resistance. Other amino acids within this extracellular region have not been adequately tested but it is possible that other positions are also involved in interacting with the drug.

The second extracellular domain H3-H4, while not responsible for conferring differential resistance among species, could nevertheless be involved in binding cardiac glycosides. To determine if this is the case, mutations have been introduced into this region of the sheep α1 isoform and the resulting constructs have been introduced into HeLa cells. The amino acid substitutions are described in Table V. No resistant colonies were observed. There are at least two explanations for such a finding. First, the region is not involved in ouabain resistance and therefore amino acid substitutions in this sequence do not alter ouabain sensitivity. The second possibility is that this region is critical for enzymatic activity and substitutions within it inactivate the enzyme. Under this latter circumstance, it would be impossible to observe resistant colonies as the enzyme would be nonfunctional. To determine which alternative is correct, the amino acid substitutions made in this region were introduced into the sheep α1 isoform cDNA which had been mutated to give substitutions conferring ouabain resistance. These are the Gln → Arg and Asn → Asp substitutions at the borders of the first extracellular region. If the amino acid substitutions in the second extracellular region inactivate the enzyme, no resistance colonies should be observed when this construct is transfected into sensitive HeLa cells. This would suggest that these residues provide an important function for the enzyme. When the constructs were introduced into HeLa cells, resistant colonies were observed. Therefore, it is likely that this region is not involved in binding ouabain, and in addition, these amino acid substitutions do not significantly affect enzymatic activity. This plus/minus selection system represents a powerful tool for examining other regions of the enzyme with respect to ouabain sensitivity. Such studies are being pursued.

Acknowledgments

This work was supported by NIH grants R01 HL-28573 and P01 HL-22619. John Orlowski is a Postdoctoral Fellow, Medical Research Council of Canada, and Elmer M. Price is a recipient of NIH postdoctoral fellowship HL-07806.

References

Breathnach, R., and P. Chambon. 1981. Organization and expression of eucaryotic split genes coding for proteins. *Annual Review of Biochemistry.* 50:349–383.

Curfman, G. D., T. J. Crowley, and T. W. Smith. 1977. Thyroid-induced alterations in myocardial sodium- and potassium-activated adenosine triphosphatase, monovalent cation active transport, and cardiac glycoside binding. *Journal of Clinical Investigation.* 59:586–590.

De Weer, P. 1985. Cellular sodium-potassium transport. *In* The Kidney: Physiology and Pathophysiology. Vol. 1. D. W. Seldin and G. Giebisch, editors. Raven Press, New York. 31–42.

Dierks, P., A. Van Ooyen, M. D. Cochran, D. Dobkin, J. Reiser, and C. Weissmann. 1983. Three regions upstream from the cap site are required for efficient and accurate transcription of the rabbit β-globin gene in mouse 3T6 cells. *Cell.* 32:695–706.

Emanuel, J. R., J. Schulz, X. M. Zhou, R. B. Kent, D. Housman, L. Cantley, and R. Levenson. 1988. Expression of a ouabain-resistant Na,K-ATPase in CV-1 cells after transfection with a cDNA encoding the rat Na,K-ATPase α1 subunit. *Journal of Biological Chemistry.* 263:7726–7733.

Fallows, D., R. B. Kent, D. L. Nelson, J. R. Emanuel, R. Levenson, and D. E. Housman. 1987. Chromosome-mediated transfer of the murine Na,K-ATPase α subunit confers ouabain resistance. *Molecular and Cellular Biology.* 7:2985–2987.

Geering, K. 1990. Topical review: subunit assembly and functional maturation of Na,K-ATPase. *Journal of Membrane Biology.* 115:109–121.

Gick, G. G., F. Ismail-Beigi, and I. S. Edelman. 1988. Hormonal regulation of Na,K-ATPase. *Progress in Clinical and Biological Research.* 268B:277–295.

Gick, G. G., J. Melikian, and F. Ismail-Beigi. 1990. Thyroidal enhancement of rat myocardial Na,K-ATPase: preferential expression of α2 activity and mRNA abundance. *Journal of Membrane Biology.* 115:273–282.

Gustafson, T. A., J. J. Bahl, B. E. Markham, W. R. Roeske, and E. Morkin. 1987. Hormonal regulation of myosin heavy chain and α-actin gene expression in cultured fetal rat heart myocytes. *Journal of Biological Chemistry.* 262:13316–13322.

Haber, R. S., and J. N. Loeb. 1988. Selective induction of high-ouabain-affinity isoform of Na,K-ATPase by thyroid hormone. *American Physiological Society.* 255:E912–E919.

Henning, S. L. 1978. Plasma concentration of total and free corticosterone during development in the rat. *American Journal of Physiology.* 235:E451–E456.

Herrera, V. L., A. V. Chobanian, and N. Ruiz-Opazo. 1988. Isoform-specific modulation of Na,K-ATPase α subunit gene expression in hypertension. *Science.* 241:221–223.

Hoh, J. F. Y., P. A. McGrath, and P. T. Hale. 1977. Electrophoretic analysis of multiple forms of rat cardiac myosin: effects of hypophysectomy and thyroxine replacement. *Journal of Molecular and Cellular Cardiology.* 10:1053–1076.

Horowitz, B., C. B. Hensley, M. Quintero, K. K. Azuma, D. Putnam, and A. A. McDonough. 1990. Differential regulation of Na,K-ATPase α1, α2, and β subunit mRNA and protein levels by thyroid hormone. *Journal of Biological Chemistry.* 265:14308–14314.

Jaunin, P., K. Richter, I. Corthesy, and K. Geering. 1990. Posttranslational processing and basic properties of a putative β3-subunit of Na,K-ATPase. *Journal of General Physiology.* 96:60*a*. (Abstr.)

Jørgensen, P. L. 1982. Mechanism of the Na,K-pump. *Biochimica et Biophysica Acta.* 694:26–68.

Kent, R. B., J. R. Emanuel, Y. B. Neriah, R. Levenson, and D. E. Housman. 1987. Ouabain resistance conferred by expression of the cDNA for a murine Na,K-ATPase α subunit. *Science.* 237:901–903.

Lin, M. H., T. Akera. 1978. Increased Na,K-ATPase concentrations in various tissues of rats caused by thyroid hormone treatment. *Journal of Biological Chemistry.* 253:723–726.

Lingrel, J. B., J. Orlowski, M. M. Shull, and E. M. Price. 1990. Molecular genetics of Na,K-ATPase. *Progress in Nucleic Acid Research and Molecular Biology.* 38:37–89.

Lo, C.-S., T. R. August, U. A. Liberman, and I. S. Edelman. 1976. Dependence of renal Na,K-adenosine triphosphatase activity on thyroid status. *Journal of Biological Chemistry.* 251:7826–7833.

Lompre, A.-M., B. Nadal-Ginard, and V. Mahdavi. 1984. Expression of the cardiac ventricular α- and β-myosin heavy chain genes is developmentally and hormonally regulated. *Journal of Biological Chemistry.* 259:6437–6446.

Long, C. S., C. P. Ordahl, and P. C. Simpson. 1989. α1-adrenergic receptor stimulation of sarcomeric actin isogene transcription in hypertrophy of cultured rat heart muscle cells. *Journal of Clinical Investigation.* 83:1078–1082.

Mernissi, G. E., and A. Doucet. 1983. Short-term effects of aldosterone and dexamethasone on Na,K-ATPase along the rabbit nephron. *Pflügers Archiv.* 388:147–151.

Noguchi, S., K. Higashi, and M. Kawamura. 1990. A possible role of the β-subunit of Na,K-ATPase in facilitating correct assembly of the α-subunit into the membrane. *Journal of Biological Chemistry.* 265:15991–15995.

Orlowski, J., and J. B. Lingrel. 1988. Tissue-specific and developmental regulation of rat Na,K-ATPase catalytic α isoform and β subunit mRNAs. *Journal of Biological Chemistry.* 263:19436–19442.

Orlowski, J., and J. B. Lingrel. 1990. Thyroid and glucocorticoid hormones regulate the expression of multiple Na,K-ATPase genes in cultured neonatal rat cardiac myocytes. *Journal of Biological Chemistry.* 265:3462–3470.

Pathak, B. G., D. G. Pugh, and J. B. Lingrel. 1990. Characterization of the 5′-flanking region of the human and rat Na,K-ATPase α3 gene. *Genomics.* 8:641–647.

Philipson, K. D., and I. S. Edelman. 1977. Thyroid hormone control of Na,K-ATPase and K^+-dependent phosphatase in rat heart. *American Journal of Physiology.* 232:C196–C201.

Price, E. M., and J. B. Lingrel. 1988. Structure-function relationships in the Na,K-ATPase α subunit: site-directed mutagenesis of glutamine-111 to arginine and asparagine-122 to aspartic acid generates a ouabain-resistant enzyme. *Biochemistry.* 27:8400–8408.

Price, E. M., D. A. Rice, and J. B. Lingrel. 1989. Site-directed mutagenesis of a conserved, extracellular aspartic acid residue affects the ouabain sensitivity of sheep Na,K-ATPase. *Journal of Biological Chemistry.* 264:21902–21906.

Price, E. M., D. A. Rice, and J. B. Lingrel. 1990. Structure-function studies of Na,K-ATPase. *Journal of Biological Chemistry.* 265:6638–6641.

Rayson, B. M., and I. S. Edelman. 1982. Glucocorticoid stimulation of Na,K-ATPase in superfused distal segments of kidney tubules *in vitro. American Journal of Physiology.* 243:F463–F470.

Schmitt, C. A., and A. A. McDonough. 1988. Thyroid hormone regulates α and α+ isoforms of Na,K-ATPase during development in neonatal rat brain. *Journal of Biological Chemistry.* 263:17643–17649.

Shüle, R., M. Muller, H. Ostuka-Murakauii, and K. Renkawitz. 1988. Cooperativity of the glucocorticoid receptor and the CACCC box binding factor. *Nature.* 332:87–90.

Shull, M. M., D. G. Pugh, and J. B. Lingrel. 1989. Characterization of the human Na,K-ATPase α2 gene and identification of intragenic restriction fragment length polymorphisms. *Journal of Biological Chemistry.* 264:17532–17543.

Shull, M. M., D. G. Pugh, and J. B. Lingrel. 1990. The human Na, K-ATPase α1 gene: characterization of the 5′-flanking region and identification of a restriction fragment length polymorphism. *Genomics.* 6:451–460.

Sinha, S. K., H. J. Rodriguez, W. C. Hogan, and S. Klahr. 1981. Mechanisms of activation of renal Na,K-ATPase in the rat: effects of acute and chronic administration of dexamethasone. *Biochimica et Biophysica Acta.* 641:20–35.

Swann, A. C., and J. D. Steketee. 1989. Subacute noradrenergic agonist infusions *in vivo* increase Na,K-ATPase and ouabain binding in rat cerebral cortex. *Journal of Neurochemistry.* 52:1598–1604.

Sweadner, K. J. 1989. Isozymes of the Na,K-ATPase. *Biochimica et Biophysica Acta.* 988:185–220.

Vasallo, P. M., W. Dackowski, J. R. Emanuel, and R. Levenson. 1989. Identification of a putative isoform of the Na,K-ATPase beta subunit: primary structure and tissue specific expression. *Journal of Biological Chemistry.* 264:4613–4618.

Walker, P., J. D. Dubois, and J. H. Dussault. 1980. Free thyroid hormone concentrations during postnatal development in the rat. *Pediatric Research.* 14:247–249.

Chapter 2

A Cell Biologist's Perspective on Sites of Na,K-ATPase Regulation

Douglas M. Fambrough, Barry A. Wolitzky, Joseph P. Taormino, Michael M. Tamkun, Kunio Takeyasu, Delores Somerville, Karen J. Renaud, M. Victor Lemas, Richard M. Lebovitz, Bruce C. Kone, Maura Hamrick, Jayson Rome, Elisabeth M. Inman, and Andrew Barnstein

Department of Biology, The Johns Hopkins University, Baltimore, Maryland 21218; Department of Molecular Genetics, Hoffmann-LaRoche Inc., Nutley, New Jersey 07110; Department of Physiology, Vanderbilt University School of Medicine, Nashville, Tennessee 37232; and Department of Physiology, The University of Virginia School of Medicine, Charlottesville, Virginia 22901

There is a natural sort of duality you experience in studying regulation of the Na,K-ATPase. On the one hand, you are motivated to learn as much as possible about this intriguing molecule. On the other hand, you find an irresistible temptation to use this molecule as a point of reference, a focus in learning as much as you can about how cells organize and regulate themselves and influence each other. This chapter is primarily an overview of regulatory mechanisms for the Na,K-ATPase, illustrating, in a somewhat personalized way, this duality.

Myogenesis and Upregulation in Tissue Culture

Let us begin with the process of myogenesis in tissue culture. As myoblasts differentiate into skeletal muscle fibers, the packing density of Na,K-ATPase molecules in the plasma membrane increases ~20-fold. The cells continue to express the α1 and β1 isoforms of the subunits, and the increase probably reflects action of muscle-specific enhancers of α1 and β1 gene expression. The steady-state level of Na,K-ATPase is soon reached in tissue culture and lasts several days. The increase in Na,K-ATPase can be visualized by labeling the myogenic cultures with a fluorescent monoclonal antibody that binds to a β1-subunit epitope on the exterior surface of the plasma membrane (Fig. 1).

The differentiating myotubes also begin to express voltage-gated sodium channels. Opening these voltage-gated sodium channels with veratridine leads to a 50–100% upregulation of the Na,K-ATPase (Fig. 2) that is dose dependent and reversible (Wolitzky and Fambrough, 1986; Taormino and Fambrough, 1990). This upregulation was accounted for quantitatively by an increased rate of biosynthesis of α–β-subunit complexes. The biosynthetic rate was estimated by pulse-labeling myotubes with [^{35}S]methionine and measuring the incorporation of radioactivity into Na,K-ATPase molecules that were isolated by immune precipitation with an anti–β-subunit monoclonal antibody. As shown in Fig. 3, when the immune precipitates were analyzed by SDS-PAGE and fluorography, it was apparent that this measure of biosynthesis was predominantly a measure of [^{35}S]methionine incorporation into α-subunits that were assembled with the immune-precipitated β-subunits. While the doubling of incorporation of [^{35}S]methionine into the α-subunit was apparent, another feature of upregulation revealed by SDS-PAGE analysis was the greatly increased labeling of the high-mannose intermediate form of the β-subunit (seen in lanes *2* and *3* of Fig. 3). In pulse–chase experiments this high-mannose form of the β-subunit disappeared rapidly. The simplest explanation of these observations is that sodium ion–induced upregulation results in a selective, marked increase in β-subunit biosynthesis; the excess β-subunits drive the assembly of more α–β complexes. β-Subunits that fail to assemble with α-subunits are nevertheless glycosylated, but then are rapidly degraded without leaving the endoplasmic reticulum. The most important primary change during upregulation appeared to be an increased rate of β-subunit biosynthesis.

At the nucleic acid level (Fig. 4), upregulation of the Na,K-ATPase involves a selective increase in transcription of the β1 gene, resulting in a tripling of β1-mRNA. This change in β1-mRNA abundance appears to be sufficient, without any change in translation rate, to account for the accelerated biosynthesis of β1-subunits. A small increase in α1-mRNA occurs quite late in the upregulation response and is seemingly of little consequence. Taken together, the nucleic acid and protein level observations

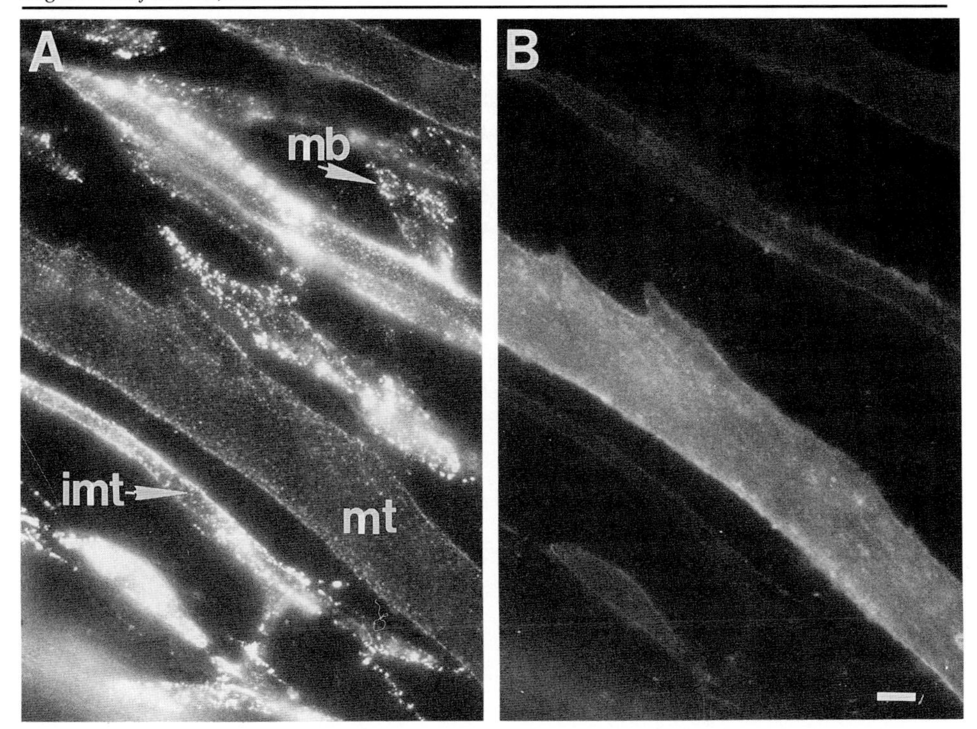

Figure 1. Fluorescence micrographs of tissue-cultured myogenic cells, showing decline in myoblast antigen (*A*) during myogenesis, while the Na,K-ATPase level (*B*) increases during myogenesis. The same field is shown in *A* and *B*. *A* is labeled with monoclonal antibody C3/1 to an early myogenic antigen (Wakshull et al., 1983) followed by rhodamine-labeled anti–mouse IgG. *B* is labeled with fluorescein-conjugated monoclonal antibody 24 to the β-subunit of the Na,K-ATPase. *mb,* myoblast; *imt,* immature myotube; *mt,* well-differentiated myotube. Bar in *B* is 20 μm.

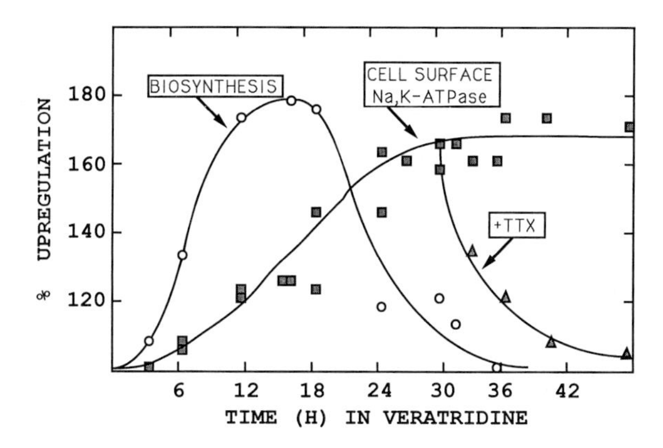

Figure 2. Kinetics of biosynthesis and upregulation of the Na,K-ATPase during stimulation of myotubes with veratridine. Primary chick skeletal muscle cultures were exposed continuously to veratridine, beginning on day 5. Number of Na,K-ATPase molecules (■) was quantified by measuring the binding of ^{125}I-labeled monoclonal antibody that recognizes an externally disposed epitope of the β-subunit of the Na,K-ATPase. The biosynthetic rate of the Na,K-ATPase was approximated by pulse-labeling myogenic cultures with [^{35}S]methionine and then quantifying the radioactivity incorporated into Na,K-ATPase molecules isolated by immunoprecipitation with an anti–β-subunit monoclonal antibody (○). In some cultures, veratridine was removed and the sodium channels were blocked with tetrodotoxin, causing a relatively rapid return to the basal state (▲). Redrawn from Wolitzky and Fambrough (1986).

suggest that during upregulation the production of Na,K-ATPase molecules is driven by accelerated β1-subunit biosynthesis.

Perhaps transcriptional activity on the β1 gene is directly responsive to elevated cytosolic sodium ion concentration. This idea is supported by preliminary tests by Takeyasu and co-workers and by Maura Hamrick in our lab. The avian β1 gene has been stably transfected into mouse myogenic cells and mouse fibroblasts, and the cells were subsequently exposed to veratridine or low potassium, or to intermediate doses of ouabain. All of these procedures, which are expected to elevate cytosolic sodium ion concentration, resulted in increases in avian β1-subunits on the mouse cell surface. Through this bioassay it may be possible to identify regions of the β1 gene that are responsive to cytosolic sodium.

In earlier experiments, encoding DNAs for avian α1- and β1-subunits were transfected into mouse cells (Takeyasu et al., 1987, 1988, 1989). Despite conferring

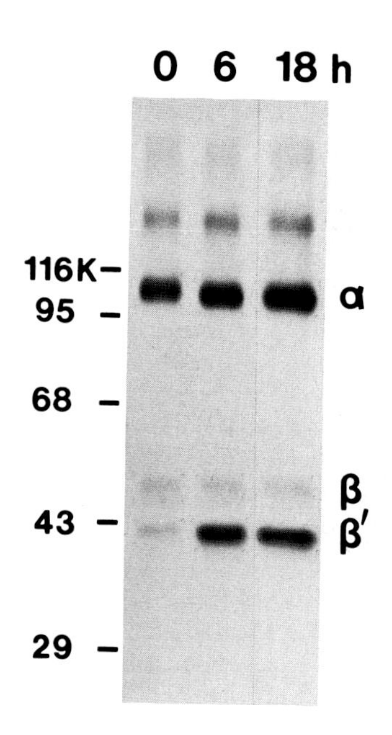

Figure 3. Effect of veratridine on the biosynthesis of Na,K-ATPase. Myogenic cells were treated for 0, 6, or 18 h with veratridine and labeled with [^{35}S]methionine for the last 2 h. Then the cells were extracted with Triton X-100 containing buffer, and immunobeads for the β-subunit of the Na,K-ATPase were used to isolate free β-subunits and α–β complexes. The precipitates were analyzed by SDS-PAGE and fluorography. Veratridine caused an approximate doubling of labeling in the α-subunit (assembled with β-subunits) and a larger increase in the labeling of the core-glycosylated form of the β-subunit (β′). There is a sixfold difference in methionine content between the α- and β-subunits; consequently, the α-subunit contains most of the [^{35}S]methionine. From Taormino and Fambrough (1990).

ouabain sensitivity on the mouse cells, overexpressed avian α1-subunits accumulated in the endoplasmic reticulum of the cells (Takeyasu et al., 1988). This suggested that the β-subunit might have as one of its functions the role of potentiating the transport of the Na,K-ATPase from the endoplasmic reticulum to the plasma membrane (Fambrough, 1988). We tested this in the mouse L cell expression system by transfecting avian cDNA encoding the β1-subunit into mouse cells expressing avian α1-subunit and vice versa. In these tests, coexpressing cell lines generally showed a high level of cell surface expression of the two avian subunits (Takeyasu et al., 1989) (Fig. 5). These results are consistent with a role of the β1-subunit in intracellular transport of the Na,K-ATPase to the plasma membrane.

To summarize, myogenesis includes a substantial increase in expression of α1- and β1-subunits of the Na,K-ATPase, supported by selective transcription of α1 and β1 genes. (Later there is a switch in α-subunit gene expression, so that in the adult, skeletal muscle expresses predominantly the α2-subunit.) In tissue culture, an increase in intracellular sodium causes a selective increase in transcription of the β1 gene, resulting in increased β1-mRNA and β1-subunit biosynthesis, driving assembly of more α1–β1 complexes (i.e., Na,K-ATPase molecules). We infer that, at least in the absence of upregulation, there is a pool of newly synthesized α-subunits that do not assemble with β-subunits but which can assemble if additional β-subunits are available.

During upregulation of the Na,K-ATPase in myogenic cells β1-mRNA falls rapidly after a period of accumulation (see Fig. 3). During the falling phase, which

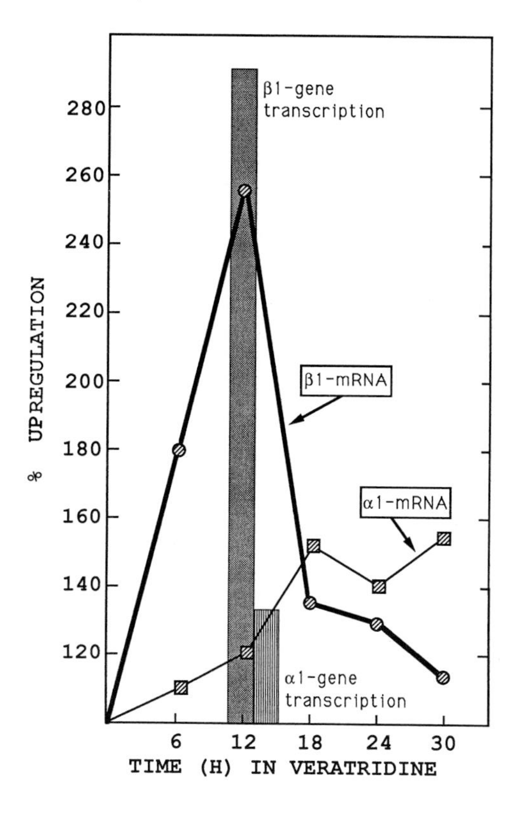

Figure 4. Effects of veratridine on mRNA levels and transcriptional rates of the Na,K-ATPase genes. Myogenic cultures were exposed to veratridine for the indicated times and then mRNA levels were quantified by RNA blot analysis. Transcriptional rates were estimated after 12 h in veratridine from nuclear run-off assays. Data redrawn from Taormino and Fambrough (1990).

occurs as the upregulatory event is nearing completion, the level of β1-mRNA decreases by two-thirds in a few hours. This is much faster than the normal turnover rate of β1-mRNA (Taormino and Fambrough, 1990). This decrease in mRNA stability may play a key role in terminating the upregulatory response. Preliminary evidence (see Taormino and Fambrough, 1991) implicates the evolutionarily conserved 3′ untranslated region of β1-mRNA as one element of potential importance in the control of β1-mRNA stability.

Fig. 6 illustrates most of the possible levels of Na,K-ATPase regulation within a single cell. This chapter has already mentioned gene selection, regulation of transcriptional rates, regulation of mRNA levels, subunit assembly, and degradation of unassembled subunits. Other early points at which regulation might be imposed include (*a*) processing of primary transcripts, (*b*) translation of mRNAs, and (*c*) integration of polypeptide chains into the membrane of the endoplasmic reticulum.

Some Other Aspects of Subunit Assembly

At this juncture a few more points of interest about the biosynthesis and assembly of Na,K-ATPase subunits deserve mention. First, pulse–chase labeling experiments

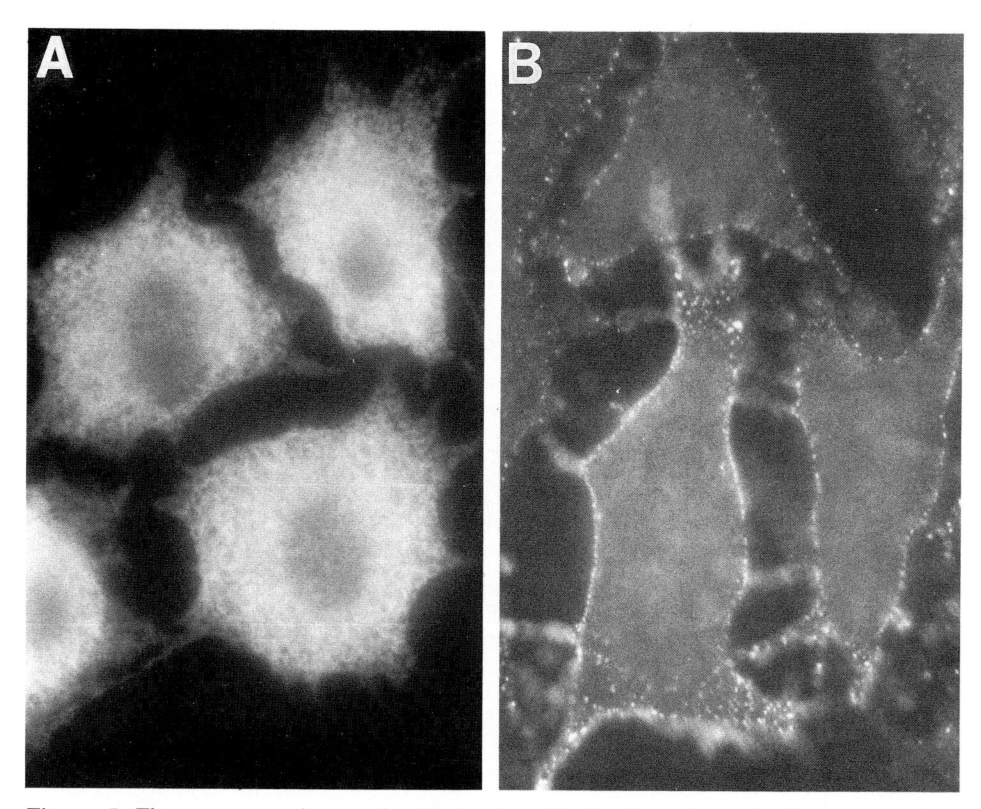

Figure 5. Fluorescence micrographs illustrating distribution of α-subunits when avian α-subunit–encoding DNA is expressed in mouse L cells without (*A*) and with (*B*) coexpression of the avian β-subunit–encoding DNA. Cells are labeled with a monoclonal antibody to the avian α-subunit (antibody 7c) followed by fluorescent-labeled goat anti–mouse IgG. Contrast the bright edge fluorescence of the cells in *B* with the reticular, intracellular pattern of fluorescence in *A*.

(Fambrough, 1983; Tamkun and Fambrough, 1986; Wolitzky and Fambrough, 1986) reveal several features of the biosynthesis/assembly process. Tissue-cultured myogenic cells labeled for as little as 5 min with [^{35}S]methionine already contain assembled α–β complexes, indicative of assembly during or immediately after polypeptide chain synthesis in the endoplasmic reticulum. The addition of high-mannose, N-linked oligosaccharide chains to the β-subunit is also fast. The assembly

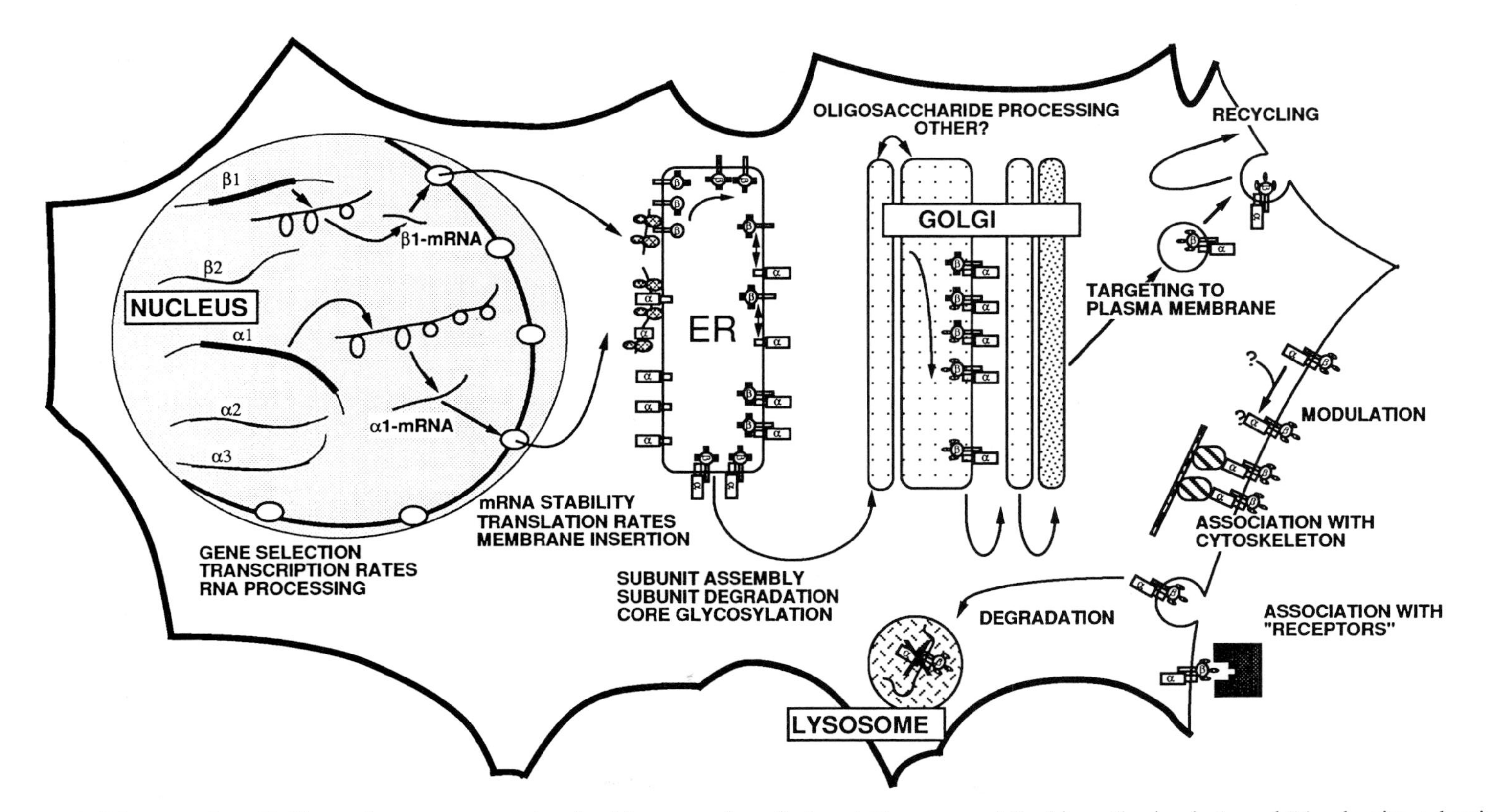

Figure 6. Diagram of a cell, illustrating some events involved in expression of α1 and β1 genes and the biosynthesis of α1- and β1-subunits, subunit assembly, post-translational processing, delivery to the plasma membrane, association with possible modulatory molecules, and degradation. Known and putative modes of regulation of Na,K-ATPase are listed near the appropriate sites in the diagram.

process precedes by nearly an hour the conversion of N-linked high-mannose oligosaccharides to complex oligosaccharides, indicating that the newly assembled α–β complexes reside for a long time in the endoplasmic reticulum after glycosylation and assembly (Fig. 7). There is currently no explanation for this long residence time.

Second, Karen Renaud has been examining regions of the β-subunit essential for α–β assembly (see Renaud and Fambrough, 1991). Expressing mutant avian β-subunit DNA in mouse L cells and scoring for assembly with the endogenous mouse α-subunit, she has found, surprisingly, that the NH_2-terminal cytosolic domain of the β-subunit is not necessary for assembly. Deletions in the membrane-spanning domain that still permit integration into the membrane also permit subunit assembly. COOH-terminal deletions of more than a few amino acids, however, fail to assemble with α-subunits. These data suggest that the major site of α–β assembly involves the ectodomains of the subunits.

Third, it appears that each isoform of the α-subunit is capable of forming α–β complexes with the β1-subunit. The experiments to demonstrate this were done as follows. Mouse L cell lines expressing the avian β1-subunit and either avian α1, α2, or

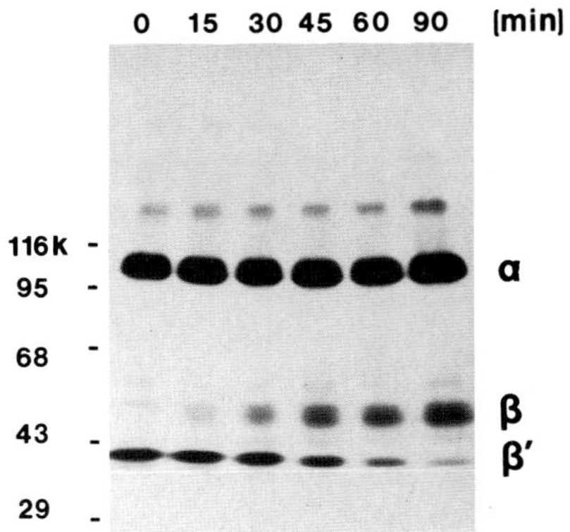

Figure 7. Kinetics of post-translational glycosylations of the Na,K-ATPase. Chick dorsal rool ganglion cultures were pulse-labeled for 30 min with [^{35}S]methionine and chased for the indicated times. At each time, cultures were extracted with Triton X-100 containing buffer, the Na,K-ATPase molecules were isolated by immune precipitation with an anti–β-subunit monoclonal antibody, and the precipitates were analyzed by SDS-PAGE and fluorography. Positions of the core-glycosylated β-subunit (β′), the mature, complex-oligosaccharide form of the β-subunit (β), and the α-subunit are indicated. Approximate molecular masses of marker proteins are also indicated at left in kilodaltons. Data from Tamkun and Fambrough (1986).

α3 were labeled with [^{35}S]methionine and the assembled α–β complexes were purified with a monoclonal antibody specific for the avian β1-subunit. The α-subunits were dissociated from the complexes in SDS and then immune precipitated with a monoclonal antibody selective for avian α-subunits. In each case, the anti–α-subunit precipitates contained the ^{35}S-labeled α-subunit (see Kone et al., 1991).

Additional Post-translational Sites of Regulation

Returning to sites of regulation in the cell, Fig. 6 depicts in cartoon fashion the late events in biosynthesis and intracellular transport to the plasma membrane. In the Golgi apparatus the β-subunit oligosaccharides are converted to complex ones that differ from one cell type to another, even though there is no difference in β-subunit polypeptide (Takeyasu et al., 1987). The Golgi apparatus is also the site of a variety of other post-translational modifications of proteins, yet none of these has been reported for the Na,K-ATPase. The question of whether or not such additional modifications occur remains unresolved, and the possible significance of such hypothetical modifications is, of course, unclear. Following modifications in the Golgi, the Na,K-ATPase moves to the *trans*-Golgi network and then is transported by vesicular traffic to the plasma membrane. The targeting of the Na,K-ATPase to basolateral membrane appears to be a direct transfer from the *trans*-Golgi network in well-polarized MDCK cells (Caplan et al., 1986). It is not known what aspects of Na,K-ATPase structure are important in targeting. A promising approach to this problem is through expression of hybrid and chimeric ATPases, composed of parts of the Na,K-ATPase and the H,K-ATPase, which are targeted to basolateral and apical membranes, respectively (Gottardi, C. J., and M. J. Caplan, personal communication). Only mature β-subunits (in the form of α–β complexes) arrive at the cell surface under normal conditions, but when N-glycosylation of the β-subunit is prevented, α–β complexes containing unglycosylated β-subunits are incorporated into the plasma membrane (Tamkun and Fambrough, 1986). In avian dorsal root ganglion cell cultures it was found that inhibition of N-glycosylation had no effect upon kinetics of α–β assembly, incorporation into plasma membrane, or degradation. If there are any effects of β-subunit glycosylation on Na,K-ATPase biogenesis and stability, they are subtle effects (see Zamofing et al., 1988). Whether β-subunit glycosylation plays any significant role in vivo is unknown.

Degradation of the Na,K-ATPase is probably accomplished by transport from the plasma membrane to the lysosomes and proteolytic destruction there. Evidence for this pathway remains indirect. Cook and colleagues (1984) found that ouabain bound to the Na,K-ATPase on the plasma membrane of tissue-cultured HeLa cells was carried slowly to lysosomes, apparently in concert with degradation of the associated Na,K-ATPase molecule. Further, as expected for this mechanism of degradation, the kinetics as measured in pulse–chase labeling experiments are first order. Degradation studies in tissue-cultured avian neurons indicate that both α- and β-subunits are degraded at the same rate, suggesting that α–β complexes are the units that are degraded (Tamkun and Fambrough, 1986). This finding also suggests (but does not prove) that α–β complexes are stable and subunit exchange does not occur. The degradation rate of the Na,K-ATPase slows after upregulation in myogenic cell cultures (Wolitzky and Fambrough, 1986). By this means the cells are able to maintain the upregulated state even though the rate of Na,K-ATPase biosynthesis is only transiently accelerated during upregulation. How this change in degradation rate is effected remains totally obscure.

There is good evidence (see Schmalzing et al., 1991*a*, *b*) for rapid removal of Na,K-ATPase molecules from the plasma membrane under certain conditions, including treatment of frog oocytes with phorbol esters. Likewise, in the avian myogenic cell cultures there is evidence for rapid removal from the plasma membrane. This occurs if, during upregulation, the entry of sodium ions is blocked (Takeyasu et al., 1989). Under these conditions, the myotubes return to the basal level of Na,K-ATPase expression in the plasma membrane, interiorizing up to half of their Na,K-ATPase molecules. The degradation rate seems to remain steady, indicating that these extra molecules become sequestered somewhere in the cytoplasm. If cells are allowed to re-upregulate, there is a rapid return to the upregulated state. We do not know whether this involves the reinsertion of Na,K-ATPase molecules into the plasma membrane. We do know, however, that this re-upregulation is accompanied by a new wave of β1-mRNA accumulation (Taormino and Fambrough, 1990). This might indicate production of new Na,K-ATPase molecules rather than, or in addition to, recycling.

Fig. 6 also indicates that there may be mechanisms for regulating Na,K-ATPase activity at the plasma membrane. One possible set of mechanisms might involve modifications of Na,K-ATPase structure. Interaction of the Na,K-ATPase with modulating ligands and covalent modifications of the Na,K-ATPase by phosphorylation or other derivitization remain possibilities. Perhaps the elusive γ-subunit will turn out to be such a ligand, analogous to the regulatory ligand phospholamban, for the SERCA2-type Ca^{2+}-ATPase (Tada and Katz, 1982).

A final class of regulatory influences depicted in Fig. 6 involves interactions between the Na,K-ATPase and the membrane cytoskeleton. The intriguing work of Nelson and Hammerton (1989) and Morrow et al. (1989) demonstrates interaction between the Na,K-ATPase and ankyrin, a protein known to link some other integral membrane proteins to spectrin. It is possible that such interactions (*a*) set levels of Na,K-ATPase expression in the plasma membrane, or (*b*) stabilize or direct a polarized placement of Na,K-ATPase molecules in polarized epithelia and neurons, or (*c*) even modulate the transport properties of the Na,K-ATPase. Similarly, interactions between the Na,K-ATPase and extracellular ligands (such as the AMOG (β2) receptor (Gloor et al., 1990) might be imagined to influence Na,K-ATPase activity and spatial distribution.

Speculations about the Role of Isoforms

Recently our lab has completed sequencing of the encoding DNAs for the α1, α2, and α3 isoforms of the avian Na,K-ATPase (Takeyasu et al., 1990). This isoform set can now be compared with that of the rat and with isoform sequences from various other mammals and some nonmammals. What emerges from comparisons is that each avian isoform has a direct homologue in mammals with 93–96% sequence identity between mammalian and avian α homologues. Moreover, each isoform has about the same tissue and cell-type distribution of expression in bird and mammal. In both the mammal and the bird, however, there is only ~82–83% sequence identity between different isoforms (α1 vs. α2, α1 vs. α3, α2 vs. α3). The differences among the isoforms are distributed throughout the primary sequence but tend to occur in the NH_2-terminal half of the α-subunit and to cluster in certain regions (Fig. 8). These sequence relationships imply that the three isoforms originated early in evolution and have been rather rigidly maintained in their primary structures through the past 200 million years (the age of the last common ancestor of birds and mammals). Maintenance of these different isoform sequences through evolution would appear to require that selection act upon the expressed protein. What aspects of Na,K-ATPase function or regulation might require several isoforms? We speculate that the isoform-specific aspects of Na,K-ATPase structure subserve functions that are restricted to particular cell types. These could be quantitative differences in transport parameters that might be important in the proper performance of specific cell types. On the other hand, the isoform-specific positions in the Na,K-ATPase molecules might be important in isoform-specific regulatory phenomena. Such phenomena remain to be discovered, but one might speculate that they would

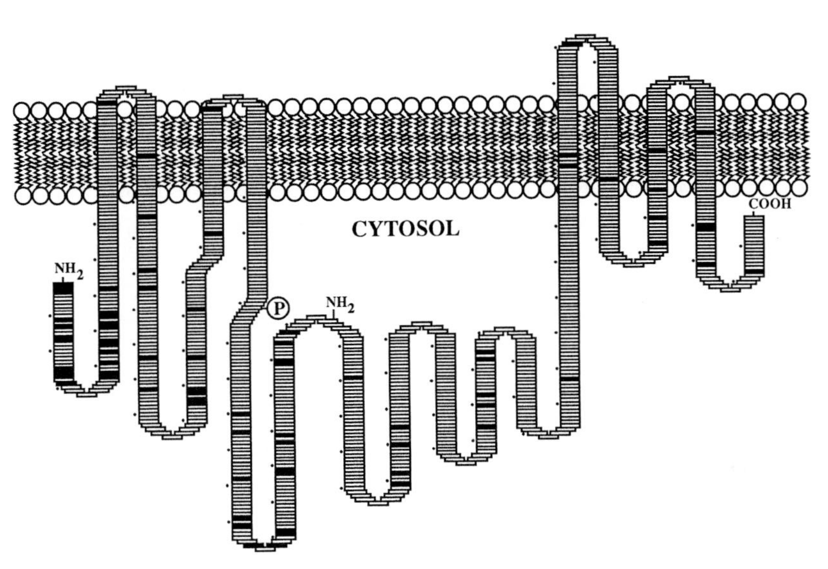

Figure 8. Cartoon of the α-subunit of the chicken Na,K-ATPase, drawn with the tramsmembrane topology suggested by Shull et al. (1985). Each amino acyl residue is symbolized by a rectangle. Dots to the left of every tenth position simplify counting. Blackened rectangles mark positions where the same amino acid occurs in rat and chicken isoform homologues (for example, in rat and chicken α1) but not all three homologue pairs have the same amino acyl residue. These positions were defined as "isoform specific" by Takeyasu et al. (1990). NH_2 and COOH termini are indicated, as well as the FITC-reactive ϵ-amino group (*NH_2*) on a Lys_{503} and the site of phosphorylation (*P*) on Asp_{371} during ion transport.

include isoform-specific interactions with cytoskeleton or with modulatory ligands. It is interesting that the single α-subunit of *Drosophila melanogaster* (Lebovitz et al., 1989), although differing in amino acid sequence about equally from the three vertebrate α-subunit isoforms, maintains very high homology, particularly with the vertebrate α3 isoform in some positions of great isoform-to-isoform divergence, such as in the region adjacent to the FITC-reactive site (Fig. 9). Evidently this sequence is critical for survival in *Drosophila* but is necessary for proper Na,K-ATPase function or regulation only in α3-expressing cells in vertebrates.

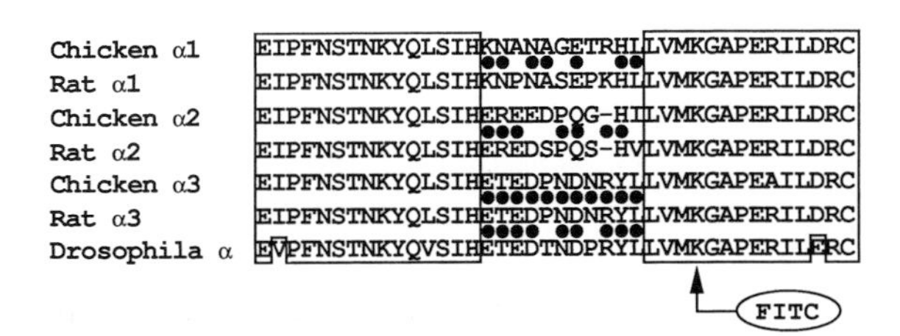

Figure 9. Aligned amino acid sequences near the FITC-reactive site of the Na,K-ATPase α-subunit isoforms of chicken and rat (rat sequences from Shull et al., 1986) and the single α-subunit found in *Drosophila melanogaster* (Lebovitz et al., 1989). Boxes surround regions of sequence identity among the vertebrates. Dots in the region of sequence variation signify positions of sequence identity between isoform homologues. At 9 of these 11 positions the *Drosophila* α-subunit sequence is identical to vertebrate α3.

Summary and Epilogue

In summary, let us take a final look at Fig. 6. We have followed most of the discrete events in the life history of the Na,K-ATPase in a single cell, pointing out the various points at which regulatory mechanisms might operate. But the picture is still one of a cell in isolation. Space has not permitted an adequate discussion of the signals that influence these cellular events. Some of the signals are well-known hormones, such as aldosterone and T3, and their sites of action appear to vary from one system to another, including influences at the level of gene selection, transcriptional rate, and translational rate. We have emphasized our work on the Na,K-ATPase in avian nerve and muscle and on expression of avian Na,K-ATPase encoding DNAs in mammalian cells. The important work of Clausen and colleagues (see Clausen, 1986, 1990) on regulation of the Na,K-ATPase in adult skeletal muscle has not been discussed here. Our work on regulation of the Na,K-ATPase in embryonic myotubes in tissue culture reveals, in some instances, mechanisms quite different from those operating in the adult. Space does not permit proper discussion of numerous examples of regulation of the Na,K-ATPase in other cell types and in other species. Given the high degree of conservation of exon/intron structure and nucleotide sequence in the genes for the Na,K-ATPase subunits and the remarkable conservation of amino acid sequences for the isoforms of the subunits in various vertebrates, one can be reasonably confident that study of any mechanism of Na,K-ATPase regulation in any organism will yield information pertinent to regulation of the Na,K-ATPase in a broad spectrum of animal species.

Finally, it should be emphasized that studies of the Na,K-ATPase are pressing at the frontiers of knowledge in many aspects of cell biology and physiology. Now we have to return to the thoughts in our opening statements: not only can we use cell biological approaches to study Na,K-ATPase regulation, but we can also use the Na,K-ATPase as a point of focus in studies of poorly understood cellular processes. For example, the changes in Na,K-ATPase degradation rate that accompany upregulation defy detailed explanation in part because so little is known about regulation of membrane protein degradation. Similarly, little is known about subunit assembly processes in the endoplasmic reticulum or about the mechanisms that determine correct targeting of membrane proteins from the *trans*-Golgi network to their proper destinations. It is our hope that in the process of elucidating the mechanisms that regulate the Na,K-ATPase, we sodium pump people will make important contributions to the larger areas of cell biology and physiology.

References

Caplan, M. J., H. C. Anderson, G. Palade, and J. D. Jamieson. 1986. Intracellular sorting and polarized cell surface delivery of (Na^+ + K^+)-ATPase, an endogenous component of MDCK cell basolateral plasma membranes. *Cell.* 46:623–631.

Clausen, T. 1986. Regulation of active Na^+-K^+ transport in skeletal muscle. *Physiological Reviews.* 66:542–590.

Clausen, T. 1990. Significance of Na^+-K^+ pump regulation in skeletal muscle. *News in Physiological Sciences.* 5:148–151.

Cook, J. S., N. J. Karin, J. B. Fishman, E. B. Tate, and L. R. Pollack. 1984. Regulation of turnover of Na^+,K^+-ATPase in cultured cells. *In* Regulation and Development of Membrane Transport Processes. J. S. Graves, editor. John Wiley & Sons, Inc., New York. 3–19.

Fambrough, D. M. 1983. Studies on the (Na^+ + K^+)-ATPase of skeletal muscle and nerve. *Cold Spring Harbor Symposium on Quantitative Biology.* 48:297–304.

Fambrough, D. 1988. The sodium pump becomes a family. *Trends in Neurosciences.* 11:325–328.

Gloor, S., H. Antonicek, K. J. Sweadner, S. Pagliusi, R. Frank, M. Moos, and M. Schachner. 1990. The adhesion molecule on glia (AMOG) is a homologue of the β subunit of the Na,K-ATPase. *Journal of Cell Biology.* 110:165–174.

Kone, B. C., K. Takeyasu, and D. M. Fambrough. 1991. Structure–function studies of Na/K-ATPase isozymes. *In* The Sodium Pump: Recent Developments. J. H. Kaplan and P. De Weer, editors. The Rockefeller University Press, New York. In press.

Lebovitz, R. M., K. Takeyasu, and D. M. Fambrough. 1989. Molecular characterization and expression of the (Na^+ + K^+)-ATPase α-subunit in *Drosophila melanogaster. EMBO Journal.* 8:193–202.

Morrow, J. S., C. D. Cianci, T. Ardito, A. S. Mann, and M. Kashgarian. 1989. Ankyrin links fodrin to the alpha subunit of Na,K-ATPase in Madin-Darby canine kidney cells and in intact renal tubule cells. *Journal of Cell Biology.* 108:455–465.

Nelson, W. J., and R. W. Hammerton. 1989. A membrane-cytoskeletal complex containing Na^+,K^+-ATPase, ankyrin, and fodrin in Madin-Darby canine kidney (MDCK) cells. Implications for the biogenesis of epithelial cell polarity. *Journal of Cell Biology.* 108:893–902.

Renaud, K. J., and D. M. Fambrough. 1991. Molecular analysis of Na,K-ATPase subunit assembly. *In* The Sodium Pump: Recent Developments. J. H. Kaplan and P. De Weer, editors. The Rockefeller University Press, New York. In press.

Schmalzing, G., H. Omay, S. Kröner, H. Appelhans, and W. Schwarz. 1991*a.* Expression of exogenous β1-subunits of Na,K-pump in *Xenopus laevis* oocytes raises pump activity. *In* The Sodium Pump: Recent Developments. J. H. Kaplan and P. De Weer, editors. The Rockefeller University Press, New York. In press.

Schmalzing, G., K. Mädefessel, and W. Haase. 1991*b.* Immunocytochemical evidence for protein kinase C–induced internalization of sodium pumps in *Xenopus laevis* oocytes. *In* The Sodium Pump: Recent Developments. J. H. Kaplan and P. De Weer, editors. The Rockefeller University Press, New York. In press.

Shull, G. E., J. Greeb, and J. B. Lingrel. 1986. Molecular cloning of three distinct forms of the Na^+,K^+-ATPase α-subunit from rat brain. *Biochemistry.* 25:8125–8132.

Shull, G. E., A. Schwartz, and J. B. Lingrel. 1985. Amino acid sequence of the catalytic subunit of the (Na^+ + K^+)-ATPase deduced from a complementary DNA. *Nature.* 316:691–695.

Tada, M., and A. M. Katz. 1982. Phosphorylation of the sarcoplasmic reticulum and sarcolemma. *Annual Review of Physiology.* 44:401–423.

Takeyasu, K., M. V. Lemas, and D. M. Fambrough. 1990. Stability of the (Na^+ + K^+)-ATPase α-subunit isoforms in evolution. *American Journal of Physiology* 259 (*Cell Physiology*)28:C619–C630.

Takeyasu, K., K. J. Renaud, J. P. Taormino, B. A. Wolitzky, A. Barnstein, M. M. Tamkun, and D. M. Fambrough. 1989. Differential subunit and isoform expression are involved in regulation of the sodium pump in skeletal muscle. *Current Topics in Membranes and Transport.* 34:143–165.

Takeyasu, K., M. M. Tamkun, K. J. Renaud, and D. M. Fambrough. 1988. Ouabain-sensitive ($Na^+ + K^+$)-ATPase activity expressed in mouse Ltk cells by transfection with DNA encoding the α-subunit of an avian sodium pump. *Journal of Biological Chemistry.* 263:4347–4354.

Takeyasu, K., M. M. Tamkun, N. R. Siegel, and D. M. Fambrough. 1987. Expression of hybrid ($Na^+ + K^+$)-ATPase molecules after transfection of mouse Ltk^- cells with DNA encoding the β-subunit of an avian brain sodium pump. *Journal of Biological Chemistry.* 262:10733–10740.

Tamkun, M. M., and D. M. Fambrough. 1986. The ($Na^+ + K^+$)-ATPase of chick sensory neurons: studies on biosynthesis and intracellular transport. *Journal of Biological Chemistry.* 261:1009–1019.

Taormino, J. P., and D. M. Fambrough. 1990. Pre-translational regulation of the ($Na^+ + K^+$)-ATPase in response to demand for ion transport in cultured chicken skeletal muscle. *Journal of Biological Chemistry.* 265:4116–4123.

Taormino, J. P., and D. M. Fambrough. 1991. Deletions in the 3′ untranslated region of the β1-subunit mRNA affect mRNA stability. *In* The Sodium Pump: Recent Developments. J. H. Kaplan and P. De Weer, editors. The Rockefeller University Press, New York. In press.

Wakshull, E., E. K. Bayne, M. Chiquet, and D. M. Fambrough. 1983. Characterization of a plasma membrane glycoprotein common to myoblasts, skeletal muscle satellite cells, and glia. *Developmental Biology.* 100:464–477.

Wolitzky, B. A., and D. M. Fambrough. 1986. Regulation of the ($Na^+ + K^+$)-ATPase in cultured chick skeletal muscle: modulation of expression by demand for ion transport. *Journal of Biological Chemistry.* 261:9990–9999.

Zamofing, D., B. C. Rossier, and K. Geering. 1988. Role of the Na,K-ATPase β-subunit in the cellular accumulation and maturation of the enzyme as assessed by glycosylation inhibitors. *Journal of Membrane Biology.* 104:69–79.

Chapter 3

Posttranslational Modifications and Intracellular Transport of Sodium Pumps: Importance of Subunit Assembly

K. Geering

Institut de Pharmacologie et Toxicologie de l'Université de Lausanne, 1005 Lausanne, Switzerland

Introduction

Active Na,K-ATPase is an integral membrane protein composed of an α-subunit and a noncovalently linked glycosylated β-subunit. Both α- and β-subunits have been cloned from several animal species, and sequence data support biochemical evidence that the α-subunit is a large multimembrane-spanning polypeptide with an important cytoplasmic domain, while the β-subunit has only one transmembrane segment and most of its mass is exposed to the extracellular side (for review see Jørgensen and Andersen, 1988; Pedemonte and Kaplan, 1990).

One of the important questions that must be answered to better understand the structure–function relationship of Na,K-APase concerns the functional role of the β-subunit. This glycoprotein is indeed an indispensable constituent of a minimal functional enzyme unit expressed at the plasma membrane (Noguchi et al., 1987; Geering et al., 1989; Horowitz et al., 1990). On the other hand, however, it is well established that the major functional domains related to the catalytic cycle such as

TABLE I
Tissue Expression and Properties of β-Isoforms

	Tissue distribution						Homology	Function	
	Kidney	Brain	Pineal gland	Heart	Liver	Embryo		Regulatory role in	
Rat							%		
β1[*]	+	+		+	–		100	?	Structural
β2[‡]	–	+	+	–	–		50	Pump	and
Xenopus								activity	functional
β1[§]	+	+					100		maturation
β3[‖]	–	+			–	+	61		
Mouse[¶]									
AMOG	–	+					40	Cell adhesion	

[*]Shyjan and Levenson, 1989; [‡]Shyjan et al., 1900; [§]Verrey et al., 1989; [‖]Good et al., 1990; [¶]Gloor et al., 1990.

cation, ATP, and phosphate binding sites are exclusively located on the α-subunit (for recent review, see Pedemonte and Kaplan, 1990) and that the affinity for cardiac glycoside binding is also largely determined by the α-subunit (Takeyasu et al., 1987, 1988; Emanuel et al., 1988; Price et al., 1989, 1990).

The functional interdependence of catalytic α- and β-subunits is best established for Na,K-ATPase, but it is interesting that most recently several groups have identified a β-subunit-like polypeptide associated with gastric H,K-ATPase (Hall et al., 1990; Okamoto et al., 1990; Shull, 1990).

The role of β-subunits remains intriguing, even more so since different putative β-isoforms have recently been identified at the gene (Lane et al., 1989; Malo et al., 1990) and at the RNA (Martin-Vasallo et al., 1989; Good et al., 1990), as well as at the protein level (Gloor et al., 1990; Good et al., 1990; Shyjan et al., 1990). The quantitative differences described in the tissue distribution of the β_1- versus β_2-isoforms in rat, or of the β_1- versus β_3-isoforms in *Xenopus laevis* (Table I), indeed point to the possibility that each isoform might have a proper specialized function.

Which properties of the enzyme should we look at to find the role of the β-subunit? As mentioned before, it is not likely that the β-subunit participates directly in the catalytic cycle. However, this fact does not exclude the possibility that the β-subunit might have an indirect regulatory role in the functioning of the active enzyme. This possibility has gained topical interest through recent data reported by Gloor et al. (1990). These authors have identified a β_2-like isoform in glial cells which expresses characteristics of an adhesion molecule and which was thus called AMOG, for adhesion molecule on glia. Antibodies against this β-isoform not only disrupt cell adhesion but also stimulate Na,K-ATPase activity at the plasma membrane. The first observation raises the interesting possibility that β-subunits associated with α-subunits might be recognition elements implicated in cell adhesion, which ultimately could mediate a fine regulation of ion transport. The second observation points to a discrete functional interplay between the two subunits in the active enzyme.

While these questions on the functional implications of β-subunits in the mature enzyme remain to be elucidated, it becomes on the other hand increasingly clear that the β-subunit, or in more general terms the assembly of α- and β-subunits, is necessary for the structural and functional maturation of the Na,K-ATPase as well as for its intracellular transport to the plasma membrane. I would like to review the experimental evidence indicating that constraints hold true for Na,K-ATPase that are similar to those for other oligomeric proteins; namely, that assembly of subunits is needed for a correct folding of proteins after their synthesis. This assembly process is essential for the acquisition of functional properties, as well as for the ability of the protein to exit the endoplasmic reticulum (ER) (for recent review, see Hurtley and Helenius, 1989).

Synthesis and Assembly of α- and β-Subunits

As is the case with most other oligomeric membrane proteins, the subunits of Na,K-ATPase are synthesized from distinct mRNA and can insert during their synthesis into ER membranes independent of each other (for review and references, see Geering, 1990). A signal recognition particle (SRP)–dependent membrane integration could so far only be demonstrated for the β-subunit (Geering, 1988; Kawakami and Nagano, 1988), but internal membrane insertion signals appear to be included in the first four transmembrane segments of the α-subunit (Homareda et al., 1988). In addition, as indicated by inhibition of membrane insertion in vitro by treatment of microsomes with *N*-ethylmaleimide, both subunits may need the signal recognition receptor for membrane integration (Cayanis et al., 1990).

Synthesis of the α- and β-subunits is apparently coordinate in certain cellular systems but not in others. In chick sensory neurons (Tamkun and Fambrough, 1986) as well as in toad bladder cells (Geering et al., 1982) the ratio between α- and β-subunits remains nearly constant during different pulse labeling periods, indicating that the basal synthesis rate of the two subunits is concurrent in these cells. In addition, upregulation of the biosynthetic rate by aldosterone, thyroid hormone, or low potassium, or downregulation by glycosylation inhibitors such as tunicamycin are similar for the two subunits in various cell types (for review, see Rossier et al., 1987; Pressley, 1988; McDonough et al., 1990). More recent reports, on the other hand, suggest that in several experimental models either β-subunit synthesis is favored over α-subunit synthesis (McDonough et al., 1990; Taormino and Fambrough, 1990) or α-subunits are expressed in excess over β-subunits (Geering et al., 1989). In any case, however, as we will discuss below, the number of functional pumps expressed appears to be determined by the number of stoichiometric α-β complexes formed in the ER.

Na,K-ATPase subunit assembly occurs during or soon after synthesis at the level of the ER (Tamkun and Fambrough, 1986; Ackermann and Geering, 1990). In this respect, Na,K-ATPase resembles most other oligomeric proteins (for review, see Hurtley and Helenius, 1989).

Nothing is known about the mechanisms that govern subunit assembly in Na,K-ATPase or other multisubunit proteins. The simplest view predicts that assembly is a random event and that successful collisions of subunits depend on numerous factors, e.g., on the absolute amount of newly synthesized subunits, the affinity between subunits, the chemical nature of the linkage bond, and the stability of newly synthesized subunits (for review, see Hurtley and Helenius, 1989). A common observation in multisubunit proteins is that a primary folding of at least one newly synthesized subunit is needed for a correct oligomeric assembly. This initial folding of subunits is thought to be facilitated by a variety of cotranslational modifications to which nascent polypeptides can be subjected; namely, glycosylation, disulfide bond formation, isomerization of prolines, and association to ER factors such as BiP (for review, see Pelham, 1989).

Co- and Posttranslational Processing of the β-Subunit of Na,K-ATPase

The β-subunit of mature Na,K-ATPase is a heavily glycosylated polypeptide. The β-subunit acquires and processes its sugars much as other glycoproteins do (for review and references, see Geering, 1990). In brief, high mannose core sugars are transferred en bloc to the nascent β-subunit from dolichol precursors. One of the prerequisites that determines the number of sugar moieties on the β-subunit is the existence of consensus asparagine residues (Asn-X-Ser(Thr)) on the protein backbone. The core sugars are then trimmed during the transport of the polypeptide to the Golgi, and finally complex-type sugars such as galactose or sialic acid are added in a *trans*-Golgi compartment.

Very little is known about the functional role of the complex-type sugars exposed to the cell exterior in the active enzyme. Neither structural properties nor the ability to perform cation-dependent conformational changes of the catalytic α-subunit are perturbed if correct terminal glycosylation is inhibited (Zamofing et al., 1988). It is interesting to note that different β-isoforms have different numbers of sugar moieties. As an example, the β_1-isoform of *Xenopus laevis* is glycosylated at three sites, while the β_3-isoform is glycosylated at four sites (Jaunin et al., 1991). In addition, since putative β_1-subunits from *Xenopus* kidney and brain have different electrophoretic mobilities (Geering, 1988), it is likely that the same β-isoform might acquire chemically distinct complex-type sugars in different cell types as described for other glycoproteins (for review, see Roth, 1987). It remains to be elucidated whether these differences in glycosylation patterns are related to distinct functions of the β-subunit isoforms in different tissues. Since carbohydrates have been implicated in cell recognition events (for review, see Anderson, 1990), it is tempting to speculate that the complex-type sugar moieties might be signals that define the specificity of the β-subunit in its putative function as a cell adhesion molecule.

In the context of our analysis of the importance of subunit assembly in the maturation and intracellular transport of the Na,K-ATPase, the acquisition and processing of the β-subunit sugars interests us for two other reasons. First, the differently glycosylated species encountered during the processing event can easily be distinguished by the difference in their molecular mass on gels and thus become valuable tools to localize the enzyme in different cellular compartments during its intracellular transport. As an illustration, Fig. 1 (right panel), shows the glycosylation processing of the newly synthesized β_1-subunit in *Xenopus laevis* kidney cells during a pulse-chase experiment from its core-glycosylated 40-kD form localized in the ER to the 39-kD trimmed form and finally to the fully glycosylated 49-kD form which has passed a *trans*-Golgi compartment.

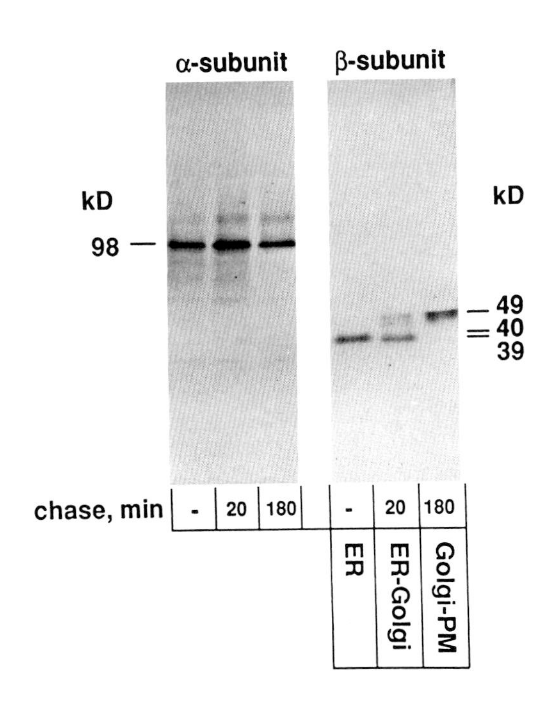

Figure 1. Intracellular routing and processing of α- and β-subunits of Na,K-ATPase. *Xenopus* kidney (A_6) cells were labeled for 7 min with [^{35}S]methionine and then chased for 20 or 180 min with an excess cold methionine. α- (*left*) and β-subunits (*right*) were immunoprecipitated with anti-α and anti-β sera and revealed by gel electrophoresis and fluorography. While no difference in molecular mass of the 98-kD α-subunit can be observed during intracellular routing, the β-subunit is processed from a 40-kD core-glycosylated form located in the ER to a 39-kD trimmed form and finally to a 49-kD fully glycosylated form which has passed a *trans*-Golgi compartment on its way to the plasma membrane (PM).

A second point of interest is related to a possible functional role of the acquisition of core sugars of the β-subunit in the subunit assembly process. Indeed, even though unglycosylated Na,K-ATPase appears to be transported to the plasma membrane and to exhibit functional properties (Tamkun and Fambrough, 1986; Takeda et al., 1988; Zamofing et al., 1989), inhibition of β-subunit core glycosylation has a pronounced effect on the cellular accumulation of both newly synthesized β- and α-subunits (Zamofing et al., 1988, 1989). This observation raises the interesting possibility that, as in other oligomeric proteins (see above), core-glycosylation might be needed for a correct initial folding of the newly synthesized β-subunit, which is possibly a prerequisite for an efficient association to the α-subunit. Nothing is known so far as to whether a similar role might be attributed to another cotranslational modification of Na,K-ATPase subunits, namely, the formation of disulfide bonds. Indeed, in β-subunits (Kirley, 1989; Miller and Farley, 1990) and in α-subunits (Miller and Farley, 1990) of purified enzyme preparations, three and two disulfide bonds, respectively, have been identified.

Posttranslational Processing of the α-Subunit of Na,K-ATPase

That the catalytic α-subunit is subjected to structural and functional modifications after its synthesis was first demonstrated in epithelial cells in culture (Geering et al., 1987). The posttranslational processing of the α-subunit during intracellular routing of the polypeptide is much more discrete than for the β-subunit and cannot be distinguished by a change in its molecular mass on gels as illustrated in Fig. 1 (left panel). Only probing by trypsinolysis reveals a posttranslational change in the structural organization of the α-subunit (Geering et al., 1987). It turns out that the

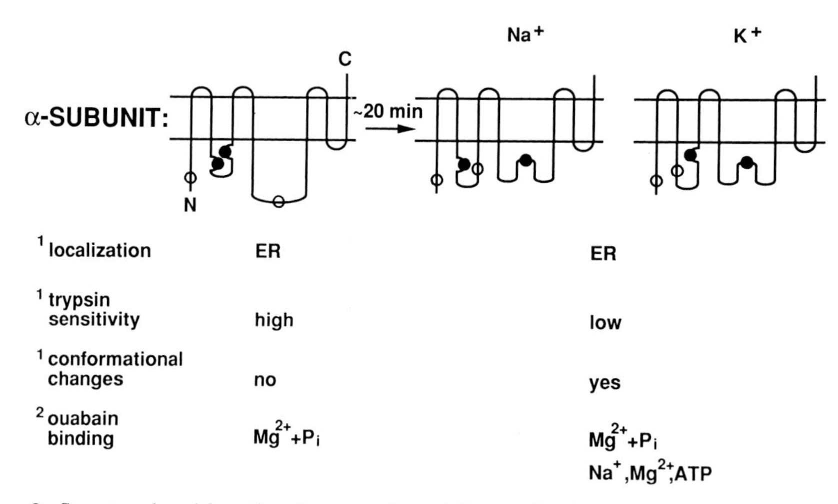

Figure 2. Structural and functional maturation of the α-subunit. For further explanations, see text. ○ and ●, exposed and occluded tryptic sites, respectively. [1]Geering et al., 1987; [2]Caplan et al., 1990.

newly synthesized α-subunits assessed in briefly pulsed cells are highly trypsin sensitive, producing upon trypsinolysis small fragments that are similar to the α-subunit synthesized in vitro in a reticulocyte lysate supplemented with ER microsomes (Geering et al., 1985). As illustrated in the model in Fig. 2, our data suggest that the newly synthesized α-subunit is in a structurally relaxed immature form, exposing several tryptic sites. However, within 20 min after synthesis, still at the level of the ER, the α-subunit becomes trypsin resistant, indicating a change in its structural organization that occludes certain tryptic sites initially exposed on the newly synthesized polypeptide. This early structural maturation of the α-subunit is likely to be a prerequisite for the α-subunit to gain at least some of its functional properties. Indeed, in parallel to increased trypsin resistance the α-subunit acquires the ability to change its conformation in response to Na^+ and K^+ (Geering et al., 1987). In addition, in the same time period the α-subunit attains the ability to bind ouabain in the presence of Na^+, Mg^{2+}, and ATP (Caplan et al., 1990).

All these observations, made in cells that synthesize α- and β-subunits concomitantly, did not lead to conclusions on the mechanisms involved in the structural and functional maturation of the α-subunit, but the fact that this process regionally and temporally coincided with subunit assembly gave the first hint about a potential role for the β-subunit. The recent introduction of several new experimental models such as transfected animal cells (Takeyasu et al., 1987, 1988; Emanuel et al., 1988) and yeast (Horowitz et al., 1990) or cRNA-injected oocytes (Noguchi et al., 1987; Geering et al., 1989) permitted us to learn more about the interdependence of α- and β-subunit synthesis.

Association to the β-Subunit Provokes a Structural Change and Stabilization in the Newly Synthesized α-Subunit

Using the *Xenopus oocyte* as an expression system, Noguchi et al. (1987) were the first to demonstrate that injection of both α- and βcRNA is needed to express functional pumps at the plasma membrane. Experiments such as the one shown in Fig. 3

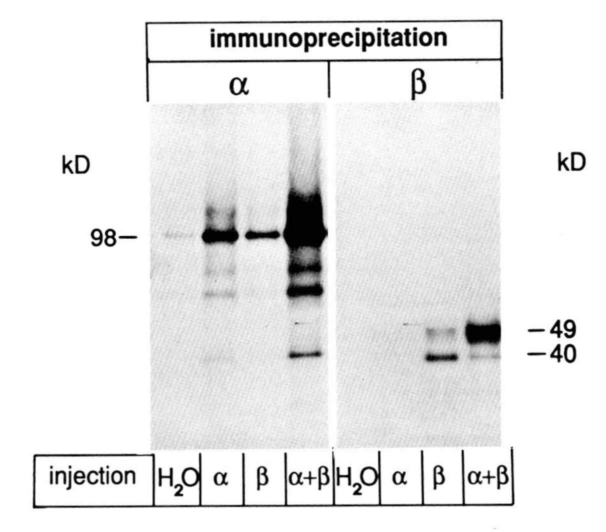

Figure 3. Expression of Na,K-ATPase subunits in *Xenopus* oocytes. Oocytes were injected with H_2O, α, β, or α- and βcRNA. cRNA was transcribed in vitro from α_1 and β_1cDNA cloned from *Xenopus* kidney cells (Verrey et al., 1989). Oocytes were labeled for 4 h with [^{35}S]methionine chased for 20 h with an excess cold methionine, and cellular extracts were immunoprecipitated with anti-α_1 (*left*) and anti-β_1 (*right*) sera. For further explanation, see text.

permitted a preliminary interpretation of this result. It turned out that the injection of αcRNA into *Xenopus* oocytes certainly increases the amount of newly synthesized α-subunits compared with water-injected controls, but not as much as injection of α- and βcRNA (Fig. 3). This result encouraged us to examine whether the degradation rate of the newly synthesized α-subunit might be influenced by the concomitant synthesis of β-subunits, which might consequently lead to an increased cellular accumulation of α-subunits. To determine the half-life of the α-subunit, *Xenopus* oocytes were injected with αcRNA alone or with α- and βcRNA and labeled for 4 h with [^{35}S]methionine, and the amount of newly synthesized α-subunit remaining was determined by immunoprecipitation after various chase periods. Such experiments clearly show that the half-life of the α-subunit synthesized together with β-subunits is considerably longer ($t_{1/2}$ at least 20 h) than the half-life of the α-subunit synthesized alone (90% degraded within 5 h with a half-life of ~2 h; Ackermann and Geering, 1990). These data suggest that in the absence of β-subunits newly synthesized α-subunits are incompletely folded and subject to rapid degradation, and that association to β-subunits provokes a structural change in the newly synthesized α-subunit which leads to its stabilization and thus to its increased cellular accumulation.

These data are in good agreement with our previous studies in the oocyte, where we looked at the effect of βcRNA injection on the structural maturation of the endogenous α-subunit (Geering et al., 1989). As illustrated in Fig. 4, it appears that *Xenopus* oocytes synthesize an excess of α-subunits over β-subunits, a result that was deduced from the fact that we could detect some β-immunoreactive material in the plasma membrane of radioiodinated oocytes, but in comparison with α-subunits we

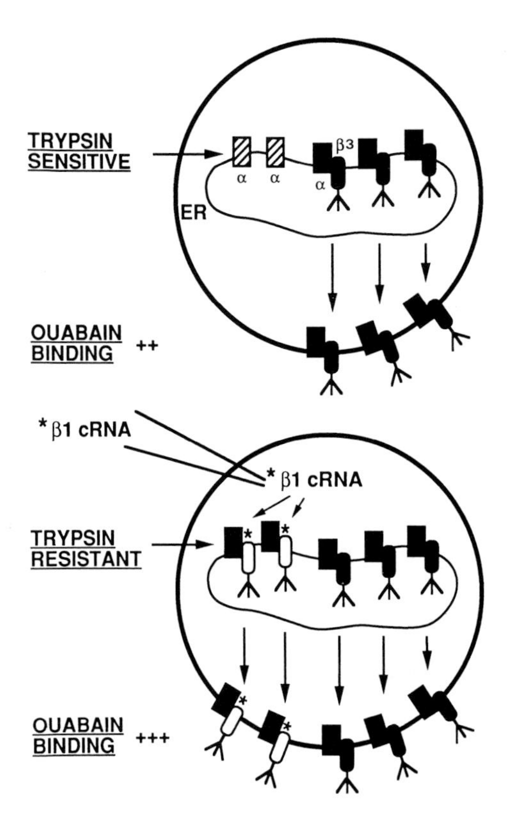

Figure 4. Maturation of *Xenopus* oocyte α-subunits by injection of βcRNA. (*Top*) *Xenopus* oocytes express Na pumps at the plasma membrane as indicated by ouabain binding to α-subunits and radioiodination of β-subunits in intact oocytes. However, oocytes snythesize more α- than β-subunits. The overexpressed α-subunit is in an immature trypsin-sensitive form. Preliminary data indicate that the *Xenopus* β-subunit might be a β_3-isoform recently cloned from *Xenopus* neurula stages (Good et al., 1990) and characterized for its basic properties (Jaunin et al., 1991). (*Bottom*) *Xenopus* β_1-subunits synthesized from injected cRNA associate to overexpressed oocyte α-subunits. This leads to an increased trypsin resistance of oocyte α-subunits and an increased number of ouabain binding sites at the plasma membrane. For further explanation, see text.

detected very few newly synthesized β-subunits in metabolically labeled oocytes (Figs. 3 and 4). Studies on the structural and functional properties of the overexpressed α-subunit in *Xenopus* oocytes then showed that, in contrast to somatic cells which synthesize stoichiometric amounts of α- and β-subunits, a large proportion of oocyte α-subunits do not acquire trypsin resistance after their synthesis. Finally, the injection of βcRNA alone into oocytes produces a trypsin-resistant membrane organization of the oocyte α-subunit and is able to perform cation-dependent conformational charges.

On the whole these data support the hypothesis that the change in trypsin sensitivity and the acquisition of at least some of the functional properties of the α-subunit that we observe in somatic cells synthesizing congruous amounts of α- and β-subunits (Fig. 2) might be due to subunit assembly. If this hypothesis is true, it is interesting to note that the newly synthesized, possibly trypsin-sensitive α-subunit in such cells already appears to possess some ouabain binding properties which are maintained throughout its intracellular routing to the plasma membrane (Caplan et al., 1990; Fig. 2). It is tempting to speculate that the expression of this functional property reflects an initial folding of the α-subunit, which is perhaps important for subunit assembly. A similar stepwise acquisition of the ultimate fully functional configuration has been described for the α-subunit of the acetylcholine receptor. The newly synthesized polypeptide is able to bind the antagonist α-bungarotoxin (Merlie, 1984), but assembly to other subunits is needed for binding of acetylcholine or for the expression of channel activity (Kurosaki et al., 1987).

As shown in Fig. 4, injection of βcRNA into oocytes not only renders the oocyte α-subunit trypsin resistant but also increases the number of functional pumps at the plasma membrane as assessed by an increased ouabain binding to intact injected oocytes compared with noninjected controls. This result encouraged us to study whether in analogy to other oligomeric proteins assembly of α- and β-subunits might be needed for the intracellular transport of the enzyme to the plasma membrane.

α- and β-Subunits of Na,K-ATPase Depend on Each Other for Their Intracellular Transport

First indications for the importance of subunit assembly for the intracellular transport of Na,K-ATPase came again from preliminary experiments performed in *Xenopus* oocytes (Fig. 3). The right panel of the figure shows the expression and processing of β-subunits synthesized from injected cRNAs. The data show that β-subunits synthesized in the presence of α-subunits are as expected in their fully glycosylated 49-kD form after a 4-h pulse and a 20-h chase. On the other hand, however, β-subunits synthesized alone are mainly in their core-glycosylated form. This observation was extended and confirmed by a detailed kinetic analysis of the glycosylation processing (Ackermann and Geering, 1990), a study that permitted us to deduce that (*a*) in *Xenopus* oocytes α-β complexes are transported to a distal Golgi compartment where full glycosylation occurs with a half-time of ~10 h, and (*b*) β-subunits synthesized alone never acquire complex-type sugars and it is thus likely that they are retained in the ER.

It is likely that the same is true for the α-subunit. Takeyasu et al. (1988) observed that in mouse cells transfected with avian αcDNA, only a limited number of α-β complexes reach the plasma membrane, while a significant proportion of avian α-subunits, probably synthesized in excess over mouse β-subunits, accumulate in an intracellular compartment. Further, if one considers the short half-life ($t_{1/2}$ ~ 2 h) of unassembled α-subunits determined in *Xenopus* oocytes (Ackermann and Geering, 1990), it is most probable that degradation of unassembled α-subunits takes place in the ER or at least in a close pre-Golgi compartment.

All data so far obtained are consistent with the hypothesis that α- and β-subunits of Na,K-ATPase are mutually dependent for the adoption of a correct configuration necessary for their transport out of the ER. It is interesting to note that in *Xenopus* oocytes, core-glycosylated unassembled β-subunits have a longer half-life ($t_{1/2}$ ~ 9 h) than unassembled α-subunits ($t_{1/2}$ ~ 2 h; Ackermann and Geering, 1990). It is possible that the longer half-life of the newly synthesized β-subunit is potentially important to assure an efficient formation of α-β complexes. This is supported by recent data which show that α-subunits newly synthesized in the oocyte can associate to preexisting β-subunits (Noguchi et al., 1990).

In this short review we have summarized the experimental evidence that defines some basic functional properties of the β-subunit rendering it an indispensable factor in the formation and expression of active Na pumps at the plasma membrane. Recently, we have shown that different β-isoforms share similar properties with respect to their ability to stabilize newly synthesized α-subunits and their dependence on the concomitant synthesis of α-subunits to leave the ER (Jaunin et al., 1991). The fine functional variations that probably exist between the different β-isoforms thus remain to be established.

Acknowledgments

I would like to thank Ms. Anne Ricou and Ms. Grazia Zorzi for preparing the manuscript, and Ms. Nancy Narbel for preparing the bibliography.

This work was supported by grant 31-26241.89 from the Swiss National Fund for Scientific Research.

References

Ackermann, U., and K. Geering. 1990. Mutual dependence of Na,K-ATPase α- and β-subunits for correct posttranslational processing and intracellular transport. *FEBS Letters.* 269:105–108.

Anderson, H. 1990. Adhesion molecules and animal development. *Experientia.* 46:2–13.

Caplan, M. J., B. Forbush III, G. E. Palade, and J. D. Jamieson. 1990. Biosynthesis of the Na,K-ATPase in Madin-Darby canine kidney cells: activation and cell surface delivery. *Journal of Biological Chemistry.* 265:3528–3534.

Cayanis, E., H. Bayley, and I. S. Edelman. 1990. Cell-free transcription and translation of Na,K-ATPase α and β subunit cDNAs. *Journal of Biological Chemistry.* 265:10829–10835.

Emanuel, J. R., J. Schulz, X. M. Zhou, R. B. Kent, D. Housman, L. Cantley, and R. Levenson. 1988. Expression of an ouabain-resistant Na,K-ATPase in CV-1 cells after transfection with a cDNA encoding the rat Na, K-ATPase α1 subunit. *Journal of Biological Chemistry.* 263:7726–7733.

Geering, K. 1988. Biosynthesis, membrane insertion and maturation of Na,K-ATPase. *In* Progress in Clinical and Biological Research. Vol. 268B. The Na^+,K^+-Pump. Part B: Cellular Aspects. J. C. Skou, J. G. Nørby, A. B. Maunsbach, and M. Esmann, editors. Alan R. Liss, Inc., New York. 19–33.

Geering, K. 1990. Subunit assembly and functional maturation of Na,K-ATPase. *Journal of Membrane Biology.* 115:109–121.

Geering, K., M. Girardet, C. Bron, J. P. Kraehenbühl, and B. C. Rossier. 1982. Hormonal regulation of (Na^+,K^+)-ATPase biosynthesis in the toad bladder: effect of aldosterone and 3,5,3′-triiodo-L-thyronine. *Journal of Biological Chemistry.* 257:10338–10343.

Geering, K., J. P. Kraehenbuhl, and B. C. Rossier. 1987. Maturation of the catalytic α-subunit of Na,K-ATPase during intracellular transport. *Journal of Cell Biology.* 105:2613–2619.

Geering, K., D. I. Meyer, M. P. Paccolat, J. P. Kraehenbühl, and B. C. Rossier. 1985.

Membrane insertion of α- and β-subunits of Na^+,K^+-ATPase. *Journal of Biological Chemistry.* 260:5154–5160.

Geering, K., I. Theulaz, F. Verrey, M. T. Häuptle, and B. C. Rossier. 1989. A role for the β-subunit in the expression of functional Na^+-K^+ATPase in Xenopus oocytes. *American Journal of Physiology.* 257:C851–C858.

Gloor, S., H. Antonicek, K. J. Sweadner, S. Pagliusi, R. Frank, M. Moos, and M. Schachner. 1990. The adhesion molecule on glia (AMOG) is a homologue of the β subunit of the Na,K-ATPase. *Journal of Cell Biology.* 110:165–174.

Good, P. J., K. Richter, and I. B. Dawid. 1990. A nervous system–specific isotype of the β subunit of Na^+,K^+-ATPase expressed during early development of Xenopus laevis. *Proceedings of the National Academy of Sciences, USA.* 87:9088–9092.

Hall, K., G. Perez, D. Anderson, C. Gutierrez, K. Munson, S. J. Hersey, J. H. Kaplan, and G. Sachs. 1990. Location of the carbohydrates present in the HK-ATPase vesicles isolated from hog gastric mucosa. *Biochemistry.* 29:701–706.

Homareda, H., K. Kawakami, K. Nagano, and H. Matsui. 1988. Membrane insertion of α and β subunits of human Na^+,K^+-ATPase. *Progress in Clinical and Biological Research.* 268B:77–84.

Horowitz, B., K. A. Eakle, G. Scheiner-Bobis, G. R. Randolph, C. Y. Chen, R. A. Hitzeman, and R. A. Farley. 1990. Synthesis and assembly of functional mammalian Na,K-ATPase in yeast. *Journal of Biological Chemistry.* 265:4189–4192.

Hurtley, S. M., and A. Helenius. 1989. Protein oligomerization in the endoplasmic retriculum. *Annual Review of Cell Biology.* 5:277–307.

Jaunin, P., P. Good, I. Corthesy, and K. Geering. 1991. Posttranslational processing and basic properties of a putative β3-subunit of Na,K-ATPase. *In* The Sodium Pump: Recent Developments. J. H. Kaplan and P. De Weer, editors. The Rockefeller University Press, New York. In press.

Jørgensen, P. L, and J. P. Andersen. 1988. Structural basis for E_1-E_2 conformational transitions in Na,K-pump and Ca-pump proteins. *Journal of Membrane Biology.* 103:95–120.

Kawakami, K., and K. Nagano. 1988. The transmembrane segment of the human Na,K-ATPase β-subunit acts as the membrane incorporation signal. *Journal of Biochemistry.* 103:54–60.

Kirley, T. L. 1989. Determination of three disulfide bonds and one free sulfhydryl in the β subunit of (Na,K)-ATPase. *Journal of Biological Chemistry.* 264:7185–7192.

Kurosaki, T., K. Fukuda, T. Konno, Y. Mori, K. Tanaka, M. Mishina, and S. Numa. 1987. Functional properties of nicotinic acetylcholine receptor subunits expressed in various combinations. *FEBS Letters.* 214:253–258.

Lane, L. K., M. M. Shull, K. R. Whitmer, and J. B. Lingrel. 1989. Characterization of two genes for the human Na,K-ATPase β subunit. *Genomics.* 5:445–453.

Malo, D., E. Schurr, R. Levenson, and P. Gros. 1990. Assignment of Na,K-ATPase β_2-subunit gene (Atpb-2) to mouse chromosome-11. *Genomics.* 6:697–699.

Martin-Vasallo, P., W. Dackowski, J. R. Emanuel, and R. Levenson. 1989. Identification of a putative isoform of the Na,K-ATPase β subunit: primary structure and tissue-specific expression. *Journal of Biological Chemistry.* 264:4613–4618.

McDonough, A. A., K. Geering, and R. A. Farley. 1990. The sodium pump needs its β subunit. *FASEB Journal.* 4:1598–1605.

Merlie, J. P. 1984. Biogenesis of the acetylcholine receptor, a multisubunit integral membrane protein. *Cell.* 36:573–575.

Miller, R. P., and R. A. Farley. 1990. β subunit of (Na^+ + K^+)-ATPase contains three disulfide bonds. *Biochemistry.* 29:1524–1532.

Noguchi, S., K. Higashi, and M. Kawamura. 1990. Assembly of the α-subunit of Torpedo californica Na^+/K^+-ATPase with its pre-existing β-subunit in Xenopus oocytes. *Biochimica et Biophysica Acta.* 1023:247–253.

Noguchi, S., M. Mishina, M. Kawamura, and S. Numa. 1987. Expression of functional (Na^+ + K^+)-ATPase from cloned cDNAs. *FEBS Letters.* 225:27–32.

Okamoto, C. T., J. M. Karpilow, A. Smolka, and J. G. Forte. 1990. Isolation and characterization of gastric microsomal glycoproteins: evidence for a glycosylated β-subunit of the H^+/K^+-ATPase. *Biochimica et Biophysica Acta.* 1037:360–372.

Pedemonte, C. H., and J. H. Kaplan. 1990. Chemical modification as an approach to elucidation of sodium pump structure-function relations. *American Journal of Physiology.* 258:C1-C23.

Pelham, H. B. 1989. Control of protein exit from the endoplasmic reticulum. *Annual Review of Cell Biology.* 5:1–23.

Pressley, T. A. 1988. Ion concentration-dependent regulation of Na,K-pump abundance. *Journal of Membrane Biology.* 105:187–195.

Price, E. M., D. A. Rice, and J. B. Lingrel. 1989. Site-directed mutagenesis of a conserved, extracellular aspartic acid residue affects the ouabain sensitivity of sheep Na,K-ATPase. *Journal of Biological Chemistry.* 264:21902–21906.

Price, E. M., D. A. Rice, and J. B. Lingrel. 1990. Structure-function studies of Na,K-ATPase: site-directed mutagenesis of the border residues from the H1-H2 extracellular domain of the α subunit. *Journal of Biological Chemistry.* 265:6638–6641.

Rossier, B. C., K. Geering, and J. P. Kraehenbuhl. 1987. Regulation of the sodium pump: How and why? *Trends in Biochemical Sciences.* 12:483–487.

Roth, J. 1987. Subcellular organization of glycosylation in mammalian cells. *Biochimica et Biophysica Acta.* 906:405–436.

Shull, G. E. 1990. cDNA cloning of the β-subunit of the rat gastric H,K-ATPase. *Journal of Biological Chemistry.* 265:12123–12126.

Shyjan, A. W., C. Gottardi, and R. Levenson. 1990. The Na,K-ATPase β2 subunit is expressed in rat brain and copurifies with Na,K-ATPase activity. *Journal of Biological Chemistry.* 265:5166–5169.

Takeda, K., S. Noguchi, A. Sugino, and M. Kawamura. 1988. Functional activity of oligosaccharide-deficient (Na,K)ATPase expressed in Xenopus oocytes. *FEBS Letters.* 238:201–204.

Takeyasu, K., M. M. Tamkun, K. Renaud, and D. M. Fambrough. 1988. Quabain-sensitive (Na^+ + K^+)-ATPase activity expressed in mouse L cells by transfection with DNA encoding the α-subunit of an avian sodium pump. *Journal of Biological Chemistry.* 263:4347–4354.

Takeyasu, K., M. M. Tamkun, N. R. Siegel, and D. M. Fambrough. 1987. Expression of hybrid (Na^+ + K^+)-ATPase molecules after transfection of mouse Ltk^- cells with DNA encoding the β-subunit of an avian brain sodium pump. *Journal of Biological Chemistry.* 262:10733–10740.

Tamkun, M. M., and D. M. Fambrough. 1986. The (Na^+ + K^+)-ATPase of chick sensory neurons: studies on biosynthesis and intracellular transport. *Journal of Biological Chemistry.* 261:1009–1019.

Taormino, J. P., and D. M. Fambrough. 1990. Pre-translational regulation of the (Na^+ + K^+)-ATPase in response to demand for ion transport in cultured chicken skeletal muscle. *Journal of Biological Chemistry.* 265:4116–4123.

Verrey, F., P. Kairouz, E. Schaerer, P. Fuentes, K. Geering, B. C. Rossier, and J. P. Kraehenbuhl. 1989. Primary sequence of Xenopus laevis Na^+-K^+-ATPase and its localization in A6 kidney cells. *American Journal of Physiology.* 256:F1034–F1043.

Zamofing, D., B. C. Rossier, and K. Geering. 1988. Role of the Na,K-ATPase β-subunit in the cellular accumulation and maturation of the enzyme as assessed by glycosylation inhibitors. *Journal of Membrane Biology.* 104:69–79.

Zamofing, D., B. C. Rossier, and K. Geering. 1989. Inhibition of N-glycosylation affects transepithelial Na^+ but not Na^+-K^+-ATPase transport. *American Journal of Physiology.* 256: C958–C966.

Chapter 4

Possible Role of the β-Subunit in the Expression of the Sodium Pump

Masaru Kawamura and Shunsuke Noguchi

Department of Biology, University of Occupational and Environmental Health, Kitakyushu, 807, Japan

Introduction

Na,K-ATPase is a membrane-bound enzyme responsible for active transport of Na^+ and K^+ across the cell membrane. It consists of two noncovalently linked subunits, a catalytic α-subunit and glycosylated β-subunit. The functional unit of the enzyme as a membrane ion transport device may be the dimer form of the heterodimeric protomer ($\alpha_2\beta_2$). Hitherto, all physiological features of the enzyme, such as location of binding sites for the transported ions (Na^+ and K^+), the energy-conferring substrate (ATP), a specific inhibitor (ouabain), and the phosphorylation site in an intermediary reaction step, have all been attributed to the larger catalytic α-subunit. No evidence indicating a specific role in ATP hydrolysis or in ion transport has been unequivocally demonstrated for this smaller glycoprotein.

However, we (Kawamura and Nagano, 1984), and recently Kirley (1990), have reported that reduction of disulfide bond(s) in the β-subunit results in the complete loss of enzyme activity. The reduction is affected by the addition of Na^+ or K^+ to the medium and not by choline ions (Kawamura et al., 1985). These results clearly indicate that the two subunits are tightly coupled.

The β-subunit has been suggested to have an anchoring role when the nascent α-subunit is incorporated into the membrane (Hiatt et al., 1984). Tamkun and Fambrough (1986) have claimed that the assembly of the α- and β-subunits occurs during or immediately after the synthesis of the polypeptides. Moreover, Noguchi et al. (1987) and Takeda et al. (1988) have reported that the expression of functionally active Na,K-ATPase in *Xenopus* oocytes can be achieved only when mRNAs for both the α- and β-subunits have been injected. These observations suggest some functional involvement for the β-subunit in the membrane insertion or the maturation of the enzyme.

In recent studies, we injected mRNAs for the two subunits of *Torpedo californica* Na,K-ATPase into *Xenopus* oocytes either separately or in combination and examined the expression, intracellular distribution, and sensitivity to trypsin of the subunits (Noguchi et al., 1990*a*). We also investigated when the assembly of the α- and β-subunits occurs or whether the β-subunit assists the α-subunit to become correctly and stably expressed in the cell (Noguchi et al., 1990*b*). The oocyte system allows the specific programmed synthesis of different subunits in a cell by alternately injecting individual mRNAs. The results obtained suggest that the β-subunit acts as a stabilizer or receptor for the newly synthesized α-subunit and facilitates the accumulation of the enzyme in the membrane.

Materials and Methods

Oocytes at stages VI and V, distinguished by size, were manually dissected from the ovarian lobes of *Xenopus laevis* and stored at 19°C in modified Barth's medium containing antibiotics. Oocytes were microinjected with mRNA (20 nl/oocyte) and incubated at 19°C for 3 d. The recombinant plasmids used for in vitro synthesis of mRNAs were constructed as described previously (Noguchi et al., 1987).

Results and Discussion

Xenopus laevis oocytes, to which mRNAα (10 ng/oocyte), mRNAβ (10 ng/oocyte), or both had been injected, were incubated for 3 d in the presence of radioactive leucine.

The homogenates of the labeled oocytes were brought to 1% with respect to Triton X-100. After centrifugation the clarified homogenate was subjected to immunoprecipitation with either antiserum to α- or β-subunit of Na,K-ATPase or with a mixture of both antisera. As shown in Fig. 1, the α-subunit detected in oocytes injected with both mRNAs simultaneously (lanes *3* and *6*) was 5- to 10-fold more abundant than

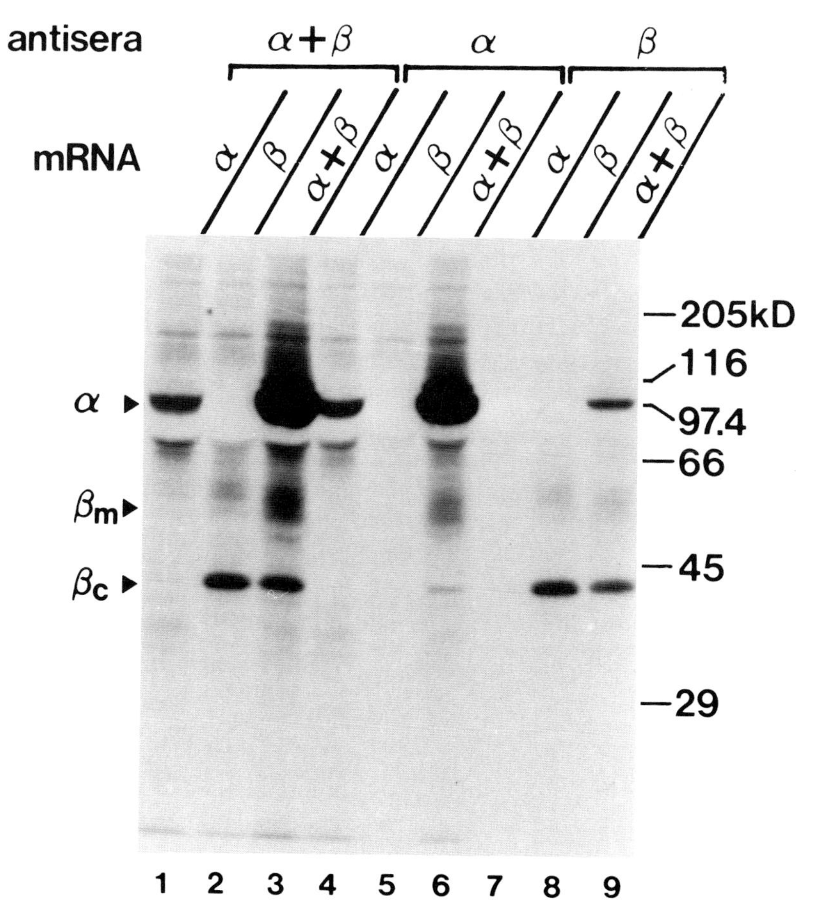

Figure 1. mRNAα alone (10 ng/oocyte) (lanes *1*, *4*, and *7*), mRNAβ alone (10 ng/oocyte) (lanes *2*, *5*, and *8*), or both (lanes *3*, *6*, and *9*) were injected into *Xenopus* oocytes, which were then incubated at 19°C for 3 d in the presence of [^{14}C]leucine. The oocytes were homogenized in the presence of 1% Triton X-100 and the labeled translation products were immunoprecipitated with antiserum to the α-subunit (lanes *4–6*) or to the β-subunit (lanes *7–9*) or with a mixture of both antisera (lanes *1–3*). The immunoprecipitates were subjected to SDS-polyacrylamide gel electrophoresis and subsequent fluorography. βm and βc represent fully glycosylated and core-glycosylated β-subunit, respectively.

that in oocytes injected with mRNAα alone (lanes *1* and *4*). On the other hand, the amount of the β-subunit synthesized under the direction of mRNAβ (lanes *2*, *3*, *8*, and *9*) was not increased, and even somewhat reduced, by coinjection with mRNAα.

The α-subunit in the oocytes injected with both mRNAs simultaneously was immunoprecipitated with antiserum to the β-subunit (lane *9*) though to a lesser extent than when immunoprecipitated in the presence of antiserum to the α-subunit (lanes *3* and *6*). This was also the case for the immunoprecipitation of the β-subunit with antiserum to the α-subunit (lanes *2*, *3*, and *6*). Partial dissociation of the αβ complex caused by Triton might be involved in these slight immunoprecipitations.

We next examined whether the translation products were stably inserted into the membrane. In this experiment, the amount of injected mRNA was increased from 10 to 60 ng per oocyte when each of the two mRNAs was injected alone to make the bands, especially the α-subunit, clearly visible on the fluorogram, whereas 10 ng each of both mRNAs were still used for coinjection. Oocytes that had been injected with mRNA(s) were homogenized with a Potter-Elvehjem-type homogenizer with a loose-fitting Teflon pestle driven at ~1,000 rpm in ice water. The homogenate was centrifuged at 160,000 *g* for 30 min. The pellet formed was suspended in the original volume of the homogenizing medium and labeled ppt-1. The supernatant was further centrifuged at 160,000 *g* for 2 h and the resulting supernatant (sup-2) and pellet (ppt-2) were saved; the latter was suspended in the original volume of the homogenizing medium. After the addition of 1% Triton X-100, the original homogenate, ppt-1, sup-2, and ppt-2 were then subjected to immunoprecipitation with a mixture of antisera to α- and β-subunit, and the results are shown in Fig. 2. When mRNAα was injected alone, the α-subunit was recovered in sup-2 as well as in ppt-1, but scarcely in ppt-2. Injection of mRNAβ alone resulted in the recovery of the β-subunit mostly in ppt-1, although a small amount of the subunit was also detected in sup-2. When mRNAβ was coinjected with mRNAα, both the α- and β-subunits were mostly recovered in ppt-1, although a small amount of the α-subunit was detected in sup-2. The recoveries of radioactivity in sup-2 fractions were 40, 10, and 10% of original homogenate for the α-subunit of oocyte injected with mRNAα alone (lane *7*), the β-subunit of oocyte injected with mRNAβ alone (lane *11*), and the α-subunit of oocytes injected with both mRNAs (lane *3*), respectively.

Fig. 3 shows the results of alkaline treatment of ppt-1. In this experiment, the ppt-1 fractions from oocytes that had been injected with mRNAα, mRNAβ, or both were treated with 0.1 M Na_2CO_3 (pH 11) on ice for 15 min and then centrifuged at 100,000 *g* for 15 min. The translation products in all the ppt-1 preparations were still tightly bound to the pellet even after the alkaline treatment. Because this treatment has been shown to remove all peripheral and adsorbed proteins from membranes it can be concluded that all the translation products associated with the ppt-1 fractions (including the α-subunit in ppt-1 from oocytes injected with mRNAα alone) were stably inserted into the membrane.

To further investigate the mode of binding of the translation products to the membrane, the ppt-1 fractions were digested with trypsin (at trypsin to ppt-1 weight rations of 1:10 and 1:100) on ice for 60 min. As shown in Fig. 4, the α-subunit in ppt-1 from oocytes injected with mRNAα alone was degraded, rapidly yielding several tryptic fragments (lane *5*, *arrowheads*). The α-subunit recovered in the sup-2 fraction was also susceptible to tryptic attack (data not shown). On the other hand, the α-subunit in ppt-1 from oocytes injected with both mRNAs was essentially resistant to trypsin. However, small amounts of tryptic fragments similar to those observed in lane *5* were detected also in this case (lane *2*, *arrowheads*), suggesting that part of the α-subunit was susceptible to trypsin even in the presence of the β-subunit. The β-subunit in ppt-1 was resistant to trypsin both in the presence and absence of the α-subunit.

Next, we examined the effects of varying amounts of injected mRNAβ on the expression of the α-subunit and on Na,K-ATPase activity of the αβ complex formed. In this experiment, the amount of mRNAα injected was 10 ng per oocyte, whereas 0,

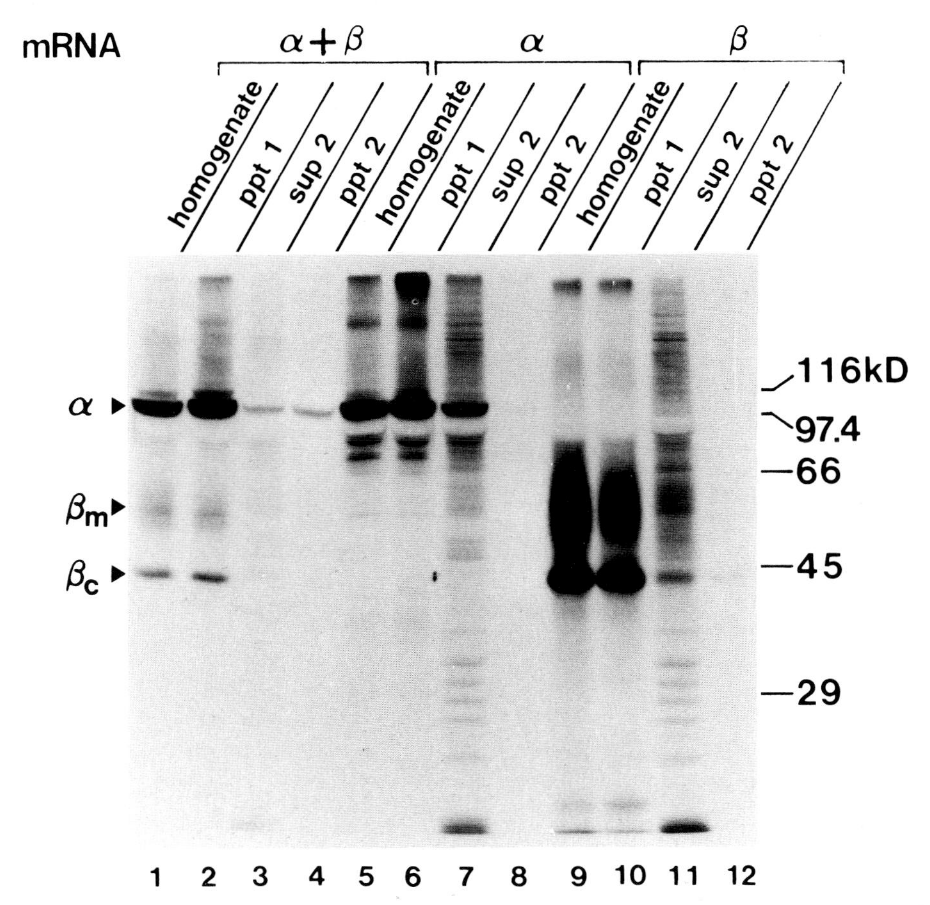

Figure 2. Insertion of translation products into membrane. Oocytes injected with both mRNAα and mRNAβ (10 ng/oocyte), mRNAα alone (60 ng/oocyte), or mRNAβ alone (60 ng/oocyte) were incubated at 19°C for 3 d in the presence of [^{3}H]leucine and fractionated into ppt-1, sup-2, and ppt-2 as described in text. Labeled translation products in each fraction were immunoprecipitated as in Fig. 1. The immunoprecipitate was analyzed by SDS-polyacrylamide gel electrophoresis followed by fluorography. Lanes *1–4*, from oocytes injected with both mRNAs; lanes *5–8*, from mRNAα-injected oocytes; lanes *9–12*, from mRNAβ-injected oocytes.

2.8, 5.6, or 11.6 ng of mRNAβ was coinjected per oocyte, which corresponded to mRNAβ to mRNAα molar rations of 0, 0.5, 1, and 2, respectively. It is clearly shown in Fig. 5, in which the radioactivities associated with the subunit bands are plotted against the mRNAβ to mRNAα molar ratio, that not only the amount of the β-subunit but also that of the α-subunit increased as the amount of mRNAβ coinjected increased. For the β-subunits, sum of core-glycosylated, βc, and mature, βm, forms was plotted. Fig. 5 also indicates that Na,K-ATPase activity of the membrane (ppt-1) fraction also increased as the amount of coinjected mRNAβ increased.

These results strongly suggest that at least one important role of the β-subunit of Na,K-ATPase is to facilitate correct assembly of the α-subunit into the membrane.

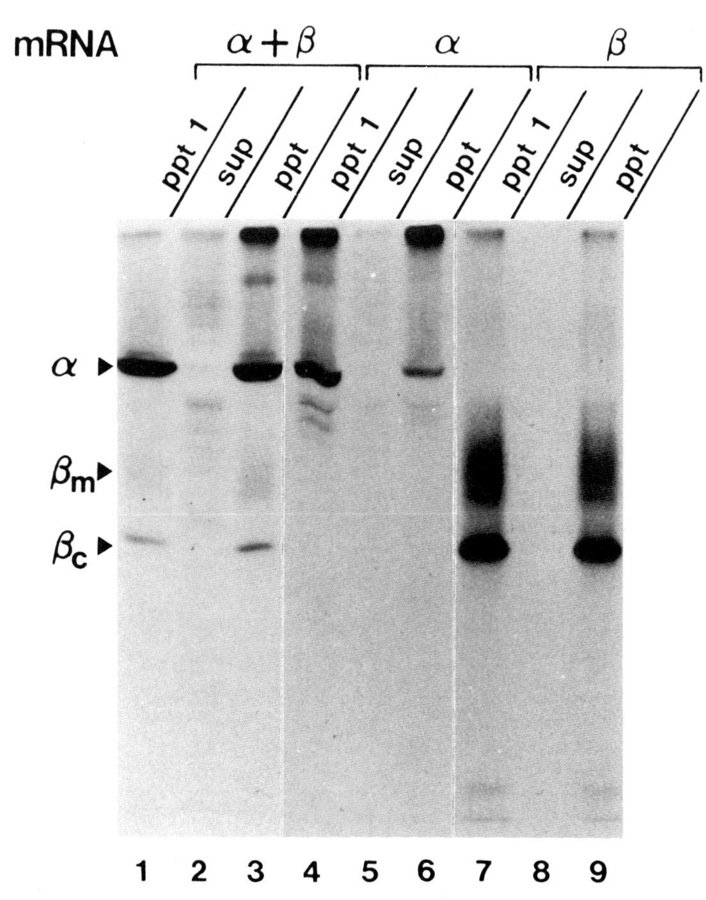

Figure 3. Alkaline treatment of the membrane (ppt-1) fractions obtained from oocytes injected with mRNAs. The ppt-1 fraction obtained as described in Fig. 2 was treated with 0.1 M Na_2CO_3 (pH 11) on ice for 15 min. After centrifugation at 100,000 *g* for 15 min, the resulting pellet (ppt) and supernatant (sup) were subjected to immunoprecipitation, and the precipitate was analyzed by SDS-polyacrylamide gel electrophoresis and subsequent fluorography. Lanes *1–3*, from oocytes injected with both mRNAs; lanes *4–6*, from mRNA α-injected oocytes; lanes *7–9*, from mRNA β-injected oocytes.

This notion is supported by the following observations. Although more than half of the α-subunit synthesized in *Xenopus* oocytes injected with mRNAα alone was recovered in the membrane fraction (Fig. 2) and not removed from the membrane by alkaline treatment (Fig. 3), this subunit as well as that recovered in the soluble (sup-2) fraction were rapidly degraded by trypsin (Fig. 4). On the other hand, when both α- and β-subunits were coexpressed in oocytes, almost all of the α-subunit was tightly bound to the membrane (Figs. 2 and 3) and was resistant to trypsin (Fig. 4). These observations indicate that, in the absence of the β-subunit, the α-subunit is inserted into the membrane in an aberrant fashion and thus exhibits no Na,K-ATPase activity (Fig. 6). In other words, correct assembly of the α-subunit into the membrane is achieved only with the aid of the β-subunit.

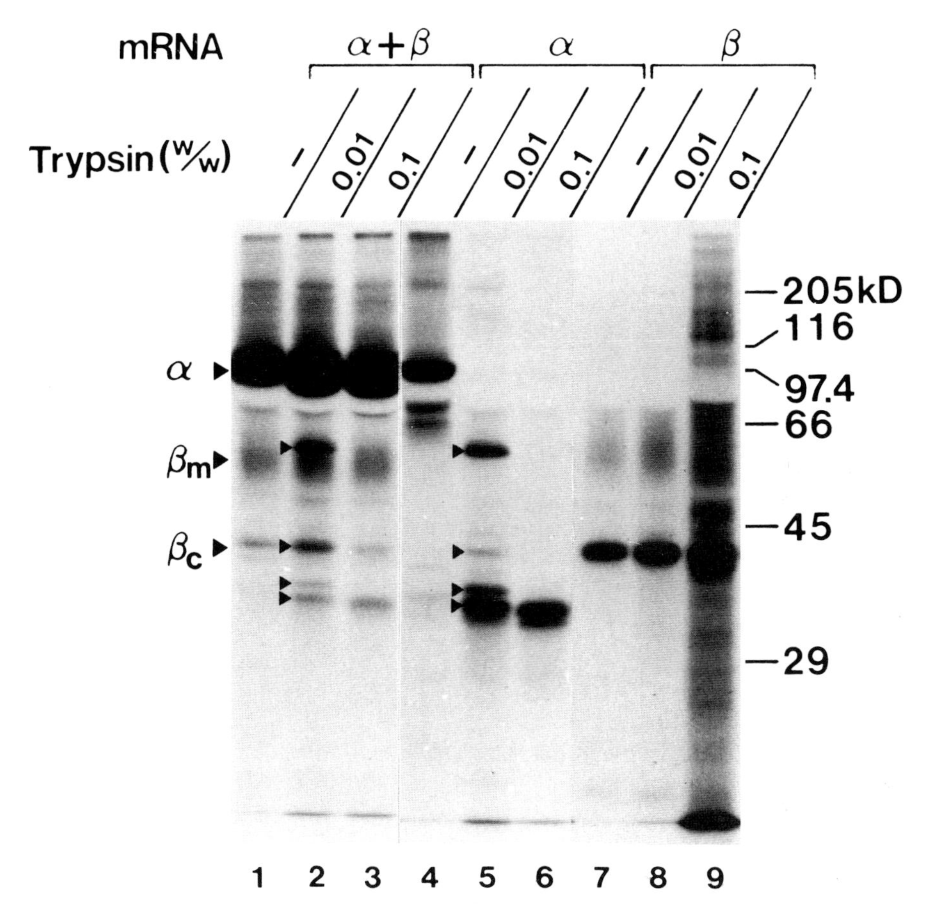

Figure 4. Trypsin sensitivity of the α- and β-subunit associated with the membrane (ppt-1) fraction. The ppt-1 fractions were prepared from oocytes injected with mRNAs and digested with trypsin (Worthington, TPCK-treated) at trypsin/protein ratios (wt/wt) of 0 (lanes *1*, *4*, and *7*), 0.01 (lanes *2*, *4*, *6*) and 0.1 (lanes *3*, *6*, and *9*) for 60 min on ice. After addition of soybean trypsin inhibitor (trypsin/inhibitor = 10 wt/wt), the membrane was analyzed by immunoprecipitation, SDS-polyacrylamide gel electrophoresis, and subsequent fluorography. Lanes *1–3*, from oocytes injected with both mRNAs; lanes *4–6*, from mRNAα-injected oocytes; lanes *7–9*, from mRNAβ-injected oocytes.

Geering et al. (1987, 1989) have reported similar results in which they claim that functional maturation of the α-subunit increases its trypsin resistance and that the endogeneous and trypsin-sensitive α-subunit present in *Xenopus* oocytes becomes trypsin resistant when oocytes are injected with mRNAβ. Geering (1990) has recently reviewed subunit assembly and functional maturation of Na,K-ATPase,

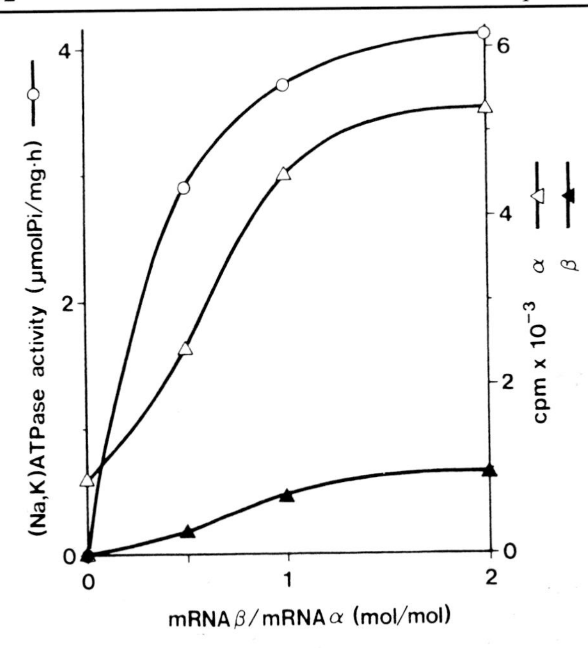

Figure 5. Effect of varying amounts of mRNAβ coinjected with a fixed amount of mRNAα on synthesis of the α- and β-subunit and Na,K-ATPase activity. To a fixed amount of mRNAα (10 ng/oocyte) was added varying amounts of mRNAβ (0, 0.28, 5.8, and 11.6 ng/oocyte). The mRNAβ to mRNAα molar ratios of these mixtures were 0, 0.5, 1, and 2, respectively. The oocytes that had received any of the mRNA mixtures were incubated at 19°C for 3 d in the presence of [^{14}C]leucine. The ppt-1 fractions that were prepared as described in Fig. 2 were subjected to immunoprecipitation, and the immunoprecipitates were analyzed by SDS-polyacrylamide gel electrophoresis and subsequent fluorography. Radioactivities associated with the α-(△) and β-subunit (βm plus βc, ▲) bands and Na,K-ATPase activities of ppt-1 fractions (○) are plotted against the molar ratios of mRNAs injected. The ppt-1 fraction from oocytes into which no mRNAβ was injected was used to assess the endogenous Na,K-ATPase activity (1.71 μmol Pi/mg per h). The Na,K-ATPase activities are plotted after subtracting the endogenous activity.

where she mentioned that assembly of α- and β-subunit is essential for expression of functional Na,K-ATPase.

The mechanism by which the β-subunit facilitates correct assembly of the α-subunit into the membrane is unclear. A clue to this problem is the observation that the quantity of the subunit synthesized in oocytes that had received only mRNAα was significantly smaller than that detected in oocytes injected with both mRNAα and mRNAβ (Fig. 1). This suggests that in oocytes the α-subunit is less stable in the absence of the β-subunit than in its presence. The stabilizing effect of the β-subunit on the α-subunit is also suggested by the finding that the amount of the α-subunit detected in oocytes that have received a fixed amount of mRNAα increases

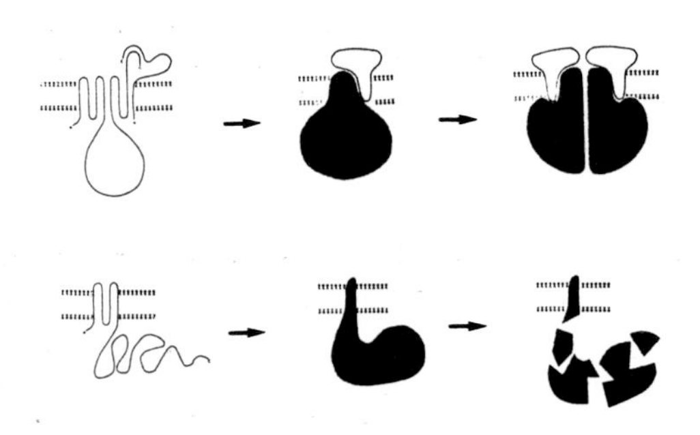

Figure 6. Membrane insertion of the α-subunit in the presence or absence of the β-subunit. In the absence of the β-subunit (*lower*), the α-subunit is inserted into membrane in an aberrant fashion and becomes unstable. In the presence of the β-subunit (*upper*), the β-subunit stabilizes the α-subunit by forming αβ complex and thereby leads to correct conformation of the α-subunit.

as the amount of coinjected mRNAβ increases (Fig. 5). It seems, therefore, that the stabilization of the α-subunit by the β-subunit is indispensable for the α-subunit to be correctly inserted into the membrane. It is likely that the β-subunit stabilizes the α-subunit by forming the αβ complex in the membrane and that only the insertion of the α-subunit in this complex, but not of the α-subunit alone, into the membrane leads to correct conformation of the α-subunit and appearance of Na,K-ATPase activity (Fig. 6).

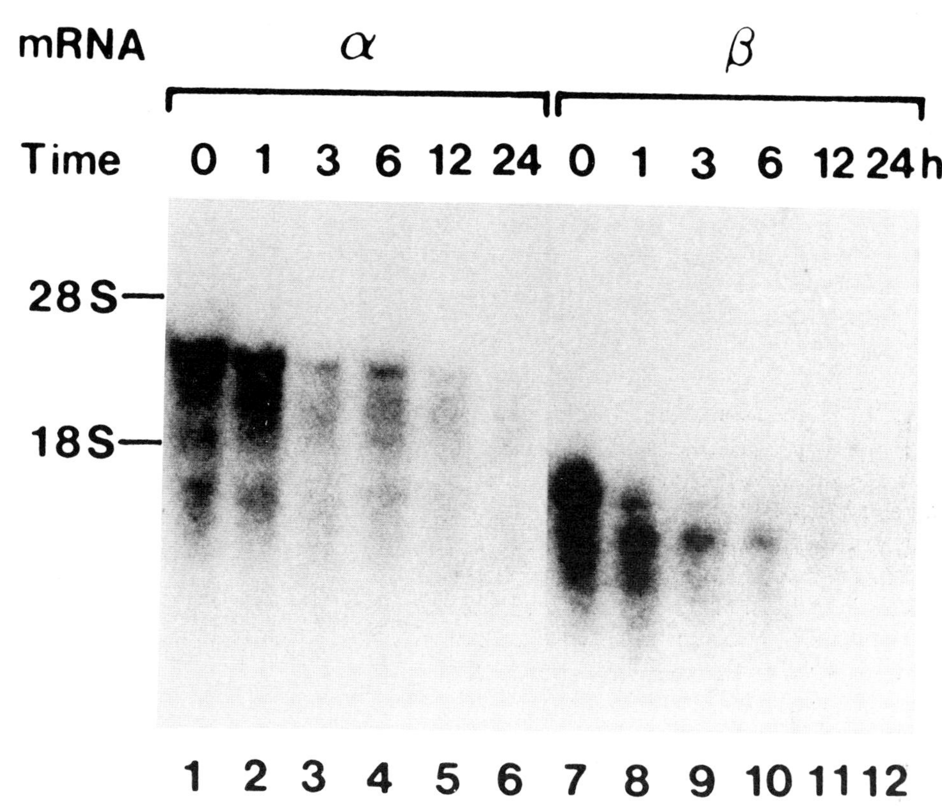

Figure 7. Blot hybridization analysis of the α- and β-subunit mRNAs of Na,K-ATPase. The oocytes were injected with either mRNA (0.5 μg/μl, 20 nl/oocyte) for the α- or β-subunit and incubated at 19°C for 3 d. The oocytes were then injected with the antisense oligonucleotide specific for either the α-subunit (1.9 μg/μl, 20 nl/oocyte) or the β-subunit (3.3 μg/μl, 20 nl/oocyte). The amounts of the antisense oligonucleotides injected corresponded to a 500 times molar excess of the respective mRNA that had been injected into an oocyte. At 0, 1, 3, 6, 12, and 24 h after injection of the antisense oligonucleotide, ~10 μg of extracted oocyte RNAs were electrophoresed, blotted, and probed with radioactive α-subunit (lanes *1–6*) or β-subunit (lanes *7–12*) cDNA probes.

It was not known whether the pre-existing β-subunit can assemble with and stabilize the α-subunit expressed later in single oocytes and whether the α-subunit inserted into the membrane in an aberrant fashion can be rescued by the β-subunit. For this purpose, the α- and β-subunits were expressed in turn in single oocytes by alternately injecting specific mRNAs for the α- and β-subunits. Dash et al. (1987) have demonstrated that injection of small oligonucleotides into *Xenopus* oocytes leads to complete degradation of complementary mRNA by means of an RNase H-like activity. To avoid the coexistence of mRNAα and mRNAβ in single oocytes, we injected a synthetic oligonucleotide specific for the mRNA first injected to degrade the first mRNA before the injection of the second mRNA. The crucial point of this experiment is whether an antisense oligonucleotide complementary to the mRNA first injected abolishes the translation of the mRNA while the second mRNA is translated. To study whether an antisense oligonucleotide could degrade mRNA for the Na,K-ATPase subunit, the oligonucleotide complementary to a part of the mRNAα [3′-(3)CCCCTTTCCCCGACGTTCACTCTTC(27)-5′] and mRNAβ [3′-(78)CCCGTCCTGGCCGTGCTCGACCAAG(102)-5′] were injected 3 d after injection of mRNAα and mRNAβ, respectively, and the time course and degree of mRNA degradation after antisense oligonucleotide injection were examined by using blot hybridization analysis. The results are shown in Fig. 7. The injected mRNA was recovered with total oocyte RNA at 1, 3, 6, 12, and 24 h after injection of

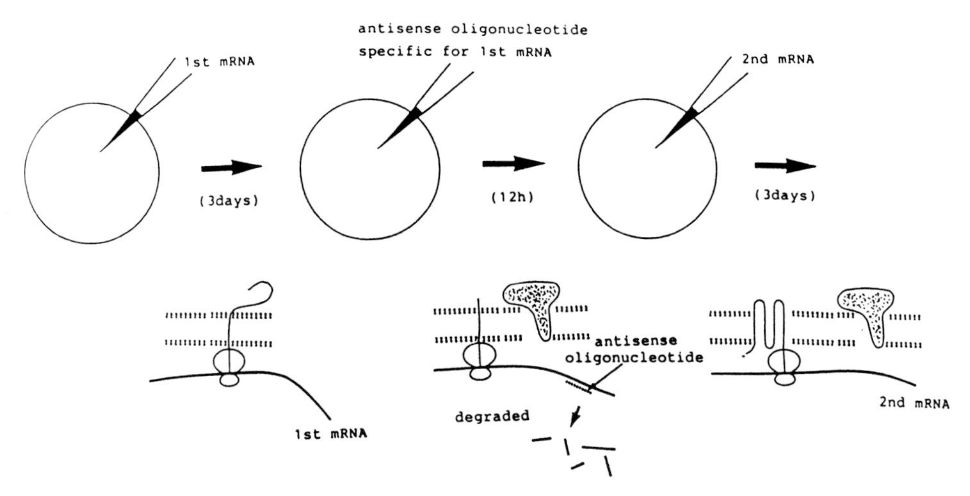

Figure 8. Sequential injection of mRNAs and antisense oligonucleotide in a single oocyte. For details, see text.

oligonucleotide, electrophoresed, and transferred to Biodyne membrane for hybridization to either α-subunit (lanes *1–6*) or β-subunit (lanes *7–12*) cDNA probes. In both cases, the mRNAs were almost completely degraded 12–24 h after the injection of the oligonucleotides.

These observations strongly indicate that the first-injected mRNA did not coexist in oocytes with the mRNA injected next as long as the latter mRNA injection was performed no earlier than 12 h after injection of the antisense oligonucleotide specific for the first mRNA. Instead, the translation products of the first-injected mRNA coexist with the second mRNA injected and hence with the nascent or newly synthesized polypeptides translated from the second mRNA (Fig. 8). We then performed sequential injections of mRNAs and antisense oligonucleotides in single oocytes. We first expressed the α-subunit followed by expression of the β-subunit in single oocytes, or in the reverse order by alternately injecting the respective mRNAs.

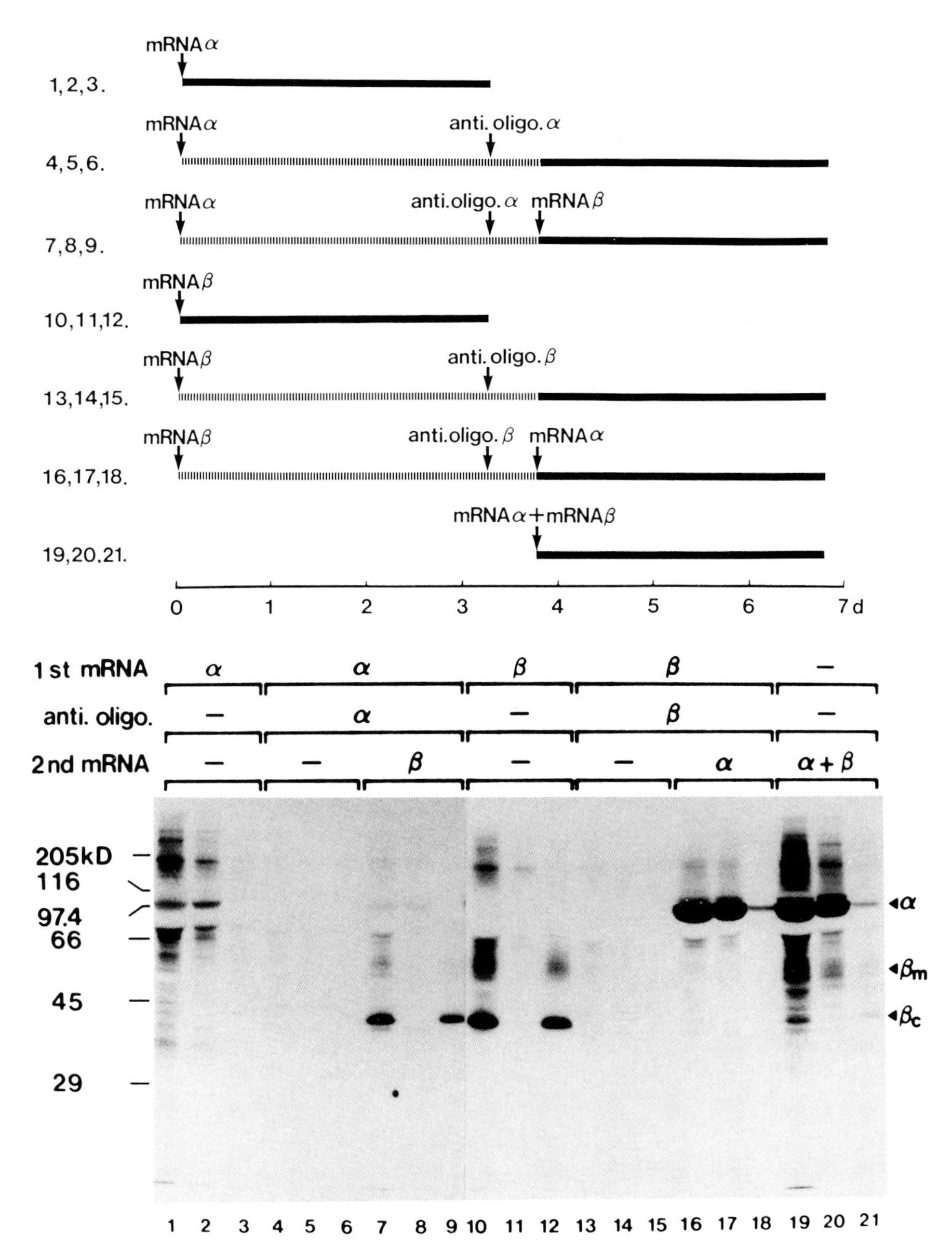

Figure 9. Expression of Na,K-ATPase subunits in mRNA-injected oocytes. (*Upper*) Time schedule for the injection of mRNAs and antisense oligonucleotides. The figures on the left side correspond to the lane numbers in the fluorograms. Broken and solid lines represent the periods when the oocytes were incubated in the absence and the presence of [^{3}H]leucine, respectively. (*Lower*) Oocytes were injected with mRNAs and antisense oligonucleotides in turn as illustrated in the upper figure. The concentrations used for injections were as follows: both mRNAs, 0.5 μg/μl each; antisense oligonucleotide for the α-subunit, 1.9 μg/μl; antisense oligonucleotide for the β-subunit, 3.3 μg/μl. 20 nl of each was injected into an oocyte. The extracts were immunoprecipitated with antiserum against the α-subunit (lanes *2*, *5*, *8*, *11*, *14*, *17*, and *20*), antiserum against the β-subunit (lanes *3*, *6*, *9*, *12*, *15*, *18*, and *21*), or with a mixture of both sera (lanes *1*, *4*, *7*, *10*, *13*, *16*, and *19*) and analyzed by gel electrophoresis followed by fluorography.

TABLE I
Na,K-ATPase and Ouabain-binding Activities in mRNA-injected Oocytes

Sample	ATPase activity	Ouabain-binding activity
	μmol Pi/mg per h	*fmol/oocyte*
1. mRNAα → anti.oligo.(α) → mRNAβ	2.65 ± 0.35	19.1 ± 3.7
2. mRNAβ → anti.oligo.(β) → mRNAα	3.71 ± 0.84	25.8 ± 3.1
3. mRNAα + mRNAβ	4.65 ± 0.09	40.5 ± 12.1
4. Noninjected	2.28 ± 0.19	17.2 ± 3.4

For samples 1 and 2, oocytes were injected with mRNAs and antisense oligonucleotides in the order indicated in the left column. Sample 3 is a positive control in which both mRNAs were injected simultaneously, and sample 4 is used to assess the endogenous activities in oocytes. The concentrations of materials and time schedule for expression were the same as indicated in Fig. 9. The values are the means of three independent experiments. Anti.oligo.(α) and anti.oligo.(β) denote the antisense oligonucleotides specific for the α- and β-subunits, respectively.

The translation of the mRNA that had been injected first was blocked by the injection of the antisense oligonucleotide specific for that mRNA before the second injection of the other mRNA. The results are shown in Fig. 9. The oocyte was first injected with mRNA for the β-subunit, then with the antisense oligonucleotide specific for the β-subunit, and finally with mRNA for the α-subunit containing a high level of the labeled α-subunit (Fig. 9, lanes *16* and *17*). Detectable labels were not incorporated into either the core-glycosylated (βc) or the fully glycosylated β-subunit (βm) of this oocyte (Fig. 9, lanes *16* and *18*), suggesting that the α-subunit but not the β-subunit was synthesized during the second 3-d incubation, which was carried out in the presence of [^{3}H]leucine. Nevertheless, the labeled α-subunit was immunoprecipitated with antiserum against the β-subunit, though the precipitated α-subunit was slight (Fig. 9, lane *18*). These results indicate that the β-subunit which has been incorporated in the membrane can assemble with the α-subunit expressed later in single oocytes. As mentioned earlier, partial dissociation of the αβ complex caused by Triton might lead to the precipitation of a small amount of the α-subunit with anti-β-subunit serum. The oocytes were first injected with the mRNA for the

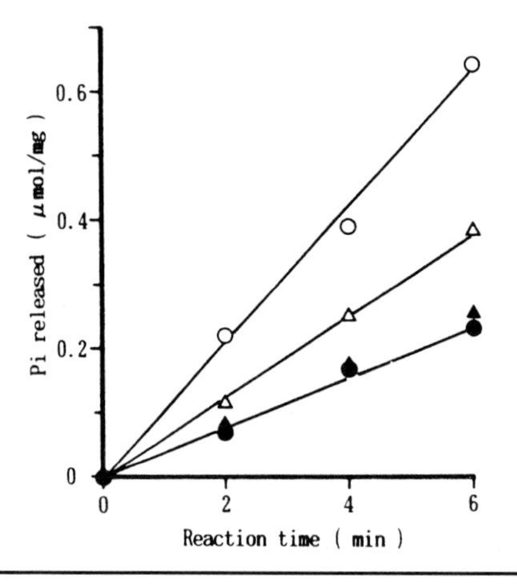

Figure 10. Na,K-ATPase activity of the complexes of the α-subunit with Cys-mutants of the β-subunit expressed in *Xenopus* oocytes. Ouabain-sensitive ATPase activity in microsomes from noninjected oocytes (●), and from oocytes injected with mRNAβ for the wild-type (○), Cys-127 to Ser (△), or Cys-278 to Ser (▲) mutant together with mRNA for the wild-type α-subunit.

α-subunit, then with an antisense oligonucleotide specific for the α-subunit, and finally with the mRNA for the β-subunit produced a high level of the β-subunit (Fig. 9, lanes *7* and *9*). Contrary to the former case, any detectable β-subunit was not immunoprecipitated with antiserum against the α-subunit (Fig. 9, lane *8*), suggesting that very little, if any, of the αβ complex is formed in this case.

The Na,K-ATPase and ouabain-binding activities of the oocytes were assayed to determine whether the αβ complexes formed through the alternate injections of mRNAs were enzymatically active. Oocytes injected with mRNAs for the α- and

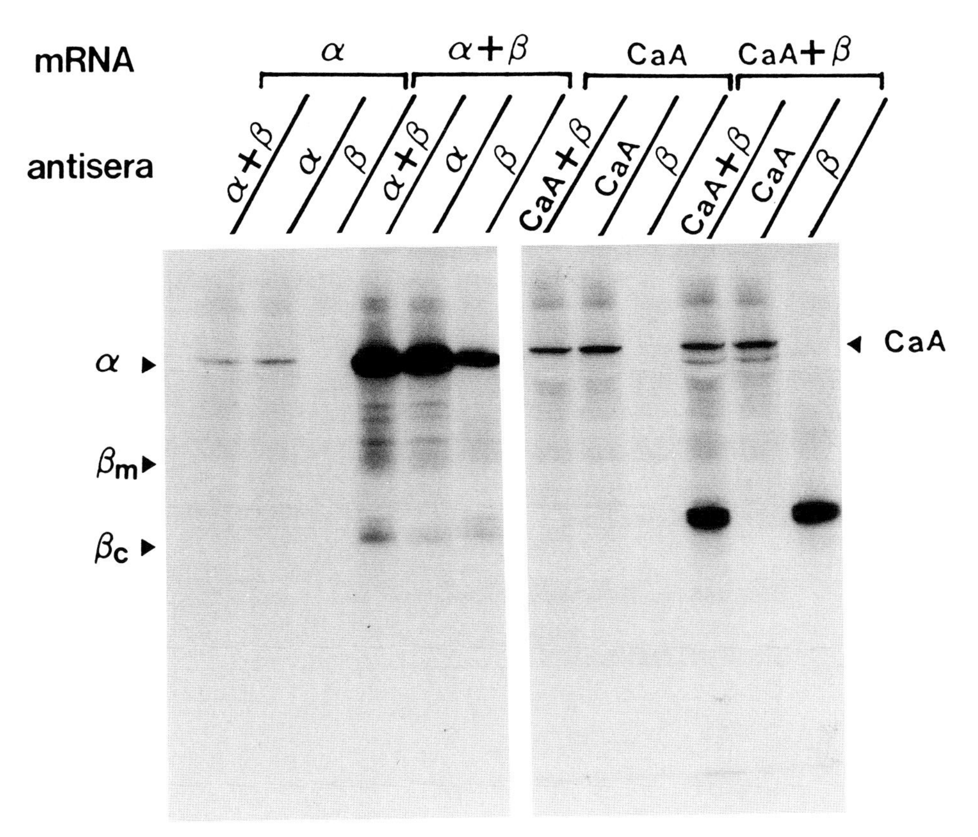

Figure 11. Coinjection of mRNA for Ca-ATPase of sarcoplasmic reticulum with mRNAβ into *Xenopus* oocytes. Oocytes injected with either mRNAα or mRNA for Ca-ATPase solely or in combination with mRNAβ were incubated at 19°C for 3 d in the presence of [^{3}H]leucine. The labeled translation products were immunoprecipitated with antisera as shown on the top of figure. CaA represents Ca-ATPase.

β-subunits simultaneously were used as positive controls, and noninjected oocytes were used as controls to assess the endogenous Na,K-ATPase activity. The results are shown in Table I. Both the ATPase and ouabain-binding activities in oocytes which were first injected with mRNA for the β-subunit, then with antisense oligonucleotide for the β-subunit, and finally with mRNA for the α-subunit were obviously higher than those in noninjected oocytes, although they were lower than those in oocytes injected with both mRNAs simultaneously. The oocytes first injected with mRNA for the α-subunit, then with antisense oligonucleotide for the α-subunit, and finally with mRNA for the β-subunit showed little increase in both activities from those of noninjected oocytes. These results indicate that the β-subunit, probably remaining in the endoplasmic reticulum, can assemble with and stabilize the newly synthesized α-subunit, leading to the active enzyme complex.

It is tempting to assume that although the α-subunit alone is inserted into the membrane, the conformation of the α-subunit thus inserted may be somehow different from that of the α-subunit in the complex with the β-subunit and that correct conformation of the α-subunit may be obtained only when nascent polypep-

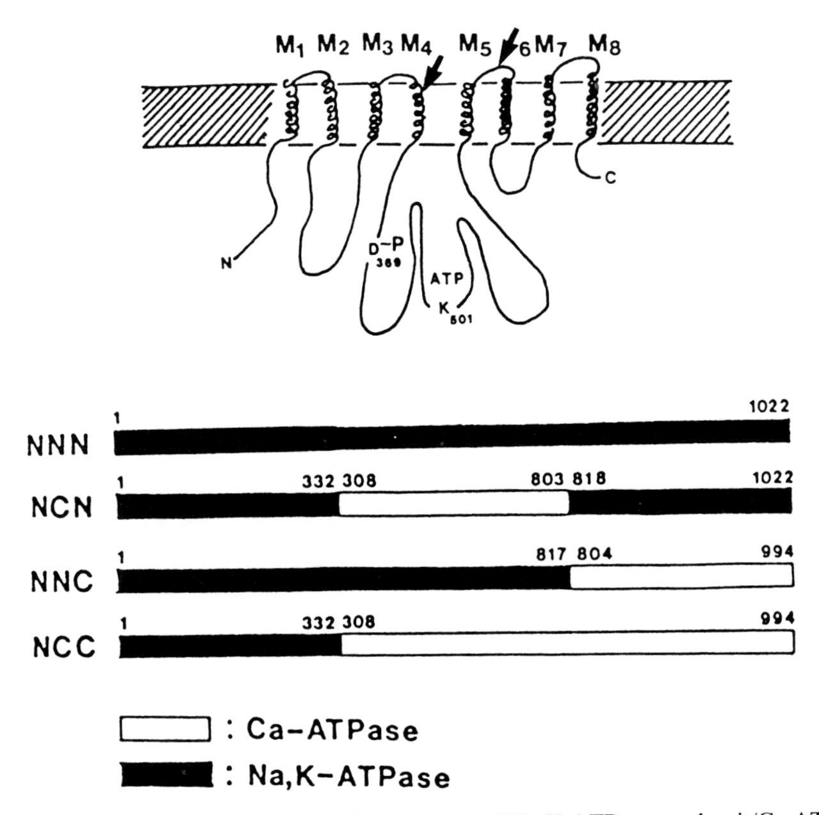

Figure 12. Schematic representation of the structures of Na,K-ATPase α-subunit/Ca-ATPase chimeric proteins. The two joining points are shown by arrowheads in the model (*upper*). N and C represent the α-subunit and Ca-ATPase, respectively. M represents transmembrane segment.

tides of the α-subunit are assembled with the β-subunit. In other words, the β-subunit is a receptor for the nascent α-subunit, as has been claimed by Hiatt et al. (1984).

Another interesting point is which part of the β-subunit interacts with the α-subunit in the αβ complex. The results obtained from site-directed mutagenesis of cysteine residues of the β-subunit (see Noguchi et al., 1991) and from expression of chimeric cDNAs between Na,K-ATPase α-subunit and Ca-ATPase of sarcoplasmic reticulum (Morohashi, M., T. Mita, and M. Kawamura, manuscript in preparation) shed light on the interaction site(s) between the α- and β-subunits. The β-subunit contains seven cysteine residues, six of which are cross-linked by disulfide bridges (for *Torpedo californica,* Cys^{127}–Cys^{150}, Cys^{160}–Cys^{176}, Cys^{215}–Cys^{278}; Kirley, 1989; Miller and Farley, 1990). Among these cysteine residues, Cys^{127} and Cys^{278} were replaced by serine by site-directed mutagenesis and resulting mutants were expressed in *Xenopus* oocytes by injecting the mutated mRNAβ together with wild-type mRNAα. As shown in Fig. 10, ouabain-sensitive ATPase activity of microsomes from oocytes injected with Cys^{127}-mutated mRNAβ obviously increased from the level of the control oocytes (none of mRNA was injected), while the ATPase activity in Cys^{278} mutant was at the control level. Immunoprecipitation experiments revealed that the αβ complex was formed in the former but not in the latter, suggesting that the Cys^{215}–Cys^{278} disulfide bridge and, hence, the conformation of the COOH-terminal region of the β-subunit are involved in the formation of the αβ complex.

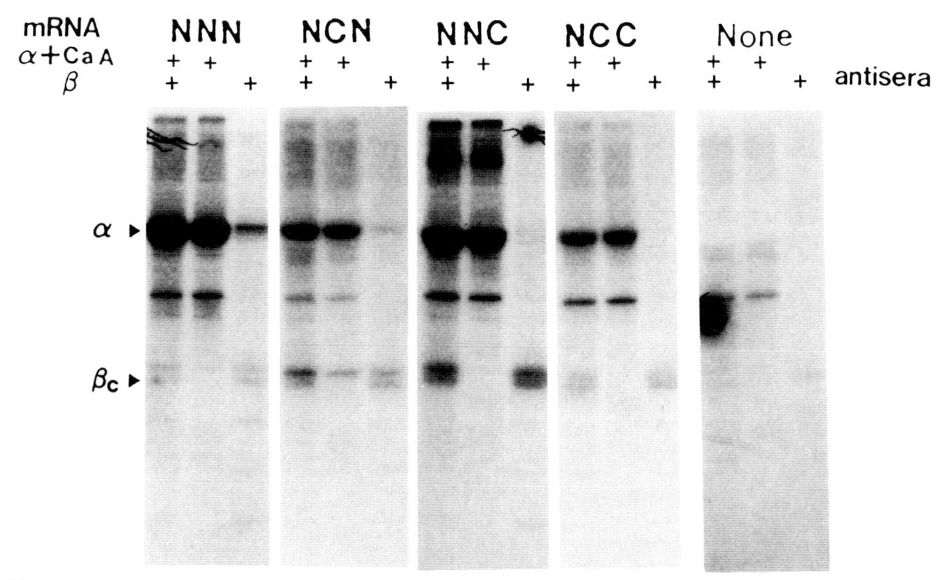

Figure 13. Expression of chimeric proteins in *Xenopus* oocytes. Chimeric mRNAs injected into oocytes and antisera used for immunoprecipitation are shown at the top of figure. mRNA for the wild-type β-subunit was simultaneously injected. For chimeric mRNAs, see Fig. 12. Injection of mRNAs, labeling with [^{3}H]leucine and immunoprecipitation were carried out as shown in Fig. 11.

The sarcoplasmic Ca-ATPase exhibits quite similar hydrophobic profiles and possesses an essentially identical transmembrane topology to that of the α-subunit of Na,K-ATPase. However, when the mRNA for Ca-ATPase and mRNAβ were coinjected into oocytes, Ca-ATPase could not be precipitated with the antiserum to the β-subunit (Fig. 11), suggesting that the Ca-ATPase could not associate with the β-subunit. Taking advantage of this fact, we constructed three chimeric cDNAs, to reveal the region of the α-subunit involved in the assembly with the β-subunit. The two joining points were set in the region conserved between the two ATPases so that the chimeric proteins would have primary structures with equal size and that structural disturbances around the joins would be minimized. The schematic representation of the structures of those three chimeric proteins is shown in Fig. 12, where

N and C represent the α-subunit of Na,K-ATPase and Ca-ATPase, respectively. These chimeric mRNAs and wild-type mRNAα were injected into oocytes together with mRNAβ.

No detectable chimeric proteins were precipitated with antiserum to the β-subunit for NNC and NCC, whose COOH-terminal one-third were Ca-ATPase, whereas chimeric proteins were precipitated with antiserum to the β-subunit for NNN and NCN, whose COOH-terminal one-third were α-subunit (Fig. 13). These results suggest that the COOH-terminal one-third of the α-subunit is indispensable to the assembly with the β-subunit. As all three chimeric cDNAs that we constructed contained the NH_2-terminal one-third of the α-subunit, it is unknown whether the NH_2-terminal one-third is required for the αβ interaction.

Construction and expression of Na and Ca pump chimeric molecules has recently been reported by Takeyasu's group (see Luckie et al., 1991). Their chimeric molecule I, consisting of NH_2-terminal two-thirds of Na pump and COOH-terminal one-third of Ca pump, is expressed exclusively inside the cells, while chimeric molecule II, containing NH_2-terminal two-thirds of Ca-pump and COOH-terminal one-third of Na pump, is transported to the plasma membrane. Because the α-subunit is transported to the plasma membrane only when it assembles with the β-subunit, it is likely that the chimeric molecule II, but not I, is associated with the β-subunit. These results also suggest that COOH-terminal one-third of the α-subunit is required for the assembly with the β-subunit.

References

Dash, P., I. Lotan, M. Knapp, K. R. Kandel, and P. Goelet. 1987. Selective elimination of mRNAs *in vivo:* complementary oligodeoxynucleotides promote RNA degradation by an RNase H-like activity. *Proceedings of the National Academy of Sciences, USA.* 84:7896–7900.

Geering, K. 1990. Subunit assembly and functional maturation of Na,K-ATPase. *Journal of Membrane Biology.* 115:109–121.

Geering, K., J. P. Kraehenbuhl, and B. C. Rossier. 1987. Maturation of the catalytic α-subunit of Na,K-ATPase during intracellular transport. *Journal of Cell Biology.* 105:2613–2619.

Geering, K., I. Theulaz, F. Verrey, M. T. Hauptle, and B. C. Rossier. 1989. A role for the β-subunit in the expression of functional (Na^+-K^+)-ATPase in *Xenopus* oocytes. *American Journal of Physiology.* 257:C851–C858.

Hiatt, A., A. A. McDonough, and I. S. Edelman. 1984. Assembly of the (Na^+,K^+)-adenosine triphosphatase. Post-translational membrane integration of the α-subunit. *Journal of Biological Chemistry.* 259:2629–2635.

Kawamura, M., and K. Nagano. 1984. Evidence for essential disulfide bonds in the β-subunit of (Na^++K^+)-ATPase. *Biochimica et Biophysica Acta.* 774:188–192.

Kawamura, M., K. Ohmizo, M. Morohashi, and K. Nagano. 1985. Protective effect of Na^+ and K^+ against inactivation of (Na^++K^+)-ATPase by high concentration of 2-mercaptoethanol at high temperature. *Biochimica et Biophysica Acta.* 821:115–120.

Kirley, T. L. 1989. Determination of three disulfide bonds and one sulfhydryl in the β subunit of (Na,K)-ATPase. *Journal of Biological Chemistry.* 264:7185–7192.

Kirley, T. L. 1990. Inactivation of (Na^++K^+)-ATPase by β-mercaptoethanol. Differential sensitivity to reduction of the three β-subunit disulfide bonds. *Journal of Biological Chemistry.* 265:4227–4232.

Luckie, D. B., K. L. Boyd, A. Mizushima, Z. Shao, A. P. Somlyo, and K. Takeyasu. 1991. Expression of functional Na and Ca pump chimeric molecules. *In* The Sodium Pump: Recent Developments. J. H. Kaplan and P. De Weer, editors. The Rockefeller University Press, New York. In press.

Miller, R. P., and R. A. Farley. 1990. β Subunit of (Na^+ + K^+)-ATPase contains three disulfide bonds. *Biochemistry.* 29:1524–1532.

Noguchi, S., K. Higashi, and M. Kawamura. 1990*a*. A possible role of the β-subunit of (Na,K)ATPase in facilitating correct assembly of the α-subunit into the membrane. *Journal of Biological Chemistry.* 265:15991–15995.

Noguchi, S., K. Higashi, and M. Kawamura. 1990*b*. Assembly of the α-subunit of *Torpedo californica* Na^+/K^+-ATPase with its pre-existing β-subunit in *Xenopus* oocytes. *Biochimica et Biophysica Acta.* 1023:247–253.

Noguchi, S., M. Mishina, M. Kawamura, and S. Numa. 1987. Expression of functional (Na^+,K^+)-ATPase from cloned cDNAs. *FEBS Letters.* 225:27–32.

Noguchi, S., Y. Mutoh, and M. Kawamura. 1991. Site-directed mutagenesis of Cys in the β-subunit of the *Torpedo californica* Na, K-ATPase. *In* The Sodium Pump: Recent Developments. J. H. Kaplan and P. De Weer, editors. The Rockefeller University Press, New York. In press.

Takeda, K., S. Noguchi, A. Sugino, and M. Kawamura. 1988. Functional activity of oligosaccharide-deficient (Na,K)ATPase expressed in *Xenopus* oocytes. *FEBS Letters.* 238:201–204.

Tamkum, M. M., and D. M. Fambrough. 1986. The (Na^+,K^+)-ATPase of chick sensory neurons. Studies on biosynthesis and intracellular transport. *Journal of Biological Chemistry.* 261:1009–1019.

Chapter 5

Overview: Subunit Diversity in the Na,K-ATPase

Kathleen J. Sweadner

Neurosurgical Research, Massachusetts General Hospital, Boston, Massachusetts 02114; and Department of Cellular and Molecular Physiology, Harvard Medical School, Boston, Massachusetts 02115

Summary

The pioneering work on the Na,K-ATPase was dominated by the concept that this enzyme was essentially the same in all tissues and organisms. Now, with the development of new molecular and immunological techniques, the Na,K-ATPase is found to be composed of several clearly different subunit types. This brief update focuses on the identification of subunits and recent ideas about their functional differences.

Introduction

Research on ion transport processes in the 1960's and 70's was dominated by the simplifying paradigm that complex tissue behavior could be understood as the sum of the actions of several essentially stereotypical transport proteins. In other words, the active transport of Na^+ and K^+ was carried out by a Na,K-ATPase that catalyzed an identical molecular reaction wherever it was found, and secondary carriers and channels determined whether a cell was involved in absorption, secretion, or action potentials. This useful concept made it possible to focus on the enzymology of the purified protein.

Biochemical, molecular, and electrophysiological techniques eventually revealed that the majority of proteins that play interesting roles in cell behavior come in multiple molecular forms, each specialized in subtle ways (Lester, 1988). The Na,K-ATPase has isoforms of both its α- and β-subunits (reviewed in Sweadner, 1989; Lingrel et al., 1990).

Alpha Isoforms: Multiple Genes Arising from One?

In vertebrates (rat, human, and chicken), it has been clearly shown that there are three independent genes for α-subunits, known as α1, α2, and α3. The deduced isoform protein sequences show unambiguous similarity even between species (see Fambrough et al., 1991; Lingrel et al., 1991), leading to the supposition that any unique functional roles they play arose early and were conserved through a substantial part of evolution.

The H,K-ATPase, which is 70% identical to the Na,K-ATPase and homologous along its entire length, can be considered to be a fourth isoform (reviewed in Sweadner, 1989). The H,K-ATPase has also been found to be complexed with a β-subunit homologous to those of the Na,K-ATPase (see below). Other members of the aspartyl-phosphate family of ion transport ATPases, such as Ca-ATPases and proton pumps, share blocks of homology with the Na,K-ATPase, but also have long stretches of completely dissimilar sequence, and have not been proven to have β-subunits.

Recent examination of Na,K-ATPase α-subunit genes from invertebrate animals has been used to assess the evolutionary origin of the Na,K-ATPase isoforms. Complete cDNA sequences for α-subunits have been determined for crustaceans *Drosophila* (Lebovitz et al., 1989) and *Artemia* (Baxter-Lowe et al., 1990), and for the coelenterate *Hydra* (Canfield et al., 1990*c*), and a partial sequence for the nematode *Caenorhabditis* (Somerville and Fambrough, 1990). In all cases there was substantial sequence homology with all three vertebrate genes, with the majority of divergence at the NH_2 terminus. Extensive restriction mapping of *Drosophila* clones did not reveal any other genes, suggesting that invertebrates may not have multiple isoform genes (Lebovitz et al., 1989). The brine shrimp *Artemia,* however, expressed Na,K-ATPase activities with demonstrable kinetic differences in different tissues, as recently detailed and reviewed by Cortas et al. (1989).

It has often been possible to partially resolve the Na,K-ATPase α-subunit isoforms from each other on SDS gels (reviewed in Sweadner, 1989). Occasionally, however, doublets are seen in preparations that ostensibly have only one isoform. Recent developments have indicated that the production of such doublets is influenced by experimental variables. Siegel and Desmond (1989) found that the proportion of alkyl sulfates of higher chain length (sodium tetradecyl sulfate) in their SDS was critical for seeing doublets in kidney preparations; Cortas et al. (1988, and manuscript submitted for publication) have shown that the NH_2-terminal sequences of the proteins in the doublets are both that of α1. A similar observation was made by Kurihara et al. (1990), who found doublets in preparations from rat submandibular gland, even in conditions where rat kidney preparations gave a single band. The new band migrated faster than either band in brain preparations. NH_2-terminal sequence analysis, however, indicated that it was (or was derived from) α1. The influence of detergent composition, heating in SDS, and reduction and alkylation upon α isoform gel mobility has now been analyzed at length (Sweadner, 1990). The conclusion was reached that anomalous electrophoretic mobility results from the incomplete unfolding of the α-subunit in detergent, and that the appearance of doublets alone is not sufficient evidence to document the existence of new isoforms.

Isoforms of β

At this writing, four different β-subunits have been identified. The prototypical Na,K-ATPase β-subunit characteristic of mammalian kidney, now known as β1, has been sequenced from several vertebrate species (reviewed in Sweadner, 1989; Lingrel et al., 1990). A mammalian β2-subunit was found by low stringency hybridization (Martin-Vasallo et al., 1989*a*), and independently cloned as a neuron–glia adhesion protein (Pagliusi et al., 1989; Gloor et al., 1990). Its sequence is now known from human, rat, mouse, and chicken (Lemas et al., 1990), and the level of identity with β1 is only ~40%. A different β-subunit type, now known as β3, has been isolated from *Xenopus* (Good et al., 1990). Finally, a β-subunit for the H,K-ATPase has been identified (Hall et al., 1990; Okamoto et al., 1990) and cloned (Canfield et al., 1990*b*; Reuben et al., 1990; Shull, 1990) from rat and rabbit (see Canfield et al., 1991). It shares 29% identity with Na,K-ATPase β1 and 37% identity with Na,K-ATPase β2 (Shull, 1990). All of the β-subunits are predicted to have a single transmembrane segment close to the NH_2 terminus, and there is conservation of cysteines. They differ, however, in the number of asparagine-linked glycosylation consensus sequences, and there are two stretches of amino acid sequence that are completely divergent when β1 and β2 are compared (Gloor et al., 1990).

Where Are α1, α2, α3, β1, and β2 Expressed?

Early work with both isoform-specific antibodies and Northern blots established the tissue specificity of α isoform expression (reviewed in Sweadner, 1989). Antibodies made against fusion proteins (Shyjan and Levenson, 1989) and NH_2-terminal sequence analysis (Arystarkhova et al., 1990) have recently been used. α1 is found nearly everywhere, while [α2] predominates in skeletal muscle. α2 is found in the adult rat heart, while α3 is expressed in the newborn rat heart (Orlowski and Lingrel, 1990; Lucchesi and Sweadner, 1991); in other species, including human, α3 is expressed in adult cardiac tissue (Amato, S., and K. J. Sweadner, unpublished observations). α3 is found predominantly in the nervous system (reviewed in Sweadner, 1989) and has been reported in ciliary epithelium (Martin-Vasallo et al., 1989*b*; Ghosh et al., 1990), pineal (Shyjan et al., 1990*b*), and in small amounts in the kidney (Arystarkhova et al., 1989). In certain human leukemia cell lines (Gilmore-Hebert et al., 1989) α3 mRNA was found to be expressed in undifferentiated cells and to change its level of expression when cells were induced to differentiate.

β1, like α1, appears to be expressed in many tissues, although its stoichiometric presence in liver and skeletal muscle remains controversial. β2 is expressed selectively in the brain (Pagliusi et al., 1989; Martin-Vasallo et al., 1989*a*; Shyjan et al., 1990*c*). It copurifies with Na,K-ATPase activity (Shyjan et al., 1990*c*) and has been purified specifically as a complex with α2 (Gloor et al., 1990). It has also been detected with α3 in the pineal (Shyjan et al., 1990*b*) and in photoreceptors (Schneider and Kraig, 1990), and with α2 or α3 in the ciliary epithelium (Coca-Prados et al., 1990).

Because all three isoforms are expressed in the brain, the nervous system has become a particular focus for investigating which cell types express which isoforms. Analysis of populations of neurons and glia in culture has been used in combination with Northern blots to demonstrate that α2 mRNA is predominant in glial cells while α3 is predominant in neuronal cells (Corthésy-Theulaz et al., 1990*b*). The heterogeneous cell populations studied, combined with the background of mRNA of each type, makes it impossible to use these data as evidence for absolute cell type specificity for each isoform; however, the data suggest a quantitative trend that has been corroborated by immunocytochemistry and in situ hybridization techniques.

In situ hybridization with isoform-specific probes has been used to assess the cellular location of the mRNAs. Schneider et al. (1988) examined cross-sections of entire rat fetuses, demonstrating characteristic organ distributions of the isoforms, and examined sections of adult rat brain at higher magnification. They found a concentration of α3 mRNA in large-diameter neurons in the cortex and cerebellum; α2 mRNA was very diffuse in its distribution. Hieber et al. (1989) used an α1 probe at relatively low stringency to compare mRNA localization with stain obtained with a crossreactive anti–Na,K-ATPase antiserum using kidney, retina, and cerebellum. As expected for the nervous system, where the plasma membrane of neurons is often very distant from the nucleus, the hybridization signal was often removed from the immunofluorescent signal. These investigators have gone on to perform high stringency hybridization with probes for α1 and α3 in the central and peripheral nervous systems (Siegel et al., 1990; Hieber et al., 1991; Mata et al., 1991). Filuk et al. (1989) employed in situ hybridization at several anatomical levels with probes for all three α isoforms using very low magnification. The low magnification images illustrate graphically how heterogeneous the isoform distribution is. In situ hybridization studies have also been used to detect the distribution of the β isoform mRNAs in different regions of the central nervous system (Filuk et al., 1989, 1990; Pagliusi et al.,

1990). In particular, a remarkable concentration of β2 is found in the granular cell layer of the cerebellum, compared with other brain regions.

Isozyme-specific antibodies have been used to demonstrate the presence of the proteins in different cell types. The most detailed studies to date have been done on nervous tissue, notably the retina, brain, and spinal cord. McGrail and Sweadner (1989) used monoclonal antibodies to α1, α2, and α3 to stain sections of the retina and optic nerve and dissociated retinal cells. Very distinct differences were seen in the distribution of the isoforms, even at low magnification. Photoreceptors expressed predominantly α3, and the stain was narrowly restricted to the inner segment, where the sodium pump powers the dark current that originates in the outer segments. Bright stain for α3 was also seen in the outer synaptic layer where photoreceptors synapse with other retinal neurons, but staining in the inner synaptic layer was relatively light, and stratified at particular levels. Stain for α1, in contrast, was very bright in the inner synaptic layer. Stain for α2 was diffuse and light, except for prominent staining of a subclass of amacrine neurons and of blood vessels. When dissociated, identified cells were examined, one type of neuron (bipolar cells) was found to express only α3, and the stain was concentrated in its dendrites. Another neuron, the horizontal cell, expressed α1 and α3 at high levels. The glial cell of the retina, the Müller cell, expressed both α1 and α2, but all of the Müller cells expressed α1, while only 15% expressed α2. The localization studies made it very clear that the Na,K-ATPase isoforms cannot be classified as "neuronal" or "glial." α1, α2, or α3 can be expressed in neurons, and α1 or α2 in glia. Investigation of the distribution of isoforms in the brain corroborated this picture of very cell-specific isoform gene expression control (McGrail et al., 1991). Cerebellar cortex, hippocampus, corpus callosum, pons, and spinal cord were examined. Many large-diameter neurons, which produce long axons, were found to express α2 and α3 in the vicinity of their cell bodies, and many white matter tracts were found to have α3 in the axolemma. There were exceptions to this generalization, however, making it impossible to draw simple conclusions.

A similar degree of cell type specificity was seen in the ciliary epithelium by Martin-Vasallo et al. (1989*b*) and Ghosh et al. (1990). This is a transporting epithelium secreting the aqueous humor of the eye, but unlike most epithelia it is anatomically complex and has two layers composed of different cell types. Nonpigmented cells of the pars plicata expressed all three isoforms, while in the pars plana they expressed α1 and α2. The pigmented cells of both pars plana and pars plicata expressed only α1.

Are the α Isoforms Functionally Different?

The difference in cardiac glycoside sensitivity of the rat Na,K-ATPase isoforms is well known, and its basis in particular amino acid residues has been reviewed elsewhere (Lingrel et al., 1990). Two recent contributions have documented that three different cardiac glycoside affinities can be detected in preparations from rat brain, presumably due to α1, α2, and α3 (Berrebi-Bertrand et al., 1990; Blanco et al., 1990). Cardiac glycoside affinity differences have been noted for other species as well, but puzzling inconsistencies between the results of different investigators continue to complicate the picture and cannot be treated in depth here. Suffice it to say that the biological role of the cardiac glycoside binding site, if any, is still not known.

One of the most fundamental questions about the Na,K-ATPase isoforms is whether or not they catalyze the same transport reaction. The H,K-ATPase differs from the Na,K-ATPase in only two significant respects: it transports H^+ instead of Na^+, and it is not detectably affected by cardiac glycosides. These two features should prove to have an interesting structural basis when the two enzymes are compared. The α1, α2, and α3 isozymes of the Na,K-ATPase have not yet been explicitly compared under identical conditions. It has been established, however, that α2 (Canfield et al., 1990*a*; Jewell and Lingrel, 1990) and α3 (Jewell and Lingrel, 1990) can support the survival of ouabain-sensitive cells in the presence of ouabain when they are modified to be ouabain resistant by site-directed mutagenesis. Transfection experiments have also established that both α3 (Hara et al., 1988) and α2 (Shyjan et al., 1990*a*) are capable of catalyzing $^{86}Rb^+$ uptake. The biochemical properties of the α3/β2 complex found in the pineal have also been studied (Shyjan et al., 1990*b*), indicating that it is essentially similar in its reaction to α1.

Other functional differences have been reported. Mixtures of α2 and α3 have a higher nucleotide affinity and broader nucleotide specificity (reviewed in Sweadner, 1989). A higher affinity for Na^+ has been known for brain Na,K-ATPase since the work of Skou (1962). This higher affinity was found to be a property of axolemma Na,K-ATPase (α2 and α3; reviewed in Sweadner, 1989), and has recently been found to be characteristic of the α3 of the pineal (Shyjan et al., 1990*b*). Anner et al. (1989) have demonstrated a difference in Na^+ affinity and maximum transport rates between the α1 isoforms of rat and rabbit, however. The difference was described as a "defect," not an affinity difference. The data raise the question of whether the isoform differences noted in the rat will be found to be applicable to other species.

Other kinetic differences between isoforms have been reported, including differences in the balance between E_1-E_2 conformation changes and the effects of ligands on K^+-pNPPase activity (reviewed in Sweadner, 1989). Since all of these experiments were performed on mixtures of α and β isoforms, there is much room for further investigation. Sensitivity to tryptic digestion, a property used to probe conformational states, has also been proven to be different. With α1 examined from a variety of animal species, cleavage has been observed at the T1, T2, and T3 sites in a manner dependent on the Na^+ or K^+ conformations of the enzyme (reviewed in Jørgensen and Andersen, 1988). The tryptic cleavage sites are conserved in the three α isoform sequences, and yet cleavage of α2 and α3 proceeds not only at different sites, but at markedly different rates (Urayama and Sweadner, 1988). Accessibility to the protease may be determined by protein packing and conformation at the surface, and these features are somehow different for the three isoforms.

Regulation of Isoforms in Intact Cells

Intriguing results have been obtained when attempts have been made to estimate the relative levels of activity of Na,K-ATPase isoforms in intact cells and synaptosomes. Ouabain-sensitive $^{86}Rb^+$ uptake has been used as a measure of pump function, taking advantage of the fact that in the rat, α1 has a much lower affinity for ouabain than α2 or α3. Resh, Lytton, and Guidotti's original work on epididymal fat pad adipocytes (reviewed in Sweadner, 1989) indicated that in untreated adipocytes the high affinity form (later shown to be α2) was virtually inactive. Preincubation of adipocytes to manipulate intracellular Na^+ concentration made it possible to demonstrate that the inactivity of α2 was due to a low apparent affinity for Na^+. Treatment of the adipocytes with insulin greatly increased the $^{86}Rb^+$ uptake sensitive to low concentrations of ouabain, and increased the apparent affinity for Na^+. When ouabain-sensitive ATP hydrolysis was measured in broken membrane preparations, the α1 and α2 isoforms were both of high Na^+ affinity, implying that the modulation of Na^+ affinity was due to a factor lost when the cells were broken.

Brodsky and Guidotti have followed up these results with two different approaches. First, Brodsky (1990*a*) examined the insulin sensitivity of a preadipocyte cell line, 3T3-F442A. Without insulin treatment, the 3T3-F442A adipocytes displayed only low–ouabain affinity $^{86}Rb^+$ uptake, like fat pad adipocytes without insulin treatment. Intracellular Na^+ was manipulated and $^{86}Rb^+$ uptake sensitive to 2 mM ouabain was measured to estimate Na^+ affinity, and the affinity was high (6.2 mM), like α1 in fat pad adipocytes. The experiments performed by Lytton on fat pad adipocytes were not replicated exactly: the ouabain sensitivity of uptake after insulin pretreatment was not examined, nor was the Na^+ affinity of uptake sensitive to 10 μM ouabain, apparently because α2 was not present at high enough levels for the experiments to be technically feasible. (α2 mRNA was present in the 3T3-F442A adipocytes, not the fibroblasts, but only as the 3.4-kb transcript. α2 protein was also present [see Brodsky's note added in proof], but the proportion relative to α1 was not known. Russo et al. [1990] similarly detected α2 in the related 3T3-L1 preadipocytes when differentiated.)

The conclusion (Brodsky, 1990*a*) was that only α1 made a significant contribution to $^{86}Rb^+$ uptake capacity in the 3T3-F442A cells, whether examined as fibroblasts or as adipocytes. When insulin was added to either fibroblasts or adipocytes, there was a greater than twofold increase in $^{86}Rb^+$ uptake, accompanied by an increase in glucose uptake. The crucial new observation was that insulin increased $^{22}Na^+$ uptake into 3T3-F442A fibroblasts, something that had been ruled out in fat pad adipocytes in earlier work. The increase was not sensitive to amiloride, and an insulin-sensitive Na^+ channel was postulated that would increase α1 pump activity by increasing substrate availability.

Brodsky and Guidotti (1990) returned to the question of modulation of Na^+ affinity using synaptosomes, which have high levels of α2 and/or α3. When the ouabain sensitivity of ATP hydrolysis was measured in synaptosomal plasma membrane preparations, 80% of the activity was high affinity and 20% was low affinity. The apparent Na^+ affinities of the high– and low–ouabain sensitivity ATPase activities in membranes were 12.5 and 9 mM, respectively. When $^{86}Rb^+$ uptake into intact synaptosomes was measured, however, the high–ouabain affinity component was only 27% of the total. If both high– and low–ouabain affinity activities had about the same Na^+ affinity, the high–ouabain affinity form(s) would have to be pumping $^{86}Rb^+$ at only 10% of maximum efficiency to explain the data. This initially appears similar to the behavior of α2 in fat pad adipocytes.

Direct estimation of Na^+ affinity for $^{86}Rb^+$ uptake by manipulating intrasynaptosomal Na^+ concentrations indicated that the low–ouabain affinity form had a Na^+ affinity of 12 mM, while the high–ouabain affinity form had a Na^+ affinity of 35 mM (Brodsky and Guidotti, 1990). Intracellular Na^+ was later shown to be 104 mM in the synaptosomes, which should not be very rate limiting. For an enzyme with a Hill coefficient of 1.5–2.0, a pump with an affinity of 35 mM should be 80–90% active at 104 mM Na^+, much more than the 10% that was observed. The reduced affinity for Na^+ is thus in the right direction, but a reduced V_{max} is indicated. The high–ouabain affinity form (α2 and/or α3) with $K_{0.5}$ of 35 mM was proposed to act as a reservoir of pump activity that could be activated either by a substantial rise in intracellular Na^+ concentration or by an increase in its Na^+ affinity. From the measured affinities and known intracellular Na^+ levels, it was calculated that only 1.4% of the high–ouabain affinity form should be active under resting conditions in vivo. Presumably it could also be activated by an increase in its V_{max}.

Brodsky (1990*b*) showed that insulin selectively activated the high–ouabain affinity $^{86}Rb^+$ uptake activity in synaptosomes; insulin was without measurable effect on Na^+ uptake processes. Very high concentrations of insulin were required, and so it is not known if the phenomenon occurs in vivo.

Do the Isoforms Have to Be Functionally Different?

Experience with isoforms of a variety of other proteins indicates that there are usually kinetic differences that contribute to differences in cellular function when one isoform is exchanged for another. In principle, this does not have to be the only reason for having isoforms, however. Two separate gene products could be required if their rates of synthesis and degradation had to be different, to be able to respond by up- or downregulation. Corthésy-Theulaz et al. (1990*a*) have reported a large difference in relative transcription rates, measured by nuclear run-on assay, and mRNA abundance for α1, α2, and α3 in cultures of central nervous system cells. The transcription rate for α3 was very slow, yet the mRNA abundance very high, when compared with the other isoforms. One might expect this of an isoform whose rate of synthesis did not need to respond to acute cellular conditions. A similar observation was made by Gilmore-Hebert et al. (1989), who observed that α3 transcription rates in the HL60 cell line were much lower than their α3 mRNA levels would suggest. Post-transcriptional regulation of α3 mRNA levels was demonstrated in the HL60 cells by the observation that DMSO induction was accompanied by a dramatic decrease in the mRNA with no change in its rate of transcription (Gilmore-Hebert et al., 1989).

Structurally different isoforms might be required to enable the cell to target the Na,K-ATPase to particular subcellular regions. There is substantial evidence for subcellular targeting now. Gottardi and Caplan (1990) have shown that Na,K-ATPase and H,K-ATPase are targeted to different membrane domains in MDCK cells. α1 and α3 appear to be in different parts of hippocampal pyramidal cells (McGrail et al., 1991). α3 is very tightly localized in photoreceptors (McGrail and Sweadner, 1989). Electron microscopy has delineated the Na,K-ATPase distribution in the photoreceptor inner segment, and has shown that the α3 is found in association with β2 (Schneider et al., 1991). The Na,K-ATPase in the inner segment is known to be anchored so that it is not free to diffuse in the membrane (Spencer et al., 1988), and spectrin has been implicated in its immobilization (Madreperla et al., 1989). Nelson and collaborators have demonstrated a direct association between the Na,K-ATPase and ankyrin (see Nelson et al., 1991).

A completely different role for an isoform of the β-subunit of the Na,K-ATPase has come from the work of Schachner and collaborators. A cDNA for β2 was cloned as a molecule mediating neuron–glial adhesion in cultures and slices of the cerebellum (Pagliusi et al., 1989; Gloor et al., 1990). The adhesion protein, called AMOG for adhesion molecule on glia, was first identified by a monoclonal antibody that blocked adhesion in in vitro assays (Antonicek et al., 1987; Antonicek and Schachner, 1988). The same antibody was used to purify the protein by affinity chromatography. It was a 46,000-M_r protein on SDS gels, accompanied by varying amounts of a 100,000-M_r protein; both subunits comigrated with purified rat brain Na,K-ATPase. Two tryptic fragments came from the sequence of α2, while three others came from no known protein. Oligonucleotide probes based on one of the tryptic fragments pulled out two clones encoding the entire sequence of mouse β2. Antibodies raised against the AMOG complex crossreacted with the Na,K-ATPase, and monoclonal antibodies specific for rat α2 recognized the 100,000-M_r protein in purified mouse AMOG preparations (Gloor et al., 1990).

The original monoclonal antibody that blocked cell–cell adhesion also stimulated $^{86}Rb^+$ uptake ~30% in intact astrocytes in culture (Gloor et al., 1990). The mechanism of stimulation is not known, but at the very least the data indicate that the antibody interacts with active Na,K-ATPase. The activity of the Na,K-ATPase is not required for the adhesion event, however. Adhesion, blocked by the AMOG antibody, occurred equally well at 4°C and in the presence of 1 mM ouabain at 37°C (Gloor et al., 1990). It appears that β2 plays a dual role: both as a subunit of the Na,K-ATPase, complexed in this case with α2, and as a component involved in cell–cell interactions. The β2 sequence is not homologous to any of the other characterized classes of adhesion molecules, although the protein isolated from mouse brain does carry the L3 carbohydrate epitope, which occurs on some other adhesion molecules. Exactly how β2 participates in adhesion and what protein acts as its receptor remains to be elucidated.

Summary

Space does not permit coverage of recent developments either in the structure of the untranslated sequences of each of the Na,K-ATPase isoforms or in the regulation of gene expression by hormones, developmental events, and pathological conditions. The explosion of interest in the isoforms promises to continue until definitive functional differences are well understood.

References

Anner, B. M., E. Imesch, and M. Moosmayer. 1989. Sodium transport defect of ouabain-resistant renal Na,K-ATPase. *Biochemical and Biophysical Research Communications.* 165:360–367.

Antonicek, H., E. Persohn, and M. Schachner. 1987. Biochemical and functional characterization of a novel neuron-glia adhesion molecule that is involved in neuronal migration. *Journal of Cell Biology.* 104:1587–1595.

Antonicek, H., and M. Schachner. 1988. The adhesion molecule on glia (AMOG) incorporated into lipid vesicles binds to subpopulations of neurons. *Journal of Neuroscience.* 8:2961–2966.

Arystarkhova, E. A., O. E. Lakhtina, N. B. Levina, and N. N. Modyanov. 1989. Immunoaffinity isolation of Na,K-ATPase alpha 3 isoform from pig kidney. *FEBS Letters.* 257:24–26.

Arystarkhova, E. A., N. M. Vladimirova, N. A. Potapenko, and N. N. Modyanov. 1990. Structural analysis of Na,K-ATPase isoforms in mammalian tissues. *Journal of General Physiology.* 96:35*a*. (Abstr.)

Baxter-Lowe, L. A., J. Z. Guo, E. E. Bergstrom, and L. E. Hokin. 1989. Molecular cloning of the Na,K-ATPase alpha-subunit in developing brine shrimp and sequence comparison with higher organisms. *FEBS Letters.* 257:181–187.

Berrebi-Bertrand, I., J. M. Maixent, G. Christe, and L. G. Lelievre. 1990. Two active Na^+/K^+-ATPases of high affinity for ouabain in adult rat brain membranes. *Biochimica et Biophysica Acta.* 1021:148–156.

Blanco, G., G. Berberian, and L. Beaugé. 1990. Detection of a highly ouabain sensitive isoform of rat brainstem Na,K-ATPase. *Biochimica et Biophysica Acta.* 1027:1–7.

Brodsky, J. L. 1990*a*. Characterization of the (Na^+ + K^+)-ATPase from 3T3-F442A fibroblasts and adipocytes: isozymes and insulin sensitivity. *Journal of Biological Chemistry.* 265:10458–10465.

Brodsky, J. L. 1990*b*. Insulin activation of brain Na^+–K^+-ATPase is mediated by alpha 2-form of enzyme. *American Journal of Physiology.* 258:C812–C817.

Brodsky, J. L., and G. Guidotti. 1990. Sodium affinity of brain Na^+–K^+-ATPase is dependent on isozyme and environment of the pump. *American Journal of Physiology.* 258:C803–C811.

Canfield, V., J. R. Emanuel, N. Spickofsky, R. Levenson, and R. F. Margolskee. 1990*a*. Ouabain-resistant mutants of the rat Na,K-ATPase alpha 2 isoform identified by using an episomal expression vector. *Molecular and Cellular Biology.* 10:1367–1372.

Canfield, V. A., C. T. Okamoto, D. Chow, J. Dorfman, P. Gros, J. G. Forte, and R. Levenson. 1990*b*. Cloning of the H,K-ATPase beta subunit. Tissue-specific expression, chromosomal assignment, and relationship to Na,K-ATPase beta subunits. *Journal of Biological Chemistry.* 265:19878–19884.

Canfield, V. A., C. T. Okamato, D. Chow, J. Dorfman, P. Gros, J. G. Forte, and R. Levenson. 1991. Identification and characterization of a gene for the H,K-ATPase β-subunit. *In* The Sodium Pump: Structure, Mechanism, and Regulation. J. H. Kaplan and P. De Weer, editors. The Rockefeller University Press, New York. 89–98.

Canfield, V. A., K.-Y. Xu, T. D'Aquila, A. W. Shyjan, and R. Levenson. 1990*c*. Molecular cloning of Na,K-ATPase alpha-subunit cDNA from *Hydra attenuata. Journal of General Physiology.* 96:19*a*. (Abstr.)

Coca-Prados, M., P. Martin-Vasallo, N. Hernando-Sobrino, and S. Ghosh. 1990. Cellular distribution and differential expression of the Na,K-ATPase alpha isoform (alpha 1, alpha 2, alpha 3), beta, and beta 2/AMOG genes in the ocular ciliary epithelium. *Journal of General Physiology.* 96:63*a*. (Abstr.)

Cortas, N., M. Arnaout, J. Salon, and I. S. Edelman. 1989. Isoforms of Na,K-ATPase in *Artemia salina.* II. Tissue distribution and kinetic characterization. *Journal of Membrane Biology.* 108:187–195.

Cortas, N., D. Elstein, and I. S. Edelman. 1988. Na,K-ATPase of renal outer medulla contains only the alpha 1 isoform appearing in multiple bands on SDS-PAGE. *Journal of Cell Biology.* 107:126*a*. (Abstr.)

Corthésy-Theulaz, I., A.-M. Merillat, P. Honegger, and B. C. Rossier. 1990*a*. Differential regulation of Na,K-ATPase isoform gene expression by thyroid hormones during rat brain development. *Journal of General Physiology.* 96:25*a*. (Abstr.)

Corthésy-Theulaz, I., A.-M. Merillat, P. Honegger, and B. C. Rossier. 1990*b*. Na^+–K^+-ATPase gene expression during in vitro development of rat fetal forebrain. *American Journal of Physiology*. 258:C1062–C1069.

Fambrough, D. M., B. A. Wolitzky, J. P. Taormino, M. M. Tamkun, K. Takeyasu, D. Somerville, K. J. Renaud, M. V. Lemas, R. M. Lebovitz, B. C. Kone et al. 1991. A cell biologist's perspective on sites of Na,K-ATPase regulation. *In* The Sodium Pump: Structure, Mechanism, and Regulation. J. H. Kaplan and P. De Weer, editors. The Rockefeller University Press, New York. 17–30.

Filuk, P. E., W. R. Anderson, A. Maki, L. M. Famiglio, and W. L. Stahl. 1990. Expression of alpha- and beta-subunit isoforms of the Na,K-ATPase in the nervous system. *Journal of General Physiology*. 96:65*a*. (Abstr.)

Filuk, P. E., M. A. Miller, D. M. Dorsa, and W. L. Stahl. 1989. Localization of messenger RNA encoding isoforms of the catalytic subunit of the Na,K-ATPase in rat brain by in situ hybridization. *Neuroscience Research Communications*. 5:155–162.

Ghosh, S., A. C. Freitag, P. Martin-Vasallo, and M. Coca-Prados. 1990. Cellular distribution and differential gene expression of the three alpha subunit isoforms of the Na,K-ATPase in the ocular ciliary epithelium. *Journal of Biological Chemistry*. 265:2935–2940.

Gilmore-Hebert, M., J. W. Schneider, A. L. Greene, N. Berliner, C. A. Stolle, K. Lomax, R. W. Mercer, and E. J. Benz, Jr. 1989. Expression of multiple Na^+,K^+-adenosine triphosphatase isoform genes in human hematopoietic cells: behavior of the novel A3 isoform during induced maturation of HL60 cells. *Journal of Clinical Investigation*. 84:347–351.

Gloor, S., H. Antonicek, K. J. Sweadner, S. Pagliusi, R. Frank, M. Moos, and M. Schachner. 1990. The adhesion molecule on glia (AMOG) is a homologue of the beta subunit of the Na,K-ATPase. *Journal of Cell Biology*. 110:165–174.

Good, P. J., K. Richter, and I. B. Dawid. 1990. A nervous system-specific isotype of the beta subunit of Na^+, K^+-ATPase expressed during early development of *Xenopus laevis*. *Proceedings of the National Academy of Sciences, USA*. 87:9088–9092.

Gottardi, C. J., and M. J. Caplan. 1990. H/K-ATPase is sorted to the apical membrane of transfected Madin Darby canine kidney (MDCK) cells. *Journal of General Physiology*. 96:26*a*. (Abstr.)

Hall, K., G. Perez, D. Anderson, C. Gutierrez, K. Munson, S. J. Hersey, J. H. Kaplan, and G. Sachs. 1990. Location of the carbohydrates present in the HK-ATPase vesicles isolated from hog gastric mucosa. *Biochemistry*. 29:701–706.

Hara, Y., A. Nikamoto, T. Kojima, A. Matsumoto, and M. Nakao. 1988. Expression of sodium pump activities in BALB/c 3T3 cells transfected with cDNA encoding alpha 3-subunits of rat brain Na^+,K^+-ATPase. *FEBS Letters*. 238:27–30.

Hieber, V., G. J. Siegel, S. T. Desmond, J. L.-H. Liu, and S. A. Ernst. 1989. Na,K-ATPase: comparison of the cellular localization of alpha-subunit mRNA and polypeptide in mouse cerebellum, retina, and kidney. *Journal of Neuroscience Research*. 23:9–20.

Hieber, V., G. J. Siegel, D. J. Fink, M. V. Beaty, and M. Mata. 1991. Differential distribution of Na,K-ATPase alpha isoforms in central nervous system. *Cellular and Molecular Neurobiology*. In press.

Jewell, E. A., and J. B. Lingrel. 1990. Characterization of the Na,K-ATPase alpha isoforms. *Journal of General Physiology*. 96:16*a*. (Abstr.)

Jørgensen, P. L., and J. P. Andersen. 1988. Structural basis for E1–E2 conformational transitions in Na,K-pump and Ca-pump proteins. *Journal of Membrane Biology.* 103:95–120.

Kurihara, K., K. Hosoi, A. Kodama, and T. Ueha. 1990. A new electrophoretic variant of alpha subunit of Na^+/K^+-ATPase from the submandibular gland of rats. *Biochimica et Biophysica Acta.* 1039:234–240.

Lebovitz, R. M., K. Takeyasu, and D. M. Fambrough. 1989. Molecular characterization and expression of the ($Na^+ + K^+$)-ATPase alpha-subunit in *Drosophila melanogaster. EMBO Journal.* 8:193–202.

Lemas, V., J. Rome, J. Taormino, K. Takeyasu, and D. M. Fambrough. 1990. Analysis of isoform-specific regions within the alpha- and beta-subunits of the avian Na^+,K^+-ATPase. *Journal of General Physiology.* 96:65*a*. (Abstr.)

Lester, H. A. 1988. Heterologous expression of excitability proteins: route to more specific drugs? *Science.* 241:1057–1063.

Lingrel, J. B., J. Orlowski, E. M. Price, and B. G. Pathak. 1991. Regulation of the α-subunit genes of the Na,K-ATPase and determinants of cardiac glycoside sensitivity. *In* The Sodium Pump: Structure, Mechanism, and Regulation. J. H. Kaplan and P. De Weer, editors. The Rockefeller University Press, New York. 1–16.

Lingrel, J. B., J. Orlowski, M. M. Shull, and E. M. Price. 1990. Molecular genetics of Na,K-ATPase. *Progress in Nucleic Acid Research and Molecular Biology.* 38:37–89.

Lucchesi, P. A., and K. J. Sweadner. 1991. Postnatal changes in Na,K-ATPase isoform expression in rat cardiac ventricle. Conservation of biphasic ovabain affinity. *Journal of Biological Chemistry.* In press.

Madreperla, S. A., M. Edidin, and R. Adler. 1989. Na^+,K^+-adenosine triphosphatase polarity in retinal photoreceptors. A role for cytoskeletal attachments. *Journal of Cell Biology.* 109:1483–1493.

Martin-Vasallo, P., W. Dackowski, J. R. Emanuel, and R. Levenson. 1989*a*. Identification of a putative isoform of the Na,K-ATPase beta subunit: primary structure and tissue-specific expression. *Journal of Biological Chemistry.* 264:4613–4618.

Martin-Vasallo, P., S. Ghosh, and M. Coca-Prados. 1989*b*. Expression of Na,K-ATPase alpha subunit isoforms in the human ciliary body and cultured ciliary epithelial cells. *Journal of Cellular Physiology.* 141:243–252.

Mata, M., G. J. Siegel, V. Hieber, M. V. Beaty, and D. J. Fink. 1991. Differential distribution of Na,K-ATPase alpha isoforms in peripheral nervous system. *Brain Research.* In press.

McGrail, K. M., J. M. Phillips, and K. J. Sweadner. 1991. Immunofluorescent localization of three Na,K-ATPase isozymes in the rat central nervous system: both neurons and glia can express more than one Na,K-ATPase. *Journal of Neuroscience.* 11:381–391.

McGrail, K. M., and K. J. Sweadner. 1989. Complex expression patterns for Na,K-ATPase isoforms in retina and optic nerve. *European Journal of Neuroscience.* 2:170–176.

Nelson, W. J., R. W. Hammerton, and H. McNeill. 1991. Role of the membrane-cytoskeleton in the spatial organization of the Na,K-ATPase in polarized epithelial cells. *In* The Sodium Pump: Structure, Mechanism, and Regulation. J. H. Kaplan and P. De Weer, editors. The Rockefeller University Press, New York. 77–87.

Okamoto, C. T., J. M. Karpilow, A. Smolka, and J. G. Forte. 1990. Isolation and characterization of gastric microsomal glycoproteins: evidence for a glycosylated beta-subunit of the H^+/K^+-ATPase. *Biochimica et Biophysica Acta.* 1037:360–372.

Orlowski, J., and J. B. Lingrel. 1990. Thyroid and glucocorticoid hormones regulate the expression of multiple Na,K-ATPase genes in cultured neonatal rat cardiac myocytes. *Journal of Biological Chemistry.* 265:3462–3470.

Pagliusi, S., H. Antonicek, S. Gloor, R. Frank, M. Moos, and M. Schachner. 1989. Identification of a cDNA clone specific for the neural cell adhesion molecule AMOG. *Journal of Neuroscience Research.* 22:113–119.

Pagliusi, S. R., M. Schachner, P. H. Seeburg, and B. D. Shivers. 1990. The adhesion molecule on glia (AMOG) is widely expressed by astrocytes in developing and adult mouse brain. *European Journal of Neuroscience.* 2:271–280.

Reuben, M. A., L. S. Lasater, and G. Sachs. 1990. Characterization of a beta subunit of the gastric H^+/K^+-transporting ATPase. *Proceedings of the National Academy of Sciences, USA.* 87:6767–6771.

Russo, J. J., M. A. Manuli, F. Ismail-Beigi, K. J. Sweadner, and I. S. Edelman. 1990. Na^+–K^+-ATPase in adipocyte differentiation in culture. *American Journal of Physiology.* 259:C968–C977.

Schneider, B. G., and E. Kraig. 1990. Na,K-ATPase of the photoreceptor: selective expression of alpha 3 and beta 2 isoforms. *Experimental Eye Research.* 51:553–564.

Schneider, B. G., A. W. Shyjan, and R. Levenson. 1991. Colocalization and polarized distribution of Na,K-ATPase alpha 3 and beta 2 subunits in photoreceptor cells. *Journal of Histochemistry and Cytochemistry.* In press.

Schneider, J. W., R. W. Mercer, M. Gilmore-Hebert, M. F. Utset, C. Lai, A. Greene, and E. J. Benz, Jr. 1988. Tissue specificity, localization in brain, and cell-free translation of mRNA encoding the A3 isoform of Na^+,K^+-ATPase. *Proceedings of the National Academy of Sciences, USA.* 85:284–288.

Shull, G. E. 1990. cDNA cloning of the beta-subunit of the rat gastric H,K-ATPase. *Journal of Biological Chemistry.* 265:12123–12126.

Shyjan, A. W., V. A. Canfield, and R. Levenson. 1990*a*. Functional properties of Na,K-ATPase alpha subunit isoforms expressed in mammalian cell lines. *Journal of General Physiology.* 96:18*a*. (Abstr.)

Shyjan, A. W., V. Cena, D. C. Klein, and R. Levenson. 1990*b*. Differential expression and enzymatic properties of the Na^+,K^+-ATPase alpha 3 isoenzyme in rat pineal glands. *Proceedings of the National Academy of Sciences, USA.* 87:1178–1182.

Shyjan, A. W., C. Gottardi, and R. Levenson. 1990*c*. The Na,K-ATPase beta 2 subunit is expressed in rat brain and copurifies with Na,K-ATPase activity. *Journal of Biological Chemistry.* 265:5166–5169.

Shyjan, A. W., and R. Levenson. 1989. Antisera specific for the alpha 1, alpha 2, alpha 3, and beta subunits of the Na,K-ATPase: differential expression of alpha and beta subunits in rat tissue membranes. *Biochemistry.* 28:4531–4535.

Siegel, G. J., and T. J. Desmond. 1989. Effects of tetradecyl sulfate on electrophoretic resolution of kidney Na,K-ATPase catalytic subunit isoforms. *Journal of Biological Chemistry.* 264:4751–4754.

Siegel, G. J., M. Mata, V. Hieber, and D. J. Fink. 1990. Alpha 1 and alpha 3 isoforms of

Na,K-ATPase are both produced in neurons and show unique distributions in the rat CNS and PNS. *Journal of General Physiology.* 96:65*a*. (Abstr.)

Skou, J. C. 1962. Preparation from mammallian brain and kidney of the enzyme system involved in active transport of Na^+ and K^+. *Biochimica et Biophysica Acta.* 58:314–325.

Somerville, D., and D. M. Fambrough. 1990. Nucleotide and deduced amino acid sequence of the alpha-subunit of the Na,K-ATPase of the nematode, *Caenorhabditis elegans. Journal of General Physiology.* 96:15*a*. (Abstr.)

Spencer, M., P. B. Detwiler, and A. H. Bunt-Milam. 1988. Distribution of membrane proteins in mechanically dissociated retinal rods. *Investigative Ophthalmology and Visual Science.* 29:1012–1020.

Sweadner, K. J. 1989. Isozymes of the Na^+/K^+-ATPase. *Biochimica et Biophysica Acta.* 988:185–220.

Sweadner, K. J. 1990. Anomalies in the electrophoretic resolution of Na^+/K^+-ATPase catalytic subunit isoforms reveal unusual protein–detergent interactions. *Biochimica et Biophysica Acta.* 1029:13–23.

Urayama, O., and K. J. Sweadner. 1988. Ouabain sensitivity of the alpha 3 isozyme of rat Na,K-ATPase. *Biochemical and Biophysical Research Communications.* 156:796–800.

Chapter 6

Role of the Membrane-Cytoskeleton in the Spatial Organization of the Na,K-ATPase in Polarized Epithelial Cells

W. James Nelson, Rachel W. Hammerton, and Helen McNeill

Department of Molecular and Cellular Physiology, Stanford University School of Medicine, Stanford, California 94305-5426

Summary

Vectorial function of polarized transporting epithelia requires the establishment and maintenance of a nonrandom distribution of Na,K-ATPase on the cell surface. In many epithelia, the Na,K-ATPase is located at the basal–lateral domain of the plasma membrane. The mechanisms involved in the spatial organization of the Na,K-ATPase in these cells are poorly understood. We have been investigating the roles of regulated cell–cell contacts and assembly of the membrane-cytoskeleton in the development of the cell surface polarity of Na,K-ATPase. We have shown that the Na,K-ATPase colocalizes with distinct components of the membrane-cytoskeleton in polarized Madin-Darby canine kidney (MDCK) epithelial cells. Significantly, we showed directly that Na,K-ATPase is a high affinity binding site for the membrane-cytoskeletal proteins ankyrin and fodrin, and that all three proteins exist in a high molecular weight protein complex that also contains the cell adhesion molecule (CAM) uvomorulin.

We have proposed that these interactions are important in the assembly at sites of cell–cell contact of the membrane-cytoskeleton, which in turn initiates the development of the nonrandom distribution of the Na,K-ATPase. To directly investigate the functional significance of these protein–protein interactions in the spatial organization of the Na,K-ATPase, we analyzed the distribution of the Na,K-ATPase in fibroblasts transfected with a cDNA encoding the epithelial CAM, uvomorulin. Our results showed that expression of uvomorulin is sufficient to induce a redistribution of Na,K-ATPase from an unrestricted distribution over the entire cell surface in nontransfected cells to a restricted distribution at sites of uvomorulin-mediated cell–cell contacts in the transfected cells; this distribution is similar to that in polarized epithelial cells. This restricted distribution of the Na,K-ATPase occurred in the absence of tight junctions, but coincided with the reorganization of the membrane-cytoskeleton. These results support a model in which the epithelial CAM uvomorulin functions as an inducer of cell surface polarity of Na,K-ATPase through cytoplasmic linkage to the membrane-cytoskeleton.

Background

Many studies of the Na,K-ATPase have focused on the regulation of expression of subunit genes, the distribution of different subunits in tissues, regulation of subunit assembly and transport to the cell surface, and the mechanisms involved in enzyme function (Jørgensen, 1982, 1991). However, in many of the major organs and tissues in the body, expression of functional Na,K-ATPase on the cell surface is not sufficient for normal cellular function. In these polarized epithelial cells, Na,K-ATPase must be localized to a specific domain of the plasma membrane (reviewed in Rodriguez-Boulan and Nelson, 1989). In the kidney, individual tubules are composed of a closed monolayer of cells that separate the lumen of the tubule from the blood supply. These cells are polarized with functionally and structurally different domains of the plasma membrane facing the lumen and blood supply. The primary function of these cells is to regulate vectorial transport of ions and solutes from the lumen of the tubule to the blood supply (Berridge and Oschman, 1972). Paracellular passage of these ions and solutes between cells is blocked by the tight junction, which is located at the boundary between the apical and basal-lateral domains of the plasma membrane (reviewed in Gumbiner, 1987). Transcellular passage of solutes is accomplished by ion channels and transporters that are localized to different domains of the plasma membrane. In the proximal kidney tubule, in which Na^+ reabsorption from the ultrafiltrate occurs, Na^+ channels and cotransporters are localized in the apical domain of the plasma membrane, where they facilitate uptake of Na^+ into the cell. Upon entry into the cell, Na^+ is rapidly pumped out into the blood supply by the Na,K-ATPase, which is localized in the basal–lateral domain of the plasma membrane. Incorrect localization of the Na,K-ATPase to the apical plasma membrane is associated with a number of kidney diseases; these include polycystic kidney disease (Wilson and Hreniuk, 1987) and ischemic damage (Molitoris et al., 1989). Taken together, these observations suggest that correct localization of the Na,K-ATPase to the basal–lateral domain of the plasma membrane is critical for normal cellular function in polarized transporting epithelia.

What is the nature of the mechanisms involved in the localization of the Na,K-ATPase to the basal–lateral domain of the plasma membrane? In the past few years, several hypotheses have been proposed: (*a*) the tight junction acts as a barrier in the lipid bilayer to the diffusion of membrane proteins between different domains of the plasma membrane; (*b*) newly synthesized proteins are targeted to the correct domain; and (*c*) proteins are linked to components of the cytoskeleton that retain them to the specific membrane domain (for detailed reviews, see Simons and Fuller, 1985, and Rodriguez-Boulan and Nelson, 1989).

We have been analyzing the role of the membrane-cytoskeleton in the polarized localization of the Na,K-ATPase and have obtained evidence for a direct linkage of the Na,K-ATPase to specific components of the membrane-cytoskeleton. This paper briefly reviews these results and proposes a working model for the role(s) of the membrane-cytoskeleton in the establishment and maintenance of the spatial distribution of the Na,K-ATPase in polarized epithelial cells.

The Membrane-Cytoskeleton

Direct evidence for the interaction of integral membrane proteins with cytoplasmic structural proteins came initially from studies on the human erythrocyte (reviewed in Bennett, 1985, and Marchesi, 1985). In reconstitution experiments using purified proteins, a membrane protein, the anion transporter (band 3), was shown to bind with high affinity to a cytoplasmic protein, termed ankyrin. Ankyrin in turn was shown to bind to another cytoplasmic protein termed spectrin, which is composed of two nonidentical subunits (M_r 240,000 and 220,000) that form a tetramer. These interactions form the basic "unit" of the membrane-cytoskeleton in the human erythrocyte: an integral membrane protein, the anion transporter, bound to an ankyrin–spectrin tetramer. These unit complexes are linked together by either direct interactions between spectrin tetramers, or indirectly through accessory proteins that include adducin, protein 4.1, and actin oligomers. These protein–protein interactions result in the formation of a dense protein meshwork that underlies the cytoplasmic surface of the plasma membrane (Bennett, 1985).

The erythroid membrane-cytoskeleton appears to be important in maintaining the structural integrity of the erythrocyte during blood circulation. Several anemias are characterized by the loss of normal expression of different cytoplasmic components of the membrane-cytoskeleton (reviewed in Marchesi, 1985). In addition, loss of expression of these cytoplasmic proteins results in an increase in the mobility of the anion transporter in the plane of the lipid bilayer (Scheetz et al., 1980), indicating that the membrane-cytoskeleton plays a direct role in immobilizing this integral membrane protein in the membrane.

The role of the erythroid membrane-cytoskeleton in immobilizing integral membrane protein in the lipid bilayer has important implications for polarized epithelial cells in which the distribution of membrane proteins is limited to specific domains of the plasma membrane (see above). Recent studies have shown that components of the membrane-cytoskeleton are not exclusive to the erythrocyte but are present in most nonerythroid cells (reviewed in Nelson and Lazarides, 1984, and Bennett, 1985). Furthermore, these proteins have been shown to have properties very similar to those of erythrocyte ankyrin and spectrin.

To investigate the possible role(s) of the membrane-cytoskeleton in polarized epithelial cells, we sought to determine the subcellular distribution of ankyrin and fodrin (the nonerythroid homologue of erythrocyte spectrin), and whether these proteins interact with integral membrane proteins. The results have provided the first evidence for the direct interaction of these membrane-cytoskeletal proteins with a membrane protein other than the anion transporter, and insight into the mechanisms involved in the establishment of cell surface polarity in epithelial cells.

Distribution and Molecular Composition of the Membrane-Cytoskeleton in Polarized Epithelial Cells

As a model of polarized epithelial cells, we have used the established cell line MDCK epithelial cells. These cells have retained many of the properties of polarized epithelial cells in situ, including the formation of structurally and functionally distinct apical and basal–lateral domains of the plasma membrane (reviewed in Simons and Fuller, 1985, and Rodriguez-Boulan and Nelson, 1989).

Antibodies were prepared against ankyrin, fodrin, and the Na,K-ATPase and used for immunofluorescence microscopy on frozen sections of polarized MDCK monolayers. Results showed that ankyrin and fodrin colocalized with the Na,K-ATPase on the basal–lateral domain of the plasma membrane; little or no staining with any of the antibodies was detected on the apical membrane domain (Nelson and Veshnock, 1986; Morrow et al., 1989; Nelson et al., 1990). We sought to determine directly whether interactions between membrane-cytoskeletal proteins and the Na,K-ATPase occurred that would result in their colocalization in the cell. Given that in the erythrocyte ankyrin binds to the anion transporter, we sought to determine whether ankyrin also binds to the Na,K-ATPase (Nelson and Veshnock, 1987*a*). In vitro reconstitution experiments showed that purified, membrane-bound Na,K-ATPase has a high affinity binding site for purified ankyrin ($K_d \sim 10^{-8}$) as demonstrated by: (*a*) cosedimentation of the proteins on sucrose gradients, (*b*) comigration of the proteins on nondenaturing polyacrylamide gels after extraction from Na,K-ATPase membranes, and (*c*) Scatchard analysis of ankyrin binding to purified Na,K-ATPase membranes (Nelson and Veshnock, 1987*a*).

The finding that Na,K-ATPase binds with high affinity to ankyrin in vitro raised the question of whether such interactions occurred in the cell. An important criterion for an analysis of this problem was that the complexes should be readily extracted from the cell under nondenaturing conditions. However, initial studies of the membrane-cytoskeleton in polarized monolayers of MDCK cells showed that

Na,K-ATPase, ankyrin, and fodrin were relatively resistant to extraction with isotonic buffers containing the nonionic detergent Triton X-100; these proteins could be extracted, but only under denaturing conditions that involved the use of ionic detergents (e.g., SDS), or chaotropic agents (e.g., 1 M KCl, urea), which resulted in disruption of the protein complexes (Nelson and Veshnock, 1986, 1987*a*, *b*; Nelson and Hammerton, 1989). However, we found that before induction of cell–cell contact these proteins were quantitatively extracted from the cells under nondenaturing conditions (for details, see Nelson and Hammerton, 1989). In addition, we showed that preexisting protein complexes present in cells before the induction of cell–cell contact were used by the cell in the assembly of the membrane-cytoskeleton. These results indicated that cells grown in the absence of cell–cell contact would be a good starting point in the search for membrane-cytoskeletal protein complexes (Nelson and Hammerton, 1989).

Monolayers of MDCK cells grown in the absence of cell–cell contact were extracted in an isotonic buffer containing Triton X-100, and the solubilized proteins were fractionated on a sucrose gradient. Analysis of the distribution of ankyrin and fodrin in the sucrose gradient by Western blotting showed that the proteins cosedimented at ~10.5 S; this region of the gradient contained <5% of the total protein, which sedimented at ~5 S (Nelson and Hammerton, 1989). Electron microscopy of the fractions containing ankyrin and fodrin (in collaboration with J. Heuser, Washington University, St. Louis, MO) showed long, rod-shaped molecules that appeared very similar to images reported for complexes of ankyrin and fodrin tetramers reconstituted in vitro from purified proteins. These results were interpreted as evidence of ankyrin–fodrin complexes in this cell extract that were similar to part of the basic unit of the membrane-cytoskeleton in the human erythrocyte (see Fig. 1).

We next determined the distribution in the sucrose gradient of Na,K-ATPase. Using Western blotting, we showed that Na,K-ATPase sedimented at ~10.5 S, similar to that of the ankyrin–fodrin complex; analysis of the sedimentation of general plasma membrane proteins that had been labeled on the cell surface with ^{125}I showed that >95% of those proteins sedimented at ~5 S (Nelson and Hammerton, 1989). To investigate whether Na,K-ATPase, ankyrin, and fodrin were cosedimenting as a protein complex, we separated the relevant sucrose gradient fractions in a nondenaturing polyacrylamide gel. Staining the resulting gel with India ink to show the distribution of all proteins revealed the presence of two bands. Western blot analysis revealed that ankyrin and fodrin were present in both protein bands, but that Na,K-ATPase was present only in the slower migrating band. Comparison of the electrophoretic mobility of these protein bands with that of purified fodrin tetramers showed that the bands containing either ankyrin–fodrin or ankyrin–fodrin–Na,K-ATPase had a slower mobility than that of purified fodrin (Nelson and Hammerton, 1989). The slower mobility of both of these bands compared with that of purified fodrin tetramers is consistent with our interpretation of the structures observed in the electron microscope as being ankyrin–fodrin tetramer complexes (see above).

We conclude from this analysis that two types of membrane-cytoskeletal complex were extracted from MDCK cells; one comprised an ankyrin–fodrin tetramer complex, and the other comprised an ankyrin–fodrin–Na,K-ATPase complex. Taken together with the in vitro binding experiments (see above), these results provide direct evidence for an interaction between an integral membrane protein, the

Na,K-ATPase, and a cytoplasmic protein complex comprising the ankyrin–fodrin tetramer. Significantly, the molecular composition of this membrane-cytoskeleton complex is similar to that described in human erythrocytes, indicating that a similar basic unit of the membrane–cytoskeleton also exists in these nonerythroid cells (Fig. 1).

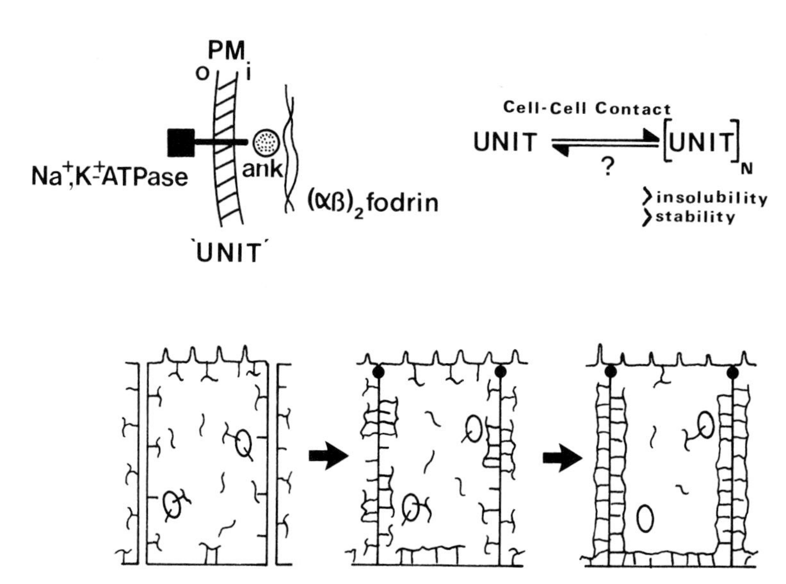

Figure 1. Identification of a membrane cytoskeletal complex in polarized epithelial cells and its role in the spatial organization of the Na,K-ATPase. Evidence from in vitro binding studies and the isolation of protein complexes from whole cells indicates that membrane-cytoskeletal complexes between Na,K-ATPase, ankyrin, and fodrin tetramers exist in polarized epithelial cells. These complexes have the same stoichiometry as the basic repeat (unit) structure of the erythrocyte membrane-cytoskeleton. We propose that these unit complexes have a random distribution on the plasma membrane before the induction of cell–cell contact. Cell–cell contact results in an aggregation of unit complexes into "$[\text{unit}]_n$". Since the CAM uvomorulin is also a component of the membrane-cytoskeleton, we suggest that this aggregation event is initiated by cell–cell contact in a manner similar to the patching of cell surface receptors that occurs in other cells in response to addition of a ligand. Due to the differential stabilization of these assembled membrane-cytoskeletal complexes at cell–cell contacts relative to units stranded in the forming apical membrane, there is a gradual accumulation of proteins at the forming basal–lateral membrane domain (we have directly measured increases in the metabolic stability and insolubility of the membrane-cytoskeleton as a function of cell–cell contact). This results in the development of the characteristic spatial distribution of the membrane-cytoskeleton and the Na,K-ATPase in polarized epithelial cells (see text for details).

What Is the Function of the Membrane-Cytoskeleton in Polarized Epithelial Cells?

We predict that the membrane-cytoskeleton has several functions in polarized epithelial cells based on knowledge of the function of these proteins in erythrocytes (see above), and from analysis of these proteins in polarized epithelial cells: (*a*) a role in maintaining protein distributions in different membrane domains, and (*b*) a role in establishing membrane domains during the development of epithelial polarity.

Inhibition of Protein Diffusion and Maintenance of the Protein Specificity of Membrane Domains

Integration of membrane proteins with the cytoskeleton is thought to result in the immobilization of those proteins in the plane of the lipid bilayer. In polarized epithelial cells, immobilization of membrane proteins is likely to be important in the regulation of distinct protein compositions of functionally different membrane domains. Proteins targeted to the appropriate membrane domain could, through linkage to the underlying cytoskeleton, be constrained from lateral diffusion in the plane of the lipid bilayer. Indeed, recent studies have shown that integral membrane proteins that were targeted to the correct membrane domain in polarized MDCK cells exhibited decreased mobility in the plane of the lipid bilayer and were relatively resistant to extraction with buffers containing Triton X-100 (Salas et al., 1988).

Earlier studies also sought to examine the lateral mobility of the Na,K-ATPase in cultures of MDCK cells (Jesaitis and Yguerabide, 1986). It was reported that the Na,K-ATPase exhibited little or no difference in lateral mobility between small and large colonies of cells. However, it is noteworthy in interpreting this result that our previous studies have shown that the Na,K-ATPase, ankyrin, and fodrin are not fully organized into the membrane-cytoskeleton until full confluency of the culture is attained (Nelson and Veshnock, 1986). Thus, while it is possible that the membrane-cytoskeleton plays a role in the maintenance of membrane protein organization in the cell, direct evidence in nonerythroid cells has not yet been obtained.

Role in the Establishment of Membrane Domains in Polarized Epithelial Cells

The results from several types of experiments indicate that the site of assembly of the membrane-cytoskeleton on the plasma membrane has an important role in initiating a remodeling of the cell surface distribution of certain integral membrane proteins (Nelson and Veshnock, 1986, 1987*b*; Nelson et al., 1990). First, indirect immunofluorescence microscopy showed that before the induction of cell–cell contact, ankyrin, fodrin, and several integral membrane proteins (e.g., Na,K-ATPase and the CAM uvomorulin [see below]) were distributed over the whole cell surface and were relatively easily extracted from the cell with Triton X-100. However, within a short time after induction of cell–cell contact (5–15 min) a portion of these proteins became localized to the zones of cell–cell contact and this portion had also become resistant to extraction with Triton X-100. We interpreted these changes in the properties and distributions of these proteins as evidence of the assembly of the membrane-cytoskeleton at those sites on the plasma membrane. Second, we showed that these changes in the distribution and properties of the proteins occurred in the presence of cyclohexamide, which inhibits protein synthesis, indicating that the membrane-cytoskeleton was being assembled from preexisting proteins (Nelson and Veshnock, 1987*b*). Third, the analysis of the molecular composition of the membrane-cytoskeleton in MDCK cells was carried out on cells grown in the absence of cell–cell contact when the membrane-cytoskeleton had not assembled into a Triton X-100–insoluble complex. Under these conditions, we detected a protein complex comprising ankyrin–fodrin tetramers and Na,K-ATPase, which, presumably, is representative of one of the classes of preexisting protein complex that becomes assembled into the membrane-cytoskeleton upon induction of cell–cell contact. Fourth, we have shown that the CAM uvomorulin is present in a complex together with ankyrin–fodrin tetramers (Nelson et al., 1990); this analysis also involved separation of proteins solubilized from MDCK cells in sucrose gradients and nondenaturing polyacrylamide gels as described above for analyzing cell extracts for complexes comprising ankyrin–fodrin–Na,K-ATPase (see above).

These results show that the basic repeating unit of the membrane-cytoskeleton, namely an ankyrin–fodrin tetramer bound to a membrane protein (in this case, Na,K-ATPase or uvomorulin), exists in whole MDCK cells before the induction of cell–cell contact, and that cell–cell contact results in the assembly of these complexes into an insoluble matrix at those contact sites. We propose that it is the assembly of the membrane-cytoskeleton at sites of cell–cell contact that initiates the remodeling of the cell surface distribution of membrane proteins, such that within a short period of time proteins bound to the membrane-cytoskeleton (e.g., Na,K-ATPase and uvomorulin) begin to accumulate at those sites (Fig. 1). The increased concentration of these proteins at those sites might cause an increase in the insolubility and metabolic stability of component proteins, properties that we have directly measured in these proteins as a consequence of the induction of cell–cell contact.

What happens to the same proteins that are trapped in the forming apical membrane domain after induction of cell–cell contact? Since there is no cell–cell contact on that membrane domain we predict that those components of the membrane-cytoskeleton trapped there would not be induced to aggregate into an insoluble protein matrix. Rather, we expect that those proteins remain as metabolically unstable unit complexes that are simply removed from the apical membrane by normal protein degradative pathways (Fig. 1). It remains a possibility that those proteins are internalized and redirected to the basal–lateral membrane, but the evidence for endogenous membrane proteins taking this pathway is not strong. Furthermore, the rapid formation of the tight junction upon induction of cell–cell contact would block simple diffusion of proteins from the apical to the basal–lateral membrane (reviewed in Gumbiner, 1987).

We propose that the apparent redistribution of certain proteins (e.g., Na,K-ATPase) to the basal–lateral membrane is due to their differential stability and hence accumulation at cell–cell contacts as a result of the assembly there of the membrane-cytoskeleton, and that the assembly of the cytoskeleton itself is initiated by the CAM uvomorulin. Uvomorulin-dependent cell adhesion is the result of homotypic interactions between uvomorulin molecules on adjacent cells. It is possible that this interaction acts like a ligand–receptor complex to induce a localized "patching" (Bourguignon and Bourguignon, 1984). This patching event may be facilitated by localized assembly on the membrane-cytoskeleton. In this regard it is noteworthy that fodrin has been shown to "co-patch" with ligand–receptor complexes in lymphocytes (Nelson et al., 1983).

This proposed role of the CAM uvomorulin as an inducer of the assembly of the membrane-cytoskeleton and associated membrane proteins at sites of cell–cell contact has been directly tested in recent studies from this laboratory (McNeill et al., 1990). The rationale for these experiments was to analyze the consequences of the expression of uvomorulin on the cell surface distribution of Na,K-ATPase and the membrane-cytoskeleton. Since uvomorulin expression is restricted to epithelial cells, we investigated this problem in fibroblasts (mouse L cells), which constitutively express ankyrin, fodrin, and Na,K-ATPase, but not uvomorulin; expression of the latter was induced by expression of exogenous uvomorulin cDNA in the cells.

A Test of the Function of the CAM Uvomorulin as an Inducer of Cell Surface Polarity of the Na,K-ATPase

Analysis of L cells expressing full-length uvomorulin showed that the cells formed tight colonies similar to those formed by epithelial cells (see also Nagafuchi et al., 1987; Mege et al., 1988; Ozawa et al., 1989). Immunofluorescence microscopy revealed that uvomorulin, Na,K-ATPase, and fodrin were localized to cell–cell contacts in a pattern indistinguishable from that in polarized epithelial cells; other proteins that were analyzed did not show a correlation between expression of uvomorulin and their cell surface distribution (for details, see McNeill et al., 1990). That there is a cytoplasmic link between uvomorulin and Na,K-ATPase, possibly through the membrane-cytoskeleton, was shown in a parallel series of experiments in which L cells were transfected with cDNAs encoding truncated uvomorulin molecules; these truncations comprised 32 or 72 amino acids from the COOH-terminal cytoplasmic domain of uvomorulin (Ozawa et al., 1989). Under these conditions, we found that redistribution of uvomorulin, fodrin, and Na,K-ATPase did not occur upon induction of cell–cell contact. We interpreted this result as a loss in cytoplasmic linkage between uvomorulin and the Na,K-ATPase due to deletion of the binding site on the cytoplasmic domain of uvomorulin to the cytoskeleton (see McNeill et al., 1990).

Although these results are consistent with the hypothesis that linkage of Na,K-ATPase and uvomorulin through the membrane-cytoskeleton plays a role in the spatial organization of these membrane proteins in the cell, we sought to determine whether the tight junction also plays a role in the redistribution of these proteins in transfected L cells. We showed that tight junctions are not expressed in these cells (for details, see McNeill et al., 1990); therefore, the establishment and maintenance of the polarized distribution of uvomorulin and the Na,K-ATPase in these cells occurred independently of the presence of tight junctions.

Conclusions

These results demonstrate that the epithelial CAM uvomorulin plays a direct role as an inducer of the reorganization and cell surface polarity of the Na,K-ATPase and perhaps other membrane proteins linked through the membrane-cytoskeleton. We suggest that the induction of cell surface polarity of constitutively expressed proteins, such as the Na,K-ATPase, by uvomorulin-mediated cell–cell contacts, in combination with the de novo formation of specialized components of the apical surface, is an integral part of the structural and functional remodeling of the cell surface in the development of the polarized epithelial cell. Once cell polarity has been established, and as proteins are turned over in the membrane, they are replaced by newly synthesized proteins that are targeted to the appropriate domain from the Golgi complex.

Acknowledgments

The studies reported from this laboratory have been supported by grants to W. J. Nelson from the National Institutes of Health (GM-35527) and the National Science Foundation (DCB 8609091 and DIR 8811434), and to R. W. Hammerton from the National Kidney Foundation. W. J. Nelson is an Established Investigator of the American Heart Association.

References

Bennett, V. 1985. The membrane skeleton of human erythrocytes and its implications for more complex cells. *Annual Review of Biochemistry.* 54:273–304.

Berridge, M. J., and J. L. Oschman. 1972. *In* Transporting Epithelia. Academic Press, New York. 91–108.

Bourguignon, L. Y. W., and G. J. Bourguignon. 1984. Capping and the cytoskeleton. *International Review of Cytology.* 87:195–223.

Gumbiner, B. 1987. The structure, biochemistry, and assembly of epithelial tight junctions. *American Journal of Physiology.* 253:C749–C758.

Jesaitis, A. J., and J. Yguerabide. 1986. The lateral mobility of the Na^+,K^+-dependent ATPase in Madin-Darby canine kidney cells. *Journal of Cell Biology.* 102:1256–1263.

Jørgensen, P. L. 1982. Mechanisms of the Na^+,K^+ pump: protein structure and conformation of the pure (Na^+ + K^+)-ATPase. *Biochimica et Biophysica Acta.* 694:27–68.

Jørgensen, P. L. 1991. Conformational transitions in the α-subunit and ion occlusion. *In* The Sodium Pump: Structure, Mechanism, and Regulation. J. H. Kaplan and P. De Weer, editors. The Rockefeller University Press, New York. In press.

Marchesi, V. T. 1985. Stabilizing infrastructure of cell membranes. *Annual Review of Cell Biology.* 1:531–562.

McNeill, H., M. Ozawa, R. Kemler, and W. J. Nelson. 1990. Novel function of the cell adhesion molecule uvomorulin as an inducer of cell surface polarity. *Cell.* 62:309–316.

Mege, R. M., F. Matzsuzaki, W. J. Gallin, J. I. Goldberg, B. A. Cunningham, and G. M. Edelman. 1988. Construction of epithelial sheets by transfection of mouse sarcoma cells with cDNAs for chicken cell adhesion molecules. *Proceedings of the National Academy of Sciences, USA.* 85:7274–7278.

Molitoris, B. A., L. K. Chan, J. I. Shapiro, J. D. Conger, and S. A. Falk. 1989. Loss of epithelial polarity: a novel hypothesis for reduced proximal tubule Na^+ transport following ischemic injury. *Journal of Membrane Biology.* 107:119–127.

Morrow, J. S., C. D. Cianci, T. Ardito, A. S. Mann, and M. Kashgarian. 1989. Ankyrin links fodrin to the alpha subunit of the Na^+,K^+-ATPase in Madin-Darby canine kidney cells and in intact renal tubule cells. *Journal of Cell Biology.* 108:455–465.

Nagafuchi, A., Y. Shirayoshi, K. Okazaki, K. Yasuda, and M. Takeichi. 1987. Transformation of cell adhesion properties by exogenously introduced E-cadherin cDNA. *Nature.* 329:340–343.

Nelson, W. J., C. A. L. S. Colaco, and E. Lazarides. 1983. Involvement of spectrin in cell-surface receptor capping in lymphocytes. *Proceedings of the National Academy of Sciences, USA.* 80:1626–1630.

Nelson, W. J., and R. W. Hammerton. 1989. A membrane-cytoskeletal complex containing

Na^+,K^+-ATPase, ankyrin and fodrin in Madin-Darby canine kidney (MDCK) cells. Implications for the biogenesis of epithelial cell polarity. *Journal of Cell Biology.* 108:893–902.

Nelson, W. J., and E. Lazarides. 1984. Assembly and establishment of membrane-cytoskeleton domains during differentiation: spectrin as a model system. *In* Cell Membranes: Methods and Reviews. Vol. 2. E. Elson, W. Frazier, and L. Glaser, editors. Plenum Publishing Corp., New York. 219–246.

Nelson, W. J., E. M. Shore, A. Z. Wang, and R. W. Hammerton. 1990. Identification of a membrane-cytoskeletal complex containing the cell adhesion molecule uvomorulin (E-cadherin), ankyrin, and fodrin in Madin-Darby canine kidney cells. *Journal of Cell Biology.* 110:349–357.

Nelson, W. J., and P. J. Veshnock. 1986. Dynamics of membrane-cytoskeleton (fodrin) organization during development of polarity in Madin-Darby canine kidney epithelial cells. *Journal of Cell Biology.* 103:1751–1766.

Nelson, W. J., and P. J. Veshnock. 1987*a*. Ankyrin binding to (Na^+,K^+)-ATPase and implications for the organization of membrane domains in polarized epithelial cells. *Nature.* 328:533–535.

Nelson, W. J., and P. J. Veshnock. 1987*b*. Modulation of fodrin (membrane skeleton) stability by cell–cell contact in Madin-Darby canine kidney epithelial cells. *Journal of Cell Biology.* 104:1527–1537.

Ozawa, M., H. Baribault, and R. Kemler. 1989. The cytoplasmic domain of the cell adhesion molecule uvomorulin associates with three independent proteins structurally related in different species. *EMBO Journal.* 8:1711–1717.

Rodriguez-Boulan, E., and W. J. Nelson. 1989. Morphogenesis of the polarized epithelial cell phenotype. *Science.* 245:718–725.

Salas, P. J. I., D. E. Vega-Salas, J. Hochman, E. Rodriguez-Boulan, and M. Edidin. 1988. Selective anchoring in the specific plasma membrane domain: a role in epithelial cell polarity. *Journal of Cell Biology.* 107:2363–2376.

Scheetz, M. P., M. Schindler, and D. E. Koppel. 1980. Lateral mobility of integral membrane proteins is increased in spherocytic erythrocytes. *Nature.* 285:510–511.

Simons, K., and S. D. Fuller. 1985. Cell surface polarity in epithelia. *Annual Review of Cell Biology.* 1:243–288.

Wilson, P. D., and D. Hreniuk. 1987. Altered polarity of Na^+,K^+-ATPase in epithelia with a genetic defect and abnormal basement membrane. *Journal of Cell Biology.* 105:176*a*. (Abstr.)

Structure of the H,K-ATPase β-Subunit and Relationship to Na,K-ATPase β-Subunits

The complete amino acid sequence of the rat gastric H,K-ATPase β-subunit (HKβ) is shown in Fig. 1. The protein consists of 294 amino acids with a predicted molecular weight of 33,689. Hydropathy analysis (Canfield et al., 1990) indicates that the protein contains a polar cytoplasmic amino terminus followed by a single transmembrane domain of 27 amino acids and a 228-residue-long extracellular carboxyl-terminal domain. When the amino acid sequence of the protein was compared with all sequences currently in the National Biomedical Research Foundation database, the H,K-ATPase β-subunit showed significant similarity only to Na,K-ATPase

```
                                                     50
HKβ    MAALQEKKSCSQRMAEFRQYCWNPDTGQMLGRTPARWVWISLYYAAFYVVMTGLFALCIYVLMQTI   66
NaKβ1  MARGKAKEEGS-----WKKFIWNSEKKEFLGRTGGSWFKILLFYVIFYGCLAGIFIGTIQVMLLTI   61
NaKβ2  MVIQKEKKSCGQVVEEWKEFVWNPRTHQFMGRTGTSWAFILLFYLVFYGFLTAMFTLTMWVMLQTV   66

                                  100
HKβ    DPYTPDYQDQLKSPGVTLRPDVYGERGLQISYNISENSSWAGLTHTLHSFLAGYTPASQQDSIN--  130
NaKβ1  SELKPTYQDRVAPPGLTQIPQI---QKTEISFRPNDPKSYEAYVLNIIRFLEKY-----KDS--AQ  117
NaKβ2  SDHTPKYQDRLATPGLMIRPKT---ENLDVIVNISDTESWDQHVQKLNKFLEPY-----NDSIQAQ  124

                             150
HKβ    --------CSSEKYFFQETFSAP--NHTKFSCKFTADMLQNCSGLVDPS-FGFEEGKPCFIIKMNR  185
NaKβ1  KDDMIFEDCGSMPSEPKERGEFNHERGERKVCRFKLDWLGNCSGLNDES-YGYKEGKPCIIIKLNR  182
NaKβ2  KNDV----CRPGRYYEQPDNGVL--NYPKRACQFNRTQLGNCSGIGDPTHYGYSTGQPCVFIKMNR  184

                               200
HKβ    IVKFLPSNNTAPR----------------VDCTFQDDPQKPRKDIEPL-QVQYYPPNG--TFSLHY  232
NaKβ1  MLGFKPKPPKNESLETYPLTMKYNPNVLPVQCTGKRDE-----DKDKVGNIEYFGMGGFYGFPLQY  243
NaKβ2  VINFYAGA--NQSMN--------------VTCVGKKDE-----DAENLGHFIMFPANG--NIDLMY  227

                   250
HKβ    FPYYGKKAQPHYSNPLVAAKFLNVPKNTQVLIVCKIMADHVTFDNPHDPYEGKVEFKLTIQK      294
NaKβ1  YPYYGKLLQPKYLQPLLAVQFTNLTLDTEIRIECKAYGENIGY-SEKDRFQGRFDVKIEV-KS     304
NaKβ2  FPYYGKKFHVNYTQPLVAVKFLNVTPNVEVNVECRINAANIATDDERDKFAARVAFKLRINKA     290
```

Figure 1. Amino acid sequences of H,K-ATPase and Na,K-ATPase β-subunits. The deduced amino acid sequence of the rat H,K-ATPase β-subunit (*HK*β, *top line*) was aligned with those of the rat Na,K-ATPase β1-(*NaK*β*1, middle line*) and β2-(*NaK*β*2, bottom line*) subunits. Dashes in the sequence allow optimal alignment for amino acid insertions/deletions. Cysteine residues are boxed; asparagine residues that are possible sites for N-linked glycosylation are underlined. The H,K-ATPase β-subunit sequence is numbered above, whereas amino acids for all three β-subunits are numbered on the right.

β-subunit isoforms. A comparison of the amino acid sequence of the rat H,K-ATPase β-subunit with the rat Na,K-ATPase β1- (Mercer et al., 1986) and β2- (Martin-Vasallo et al., 1989) subunits is presented in Fig. 1. Dashes in the sequences allow optimal alignment for amino acid insertions and/or deletions. Cysteine residues are enclosed in boxes, and sites for potential N-linked glycosylation are underlined. The amino acid sequence of the rat H,K-ATPase β-subunit exhibits 41% identity with the rat Na,K-ATPase β2-subunit and 35% identity with the rat β1-subunit. A putative transmembrane segment is located between residues 40 and 66 in the H,K-ATPase β-subunit and 40 and 67 in the Na,K-ATPase β2-subunit. This region is highly conserved between the two proteins. Of the 27 amino acid residues compared, 11 are identical and 9 represent conservative substitutions. There are six cysteine residues within the presumed extracellular domain of the H,K-ATPase β-subunit (positions 131, 152, 162, 178, 201, and 266). In the computer-aligned sequences, these cysteine residues appear to be highly conserved among the H,K-ATPase and Na,K-ATPase β-subunits. These residues could play an important role in maintaining β-subunit structure, possibly forming several folded subdomains each cross-linked by disulfide bonds.

As shown in Fig. 1, there are seven potential sites for N-linked glycosylation in the rat H,K-ATPase β-subunit, seven in the rat Na,K-ATPase β2-subunit, and three in the β1-subunit polypeptide. One of the predicted N-linked glycosylation sites in the H,K-ATPase β-subunit (Asn_{161}) is exactly conserved relative to one of the

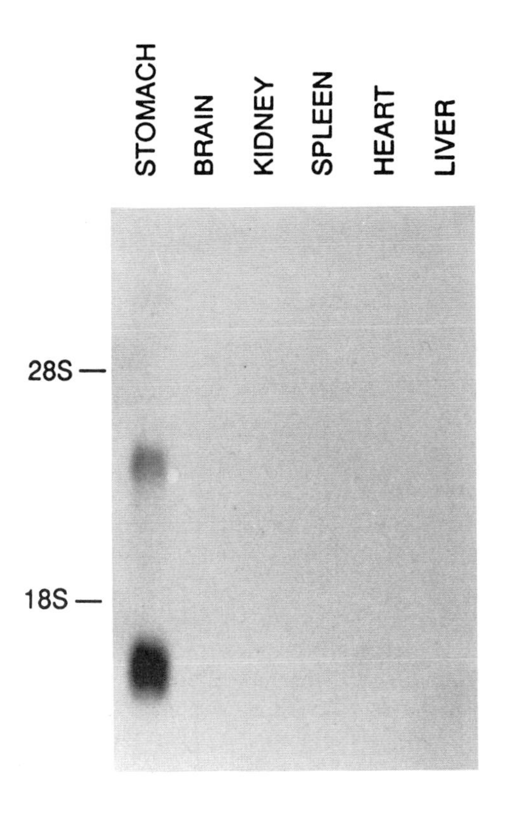

Figure 2. Expression of H,K-ATPase β-subunit mRNA in rat tissues. RNA was prepared from the adult rat tissues indicated. Total cellular RNA (25 μg) was fractionated by electrophoresis through a 1% agarose-containing formaldehyde gel, transferred to a Zetabind filter, and hybridized with a cDNA probe for the H,K-ATPase β-subunit. The positions of the 28S and 18S markers are indicated at the left.

predicted N-linked glycosylation sites found in the rat β1- and β2-subunits, whereas a second (Asn_{193}) is located three residues from a predicted N-linked glycosylation site that is conserved between the β1- and β2-subunits. A third asparagine residue (Asn_{100}) is exactly conserved relative to an N-linked glycosylation site found in the rat β2-subunit, while a fourth asparagine residue (Asn_{255}) is exactly conserved relative to the position of an N-linked glycosylation site conserved between the β1- and β2-subunits. However, Asn_{255} is not contained within a consensus glycosylation sequence. Amino acid sequence analysis of deglycosylated H,K-ATPase β-subunits (Canfield et al., 1990; Forte, J. G., C. T. Okamoto, D. Chow, V. A. Canfield, and R. Levenson, unpublished observations) reveals that aspartate is present in positions 193 and 225 instead of asparagine residues predicted from the cDNA sequence.

Since deglycosylation with *N*-glycanase F deaminates asparagine to aspartic acid (while releasing oligosaccharide), we conclude that Asn_{193} and Asn_{225} are likely to be glycosylated in situ. Together, the structural similarities between H,K- and Na,K-ATPase β-subunits strongly indicate that these polypeptides evolved from a common ancestral gene and comprise a family of related P-type ion transport protein subunits.

Tissue-specific Expression of the H,K-ATPase β-Subunit

To analyze expression of the H,K-ATPase β-subunit gene, β-subunit cDNA was hybridized to an RNA transfer blot containing rat tissue RNA. The results are shown in Fig. 2. Of the rat tissues analyzed (brain, heart, lung, kidney, liver, spleen, and stomach), β-subunit mRNA was detected only in stomach. The β-subunit gene appears to encode two transcripts, ~1.7 and 3.7 kb in size. The 1.7-kb transcript is

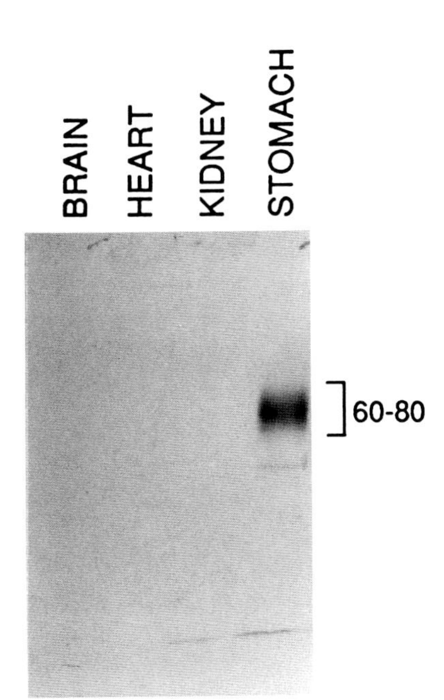

Figure 3. Western blot analysis of H,K-ATPase β-subunit distribution. Crude microsomal membrane fractions were prepared from rat tissues. Solubilized membrane proteins (80 μg/lane) were fractionated by electrophoresis through an SDS-containing 10% polyacrylamide gel, transferred to a nitrocellulose filter, and probed with an anti–gp 60–80 monoclonal antibody. The apparent molecular weight of the glycosylated H,K-ATPase β-subunit is indicated at the right.

the predominant mRNA species and is ~20-fold more abundant than the 3.7-kb transcript. We do not yet know whether the two β-subunit mRNAs arise from differential splicing, utilization of alternative polyadenylation signals, or if the larger transcript represents an unprocessed nuclear precursor.

To determine the tissue distribution of H,K-ATPase β-subunit polypeptides, we probed a Western blot of rat microsomal membrane fractions with the H,K-ATPase β-subunit monoclonal antibody. As shown in Fig. 3, the antibody reacted with a broad band of ~60 to ~80 kD in rat stomach. In contrast, no β-subunits were detectable in kidney, brain, and heart. These results suggest that the H,K-ATPase β-subunit is expressed exclusively in stomach. In addition, we have found that the H,K-ATPase β-subunit is expressed in the stomach of a wide variety of animal species including rat, rabbit, hog, cow, mouse (Canfield et al., 1990), and frog

(Beesley and Forte, 1973). These results are consistent with the view that the β-subunit is expressed in all vertebrates capable of gastric HCl secretion.

Chromosomal Localization and Organization of the H,K-ATPase β-Subunit Gene

We have used segregation of restriction fragment length polymorphisms (RFLPs) among recombinant inbred strains of mice to identify the chromosomal location of the mouse gene (*Atp4b*) encoding the gastric H,K-ATPase β-subunit. As shown in Fig. 4, mouse genomic β-subunit DNA sequences were identified by hybridizing Southern blots of Taq I–digested mouse DNA with β-subunit cDNA as probe. This probe hybridizes to three major genomic fragments, 5.0, 3.2, and 2.4 kb long. The 3.2-kb fragment is common to the six strains of mice tested, whereas the 5.0-kb fragment is specific to strains 1, 2, 3, 4, and 6 and the 2.4-kb fragment is specific to strain 5. Analysis of the strain distribution pattern of 2/5 recombinant mouse strains (Canfield et al., 1990) reveals linkage of the β-subunit gene with chromosome 8

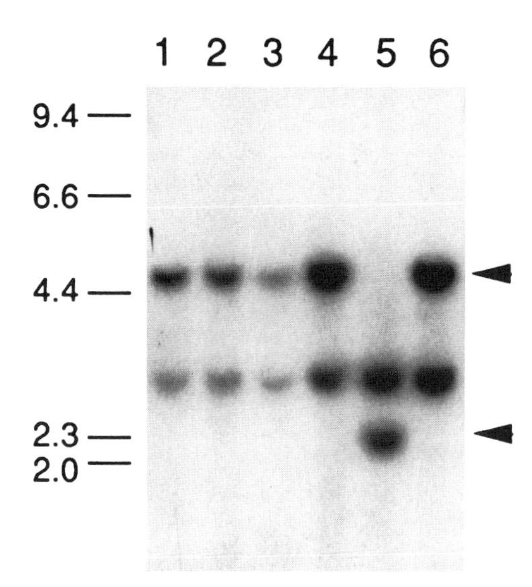

Figure 4. Restriction fragment length polymorphism for the H,K-ATPase β-subunit. Genomic DNA was prepared from six inbred mouse strains. Genomic DNA was digested to completion with Taq I, separated on a 1% agarose gel, blotted onto a hybridization membrane, and probed with H,K-ATPase β-subunit cDNA. Positions of DNA size markers are shown at the left. Arrows indicate the position of polymorphic hybridizing DNA fragments.

markers (Table I). Previous chromosome mapping experiments have demonstrated that the Na,K-ATPase β1- and β2-subunit genes are located on murine chromosomes 1 and 11, respectively (assignments of all H,K- and Na,K-ATPase subunit genes are presented in Table I). The genomic distribution of H,K-ATPase and Na,K-ATPase β-subunit genes indicates that they form a multigene family. Chromosomal mapping of the β-subunit genes reveals their dispersion in the murine genome; therefore, the existence of a common *cis*-acting mechanism regulating expression of the β-subunit genes is ruled out.

The isolation of genomic sequences overlapping the H,K-ATPase β-subunit gene would permit the analysis of putative *cis*- and *trans*-acting elements participating in the regulation of parietal cell-specific expression of the β-subunit gene. Further, elucidation of the genomic organization of the β-subunit gene would permit an analysis of the evolutionary history of a gene that shares considerable sequence homology with a group of genes that participate in related yet distinct transport

TABLE I
Chromosomal Assignment of H,K- and Na,K-ATPase Genes

ATPase subunit	Mouse locus	Mouse chromosome	Linked loci	Reference
Na,K α1	*Atp1a1*	3	*Egf*	Kent et al., 1987
Na,K α2	*Atp1a2*	1	*Spna-1*	Kent et al., 1987
			Fcelα, Sap, Crp	Kingsmore et al., 1989
Na,K α3	*Atp1a3*	7	*P-450b/Coh*	Kent et al., 1987
H,K α	*Atp4a**	Unknown		
Na,K β1	*Atp1b1*	1	*Spna-1*	Kent et al., 1987
Na,K β2	*Atp1b2*	11	*Zfp3-1, Asgr-1/2, Akv-4, Evi-2*	Malo et al., 1990
H,K β	*Atp4b*	8	*Xmv-26, Xmv-12*	Canfield et al., 1990

Descriptions of loci are as follows: *Egf,* epidermal growth factor; *Spna-1,* α spectrin; *Fce1α,* Fc receptor for IgE; *Sap,* serum amyloid P-component; *Crp,* C-reactive protein; *P450b/Coh,* coumarin hydroxylase; *Zfp3,* zinc finger protein; *Asgr-1/2,* asialoglycoprotein receptor; *Akr-4, Evi-1,* ecotropic proviruses; *Xmv-12, Xmv-26,* xenotropic leukemia proviruses.

*Reserved name.

properties. To isolate the H,K-ATPase β-subunit gene, a full-length rat H,K-ATPase β-subunit cDNA was used as a probe to identify mouse H,K-ATPase β-subunit clones in a pWE15 cosmid library constructed with genomic DNA from Balb/3T3 fibroblasts. A restriction map of the ~40-kb genomic DNA domain encompassed by the inserts of three cosmid clones is shown in Fig. 5. Exon/intron boundaries were established by DNA sequence comparison of cDNA and genomic subclones. The schematic map shows that the gene spans ~12 kb of genomic DNA consisting of ~11.1 kb of intervening sequences and ~0.9 kb of coding sequences. The gene is interrupted by six introns ranging in size from 150 bp to 3 kb, while the length of the seven exons varies from 57 to 560 bp. Exon 1 contains the first 175 untranslated nucleotides at the 5′ end of the gene, the ATG translation initiation codon, and the first 112 nucleotides of coding sequence corresponding to the amino-terminal intracellular domain of the β-subunit protein. Exon 2 contains sequences encoding the putative transmembrane domain. Exons 3–7 encode the entire carboxyl-terminal extracellular domain. Exon 7 contains the last 159 nucleotides of coding sequence,

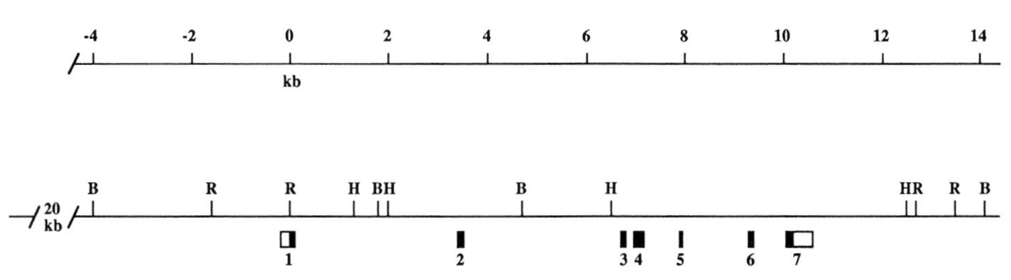

Figure 5. Physical map of the mouse H,K-ATPase β-subunit gene. The structure of the H,K-ATPase β-subunit gene is schematically represented. Exons are represented by boxes (numbered 1–7). Solid boxes contain coding sequence; open boxes represent untranslated regions. The direction of transcription is from left to right. The scale (every 2 kb) is shown above the gene. Recognition sites for restriction enzymes are designated by single letters. *B,* Bam HI; *R,* Eco RI; *H,* Hind III.

the translation stop codon, and ~400 bp of 3′ untranslated nucleotides. In aggregate, the cosmid clones contain >20 kb of sequence flanking the 5′ end of the β-subunit gene. The availability of these cosmid clones thus provides a basis for the identification and characterization of control elements mediating the tissue-specific expression of the β-subunit gene.

Conclusions and Discussion

The isolation of cDNA clones for the β-subunit of the gastric H,K-ATPase reveals a polypeptide that is remarkably similar to Na,K-ATPase β-subunit isoforms. The rat H,K-ATPase β-subunit exhibits 41% amino acid sequence identity to the rat Na,K-ATPase β2-subunit and, like Na,K-ATPase β-subunits, contains a short intracellular amino-terminal domain, a single transmembrane segment, and a large extracellular carboxyl-terminal domain containing multiple sites for N-linked glycosylation. Our results suggest that the H,K- and Na,K-ATPase β-subunit genes comprise a multigene family. Chromosomal mapping studies show that the H,K-ATPase β-subunit gene is located on mouse chromosome 8, the Na,K-ATPase β1-subunit gene resides on mouse chromosome 1, and the β2-subunit gene maps to mouse chromosome 11. Dispersion of the β-subunit genes argues in favor of the idea that the polypeptide encoded by each gene may have properties selected in response to different physiological demands. Genomic clones encompassing the coding and 5′ flanking regions of the mouse H,K-ATPase β-subunit gene have been isolated and characterized. The gene is composed of seven exons and spans ~12 kb of genomic DNA. Analysis of the 5′ flanking sequence should provide insight into control elements that regulate the cell-specific expression of the H,K-ATPase β-subunit gene.

Identification of the gene for the gastric H,K-ATPase β-subunit raises a number of intriguing questions. Is it possible for the H,K-ATPase β-subunit to associate with the Na,K-ATPase α-subunit (and vice versa), and if so, to produce a molecule with enzymatic activity? Both H,K-ATPase and Na,K-ATPase are expressed in the parietal cell. However, H,K-ATPase is sorted to the tubulovesicular membrane compartment, while Na,K-ATPase is targeted to the basolateral domain. Distribution of H,K-ATPase and Na,K-ATPase activities to separate membrane compartments suggests that there may be regulatory elements that govern the specific association of H,K- and Na,K-ATPase β-subunits with their respective α-subunits. It should be pointed out, however, that parietal cells express the Na,K-ATPase β1-subunit isoform, and the H,K-ATPase β-subunit exhibits greater sequence similarity to the Na,K-ATPase β2-subunit than to the β1-subunit. This raises the possibility that the H,K-ATPase β-subunit might be capable of substituting for the Na,K-ATPase β2-subunit, especially in situations in which β-subunit expression is required to maintain cellular viability. The ability to specifically "knock out" β2-subunit gene expression, then rescue a cell via introduction of the cloned H,K-ATPase β-subunit gene could provide an approach for addressing this question.

It will clearly be of interest to determine how the sequence of each β-subunit leads to specificity regarding interaction with the H,K- or Na,K-ATPase α-subunit. The construction and expression of chimeric cDNA molecules between H,K-ATPase and Na,K-ATPase β-subunit cDNAs should permit identification of sites within a given β-subunit that interact with the corresponding α-subunit. This type of approach should also be useful for identifying other functional domains within the H,K- and Na,K-ATPase β-subunits. A further issue that should be raised in this context is the elucidation of the function of the β-subunit for H,K- or Na,K-ATPase activity. It will clearly be of interest to determine if expression of H,K/Na,K β-subunit chimeras affects the biochemical and/or enzymatic properties of H,K- or Na,K-ATPases.

Our data indicate that the H,K-ATPase β-subunit is expressed exclusively in the stomach of a wide variety of mammalian species. What factors govern the tissue-specific expression of the H,K-ATPase β-subunit gene? Isolation of genomic DNA sequences encompassing the β-subunit gene provides a framework for addressing this question. For example, testing the transcriptional activities of 5′ deletion fragments from the promoter (linked to a reporter gene) in cultured parietal cells should permit us to define *cis*-acting elements implicated in transcriptional regulation of the mouse H,K-ATPase β-subunit gene.

Characterization of positive (and negative) *cis*-acting 5′ regulatory elements would provide a basis for the identification and analysis of *trans*-acting factors mediating stomach-specific expression of the β-subunit gene. Elucidation of this mechanism could have important implications for understanding H,K-ATPase biogenesis in particular, and stomach physiology in general.

Acknowledgments

We thank Dr. Andrew Shyjan who originally suggested this project to us. We are grateful to Dr. Edwin Geissler (Beth Israel Hospital, Boston, MA) for providing the mouse cosmid DNA library.

This work was supported by grants from the Medical Research Council (MRC) of Canada to P. Gros and the National Institutes of Health to J. G. Forte and R. Levenson.

References

Beesley, R. C., and J. G. Forte. 1973. Glycoproteins and glycolipids of oxyntic cell microsomes. *Biochimica et Biophysica Acta.* 307:372–385.

Canfield, V. A., C. T. Okamoto, D. Chow, J. Dorfman, P. Gros, J. G. Forte, and R. Levenson. 1990. Cloning of the H,K-ATPase β subunit: tissue specific expression, chromosomal assignment, and relationship to Na,K-ATPase β subunits. *Journal of Biological Chemistry.* 265:19878–19884.

Farley, R. A., and L. D. Faller. 1985. The amino acid sequence of an active site peptide from the H,K-ATPase of gastric mucosa. *Journal of Biological Chemistry.* 260:3899–3901.

Frohman, M. A., M. K. Dush, and G. R. Martin. 1988. Rapid production of full-length cDNAs from rare transcripts: amplification using a single gene-specific oligonucleotide primer. *Proceedings of the National Academy of Sciences, USA.* 85:8998–9002.

Kent, R. B., D. A. Fallows, E. Geissler, T. Glaser, J. R. Emanuel, P. A. Lalley, R. Levenson, and D. E. Housman. 1987. Genes encoding α and β subunits of the Na,K-ATPase are located on three different chromosomes in the mouse. *Proceedings of the National Academy of Sciences, USA.* 84:5369–5373.

Kingsmore, S. F., M. L. Watson, T. A. Howard, and M. F. Seldin. 1989. A 6000 kb segment of chromosome 1 is conserved in human and mouse. *EMBO Journal.* 8:4073–4080.

Malo, D., E. Schurr, R. Levenson, and P. Gros. 1990. Assignment of the Na,K-ATPase β2-subunit gene (*Atpb*-2) to mouse chromosome 11. *Genomics.* 6:697–699.

Martin-Vasallo, P., W. Dackowski, J. R. Emanuel, and R. Levenson. 1989. Identification of a putative isoform of the Na,K-ATPase β subunit. *Journal of Biological Chemistry.* 264:4613–4618.

Mercer, R. W., J. W. Schneider, A. Savitz, J. R. Emanuel, E. J. Benz, Jr., and R. Levenson. 1986. Rat-brain Na,K-ATPase β-chain gene: primary structure, tissue-specific expression, and amplification in ouabain-resistant C^+ cells. *Molecular and Cellular Biology.* 6:3884–3890.

Okamoto, C. T., J. M. Karpilow, A. Smolka, and J. G. Forte. 1990. Isolation and characterization of gastric microsomal glycoproteins: evidence for a glycosylated β-subunit of the H^+/K^+-ATPase. *Biochimica et Biophysica Acta.* 1037:360–372.

Shull, G. E., and J. B. Lingrel. 1986. Molecular cloning of rat stomach ($H^+ + K^+$)-ATPase. *Journal of Biological Chemistry.* 261:16788–16791.

Shyjan, A. W., C. Gottardi, and R. Levenson. 1990. The Na,K-ATPase β2 subunit is expressed in rat brain and copurifies with Na,K-ATPase activity *Journal of Biological Chemistry.* 265:5166–5169.

Structure

Chapter 8

Architecture of the Sodium Pump Molecule: Probing the Folding of the Hydrophobic Domain

Nikolai Modyanov, Svetlana Lutsenko, Elena Chertova, and Roman Efremov

Shemyakin Institute of Bioorganic Chemistry, USSR Academy of Sciences, 117871, Moscow, USSR

Introduction

Current ideas on the mechanism of such a complex biological machine as the active transport system for cations are still primitive. Evidently, to understand the molecular events that underlie coupled ATP hydrolysis, energy transduction, and cation transport, thorough knowledge of the three-dimensional structure of the Na,K-ATPase molecule and the dynamics of its changes is absolutely necessary. As first steps in this direction, a combination of various experimental approaches has been used in our laboratory to obtain specific information on spatial localization of certain structural elements of the subunit polypeptide chains and to localize components of the ATP-hydrolyzing center and other functional domains.

This paper consists of two parts. The first briefly summarizes the earlier published data on the Na,K-ATPase structural organization. The second considers in detail the current status of spatial structure analysis of the intramembrane portions of the sodium pump, directed toward identification of the protein regions involved in the formation of ion-conducting pathways and establishment of their spatial relationships.

Transmembrane Arrangement of Polypeptide Chains

Determination of a complete primary structure of Na,K^+-ATPases from diverse sources (Kawakami et al., 1985; Ovchinnikov et al., 1985*a*, 1986*b*; Shull et al., 1985, 1986; Noguchi et al., 1986) initiated a new stage in studying the molecular organization of the enzyme. A first urgent task is to reliably identify exposed and intramembrane domains; i.e., to design a two-dimensional scheme for the transmembrane protein folding.

To reveal extramembrane regions of the enzyme molecule we developed a procedure of stepwise limited trypsinolysis of the native membrane-bound Na,K-ATPase. This assumes that the intramembrane moiety is protected against protease attack by the lipid environment. A single hydrolysis under certain conditions provided exhaustive digestion of exposed regions of the α-subunit with complete retention of glycoprotein integrity. Before the second step of trypsinolysis, to overcome the β-subunit resistance to proteolysis, its disulfide bonds were reduced by β-mercaptoethanol treatment. The results of detailed structural analysis of water-soluble peptides isolated from both hydrolyzates (in total ~600 and 180 amino acid residues of α- and β-subunits, respectively) provided evidence of the extramembrane localization of the corresponding regions of the polypeptide chains. This enabled us to propose a scheme for subunit folding in the plasma membrane. According to the model, α- and β-subunits have seven and one transmembrane segments, respectively. So amino- and carboxy-termini of both polypeptides are situated on opposite sides of the membrane (Ovchinnikov et al., 1985*a*, 1986*b*, 1987*a*).

Immunochemical analysis was the first method used to test this model. The proposed spatial disposition of the hydrophilic loops linking membrane segments V, VI, and VII of the catalytic subunit was tested using monoclonal antibodies IIC_9 and Vg_2, which recognize the corresponding amino acid sequences, upon immunofluorescent analysis of viable cells or smears of a pig kidney embryonic (PKE) cell line (Modyanov et al., 1987; Ovchinnikov et al., 1988*b*). The extracellular orientation of the COOH-terminal region of the α-subunit was confirmed using affinity purified antibodies (α-p999) to the synthetic peptide of the corresponding sequence. The presence of the α-p999 epitope on the outer surface of the plasma membrane was demonstrated in experiments with viable PKE cells by flow cytofluorimetry measurements. In addition, it was shown that trypsin treatment of PKE cells considerably reduced the density of the α-p999 epitope on the cell surface and that perforation of the trypsinized cells did not increase the fluorescence intensity (Modyanov et al., 1988*a;* Ovchinnikov et al., 1988*b*). The proposed disposition of the COOH-terminal region of the α-subunit was confirmed recently by Bayer (1990).

Analysis of the secondary structure of hydrophilic and hydrophobic domains of Na,K-ATPase was performed by Raman spectroscopy of the native enzyme and membrane-bound products of stepwise proteolysis (Ovchinnikov et al., 1988*a*). Spectral data demonstrate unambiguously that all components of the intramembrane moiety of Na,K-ATPase are in the α-helical conformation. The estimated content of regular secondary structures in the cytoplasmic domain of the α-subunit and in the exoplasmic domain of the β-subunit was used to calculate the secondary structure organization of the Na,K-ATPase molecule by statistical prediction techniques. Large reversible conformational changes of the enzyme molecule associated with the $E_1 \leftrightarrow E_2$ transition were detected by the same spectroscopic method (Nabiev et al., 1988).

We used affinity modification with two substrate analogues to study the active site topography. An alkylating ATP analogue, γ-[4-*N*-2-chloroethyl-*N*-methyl-amino)]benzylamide-ATP (ClRATP), covalently binds to the α-subunit, yielding a product resistant to hydrolysis by the enzyme and inhibiting the ATPase activity (Dzhandzhugazyan and Modyanov, 1985). Sequence determination of the modified peptides allowed us to identify Asp^{710} and Asp^{714} as the targets of modification in E_1 and E_2 enzyme forms, respectively (Dzhandzhugazyan et al., 1986, 1988; Ovchinnikov et al., 1987*b*). This finding directly demonstrates a rearrangement of the molecular structure of the active site upon the $E_1 \leftrightarrow E_2$ transition. We performed the affinity modification of Na,K-ATPase with purified dialdehyde ATP derivative (oATP) to reveal the components of the active site interacting with the substrate ribosyl moiety. A 10-kD modified fragment of the α-subunit localized in the COOH-terminal part of the large cytoplasmic loop between the fourth and fifth membrane segments was isolated from the tryptic hydrolyzate (Bernikov et al., 1990).

Determination of the complete sequence of a human gene for the catalytic subunit of Na,K-ATPase allowed analysis of a probable correlation between the exon–intron organization of gene and protein structure–function features (Ovchinnikov et al., 1988*c;* Broude et al., 1989). In five cases the positions of the introns coincided with the boundaries of transmembrane segments. A preferred disposition of introns in the gene moiety encoding protein surface regions has been proposed by Craik et al. (1982). This is true for all fragments of the polypeptide chain identified on the surface of the Na,K-ATPase molecule by immunochemical methods (Modyanov et al., 1987; Ovchinnikov et al., 1988*b*). Furthermore, the intron positions coincide well with characteristic points of limited proteolysis of the native enzyme (Jorgensen and Collins, 1986). This correlation enables us to consider the intron positions as markers of the protein surface.

Comparison of the sequences of different E_1-E_2 ATPases revealed a number of highly conserved regions, which included all the known components of the ATP binding and hydrolysis sites (for a review, see Serrano, 1989). The genes for sarcoplasmic reticulum Ca^{2+}-ATPase (Korczak et al., 1988) and Na,K-ATPase α-subunit have similar exon–intron structures in corresponding regions (Broude et al., 1989). The data support the existence of a hypothetical common ancestor gene of ion pumps, duplicated and diverged during the evolution into various forms with different specificities. Usually, all E_1-E_2 ATPases are considered as a single family of related enzymes. However, recent data suggest a more detailed classification. The discovery of the second subunit for H^+,K^+-ATPase related to the β-subunit of the Na,K-ATPase (Hall et al., 1990; Okamoto et al., 1990; Shull, 1990) and a high level of sequence homology for the catalytic subunits suggest that these enzymes are the most closely related among ion pumps of eukaryotic cells. Even the exon–intron structures of their genes are almost the same (Ovchinnikov et al., 1988*c*; Maeda et al., 1990).

On the other hand, current models of the transmembrane organization of the Ca^{2+}-ATPase and the α-subunit of Na,K-ATPase have different folding of the COOH-terminal parts of polypeptide chains starting from the fifth membrane segment (Brandl et al., 1986; Modyanov et al., 1988*a*; Shull and Greeb, 1988). The carboxy-termini of these proteins were located by site-directed antibodies on opposite membrane surfaces (Ovchinnikov et al., 1988*b*; Matthews et al., 1989). Cytoplasmic orientation of the COOH-terminal region can be regarded as proven for the Ca^{2+}-ATPase of the plasma membrane, since it contains the calmodulin binding site (James et al., 1988).

Thus, all known ion pumps of animal cells could be divided into two groups differing essentially in structure and transport properties. The first group includes all known Ca^{2+}-ATPases, which consist of a single polypeptide and transfer only one type of cation. The second group includes Na,K-ATPase and H^+,K^+-ATPases which are composed of two different subunits and counter-transport ions of two types across the plasma membrane. Functional differences between these two groups are apparently based on the peculiarities of the transmembrane folding of the catalytic polypeptides.

Structural Analysis of Intramembrane Moiety

Hydrophobic Labeling of E_1 and E_2 Enzyme Forms

Complete understanding of the cation transport mechanism is impossible without a detailed knowledge of the spatial organization of the intramembrane moiety. The crucial questions here are how transmembrane segments are interlocated and oriented relative to the lipid bilayer and how their structures could reversibly rearrange during the pump functioning. Recent success in crystallization of sarcoplasmic reticulum Ca^{2+}-ATPase (Stokes and Green, 1990) promises more information on the three-dimensional structure of one of the E_1-E_2 ATPases in the near future. Data available on electron microscopy of two-dimensional crystals of Na,K-ATPase provide only a general idea of the topography of the membrane complex (Ovchinnikov et al., 1985*b*).

In our opinion a method based on selective modification with hydrophobic reagents is a powerful tool for the structural characterization of the intramembrane part of Na,K-ATPase. The carbene-generating reagents 3-(trifluoromethyl)-3 (m-[^{125}I]iodophenyl)diazarine)[^{125}I]TID) and 1-palmitoyl-2-[11-[4-[3-(trifluoromethyl)diazirinyl]phenyl][2-^{3}H]undecanoyl]-*sn*-glycero-3-phosphorylcholine ([^{3}H]PTPC/11) were designed as highly selective markers of the membrane-embedded protein regions: these reagents preferentially modify the lipid-contacting amino acid residues (Brunner and Semenza, 1981; Meister et al., 1985; Harter et al., 1988; White and Cohen, 1988). It was demonstrated that [^{125}I]TID can react with any amino acid residue, including aliphatic chains (i.e., labeling with this reagent does not depend on the amino acid composition of a transmembrane fragment and the modification level is determined mainly by its exposure to the lipids). Photoactivable phospholipid derivatives could react only with inner membrane components, so the specificity of their interaction is higher than that of the other hydrophobic reagents (Brunner, 1989).

The functioning of Na,K-ATPase is associated with a structural rearrangement of the enzyme (Jorgensen, 1975). To examine the involvement of the intramembrane portion in these conformational changes, the modification of Na,K-ATPase with hydrophobic reagents was performed with the enzyme stabilized in different conformational states. The experimental scheme includes modification of the Na,K-ATPase, stabilized in E_1 or E_2 forms with hydrophobic reagents, separation of the labeled preparation by gel-permeation chromatography or polyacrylamide gel electrophoresis (PAGE), and digestion of isolated subunits with trypsin. Tryptic hydrolyzates were then separated by reverse-phase chromatography and amino acid sequences of the peptides were determined. The Na,K-ATPase modification with [^{125}I]TID was performed according to Chertova et al. (1989) in the presence of glutathione to avoid unspecific binding of the reagent to the extramembrane protein moiety. The membrane-bound Na,K-ATPase was labeled with the photoactivated phospholipid derivative [^{3}H]PTPC/11, using enzyme (1 mg/ml) in 50 mM imidazol-HCl, pH 7.5, containing 2 mM β-mercaptoethanol and 150 mM NaCl or 150 mM KCl and incubating with 60 nmol of the reagent (15 Ci/mmol) in the presence of the phospholipid-exchanger protein at 37°C for 30 min. The excess of the unbound reagent was removed by centrifugation through a step of 10% sucrose. The pellet was resuspended in the same buffers and UV-irradiated. The specificity of [^{3}H]PTPC/11 labeling was checked in control experiments: in the absence of the phospholipid-exchanger protein no radioactivity was detected in the α- and β-subunits (manuscript in preparation).

The modified proteins were analyzed by gel-permeation chromatography on an Ultrapore TSK-3000 SW column and by 7.5% PAGE according to Laemmli (1970). Specific radioactivity incorporated into the subunits was determined by comparing densitograms of the polyacrylamide gel with its autoradiogram.

Irradiation of the membrane-bound Na,K-ATPase both in Na^+ and K^+ forms in the presence of [^{125}I]TID and [^{3}H]PTPC/11 resulted in covalent labeling of lipids and both of the enzyme subunits (Fig. 1). Incorporation of the reagents into the protein moiety of Na,K-ATPase preparations was 2–3% and 0.1–0.2% of the total covalently bound radioactivity for [^{125}I]TID- and [^{3}H]PTPC/11-treated samples, respectively. The labeling of both the α- and β-subunits coincides with results previously obtained for Na,K-ATPase modification by [^{125}I]TID (Jorgensen and Brunner, 1983) and by two photoactivated phosphatidylcholine derivatives, 1-myristoyl,2-[12-amino(4*N*-3-nitro-1-azidophenyl)]-dodecanoyl-*sn*-glycero-3[^{14}C]phosphocholine (PL-1) bearing its reactive group in the central membrane part like [^{3}H]PTPC/11, and 1-palmitoyl,2-(2-azido-4-nitro)benzoyl-*sn*-glycero-3-[^{3}H]phosphocholine (PL-2) (Montecucco et al., 1981). Both enzyme subunits appear to be labeled in these experiments. On the contrary, modification of the Na,K-ATPase with iodonaphthylazide (INA) (Jorgensen et al., 1982) and adamantanyldiazarine (AD) (Farley et al., 1980) resulted in hydrophobic reagent incorporation in the α-subunit only. This difference is perhaps due to chemical and steric peculiarities of the reagents (for review, see Brunner, 1989).

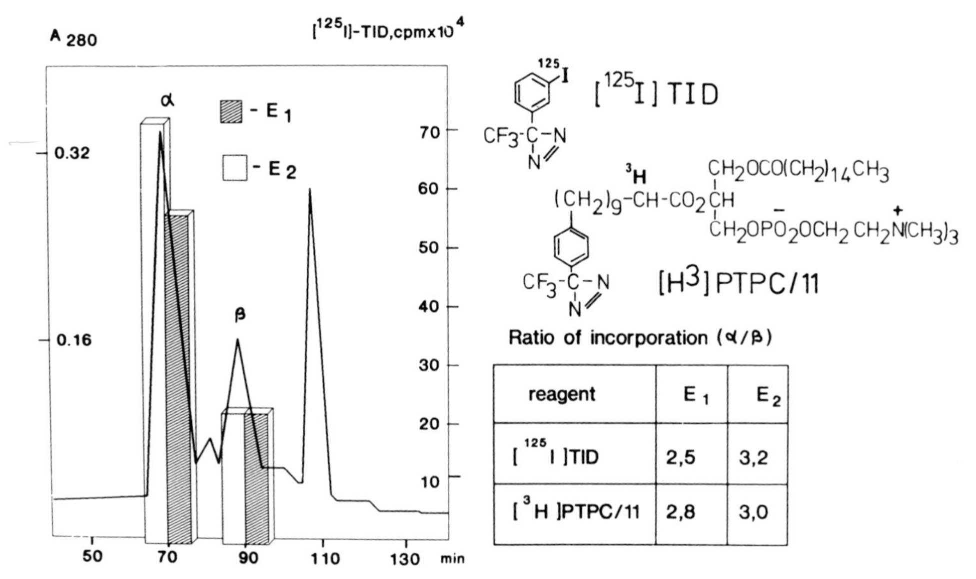

reagent	E_1	E_2
[^{125}I]TID	2,5	3,2
[^{3}H]PTPC/11	2,8	3,0

Figure 1. Determination of the incorporation of the hydrophobic reagents into the Na,K-ATPase subunits. Separation of Na,K-ATPase modified with hydrophobic reagents in E_1 and E_2 enzyme forms (*left*). The preparation of Na,K-ATPase was solubilized in 0.1 M Na-acetate buffer, pH 5.6 (buffer A), containing 5% SDS, 5% β-mercaptoethanol, then separated on the UltroPac TSK-3000 SWG column in buffer A with 0.2% SDS (flow rate 1 ml/min); the protein absorbance was monitored at 280 nm. The height of the blocks indicates the value of radioactivity incorporated into the subunits in the E_1 and E_2 conformations. The value of the ratio of reagent incorporation according to data of gel-permeation chromatography and densitometry of PAGE and autoradiography are summarized in the table (*right*).

Analysis of the β-subunit samples isolated from the [^{125}I]TID-modified Na,K-ATPase reveals the same value of specific radioactivity incorporation in E_1 and E_2 forms (Fig. 1). Similar results were obtained in the experiments with [^{3}H]PTPC/11. This finding indicates that the level of the β-subunit labeling does not depend on a conformational state of the enzyme; i.e., the area of its contact with lipids does not change upon the $E_1 \rightarrow E_2$ transition. Unlike the β-subunit, the labeling of the α-subunit increases upon the $E_1 \rightarrow E_2$ transition for both reagents. The ratio of label incorporation (α/β) increases from 2.3 for [^{125}I]TID and 2.8 for [^{3}H]PTPC/11 (in the Na-(E_1) enzyme form) to 3.5 for [^{125}I]TID and 3.2 for [^{3}H]PTPC/11 (in K^+-(E_2) enzyme form).

These data clearly demonstrate the changes in accessibility of membrane-bound fragments of the α-subunit for hydrophobic reagents in E_1 and E_2 forms; i.e., the intramembrane moiety of Na,K-ATPase (α- but not β-subunit) is involved in a reversible structural rearrangement during enzyme functioning. Analysis of the INA interaction with Na,K-ATPase in different enzyme conformational states also revealed more extensive labeling of the α-subunit in the K^+ form (by 10–15%) as compared with the Na^+ form (Jorgensen et al., 1982). The difference in the ratio of reagent incorporation into the subunits in distinct enzyme forms is greater for $[^{125}I]$TID as compared with $[^3H]$PTPC/11 (Fig. 1). Hence, the contact area of the α-subunit with lipids in the central part of the bilayer does not change drastically during the E_1-E_2 transition. These differences in the case of PTPC/11 could be explained by changes in mutual dispositions of the transmembrane rods and/or distance between the protomers. An additional increase of the hydrophobic labeling revealed by $[^{125}I]$TID indicates the rearrangement of other hydrophobic portions of the α-subunit, namely, the regions located near the phospholipid heads or forming extramembrane hydrophobic pockets. The proposal, that the α-subunit has an extended segment partially embedded in the membrane, is supported by earlier data of Montecucco et al. (1981). These authors showed that in the case of modification by the phosphocholine derivative PL-2, with a reactive group near the head groups of phospholipids, the ratio of label incorporation α/β increased up to a value of 5 as compared with a ratio of 3 for PL-1 carrying an azido group at the end of the fatty acid chain.

Calculation of the hydropathy plots of the α- and β-subunits and analysis of the enzyme topography leads to a ratio of 7:1 for the mass of the membrane bound portions (Ovchinnikov et al., 1987). Thus the specific radioactivity incorporation into the intramembrane portion of the β-subunit upon hydrophobic labeling is threefold higher than that of the corresponding α-subunit region. On the one hand, this finding points to the peripheral location of the β-subunit in the oligomeric membrane complex. On the other hand, it suggests the lack of contact of some α-subunit membrane fragments with lipids. This can be explained by the tight interaction of the α-subunits within an oligomer and/or possible location of one or two rods in the center of the αβ-protomer membrane moiety. Protection of the α-subunit by the tightly bound lipids or low molecular weight proteins cannot be excluded.

Isolation and Characterization of Labeled Fragments

Evidently $[^3H]$PTPC/II is a more selective reagent than $[^{125}I]$TID, which is also capable of modifying clusters of hydrophobic residues in extramembrane protein domains (Brunner, 1989). Nevertheless, the ratios of radioactivity incorporation into the subunits for these two reagents are similar (Fig. 1). This testifies to the specific interaction of $[^{125}I]$TID with the membrane sector of the enzyme. The specific radioactivity of $[^{125}I]$TID-modified preparations is higher compared with those treated with $[^3H]$PTPC/11 and isolation of the $[^{125}I]$TID-labeled peptides is not as complicated. Besides the α- and β-subunits, membrane-bound preparations of Na,K-ATPase contain some low molecular weight components capable of $[^{125}I]$TID incorporation. Because of this, modified α and β peptides were separated and identified after isolation of the enzyme subunits. The purified α-subunit of Na,K-ATPase was treated with trypsin. Gel-permeation chromatography of the tryptic hydrolyzate on a TSK-2000SW column (Fig. 2) showed that the two reagents were incorporated exclusively into high molecular weight fragments. Low molecular weight products from the exhaustive tryptic hydrolysis of regions corresponding to the extramembrane segments of the subunits were free of the label. Similar results were obtained with the β-subunit (Chertova et al., 1989). According to calculation, only the hydrophobic fragments have molecular masses > 3,000 D. Thus the absence of the label in the region of low molecular weight peptides indirectly points to a highly selective labeling.

The modified fragments were separated on a reverse-phase column (Nucleosil 5C4—300). Determination of the sequence of the β-subunit–modified fragments

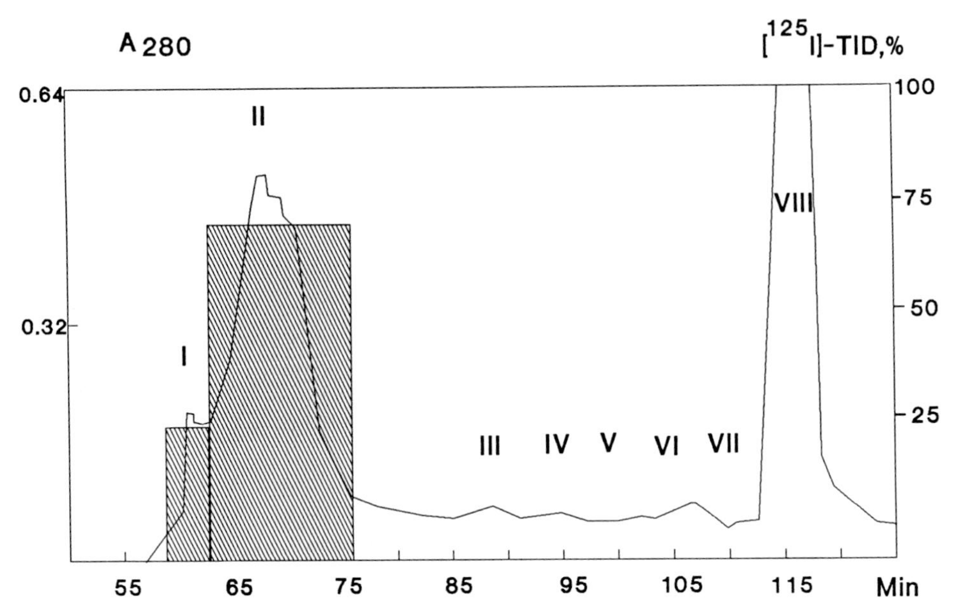

Figure 2. Separation of tryptic hydrolyzate of the α-subunit by gel-permeation chromatography. Tryptic hydrolyzate was lyophilized, then dissolved in buffer A containing 5% SDS and 5% β-mercaptoethanol and applied to the UltroPac TSK-2000SW column (7.5 × 600 mm) equilibrated in buffer A with 0.2% SDS (flow rate 0.2 ml/min). The peptide absorbance was detected at 280 nm and the position of the labeled fragments was determined by the yield of radioactivity. The values of radioactivity as a percent of the applied are shown.

revealed labeling of only one peptide, Thr^{27}-Arg^{71}. The positions of amino acid residues preferentially modified with [^{125}I]TID were identified and indicated that fragment Thr^{27}-Arg^{71} of the β-subunit is bound to the membrane (Chertova et al., 1989).

The same procedures were applied to separate tryptic hydrolyzates of the enzyme α-subunit modified in E_1 and E_2 conformations (Fig. 3). Though separation and tryptic hydrolysis of the subunits as well as isolation of the fragments were performed under the same conditions, the radioactivity distribution on chromatograms is different for the E_1 and E_2 forms. All the fractions containing radioactive material were analyzed for both enzyme conformations. A set of labeled peptides was shown to be the same for E_1 and E_2 forms:

Asp^{68}—Gly—Pro—Asn—Ala—Leu—Thr—Pro—Pro—Pro—Thr—Thr—Pro—Glu—Trp—Val—Lys— —Phe—Cys—Arg—Gln—Leu—Phe—. . .—Lys^{142}

Ile^{265}—Ala—Thr—Leu—Ala—Ser—Gly—. . .—Lys^{341}

Val^{545}—Leu—Gly—Phe—Cys—His—Phe—Leu—Pro—Asp—. . .—Arg^{589}

Ser^{770}—Ile—Ala—Tyr—Thr—Ser—Asn—Ile—Pro—Glu—Ile—. . .—Lys^{826}
Leu^{842}—Ile—Ser—Met—Ala—Tyr—Gly—Gln—Ile—Gly—Met—Ile—Gln—Ala—Leu—Gly—Gly—Phe—Phe—Thr—Tyr—Phe—Val—Ile—Leu—Ala—Glu—Asn—Gly—Phe—. . .—Arg^{880}

Asn^{936}—Ser—Val—Phe—Gln—Gln—Gly—Met—Lys—Asn—Lys—Ile—Leu—Ile—Phe—. . .—Arg^{972}

Met^{973}—Tyr—Pro—Leu—Lys—Pro—Thr—. . .—Arg^{999}

Thus the difference in intensity of labeling of the α-subunit in the E_1 and E_2 conformations can be explained by a change in the intensity of labeling of the specific peptides rather than by the labeling of different peptides. The majority of fractions analyzed contained a mixture of two or three fragments. We are currently attempting to purify all of the fragments to determine the extent of labeling and to establish the sidedness of modification. Modification of all the fragments with [^{125}I]TID agrees with data on limited proteolysis of [^{125}I]TID-labeled Na,K-ATPase (Jorgensen and Brunner, 1983), which showed the label to be evenly distributed along the polypeptide chain of the α-subunit.

Mapping of the membrane sector with a series of hydrophobic reagents also showed label incorporation into other portions of the α-subunit. Iodonaphthylazide labels the NH_2-terminal 12-kD fragment (Jorgensen et al., 1982), including peptide Asp^{68}-Lys^{142}, which was also isolated. [^{3}H]Adamantanyldiazirin ([^{3}H]AD) modifies the COOH-terminal portion of the α-subunit (Farley et al., 1980) which includes peptides Val^{545}-Arg^{589}, Ser^{770}-Lys^{826}, Leu^{842}-Arg^{880}, Asn^{936}-Arg^{972}, and Met^{973}-Arg^{999} in the labeled fractions. Nicholas (1984) provided extensive structural information on modification of Na,K-ATPase with [^{3}H]AD. [^{3}H]AD-labeled peptides were isolated and their NH_2-terminal amino acid sequences were determined. Peptides Ile-Ala-Thr-Leu- and Val-Leu-Gly-Phe- were identified as transmembrane fragments. These sequences were determined before the complete primary structure of Na,K-ATPase was known. It is now possible to identify these peptides as Ile^{265}-Ala-Thr-Leu- and Val^{545}-Leu-Gly-Phe- (we identified both peptides as [^{125}I]TID-labeled; see above). Amino acid sequences Leu-Ile-X-Leu-Ala and Met-Tyr-Leu-Pro- presumably, correspond to fragments Leu^{842}-Lle-Ser-Met-Ala- and Met^{974}-Tyr-Pro-Leu of the α-subunit (Kyte et al., 1987) and they are also isolated among [^{125}I]TID-labeled peptides.

We were able to isolate α-subunit fragment Leu^{842}-Arg^{880} and β-subunit fragment Thr^{27}-Arg^{71} and to determine the positions of preferential labeling. The first target of modification in the β-subunit intramembrane portion is Phe^{41}, which may be a result of masking the NH_2-terminal highly hydrophobic sequence Leu^{33}-Leu^{40} by the α-subunit or exposure of this region of the β-subunit in cytoplasm. On the other hand, the sixth α-subunit transmembrane segment evidently also includes the

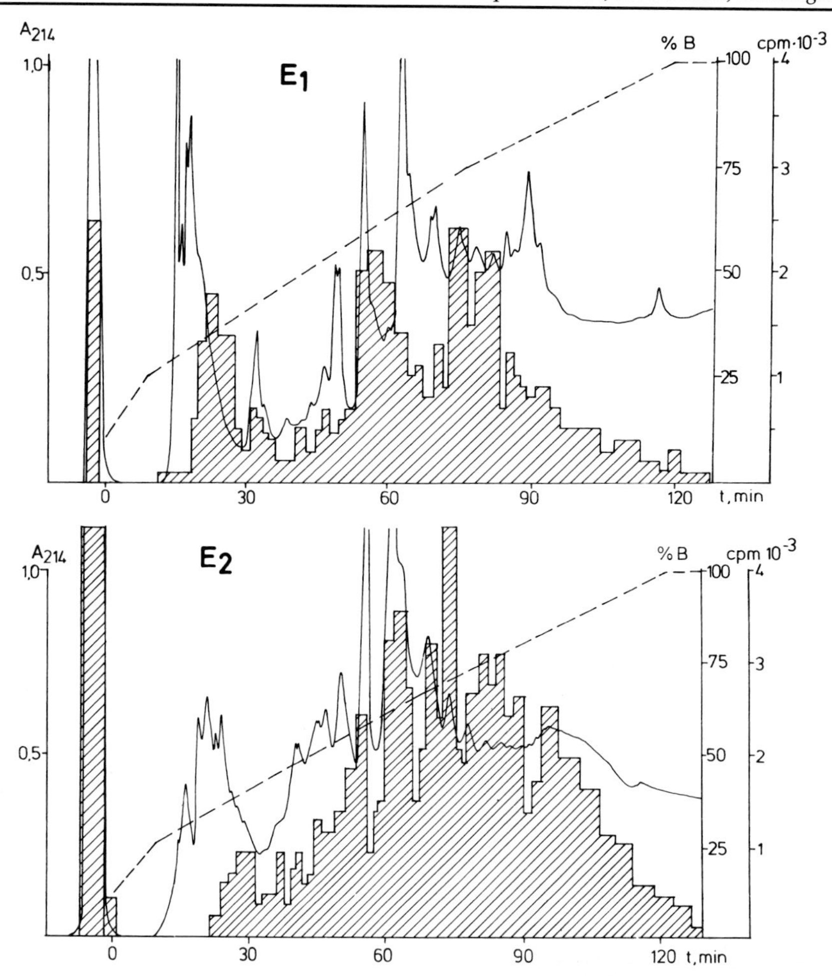

Figure 3. Separation of [^{125}I]TID-labeled peptides of the α-subunit modified in E_1 (*top*) and E_2 (*bottom*) conformations on Nucleosil 5C_4-300. The radioactive pools from gel filtration (Fig. 2) were lyophilized and redissolved in buffer B (10% mixture of acetonitrile:isopropanol 2:1 in 0.1% trifluoroacetic acid) containing 5% SDS and applied to Nucleosil 5C_4-300 column (4.6 × 100 mm) equilibrated with buffer B in the gradient of buffer C (acetonitrile:isopropanol 2:1, 0.1% TFA): 0–25% C for min, 25–50% C for 75 min, 50–75% C for 50 min, 75–100% C for 25 min, 100% C for 25 min (change in percentage of acetonitrile is shown by hatched bars). Flow rate, 0.5 ml/min. The peptide yield was determined by the absorption at 214 nm. Histograms show the radioactivity profile.

[^{125}I]TID-modified residue of Glu^{868}, which followed a predicted hydrophobic region. These findings allowed us to refine the position of boundaries of transmembrane rods proposed from hydropathy plots.

Purification of other fragments labeled with [^{125}I]TID in different enzyme conformations and identification of modification points are now in progress. These results will clarify how the conformational transition $E_1 \rightarrow E_2$ influences the position of these peptides relative to the lipids.

Spatial Organization of Intramembrane Moiety

The results of identification of the membrane-bound fragments of Na,K-ATPase coincide in general with the subunit transmembrane arrangement proposed earlier (Ovchinnikov et al., 1987). As shown in Fig. 4, in addition to the predicted transmembrane fragments, the rather hydrophobic fragments Val^{545}-Arg^{590} and Met^{973}-Arg^{999} appeared to be modified by [^{125}I]TID. The same peptides were also shown to be labeled with AD (Nicholas, 1984). On the other hand, both these fragments were identified in the supernatant after precipitation of the membrane preparation of the Na,K-ATPase subjected to limited tryptic hydrolysis (Ovchinnikov et al., 1987*a*).

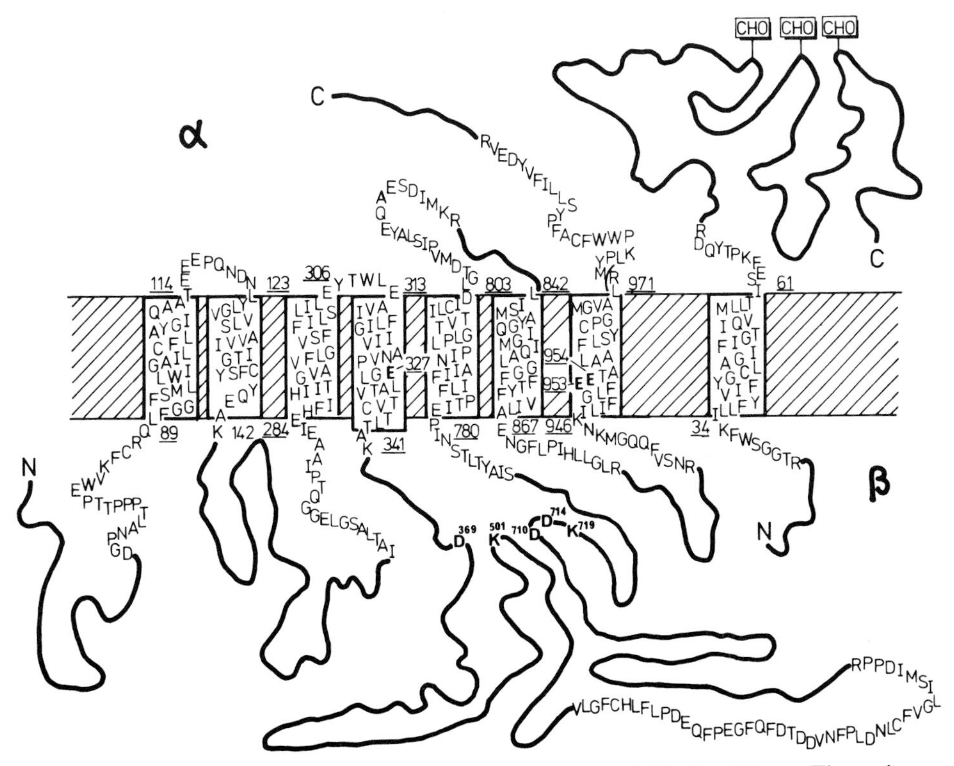

Figure 4. Scheme for the transmembrane arrangement of Na,K-ATPase. The primary structures of [^{125}I]TID-labeled fragments are shown.

These findings indicate that the α-subunit regions Val^{545}-Arg^{590} and Met^{973}-Arg^{999} are either partially embedded in the membrane or lie on the surface of the intramembrane moiety of the Na,K-ATPase. There is still a possibility that these fragments form very hydrophobic "pockets" capable of selective binding of the lipophilic reagents. As mentioned above, changes in the modification level of the α-subunit in various conformational states can be explained by a change of the contact area between the subunits in the oligomer and by a different depth in the lipid bilayer of the fragments. Peptides Val^{545}-Arg^{590} and Met^{973}-Tyr are the best candidates for this role.

Thus, a combination of different experimental approaches allowed us to reveal the following structural features of the intramembrane portion of the sodium pump. The α- and β-subunits have seven and one transmembrane segments, respectively (Ovchinnikov et al., 1987*a*). They are all in the α-helical conformation (Ovchinnikov, 1988*a*). Present data on hydrophobic labeling indicate that first and/or second, third and/or fourth, as well as fifth, sixth, and seventh membrane segments of the α-subunit are in direct contact with the lipid bilayer. The transmembrane segment of the β-subunit is located on the periphery of the αβ-complex.

This information was used to propose the first model of the spatial organization of the intramembrane domain of the Na,K-ATPase (Figs. 4 and 5). The hypothetical three-dimensional structure of all transmembrane helices was calculated by using the values of the torsion angles *j* and *v* classically attributed to an α-helix and the side chains corresponding to each peptide. The structures obtained were visualized with the PC-CHEMMOD Molecular Modeling System (U-Micro Ltd., Cheshire, UK). The axial projections of the models for the intramembrane part of the β-subunit and

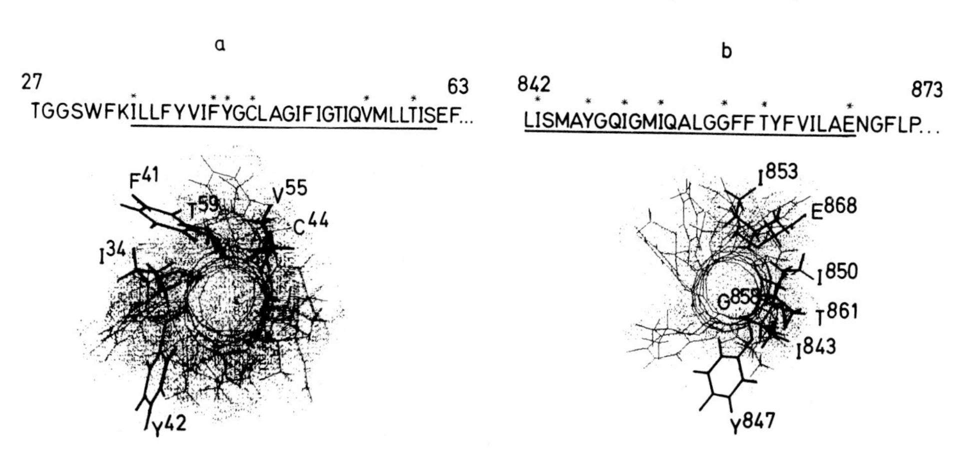

Figure 5. The amino acid sequences and axial projections of the intramembrane fragment of the β-subunit Lys^{33}-Ile^{61} (*a*) and sixth transmembrane segment of the α-subunit Leu^{842}-Ala^{867} (*b*). The [^{125}I]TID-labeled amino acid residues are displayed.

the sixth transmembrane segment of the α-subunit (Fig. 5) clearly demonstrate an asymmetrical distribution of the TID-labeled amino acid residues. Thus, parts of both helices accessible to the hydrophobic reagent are, obviously, exposed to the lipid matrix, so other portions of the membrane segments are located inside the protein. We examined the spatial relationships of the intramembrane components of the Na,K-ATPase by comparison with a set of the above α-helical models with the corresponding part of the three-dimensional structure of an αβ-protomer revealed by electron microscopy of two-dimensional crystals (Ovchinnikov et al., 1985*b*). The Van der Waals surfaces of all helices were calculated. The following assumptions were made for the fitting: (*a*) segments 1 and 2 as well as 3 and 4 connected by short hydrophilic sequences are associated in pairs; (*b*) critical charged residues (Glu^{327}, Glu^{953}, and Glu^{954}) and other potential components of ion-conducting pathways are located inside the protein; and (*c*) residues accessible to a hydrophobic reagent are in contact with the lipid environment. The resulting map of possible arrangements of transmembrane helices fitted into the cross-sections of the membrane part of the three-dimensional model of Na,K-ATPase is shown in Fig. 6. The contours of cross-sections parallel to the membrane plane were calculated near the extracellular and cytoplasmic surfaces and in the center of the lipid bilayer. The most probable position for the β-subunit on the map was selected by comparison of the three-dimensional models of the αβ-protomer and individual β-subunit (Ovchinnikov, 1985*b*, 1986*a*; Modyanov, 1988*b*).

Obviously, this version of the disposition of α-subunit helices is only one of many possible arrangements. We hope that results of the studies on hydrophobic labeling might restrict this variety and lead to a more realistic model of the structural organization of the intramembrane part of the Na,K-ATPase. The challenging problem remaining is the determination of which protein fragments are involved in structural rearrangements during pumping. This information is essential for understanding the cation transport mechanism.

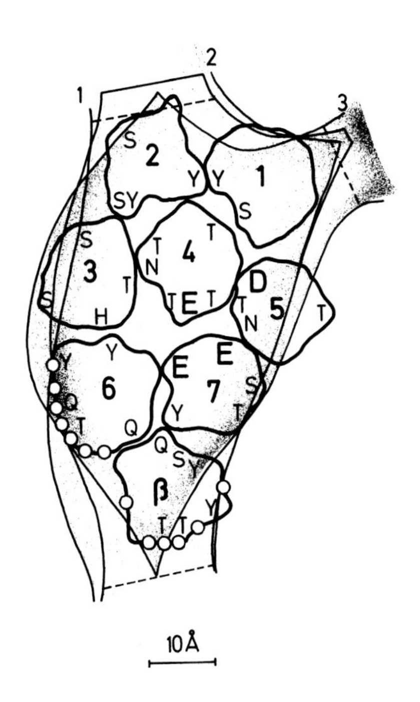

Figure 6. Hypothetical arrangement of transmembrane α-helices fitted into the cross-sections of the intramembrane part of the three-dimensional model of Na,K-ATPase revealed by electron microscopy and two-dimensional crystals. The contours of cross-sections parallel to the membrane plane were calculated near the extracellular (*1*) and cytoplasmic (*3*) surfaces and in the middle of the lipid bilayer (*2*). Circles indicate the position of the [^{125}I]TID-labeled residues. Hydrophilic amino acid residues are designated.

Acknowledgments

We thank Dr. J. Brunner (Swiss Federal Institute of Technology, Zurich) for valuable advice and for providing [^{3}H]PTPC/11 and the phospholipid protein exchanger. We also thank Dr. I. Nabiev for fruitful discussions and Drs. V. Demin, A. Barnakov, and A. Kuzin for recalculations of the three-dimensional model of the Na,K-ATPase crystal unit cell.

References

Bayer, R. 1990. Topological disposition of the sequences -QRKIVE- and -KETYY in native (Na^+ + K^+)-ATPase. *Biochemistry.* 29: 2251–2256.

Bernikov, L. R., K. N. Dzhandzhugazyan, S. V. Lutsenko, and N. N. Modyanov. 1990. Dialdehyde ATP derivative as a modifier of the Na^+,K^+-ATPase active site. *European Journal of Biochemistry.* 194:413–421.

Brandl, C. J., N. M. Green, B. Korczak, and D. H. MacLennan. 1986. Two Ca^{2+} ATPase genes: homologies and mechanistic implications of deduced amino acid sequences. *Cell.* 44:597–607.

Broude, N. E., N. N. Modyanov, G. S. Monastyrskaya, and E. D. Sverdlov. 1989. Advances in Na, K-ATPase studies: from protein to gene and back to protein. *FEBS Letters.* 257:1–9.

Brunner, J. 1989. Photochemical labeling of apolar phase of membranes. *Methods in Enzymology.* 172:628–687.

Brunner, J., and G. Semenza. 1981. Selective labeling of the hydrophobic core of membranes with 3-(trifluoromethyl)-3-(m-[^{125}I]iodophenyl) diazirine, a carbene generating reagent. *Biochemistry.* 20:7174–7181.

Chertova, E. N., S. V. Lutsenko, N. B. Levina, and N. N. Modyanov. 1989. Probing the topography of the intramembrane part of Na^+,K^+-ATPase by photolabelling with 3-(trifluoromethyl)-3-(m-[^{125}I]-iodophenyl)-diazirine: analysis of the hydrophobic domain of the β-subunit. *FEBS Letters.* 254:13–16.

Craik, C. S., S. Sprang, R. Fletterick, and W. J. Rutter. 1982. Intron-exon splice junctions map at protein surfaces. *Nature.* 299:180–182.

Dzhandzhugazyan, K. N., S. V. Lutsenko, and N. N. Modyanov. 1986. Na^+,K^+-activated adenosine triphosphatase from pig kidney. IV. Structure of the α-subunit active site fragment modified with alkylating ATP analog. *Biological Membranes (USSR).* 3:858–868.

Dzhandzhugazyan, K. N., S. V. Lutsenko, and N. N. Modyanov. 1988. Target-residues of the active site affinity modification are different in E_1 and E_2 forms. *In* The Na^+,K^+-Pump. Part A: Molecular Aspects. J. C. Skou, J. G. Norby, A. B. Maunsbach, and M. Esmann, editors. Alan R. Liss, Inc., New York. 268A:181–188.

Dzhandzhugazyan, K. N., and N. N. Modyanov. 1985. Chemical and immunochemical approaches to the structure of membrane-bound Na^+,K^+-ATPase. *In* The Sodium Pump. I. Glynn, and C. Ellory, editors. The Company of Biologists Ltd., Cambridge, UK. 129–134.

Farley, R., D. Goldman, and H. Bayley. 1980. Identification of regions of the catalytic subunit of (Na^+,K^+)-ATPase embedded within the cell membrane. *The Journal of Biological Chemistry.* 255:860–864.

Hall, K., G. Perez, D. Anderson, C. Gtierrez, K. Munson, S. J. Hersey, J. H. Kaplan, and G. Sachs. 1990. Location of the carbohydrates present in the H^+,K^+-ATPase vesicles isolated from dog gastric mucosa. *Biochemistry.* 29:701–706.

Harter, C., T. Bachi, G. Semenza, and J. Brunner. 1988. Hydrophobic photolabeling identifies BHA2 as the subunit mediating the interaction of bromelain-solubilized influenza virus hemagglutinin with liposomes at low pH. *Biochemistry.* 27:1856–1864.

James, P., M. Maeda, R. Fisher, A. K. Verma, J. Krebs, J. Penniston, and E. Carafoli. 1988. Identification and primary structure of a calmodulin binding domain of the Ca^{2+} pump of human erythrocytes. *The Journal of Biological Chemistry.* 263:2905–2910.

Jorgensen, P. L. 1975. Purification and characterization of (Na^+,K^+)-ATPase. V. Conformational changes in the enzyme: transitions between the Na-form and the K-form studied with tryptic digestion as a tool. *Biochimica et Biophysica Acta.* 401:399–415.

Jorgensen, P. L., and J. Brunner. 1983. Labeling of the intramembrane segments of the α-subunit and β-subunit of pure membrane-bound (Na^+,K^+)-ATPase with 3-(trifluoromethyl)-3-(m-[^{125}I]iodophenyl) diazirine. *Biochimica et Biophysica Acta.* 735:291–296.

Jorgensen, P. L., and J. H. Collins. 1986. Tryptic and chymotryptic cleavage sites in the sequence of α-subunit of Na,K-ATPase from outer medulla of mammalian kidney. *Biochimica et Biophysica Acta.* 860:570–576.

Jorgensen, P. L., S. J. Karlish, and C. Gitler. 1982. Evidence for the organization of the transmembrane segments of (Na,K)-ATPase based on labeling lipid-embedded and surface domains of the α-subunit. *The Journal of Biological Chemistry.* 257:7435–7441.

Kawakami, K., S. Nogushi, M. Noda, H. Takahashi, T. Ohta, M. Kawamura, H. Nojima, K. Nagano, T. Hirose, S. Inayama, et al. 1985. Primary structure of the α-subunit of *Torpedo californica* Na^+,K^+-ATPase deduced from cDNA sequence. *Nature.* 316:733–736.

Korczak, B., A. Zarain-Herzberg, C. J. Brandl, C. J. Ingles, N. M. Green, and D. H. MacLennan. 1988. Structure of the rabbit fast-twitch skeletal muscle Ca^{2+}-ATPase gene. *The Journal of Biological Chemistry.* 263:4813–4819.

Kyte J., K. Xu, and R. Bayer. 1987. Demonstration that lysine-501 of the polypeptide of native sodium and potassium ion activated adenosinetriphosphatase is located on its cytoplasmic surface. *Biochemistry.* 26:8350–8360.

Laemmli, U. K. 1970. Cleavage of structural proteins during assembly of the head of bacteriophage T_4. *Nature.* 227:680–685.

Maeda, M., Ko.-I. Oshiman, S. Tamura, and M. Futai. 1990. Human gastric (H^+ + K^+)-ATPase gene. *The Journal of Biological Chemistry.* 265:9027–9032.

Matthews, I., J. Colyer, A. M. Mata, N. M. Green, R. P. Sharma, A. G. Lee, and J. M. East. 1989. Evidence for the cytoplasmic location of the N- and C-terminal segments of the sarcoplasmic reticulum (Ca^{2+}-Mg^{2+})-ATPase. *Biochemical and Biophysical Research Communications.* 161:683–688.

Meister, H., R. Bachofen, G. Semenza, and J. Brunner. 1985. Membrane topology of light-harvesting protein B870-α of *Rhodospirillum rubrum* G-9^+. *The Journal of Biological Chemistry.* 260:16326–16331.

Modyanov, N. N., E. A. Arystarkhova, N. M. Gevondyan, N. M. Arzamazova, R. G. Efremov, I. R. Nabiev, N. E. Broude, G. S. Monastyrskaya, and E. D. Sverdlov. 1988*a*. Sodium pump: current view on structural organization. *In* Molecular Basis of Biomembrane Transport. F. Palmieri and E. Quagliariello, editors. Elsevier Science Publishers, Amsterdam. 229–237.

Modyanov, N. N., E. A. Arystarkhova, and S. A. Kocherginskaya. 1988*b*. Structural basis of Na^+,K^+-pump functioning. *Biological Membranes, USSR.* 5:341–385.

Modyanov, N. N., N. M. Arzamazova, E. A. Arystarchova, N. M. Gevondyan, E. E. Gavrilyeva, K. N. Dzhandzhugazyan, S. V. Lutsenko, N. M. Luneva, G. I. Shafieva, and E. N. Chertova. 1987. Na^+K^+-pump: structural organization. *In* Receptors and Ion Channels. Yu. A. Ovchinnikov and F. Hucho, editors. Walter de Gruyter, New York, Berlin. 287–294.

Montecucco, C., R. Bisson, C. Gache, and A. Johannsson. 1981. Labelling of the hydrophobic domain of the Na^+,K^+-ATPase. *FEBS Letters.* 128:17–21.

Nabiev, I. R., K. N. Dzhandzhugazyan, R. G. Efremov, and N. N. Modyanov. 1988. Binding of monovalent cations induces large changes in the secondary structure of Na,K-ATPase as probed by Raman spectroscopy. *FEBS Letters.* 236:235–239.

Nicholas, R. A. 1984. Purification of the membrane-spanning tryptic peptides of the α polypeptide from sodium and potassium ion activated adenosinetriphosphatase labeled with 1-tritiospiro-[adamantane-4,3′-diazirine]. *Biochemistry.* 23:888–898.

Noguchi, S., M. Noda, H. Takahashi, K. Kawakami, T. Ohta, K. Nagano, T. Hirose, S.

Inayama, M. Kawamura, and S. Numa. 1986. Primary structure of the β-subunit of *Torpedo californica* (Na^+,K^+)-ATPase deduced from the cDNA sequence. *FEBS Letters*. 196:315–320.

Okamoto, C. T., J. M. Kaprilov, A. Smolka, and J. C. Forete. 1990. Isolation and characterization of gastric microsomal glycoproteins: evidence for a glycosylated β-subunit of the H^+,K^+-ATPase. *Biochimica et Biophysica Acta*. 1037:360–372.

Ovchinnikov, Yu. A., S. G. Arsenyan, N. E. Broude, K. E. Petrukhin, A. V. Grishin, N. A. Aldanova, N. M. Arzamazova, E. A. Arystarkhova, A. M. Melkov, Yu. V. Smirnov et al. 1985*a*. Nucleotide sequence of cDNA and primary structure of the α-subunit of pig kidney Na^+,K^+-ATPase. *Proceedings of the Academy of Sciences, USSR*. 285:1491–1495.

Ovchinnikov, Yu. A., E. A. Arystarkhova, N. M. Arzamazova, K. N. Dzhandzhugazyan, R. G. Efremov, I. R. Nabiev, and N. N. Modyanov. 1988*a*. Differentiated analysis of the secondary structure of hydrophilic and hydrophobic regions in alpha- and beta-subunits of Na,K-ATPase by Raman spectroscopy. *FEBS Letters*. 227:235–239.

Ovchinnikov, Yu. A., N. M. Arzamazova, E. A. Arystarkhova, N. M. Gevondyan, N. A. Aldanova, and N. N. Modyanov. 1987*a*. Detailed structural analysis of exposed domains of membrane-bound Na^+,K^+-ATPase. *FEBS Letters*. 217:269–374.

Ovchinnikov, Yu. A., V. V. Demin, A. N. Barnakov, A. P. Kuzin, A. V. Lunev, N. N. Modyanov, and K. N. Dzhandzhugazyan. 1985*b*. Three-dimensional structure of (Na^+ + K^+)-ATPase revealed by electron microscopy of two-dimensional crystals. *FEBS Letters*. 190:73–76.

Ovchinnikov, Yu. A., V. V. Demin, A. N. Barnakov, E. A. Svetlichni, N. N. Modyanov, and K. N. Dzhandzhugazyan. 1986*a*. Three-dimensional structure of Na^+,K^+-ATPase β-subunit revealed by electron microscopy of two-dimensional crystals. *Biological membranes (USSR)*. 3:537–541.

Ovchinnikov, Yu. A., K. N. Dzhandzhugazyan, S. V. Lutsenko, A. A. Mustayev, and N. N. Modyanov. 1987*b*. Affinity modification of E_1-form of Na^+,K^+-ATPase revealed Asp-710 in the catalytic site. *FEBS Letters*. 217:111–116.

Ovchinnikov, Yu. A., N. M. Luneva, E. A. Arystarkhova, N. M. Gevondyan, N. M. Arzamazova, A. T. Kozhich, V. A. Nesmeyanov, and N. N. Modyanov. 1988*b*. Topology of Na^+,K^+-ATPase: identification of the extra- and intracellular hydrophilic loops of the catalytic subunit by specific antibodies. *FEBS Letters*. 227:230–234.

Ovchinnikov, Yu. A., N. M. Modyanov, N. E. Broude, K. E. Petrukhin, A. V. Grishin, N. I. Kiyatkin, N. M. Arzamazova, N. A. Aldanova, G. S. Monastyrskaya, and E. D. Sverdlov. 1986*b*. Pig kidney Na^+,K^+-ATPase: primary structure and spatial organization. *FEBS Letters*. 201:237–245.

Ovchinnikov, Yu. A., G. S. Monastyrskaya, N. E. Broude, Yu. A. Ushkaryov, A. M. Melkov, Yu. V. Smirnov, I. V. Malyshev, R. L. Allikmets, M. B. Kostina, I. E. Dulubova, et al. 1988*c*. Family of human Na^+,K^+-ATPase genes: structure of the gene for the catalytic subunit (αIII-form) and its relationship with structural features of the protein. *FEBS Letters*. 233:87–94.

Serrano, R. 1989. Structure and function of plasma membrane ATPase. *Annual Review of Plant Physiology and Plant Molecular Biology*. 40:61–94.

Shull, G. E. 1990. cDNA cloning of the β-subunit of the rat gastric H^+,K^+-ATPase. *Journal of Biological Chemistry*. 265:12123–12126.

Shull, G. E., and J. Greeb. 1988. Molecular cloning of two isoforms of the plasma membrane Ca^{2+}-transporting ATPase from rat brain. *Journal of Biological Chemistry*. 263:8646–8657.

Shull, G. E., L. K. Lane, and J. B. Lingrel. 1986. Amino-acid sequence of the β-subunit of the $(Na^+ + K^+)$-ATPase deduced from cDNA. *Nature.* 321:429–431.

Shull, G. E., A. Schwartz, and J. B. Lingrel. 1985. Amino-acid sequence of the catalytic subunit of the $(Na^+ + K^+)$-ATPase deduced from a complementary DNA. *Nature.* 316:691–695.

Stokes, D. L., and N. M. Green. 1990. Three-dimensional crystals of CaATPase from sarcoplasmic reticulum. Symmetry and molecular packing. *Biophysical Journal.* 57:1–14.

White, B. H., and J. B. Cohen. 1988. Photolabeling of membrane-bound *Torpedo* nicotinic acetylcholine receptor with the hydrophobic probe 3-(trifluoromethyl)-3-(m-[^{125}I]iodophenyl)diazirine. *Biochemistry.* 27:8741–8751.

Chapter 9

Localization of Ligand Binding from Studies of Chemical Modification

Jack H. Kaplan

Department of Physiology, University of Pennsylvania, Philadelphia, Pennsylvania 19104-6085

Introduction

Systematic approaches to defining structure–function relations for the sodium pump protein in terms of specific amino acids date from the relatively recent acquisition of α-subunit primary sequence from gene cloning techniques (Kawakami et al., 1985; Shull et al., 1985). The absence of any direct evidence linking a catalytic role with the β-subunit has focused attention on identifying which regions of the α-subunit can be associated with specific catalytic and transport functions of the sodium pump reaction cycle. Along with such considerations goes the difficult step of transforming the amino acid sequence data obtained from cloning into a meaningful and reliable topographic model for the transmembrane disposition of the α-subunit, as well as a helpful picture for the protein secondary structure (i.e., folding). Recent studies with the sodium pump α-subunit and its relatives in the family of P-type ATPases, which include the sarcoplasmic reticulum (SR) and plasma membrane (PM) calcium pumps, the proton pump or H,K-ATPase from gastric mucosa, and the other transport ATPases, have led to a more or less unified picture. Obviously, these pumps will not be the same in detail; however, it is now generally accepted that similarities in function, enzymatic reaction cycle, and regions of primary sequence homology will be reflected in similarities in structure and the ways in which the different proteins interact with their ligands. Currently, each of these specific areas is the subject of intense activity (and speculation). At the moment, models exist that propose seven (Ovchinnikov et al., 1988), eight (Kawakami et al., 1985; Shull et al., 1985), or ten (Maclennan et al., 1985) transmembrane helices for the sodium or calcium pumps. Obviously, if this is an even number then the carboxy terminus of the α-subunit is in the cytoplasm, as is true for the calmodulin-binding carboxy terminus of the PM calcium pump (James et al., 1988).

The predominant mass of the α-subunit is contained in a large cytoplasmic loop between transmembrane helices 4 and 5 (numbered from the amino terminus). This loop, running approximately from residue 340 to 780 (in sheep kidney Na,K-ATPase), contains sequences that are homologous to the other ion pumps and also contains stretches that show similarities with other ATP binding proteins (Taylor and Green, 1989). These homologies have led to a consensus structure for ion pumps based on data from mutagenesis studies, chemical modification, and proteolytic cleavage (Green et al., 1988). A sketch of the overall organization of this structure is shown in Fig. 1. Major portions of the cytoplasmic loop contribute to the ATP binding domain and the phosphorylation region. In the model developed for the calcium pump, the large cytoplasmic loop is linked to the membrane-embedded transmembrane helices by a stalk region. Early suggestions for the locus of the cation binding sites of the Ca pump placed the significant carboxyl residues (see below) in this stalk region (Brandl et al., 1986); however, the results of studies on transiently expressed mutagens make this unlikely (Clark et al., 1989*b*). More recent data from mutagenesis studies place the putative cation-associated carboxyl residues in the transmembrane helices (Clarke et al., 1989*a,* 1990). Some of these are contributed from the helices which are carboxy-terminal to the cytoplasmic loop (where the various family members of the P-type ATPases show rather less homology), while others come from more amino-terminal transmembrane helices. If this model is correct, it brings together in close spatial proximity the transmembrane helices which in the linear amino acid sequence are greatly separated on either side of the large cytoplasmic loop.

These models rely on a combination of information from mutagenesis studies of the calcium pump (referred to above), high resolution data from electron diffraction (Stokes and Green, 1990), and plausibility arguments. The latter are based on the expectation that hydrophilic and highly charged substrates (ATP, ADP, etc.) bind to cytoplasmically exposed domains of the protein. Phosphorylation of the protein occurs in this region and the information is transferred to some portion of the protein that forms a pathway for the cations through the low dielectric membrane bilayer. The task confronting contemporary workers who wish to understand how this protein works is to identify the specific regions involved in ligand binding and to learn how they are spatially related and how these relationships change during the reaction cycle.

Although the reaction mediated by the pump is complex and many intermediate states exist, this complexity can be used as an aid in protein modification studies. The

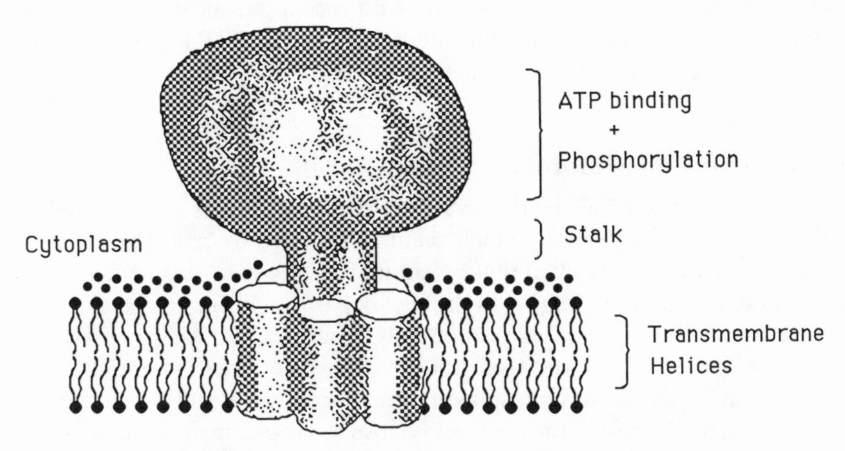

Figure 1. Schematic of α-subunit topography.

greater the number of properties of the protein that can be measured and the more intermediates characterized, the more confidence can be placed in the specificity and limits of the effects of modifications. For example, if major dephosphoenzyme conformations are characterized in some way, nucleotide binding measured, phosphorylation level assayed, cation binding or occlusion as well as activity measured, then if a reagent is hypothesized to act at a particular ligand binding site its specificity of action can be assessed. In such a complex enzymatic pathway as that of the sodium pump, the loss of overall enzymatic activity (ATPase) is by itself not a very useful criterion for the evaluation of the basis of the effects of a specific alteration (via mutagenesis or chemical modification).

Currently (Fall of 1990 or just after), there are no expression systems available that will produce significant or adequate quantities of functioning sodium pumps, so that systematic studies on the functional effects of mutations, deletions, and so on cannot be performed. We are thus limited to an amino acid–based strategy rather than a nucleotide-based strategy for enzyme modification. The approach used in chemical modification studies has been described and discussed recently in a review of this field (Pedemonte and Kaplan, 1990) and will not be examined in detail here. We hope that by the development of new specific reagents for the inactivation and modification of the Na,K-ATPase and identification of their sites of action we can learn more about structure–function relations in this protein and also provide rational targets for a mutagenesis program when such an approach becomes viable.

It should be borne in mind that the intrinsic ambiguity in interpreting the data from modification studies cannot be avoided. Whether the modification used is a specific chemical alteration of an intrinsic amino acid or the replacement of one endogenous amino acid by another, we cannot know until high resolution structural data become available whether the effects we observe are the result of direct actions at a site between the modified residue and a ligand or some other allosteric or through-space interaction mediated by the macromolecule. An undeniable power of a nucleotide-based strategy is the ability to make such modifications at predetermined locations with a range of more or less conservative substitutions in bulk, charge, or some other property.

The remainder of this article will deal with recent chemical attempts to obtain information on two of the functional protein regions, the ATP binding domain, and the monovalent cation binding domain.

The ATP Binding Domain

Studies in the last several years have identified regions of the α-subunit, and sometimes specific amino acids, which seem to be close to or at the ATP binding region of the enzyme. So far all of these residues are between transmembrane helices 4 and 5 and thus within the large cytoplasmic loop of ~440 amino acids (see Fig. 2). This is the region that has figured prominently in secondary structure predictions of the P-type ATPases (Taylor and Green, 1989). These residues include the aspartate (D-369), which forms the acyl phosphoenzyme intermediate and which presumably, at least in NaE_1ATP forms of the enzyme, is close to the gamma phosphate residue of ATP. Lysine 501, the major FITC-binding lysine, is in this region along with two aspartates (D-710 and D-714) which have been reported to be modified by a reagent that contains an alkylating moiety attached to the gamma phosphate of ATP (Dzhandzhugazyan and Modyanov, 1985; Ovchinnikov et al., 1987). This implies that region 710–714 is probably close to D-369 in NaE_1ATP. Lysine 719 and cysteine 656, both in the carboxy-terminal portion of the cytoplasmic loop, have been suggested to be the ATP-protectable targets of inactivation by FSBA (Ohta et al., 1986). The sources of possible error and controversial aspects of these labeling studies have been discussed in detail elsewhere (Pedemonte and Kaplan, 1990). More recent work by Hinz and Kirley (1990) has used pyridoxal phosphate and pyridoxal-5′-diphospho-5′-adenosine (ADP-pyridoxal) as ATP analogues. These reagents had previously been used in studies on the gastric H,K-ATPase (Tamura et al., 1989) and the SR Ca-ATPase (Yamamato et al., 1988). Evidence was obtained by Hinz and Kirley (1990) for the involvement of a lysine (K-480) in the α-subunit of lamb kidney Na,K-ATPase. This residue was contained in a sequence which ran from I-470 to K-487 which is highly conserved in sodium pumps from many species as well as in the gastric H,K-ATPase. Interestingly, this sequence corresponds to the radiolabeled fragment previously identified with the ATP-binding region from studies with

8-azido-ATP (Tran et al., 1988). Another segment of this loop which has been tentatively proposed to be involved in ATP binding is the sequence around K-598 to K-605, which was identified from sequencing of a labeled peptide after H_2DIDS inactivation of the sodium pump (Pedemonte and Kaplan, 1988*a,* 1990). The identification was not conclusive, since sequencing did not extend sufficiently far to identify the modified lysine. Recent work on the *Neurospora* proton pump with another reagent also suggests that this segment is important for ATP binding. In this work, 2-azido-ATP was used to photochemically inactivate the protein and a labeled fragment that corresponds to the region G-597 to K-622 in the sheep kidney sodium pump was sequenced. This region shows homology among the *Neurospora* proton

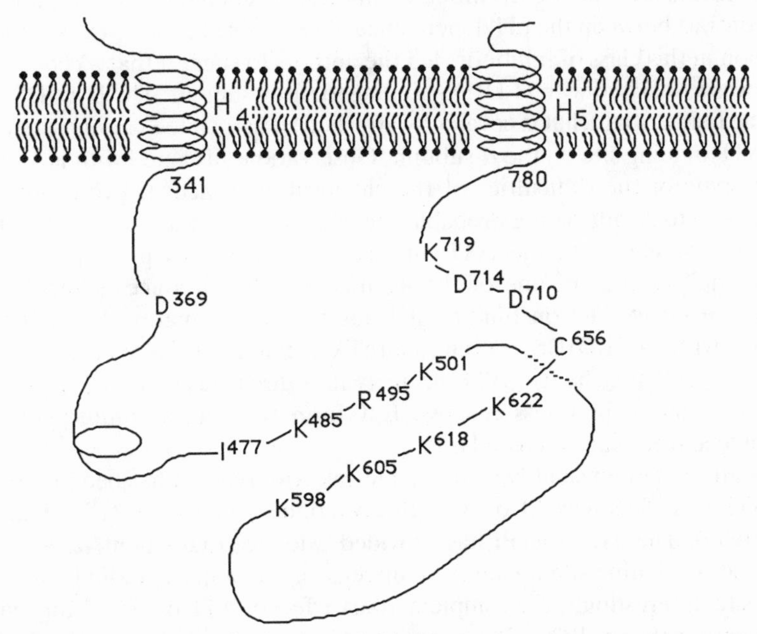

Figure 2. The cytoplasmic loop of the α-subunit. The fragment shows the various residues and stretches which have been associated with the ATP-binding domain following the results of chemical modification studies.

pump, the calcium pump, and the sodium pump, and contains a sequence with homology to a consensus sequence in phosphofructokinase (Davis et al., 1990).

Recently, Xu (1989) reported that in addition to K-501, the major lysine modified by FITC (in an ATP-protectable fashion), minor labeling with similar characteristics was also observed at K-480 and K-766. The former of these is within the sequence labeled by 8-azido-ATP (mentioned above). The latter, if it is part of the ATP binding region in close proximity to ATP, means that the folding of this loop to form the ATP binding site includes residues that are very close to the membrane–cytoplasm interface (see Fig. 2).

Until now, most of the amino acids identified by direct labeling as being involved in the ATP binding domain have been lysine residues. This is probably a result of the greater availability of lysine-selective reagents rather than a particular concentration of lysine residues in this region. Recent studies using *N*-acetylimidazole provide evidence for a residue in the α-subunit which when modified (acetylated) results in a loss of enzyme activity. This loss is due to the disruption of high affinity ATP binding by the enzyme. The inactivation is prevented by the simultaneous presence of ATP and, importantly, can be reversed by treatment with hydroxylamine (Argüello and Kaplan, 1990). Taken together, these data provide evidence for a tyrosine residue in the ATP binding region of the α-subunit.

The observations of Xu (1989) referred to above suggest that lysines 501, 480, and 766 cluster in some way so that they may all be directly associated with ATP binding to the α-subunit. Earlier evidence for a clustering of such titrable, positively charged side-chains was obtained from studies of enzyme inactivation by H_2DIDS and other reversible stilbene disulfonate inhibitors (Pedemonte and Kaplan, 1988*a, b*). The contrast between the pH dependence of reversible inhibition and irreversible inactivation in this class of inhibitors led the authors to suggest that a cluster of such residues was associated with H_2DIDS binding to the α-subunit. The H_2DIDS binding site (or the lysine to which it is covalently attached) is associated closely with the high affinity ATP binding site on the α-subunit. Observations obtained with H_2DIDS also illustrate some of the difficulties of the chemical approach. If protection studies alone are used to decide on the probable site of action, it is just as likely that H_2DIDS binds to the K site as to the ATP site, since both ligands prevent attachment. However, the presence of one of the ligands probably produces protection via a conformation effect; i.e., on binding it leads to a conformation less reactive (or unreactive) with the H_2DIDS. It seems more likely that H_2DIDS binds to the E_1 form close to the ATP site; hence, ATP protects in a direct fashion and K protects by holding the enzyme in a less reactive E_2K form (i.e., via a conformation effect [Pedemonte and Kaplan, 1988*a, b*]).

Recently, a new covalently acting lysine-selective reagent has been reported for P-type ATPases. This is NIPI or *N*-isothiocyanophenylimidazole (Ellis-Davies and Kaplan, 1990). The use of NIPI has provided evidence that when Na binds to the α-subunit at its binding site it causes an increase in the apparent affinity for ATP at the ATP site. Interestingly, the complementary effect of ATP on Na affinity had been reported earlier (Skou, 1979). It was observed only at elevated pH (pH 8.2), exactly the conditions under which NIPI modification of the enzyme is performed (Ellis-Davies and Kaplan, 1990). These observations suggest that ATP binding in this cytoplasmic loop region results in a structural change (corresponding to $E_1 \rightarrow E_1$ ATP) which is transmitted to the side-chains involved in Na coordination elsewhere in the molecule (see below).

In the sodium pump reaction cycle ATP acts with two different apparent affinities, ~0.1 μM in the catalytic site of phosphorylation (NaE_1ATP) and ~100 μM in stimulating the release of occluded K (E_2KATP). It has been argued that these different effects of ATP may be due to ATP binding at different regions in the protein. Leaving aside issues of active dimeric or oligomeric enzyme forms (of the $\alpha_2\beta_2$ type, for example), it seems more likely that the effects are due to subtle changes in a single ATP binding domain as the protein progresses through the E_1Na and E_2K states. Observations with NIPI show that modification of a single lysine residue in the α-subunit results in the loss of both ATP functions: the high affinity binding in the presence of Na ions and the low affinity deocclusion of Rb ions (Ellis-Davies and Kaplan, 1990). This makes it most likely that a single protein domain is involved in both functions. It is anticipated that studies underway using [^{3}H]NIPI will identify the precise lysine residue involved in the modification. The specific residues and structures that have been located in the primary structure of the cytoplasmic loop and are thought to be directly involved in ATP binding are shown in Fig. 2. These residues have been identified as a result of their identification as the locus of attachment of covalent inactivators of the pump. In addition to the difficulty in deciding whether their effects are direct (as discussed above), another potential pitfall is always present in such studies. This has been discussed in a recent review (Pedemonte and Kaplan, 1990) and is illustrated in some recent studies by Chang and Slayman (1990) of the *Neurospora* plasma membrane proton pump. Another reason for inactivation by a reagent in the ATP site, for example, may not be because the labeled amino acid is critically important for coordination to ATP, but because the space close to the residue is important for function and its occupancy by a bulky reagent may lead to inactivation. Earlier work by Pardo and Slayman (1989) had established that modification of the specific cysteine residue (C-532) by *N*-ethylmaleimide resulted in inactivation of H^+-ATPase activity and that the inactivation (and labeling) could be prevented by MgADP. A conclusion might be that C-532 was involved in ADP (hence ATP) binding. The subsequent work by Chang and Slayman (1990) shows that modification of C-532 with methyl methanethosulfonate, a small sulfhydryl-reactive reagent, results in neither loss of activity nor any detectable structural change. This removes any basis for the assumption that C-532 plays an essential functional in catalysis. It would be interesting in the case of the alkylating derivatives of ATP discussed above if the modified protein were treated enzymatically or under mild chemical conditions to remove the adenosine phosphate moieties from the covalent attachment. The resulting protein would have relatively less bulky groups attached at the apparently essential amino acids, say D-710 or D-714. If this protein were inactivated and unable to bind ATP, a stronger case would be made for the proposed role of D-710 or D-714 in ATP binding.

The Monovalent Cation Binding Domain

Compared with the wealth of data available on the ATP binding domain, far less is known about the cation binding domain. The extent to which the ATP binding domains have been conserved through many enzyme systems has been of great help in advancing our understanding of a variety of ATPases, kinases, etc. However, it is probably the evolutionary development of a cation binding site that enables these ATP hydrolyzing proteins to smuggle cations via a nonleaky pathway through the membrane which gave the P-type ATPases their special character. It has seemed plausible for many years that if monovalent cation binding sites on the sodium pump were to satisfy local charge interactions, a negatively charged amino acid side-chain might be directly involved. This reasoning gave rise to studies with carbodiimides, well-characterized chemical reagents which react with carboxyl residues. The early work on carboxyl group modification was initiated by Robinson, who showed that dicyclohexylcarbodiimide (DCCD) inactivated the sodium pump in a characteristic way, which was prevented by the simultaneous presence of sodium ions (Robinson, 1974). This supported the idea that carboxyl side-chains may play a role in Na^+ or K^+ ion coordination by the enzyme. Similar subsequent work on the sodium and calcium pumps has been reviewed (Pedemonte and Kaplan, 1990). The fundamental problem associated with the use of carbodiimides lies in the complex alternative pathways available for the initially formed carbodiimide–protein carboxyl adduct (see Fig. 3). Two main fates are available: either (*a*) condensation, with elimination of the carbodiimide as a urea and formation of a peptide bond (intramolecularly with an endogenous amine as shown or with an amine of another peptide to form an intermolecular cross-link), or (*b*) rearrangement. Only in the latter case does the extent of incorporation of the carbodiimide into the protein reflect the extent of inactivation and involvement of an essential carboxyl. Indeed, for the incorporation and inhibition to be related in a simple way, the condensation pathway must not inactivate at all. Recent studies by Pedemonte and Kaplan (1986*a, b*) showed with both water-soluble and hydrophobic carbodiimides that inactivation was almost certainly due, under their reaction conditions, to condensation. The most telling evidence came from the observation that addition of exogenous nucleophiles led to nucleophile incorporation but fully active enzyme (Pedemonte and Kaplan, 1986*a*). In other words, inactivation could not be due to the modification of an essential carboxyl (since the carboxyl was modified in the fully active enzyme obtained with incorporated nucleophile), but was probably due to cross-link formation and production of a rigid, fixed protein or the removal by condensation of an essential amino side-chain. It is ironic that the use of a carboxyl-selective reagent might provide evidence for the involvement of an essential amino side-chain. Different results were obtained by Shani-Sekler et al. (1988), who have reported DCCD inactivation of the pig kidney sodium pump enzyme which was prevented by monovalent cations (in agreement with Pedemonte and Kaplan, 1986*a*) but was not prevented by the simultaneous presence of exogenous nucleophiles. Goldshleger et al. (1990) obtained evidence for the incorporation of radiolabeled DCCD which corresponded to the modification of carboxyl residues that are assumed to be at the monovalent cation binding site. The reason for the difference in observations is not clear but some possibilities have been discussed previously (Shani-Sekler et al., 1988; Pedemonte and Kaplan, 1990).

To avoid the potential complications of using carbodiimides we recently employed the stabilized diazomethane analogue 4-(diazomethyl)-7-(diethylamino)-coumarin (DEAC). Reaction of a diazomethane with a carboxyl residue results simply in esterification of the carboxyl (see Fig. 3). Initial findings with DEAC have recently been reported (Argüello and Kaplan, 1991*a, b*). The basic findings are that a K- or Na-preventable inactivation results from treatment of the enzyme with this reagent. Complete inactivation of the enzyme results from the K-preventable incorporation of one or two molecules per active α-subunit. The modified carboxyl residues are not in peptides released from the membranes on extensive proteolytic treatment of modified enzyme. Thus, it is likely that the esterified carboxyl residues are located in or close to the transmembrane segments of the α-subunit. The modified enzyme has unaltered nucleotide binding properties but less than 10% of the native proteins' capacity to occlude Rb ions. It is also apparent that the DEAC-modified enzyme is still able to undergo $E_1 \rightleftharpoons E_2$ transitions (Argüello and Kaplan, 1991*b*). The precise localization and identification of the DEAC-modified aspartate or glutamate residues is now underway. We anticipate that some of the amino acids involved in the coordination of K^+ or Na^+ ions with the α-subunit will soon be identified. The identification of other residues which are not glutamates or

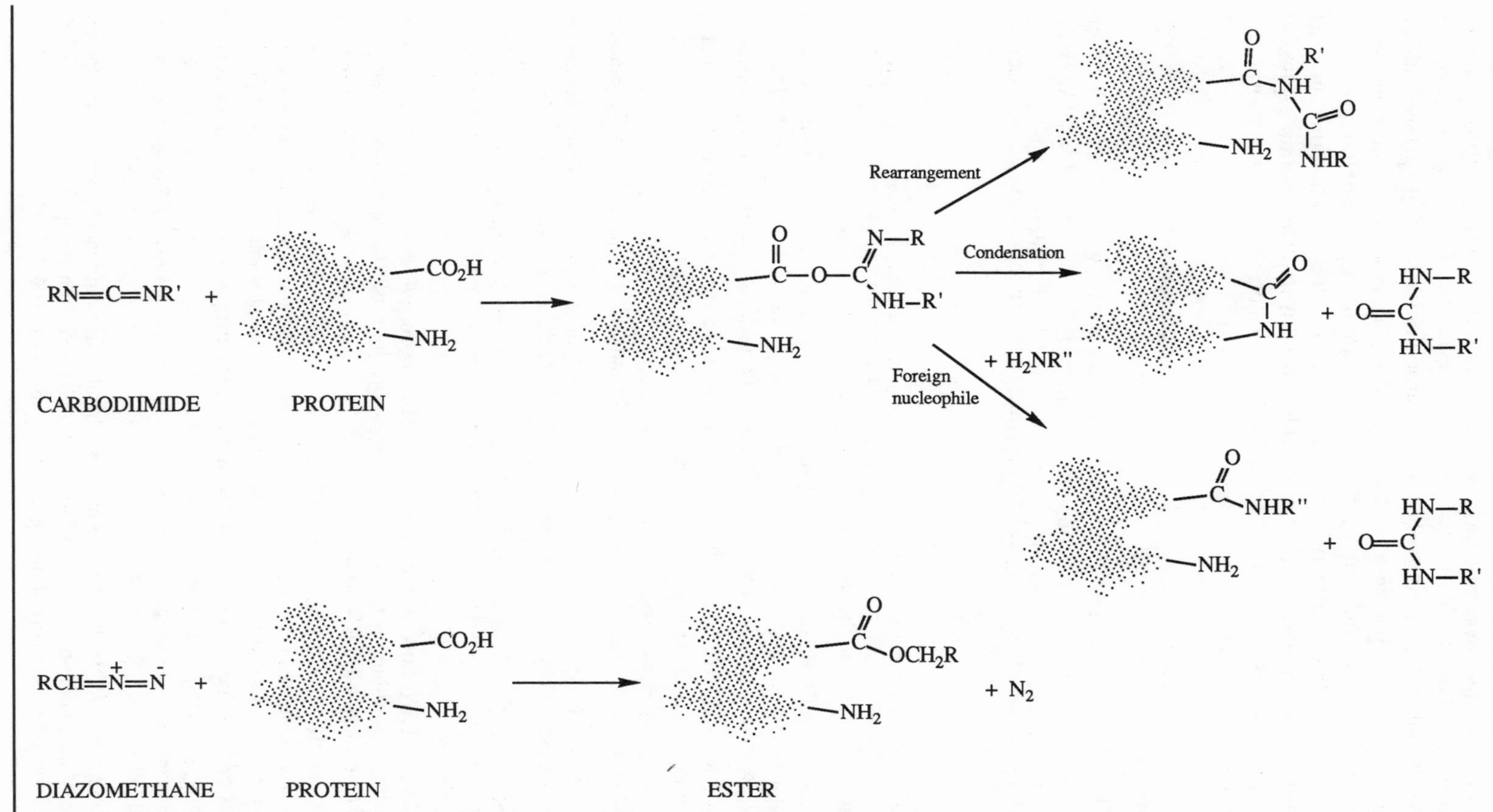

Figure 3. Modification of protein carboxyl side-chains. A comparison of the reaction pathway and products of side-chain carboxyl residues with either carbodiimides or diazomethanes.

aspartates in the monovalent cation coordination will require the development of other chemical reagents.

In summary, then, the systematic application of chemical labeling approaches is beginning to identify specific amino acids that are likely to be involved in ligand coordination by the α-subunit of the sodium pump. An understanding of the mechanism of action of our favorite transport protein will come when this approach can be used in combination with a nucleotide-based strategy for mutant expression and when techniques are available to obtain high resolution structural information from the crystalline protein.

Acknowledgments

I would like to acknowledge with gratitude my collaborators who appear as coauthors in the cited work, Drs. Carlos Pedemonte, Graham Ellis-Davies, and José Argüello.

The work described here that was performed in the author's laboratory was supported by NIH grants HL-30315 and GM-39500.

References

Argüello, J. M., and J. H. Kaplan. 1990. *N*-Acetylimidazole inactivates renal Na,K-ATPase by disrupting ATP binding to the catalytic site. *Biochemistry.* 29:5775–5782.

Argüello, J. M., and J. H. Kaplan. 1991*a.* Esterification of a carboxyl residue in the Na pump α-subunit abolishes K sensitivity. *In* The Sodium Pump: Recent Developments. J. H. Kaplan and P. De Weer, editors. The Rockefeller University Press, New York. 000–000.

Argüello, J. M., and J. H. Kaplan. 1991*b.* Evidence for essential carboxyls in the cation binding domain of the Na,K-ATPase. *Journal of Biological Chemistry.* In press.

Brandl, C. J., N. M. Green, B. Korczak, and D. H. Maclennan. 1986. Two Ca^{2+} ATPase genes: homologies and mechanistic implications of deduced amino acid sequences. *Cell.* 44:597–607.

Chang, A., and C. W. Slayman. 1990. A structural change in the Nemospora plasma membrane [H^+]-ATPase induced by N-ethylmaleimide. *Journal of Biological Chemistry.* 265:15531–15536.

Clarke, D. M., T. W. Loo, and D. H. Maclennan. 1990. Functional consequences of alterations to probe amino acids located in the transmembrane domain of the Ca^{2+} ATPase of sarcoplasmic reticulum. *Journal of Biological Chemistry.* 265:6262–6267.

Clarke, D. M., T. W. Loo, G. Inesi, and D. H. Maclennan. 1989*a.* Location of high affinity Ca^{2+}-binding sites within the predicted transmembrane domain of the sarcoplasmic reticulum Ca^{2+}-ATPase. *Nature.* 339:476–478.

Clarke, D. M., K. Marayaua, T. W. Loo, E. Lebeser, G. Inesi, and D. H. Maclennan. 1989*b.* Functional consequences of glutamate, aspartate, glutamine and asparagine mutations in the stalk sector of the Ca^{2+}-ATPase of sarcoplasmic reticulum. *Journal of Biological Chemistry.* 264:11236–11251.

Davis, C. B., K. E. Smith, B. N. Campbell, Jr., and G. G. Hammes. 1990. The ATP binding site of the yeast plasma membrane proton-translocating ATPase. *Journal of Biological Chemistry.* 265:1300–1305.

Dzhandzhugazyan, K., and N. Modyanov. 1985. Chemical and immunochemical approaches to structure of membrane-bound Na^+-K^+-ATPase. *In* The Sodium Pump. I. M. Glynn and C. Ellory, editors. The Company of Biologists, Ltd., Cambridge, UK. 129–134.

Ellis-Davies, G. C. R., and J. H. Kaplan. 1990. Binding of Na^+ ions to the Na,K-ATPase increases the reactivity of an essential residue in the ATP binding domain. *Journal of Biological Chemistry.* 265:20570–10576.

Goldshleger, R., S. J. D. Karlish, W. D. Stein, and D. M. Tal. 1990. Rubidium occlusion occurs in membrane-embedded tryptic fragments of renal Na,K-ATPase. *Biophysical Journal.* 57:17*a*. (Abstr.)

Green, N. M., W. R. Taylor, and D. H. Maclennan. 1988. A consensus structure for ion pumps. *Progress in Clinical and Biological Research.* 273:15–24.

Hinz, H. R., and T. L. Kirley. 1990. Lysine 480 is an essential residue in the putative ATP site of lamb kidney (Na,K)-ATPase. *Journal of Biological Chemistry.* 265:10260–10265.

James, P., M. Maeda, R. Fisher, A. K. Verma, J. Krebs, J. Penniston, and E. Carafoli. 1988. Identification and primary structure of a calmodulin binding domain of the Ca^{2+} pump of human erythrocytes. *Journal of Biological Chemistry.* 263:2904–2910.

Kawakami, K., S. Noguchi, M. Noda, H. Takahashi, T. Ohta, M. Kawamura, H. Nojima, K. Nagano, T. Hiroshe, S. Inayama, H. Hayashida, T. Miyata, and S. Numa. 1985. Primary structure of the α-subunit of *Torpedo californica* ($Na^+ + K^+$)-ATPase deduced from cDNA sequence. *Nature.* 316:733–736.

Maclennan, D. H., C. J. Brandl, B. Korczak, and N. M. Green. 1985. Aminoacid sequence of a Ca-ATPase from rabbit muscle sarcoplasmic reticulum deduced from its complementary DNA sequence. *Nature.* 316:696–700.

Ohta, T., K. Nagano, and M. Yoshida. 1986. The active site structure of Na^+/K^+-transporting ATPase: location of the 5-(*p*-fluorosulfonyl)benzoyladenosine binding site and soluble peptides released by trypsin. *Proceedings of the National Academy of Sciences, USA.* 83:2071–2075.

Ovchinnikov, Y. A., N. K. Dzhandzugazyan, S. V. Lutsenko, A. A. Mustayev, and N. N. Modyanov. 1987. Affinity modification of E1-form of $Na^+ + K^+$-ATPase revealed Asp-710 in the catalytic site. *FEBS Letters.* 217:111–116.

Ovchinnikov, Y. A., N. M. Luneva, E. A. Arystarkhova, N. M. Gevondyan, N. M. Arzamazova, A. T. Kozhich, V. A. Nesmeyanov, and N. N. Modyanov. 1988. Topology of Na^+, K^+-ATPase: identification of the extra- and intracellular hydrophilic loops of the catalytic subunit by specific antibodies. *FEBS Letters.* 227:230–234.

Pardo, J. P., and C. W. Slayman. 1989. Cysteine 532 and cysteine 545 and the N-ethylmaleimide-reactive residues of the Neurospora plasma membrane H^+-ATPase. *Journal of Biological Chemistry.* 2663:9373–9379.

Pedemonte, C. H., and J. H. Kaplan. 1986*a*. Carbodiimide inactivation of Na^+-K^+-ATPase. A consequence of internal cross-linking and not carboxyl group modification. *Journal of Biological Chemistry.* 261:3632–3639.

Pedemonte, C. H., and J. H. Kaplan. 1986*b*. Carbodiimide inactivation of Na^+-K^+-ATPase, via intramolecular cross-link formation, is due to inhibition of phosphorylation. *Journal of Biological Chemistry.* 261:16660–16665.

Pedemonte, C. H., and J. H. Kaplan. 1988*a*. Inhibition and derivatization of the renal Na^+-K^+-ATPase by dihydro-4,4′-diisothiocyanatostilbene-2,2′-disulfonate. *Biochemistry.* 27: 7966–7973.

Pedemonte, C. H., and J. H. Kaplan. 1988*b*. Interaction of the Na,K-ATPase with H_2DIDS. *In* The Na^+,K^+-Pump, Part A: Molecular Aspects. J. C. Skou, J. G. Norby, A. B. Maunsbach, and M. Esmann, editors. Alan R. Liss, Inc., New York. 327–334.

Pedemonte, C. H., and J. H. Kaplan. 1990. Chemical modification as an approach to elucidation of sodium pump structure-function relations. *American Journal of Physiology.* 258 (*Cell Physiology* 27):C1–C23.

Robinson, J. D. 1974. Affinity of the (Na^+-K^+)-dependent ATPase for Na^+ measured by Na^+-modified enzyme inactivation. *FEBS Letters.* 38:325–328.

Shani-Sekler, M., R. Goldshleger, D. M. Tal, and S. J. D. Karlish. 1988. Inactivation of Rb^+ and Na^+ occlusion on (Na,K)-ATPase by modification of carboxyl groups. *Journal of Biological Chemistry.* 263:19331–19341.

Shull, G. E., A. Schwartz, and J. B. Lingrel. 1985. Amino-acid sequence of the catalytic subunit of the (Na^+ + K^+)-ATPase deduced from complementary DNA. *Nature.* 316:691–695.

Skou, J. C. 1979. Effects of ATP on the intermediary steps of the reaction of the (Na^+ + K^+)-ATPase. *Biochimica et Biophysica Acta.* 867:421–435.

Stokes, D. L., and N. M. Green. 1990. Structure of Ca-ATPase: electron microscopy of frozen-hydrated crystals at 6A resolution in projection. *Journal of Molecular Biology.* 213:529–539.

Tamura, S., M. Tagaya, M. Maeda, and M. Futai. 1989. Pig gastric (H^+ + K^+)-ATPase lysine 497 conserved in cation transporting ATPases is modified with pyridoxal-5′ phosphate. *Journal of Biological Chemistry.* 264:8580–8584.

Taylor, W. R., and N. M. Green. 1989. The predicted secondary structure of the nucleotide-binding sites of six cation-transporting ATPase lead to a probable tertiary fold. *European Journal of Biochemistry.* 179:241–248.

Tran, C. M., G. Scheiner-Bobis, W. Schoner, and R. A. Farley. 1988. The amino acid sequence of a peptide from the ATP-bindings site of Na^+-K^+-ATPase labeled by 8-N_3-ATP. *Biophysical Journal.* 53:343*a*. (Abstr.)

Yamamoto, H., M. Tagaya, T. Fukai, and M. Kawakita. 1988. Affinity labeling of the ATP-binding site of Ca^{2+} transporting ATPase of sarcoplasmic reticulum by adenine triphosphopyridoxal: identification of the native lysyl residue. *Journal of Biochemistry.* 103:452–457.

Xu, K. 1989. Any of several lysines can react with 5′-isothiocyanateofluoroscein to inactivate sodium and potassium ion activated adenosinetriphosphatase. *Biochemistry.* 28:5764–5772.

Chapter 10

Structure of the Cation Binding Sites of Na/K-ATPase

S. J. D. Karlish, R. Goldshleger, D. M. Tal, and W. D. Stein

Department of Biochemistry, Weizmann Institute of Science, Rehovot, Israel; and Institute of Life Science, Hebrew University, Jerusalem, Israel

The question "How does a cation pump, pump?" requires knowledge as to the function and structure of both ATP and cation sites and their detailed interactions. The structure of the cation sites is the least understood but arguably the most significant aspect of the different pumps, since herein lies their functional specificity.

A necessary framework for a description of the sites are the models of *trans*-membrane arrangement of the α chains of the related pumps, Na/K, Ca, etc. There are several models, which differ mainly in the predicted number of *trans*-membrane segments in the COOH-terminal half of the chain and orientation of the COOH terminus, while the first four NH_2-terminal *trans*-membrane segments and the position of the first and the major second cytoplasmic loops are less controversial. The models of Shull et al. (1985) and Kawakami et al. (1985) for the α chain of Na/K-ATPase predict eight *trans*-membrane segments (four in the COOH-terminal domain) with a cytoplasmically oriented COOH terminus, while that of Ovchinnikov et al. (1986) proposes seven *trans*-membrane segments (three in the COOH-terminal domain) and an extracellular COOH terminus. The model of the sarcoplasmic reticulum Ca pump supposes that there are ten *trans*-membrane segments (six in the COOH-terminal domain) with a cytoplasmic COOH terminus (Maclennan et al. 1985). The hydropathy profiles of the different pumps are similar and it is therefore unlikely that there are in reality different numbers of *trans*-membrane segments. Thus the precise chain folding, particularly at the COOH terminus, is the subject of detailed investigation by a variety of techniques. As discussed below, this region turns out to be crucial for the cation sites. By comparison with ATP binding sites, which are recognizable within the major cytoplasmic loop by residues at or near the site (see Jorgensen and Andersen, 1988), the information leading to detailed folding models (Taylor and Green, 1989), the nature of the cation residues is still largely speculative. One possibility is that negatively charged groups (glutamates or aspartates) within *trans*-membrane segments are involved (Ovchinnikov, 1987), but there are also suggestions in the literature that cation and ATP sites are in close proximity (Klemens et al., 1988), which might seem paradoxical. In addition, glutamates at the first extracellular loop of Na/K-ATPase (Shull et al., 1985), or, in the case of Ca-ATPase, in the first cytoplasmic loop, the so-called "stalk" (Maclennan et al., 1985), were also suggested to bind cations, and the NH_2-terminal region of Na/K-ATPase, rich in positively charged residues, was proposed to serve as an ion-selective gate controlling passage of Na and K to and from the sites (Shull et al., 1985; Lingrel et al., 1990).

Known structures of cation binding proteins and ionophores are useful precedents and suggest that bound or occluded cations will be largely dehydrated and ligated with six to eight oxygen-containing groups (including carboxyl, and/or neutral groups such as carbonyl, serine, threonine, or phenolic hydroxyl, and/or remaining solvation water). Residues could reside on either continuous or distant segments of the polypeptide (see Glynn and Karlish, 1990). But of course a cation "site" on a pump is not static. Thus different residues could be involved at the opposing surfaces of the protein or in more deeply embedded occluded states. Movement of cations between the surface and occluded states may occur in narrow crevices or so-called "ion wells."

In the absence of a high resolution structure, some insight into cation sites in pumps is provided by studies of chemical modification of carboxyl groups, for gastric H/K-ATPase with a K-site affinity label, by site-directed mutagenesis in the case of

Ca-ATPase, from observations on charge transfer properties or voltage dependence of pumps, and also by using cation analogues as probes of sites in fluorescence or NMR studies. We have recently reviewed the state of this knowledge up to the Summer of 1989 (Glynn and Karlish, 1990). This article will therefore summarize very recent experiments using tryptic digestion of renal Na/K-ATPase to identify relevant peptide segments, and modification by dicycolhexylcarbodiimide (DCCD). The findings will then be compared with those from other techniques, particularly the mutation experiments of Clarke et al. (1989*a*, *b*) on Ca-ATPase.

The questions to be addressed include (*a*) the location of the sites relative to *trans*-membrane segments and the ATP binding domain, (*b*) the nature of the ligating groups, especially the possible role of carboxyl residues, and (*c*) whether Na and K bind at the same sites. Na and K ions compete, and so a common path of binding and transport would be the simplest possibility. Of course only two Na and two K sites could be the same. Presumably the third Na site would be distinct.

For any chemical or genetic manipulation intended to modify cation sites, it is highly desirable to assay cation binding or occlusion directly, rather than cation-dependent reactions, to minimize the possibility that observed effects are indirect. For this type of study a simple but accurate manual assay of Rb or Na occlusion was devised by Shani et al. (1987). It is based on the great stability of cation–enzyme

TABLE I
Stoichiometry of Rb and Na Occlusion

Phosphoenzyme	Na occlusion	Rb occlusion	Rb/EP	Na/EP
	nmol/mg protein ± SE			
2.01 ± 0.068	6.34 ± 0.17	3.93 ± 0.047	1.96 ± 0.036	3.15 ± 0.14

Reproduced from Shani-Sekler et al. (1988).

complexes at 0°C, which allows slow elution on Dowex-50 cation exchange columns for efficient removal of free isotope. It was designed to allow measurement of equilibrium levels of occlusion at high concentrations of the cations, which is an important feature. Table I shows the classical stoichiometries of 2Rb or 3Na ions occluded per pump molecule, when measured at saturating Rb or Na concentrations (Shani-Sekler et al., 1988). Na occlusion is induced with oligomycin as described by Esmann and Skou (1985).

Location

One might ask oneself how much of the enzyme can be digested away from the membrane with trypsin without destroying occlusion. The answer turns out to be a surprisingly large amount (Karlish et al., 1990)! Fig. 1 *A* shows a 10% Tricine gel (Schagger and von Jagow, 1987) of fragments remaining in the membrane after digestion of renal Na/K-ATPase with a massive, equimolar concentration of trypsin in the presence of Rb ions and EDTA, and below the fraction of Rb occlusion remaining. As described by Jorgensen (1975), the α chain is rapidly digested and fragments appear and disappear. But the important point is that after 1 h a stable 19-kD fragment has accumulated, with (in this gel) some poorly resolved smaller fragments, while the β chain is largely intact. But the striking thing is that Rb occlusion is essentially normal. About half of the membrane protein is solubilized. Clearly the Rb occlusion domain is protected from the trypsin, probably within *trans*-membrane segments.

The remarkable resistance of the 19-kD fragment to further digestion depends on the presence of Rb and absolutely on the absence of Ca ions (Fig. 1 *B*). In the absence of Rb or the presence of Ca the fragment does not accumulate and occlusion

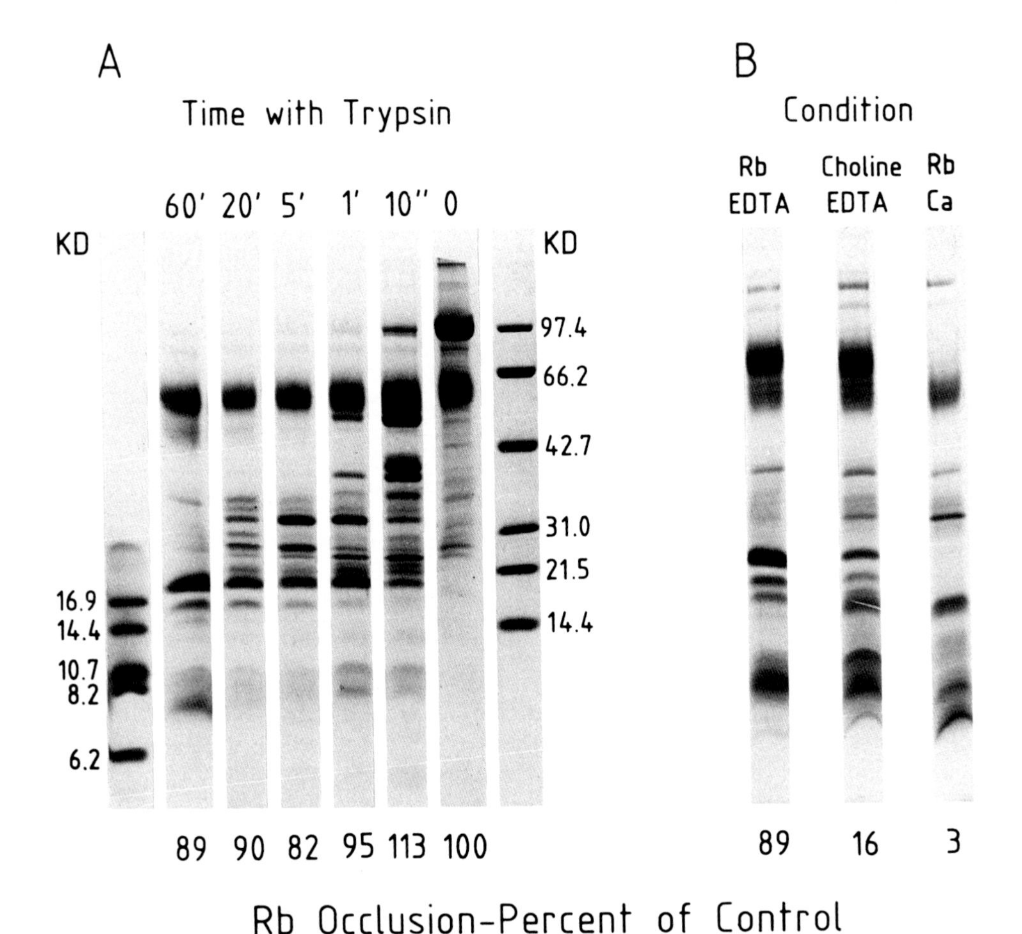

Figure 1. Tryptic digestion of renal Na/K-ATPase, with accumulation of the 19-kD fragment. Reproduced from Karlish et al. (1990).

is lost. Unpublished experiments (Capasso, J., S. Hoving, and S. Karlish) show that K or any of its congeners, but also Na ions, protect the 19-kD fragment. In Fig. 1 *B* one can also see partial digestion of β into fragments of 40–50 kD and 16 and 14 kD (identified as β fragments by sequencing; see Table II), and again poorly resolved smaller fragments. When digested at pH 6.5 the β chain can be completely cut into the fragments of 40–50 and 16 kD without loss of occlusion (unpublished experiments). Thus an intact β is not required.

The experiment in Fig. 1 suggests an intimate connection between the 19-kD fragment and occlusion. To prove this, a quantitative correlation was made by preparing 19-kD membranes, as we call them, digesting with trypsin in the presence of Ca, and at different times stopping the digestion with tryptic inhibitor. The correlation coefficient between the amount of 19-kD fragment on gels and Rb occlusion is ~0.95, suggesting that the 19-kD is essential for occlusion. This does not necessarily mean that the fragment directly binds Rb, but in fact we think it does (see below).

TABLE II
NH_2-terminal Sequences of Tryptic Fragments of α and β Chains

α fragments		
19 kD	Found	NPKTDKLVNERLISMA
		830
	Pig α	NPKTDKLVNERLISMA
≈10 kD	Found	SGPNALTPPGTT
		68
	Pig α	DGPNALTPPPTT
≈9 kD	Found	SGPNALTPPGTTPEN
		68
	Pig α	DGPNALTPPPTTPEN
≈8 kD	Found	ATLDMGLEXGQ
		264
	Pig α	ATLASGLEGGQ
≈7 kD	Found	LIVTGVEEGELIFD
		749
	Pig α	SIVTGVEEGRLIFD
β fragments		
16 kD	Found	AKEEGSIWKF
		5......................
	Pig β	AKEEGSWKKF
14 kD	Found	XIRNXEKXEFLGRTHG
		14....................................
	Pig β	FIWNSEKKEFLGRTGG
≈50 kD	Found	GERKVKRFXLEXLGR
		143................................
	Pig β	GERKVCRFRLEWLGN

To define which peptides actually bind the ions, it is necessary to define more precisely the components of these truncated pumps. This has been done by separating the SDS-solubilized peptides on a size-exclusion HPLC column, concentrating the protein peaks, and then running on a 16% Tricine gel which resolves below 5 kD. On a TSK-G3000 SW column under nonreducing conditions, four characteristic peaks appear. The first corresponds to high molecular weight contaminants or aggregates. Peak two contains the β chain and the 40–50- and 16-kD fragments seen in Fig. 1 *B* connected by an S-S bridge. Peak three contains the 19-kD fragment about

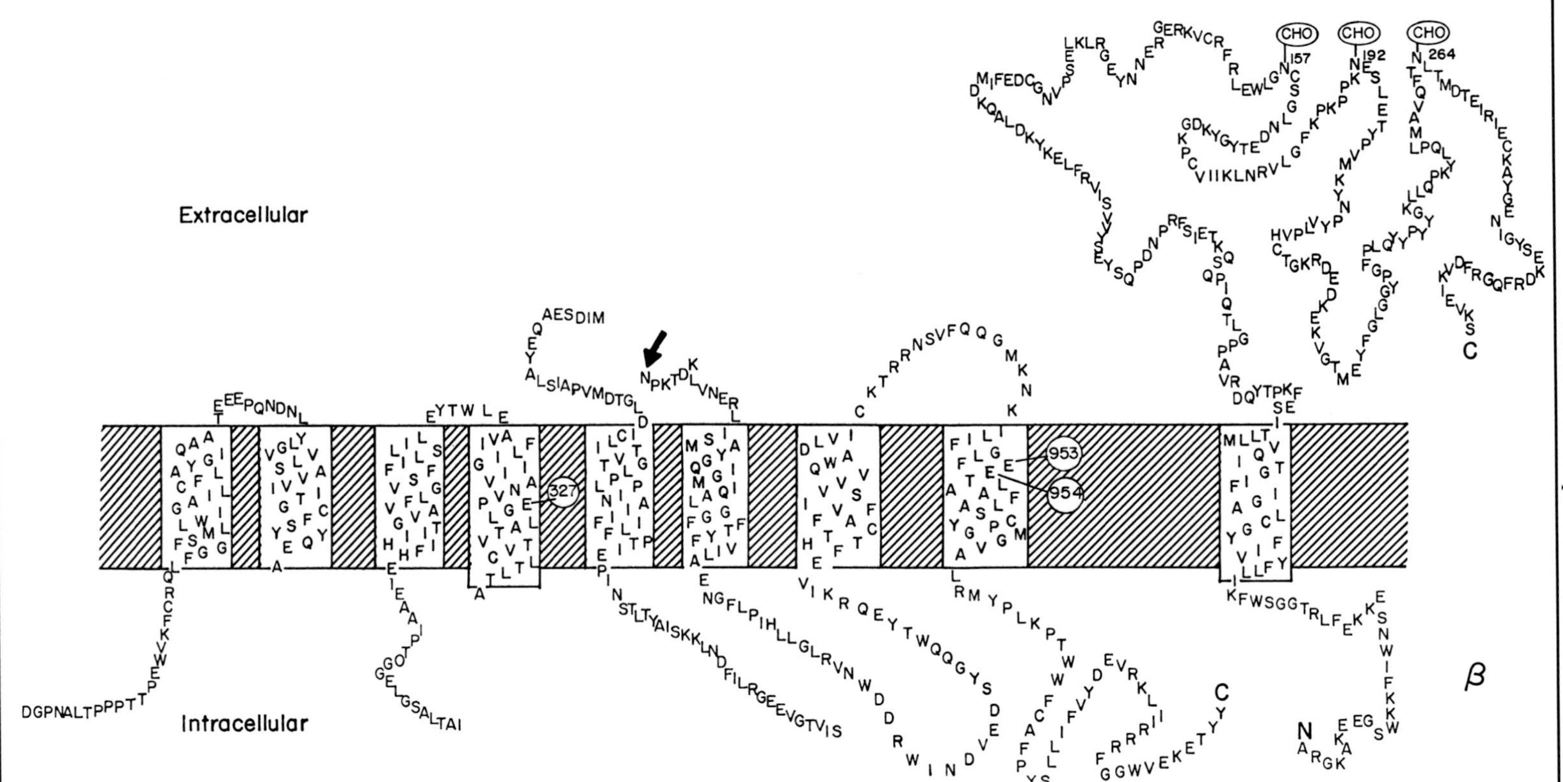

Figure 2. Hypothetical arrangement of tryptic peptides in 19-kD membranes.

one-third pure. Peak four contains a family of four or five peptides of 7–11 kD, reasonably but not perfectly separated.

The NH_2 termini of these peptides have now been sequenced after transfer to Immobilon (Matsudaira, 1987) and located in the primary sequence of α or β chains (Table II). The NH_2 terminus of the 19-kD fragment is unambiguously identified as residue 830, extending toward the COOH terminus. The sequences of the family of smaller fragments (peak four) are as follows. The first two of molecular mass ~9 and ~10 kD have the same NH_2 terminus, beginning at residue 68, but are different in length. They contain putative M1 and M2. Fragment three, ~8 kD, is mainly the peptide beginning at residue 264, containing putative M3 and M4, but here a minor sequence corresponding to M5 is also seen. Fragment four had a sequence which, in membranes digested at pH 7.0, could not be identified in either α or β. In membranes digested at pH 8.3, this ~7-kD fragment was identified as beginning at residue 750

TABLE III
Functional Properties of Control and 19-kD Fragment–containing Membranes

	Control	19-kD fragment–containing membranes
Cation occlusion	*nmol/mg*	*protein* ± *SE*
1. Rb occlusion	3.75 ± 0.12	7.74 ± 0.15
2. Na occlusion	5.22 ± 0.22	11.86 ± 0.12
3. Ratio Na/Rb	1.39 ± 0.074	1.53 ± 0.035
ATP-dependent functions		
1. ATPase activity, *μmol/min per mg protein*	15–21	<2% of control
2. Phosphorylation by ATP, *nmol/mg protein*	1.6–2.2	<1% of control
3. ATP-displaceable eosin binding, *% of total fluorescence*	11.8%	<0.5%
4. Inhibition of Rb occlusion Rb(0.1 mM ± ATP [1 mM]), *% inhibition*	70%	Undetectable

Reproduced from Karlish et al. (1990).

and containing putative M5. Improved separations of these peptides would probably clean up the picture, but this information is sufficient to propose the following model of 19-kD membranes.

Fig. 2 is based on the folding models of the α chain of Shull et al. (1985) and Kawakami et al. (1985). (See Karlish et al. [1990] for a model based on that proposed by Ovchinnikov [1987].) The 19-kD fragment includes putative M6, M7, and M8. It is shown to extend to the COOH terminus, but could be a little shorter. The NH_2 terminus of the smaller peptides is either close to or at the theoretical limit for tryptic digestion. The COOH termini shown are the theoretical tryptic limit; they could be a little longer. In one case, the sequence of the M5 fragment appears to be a chymotryptic split. The finding of all these sequences is therefore compatible with, but does not prove, this model. A major uncertainty is that the location of the COOH terminus, cytoplasmic or extracellular, has not been unequivocally demonstrated.

Studies of binding of COOH terminus–specific antibodies to whole cells (extracellular) (Ovchinnikov et al., 1988) or kidney vesicles (cytoplasmic) (Kramer et al., 1990) have given contradictory results. It also remains to be seen if the NH_2 terminus of the 19-kD fragment is really extracellular as predicted here. What is clear is that $>50\%$ of the α chain is removed, including the major cytoplasmic loops, but the predicted *trans*-membrane segments appear intact. The finding of fragments beginning at residue 68 (Table II) in membranes occluding both Rb and Na (Table III) seems to exclude a crucial role for the NH_2 terminus of the α chain in determining cation binding and specificity; this region is probably involved in control of the E_1-E_2 conformational equilibrium (Jorgensen and Collins, 1986). Clearly the big question is how the *trans*-membrane segments of α are arranged in the plane of the membrane, and which ones form the cation occlusion cage.

Functional Properties of 19-kD Membranes

Table III compares capacities for Rb and Na occlusion and ATP-dependent functions in control and 19-kD membranes (Karlish et al., 1990). The top half shows that Na occlusion capacity is intact. Specific Rb and Na occlusion are both about double that of the control due to removal of half of the protein. The ratio between Rb and Na is close to 1.5 in both cases. Thus all three Na (i.e., all cation sites) are preserved in the 19-kD membranes. By contrast, ATP-dependent activities are totally destroyed. Thus the ATP and cation occlusion domains have been physically separated. In normal active transport the interactions between these domains must therefore be indirect.

Despite the maintenance of occlusion capacity, 19-kD and control membranes show some differences in occlusion properties. (*a*) For example, the apparent affinity of Rb for occlusion is somewhat (pH 7.0) or decidedly (at pH 8.5) higher for 19-kD membranes compared with control, while that of Na for its occlusion is significantly lower and more cooperative than in control. (*b*) Deocclusion of ^{86}Rb into a Rb-containing medium is clearly biphasic for 19-kD membranes, while only monophasic dissociation is observed for control enzyme. Biphasic dissociation in Rb media containing phosphate has already been described by Glynn et al. (1985) and Forbush (1987) for undigested enzyme, and explained by ordered dissociation of the two Rb ions. Apparently dissociation of Rb from 19-kD membranes can also be ordered. On the "flickering gate" model of Forbush (1987), the "gate" in 19-kD membranes is closed more often than in the control enzyme.

An obvious question was whether the entire cation transport pathway was intact in the 19 kD-containing membranes. Proteoliposomes loaded with RbCl were prepared from 19-kD membranes by freeze–thaw sonication. ^{86}Rb uptake was measured. A slow passive Rb/Rb exchange, in the absence of ATP, which we have described for reconstituted control enzyme (Karlish and Stein, 1982), was readily detected for vesicles prepared with 19-kD membranes (Karlish et al., 1990). Again there were some quantitative differences from the flux in control vesicles. But the important conclusion is that the entire Rb transport path remains intact in 19-kD membranes.

Modification and Labeling with DCCD

The use of DCCD is not simple due to lack of selectivity and the complex reaction mechanism. After the initial carboxyl activation, cross-linking to neighboring nucleophiles can occur, as well as the rearrangement to the stable *N*-acylurea. The *O*-acylurea can also be displaced by external nucleophiles. Pedemonte and Kaplan (1986) have shown that in some conditions inactivation of Na/K-ATPase is caused by cross-linking. By contrast, inactivation of cation occlusion in both control and 19-kD membranes appears to be caused by direct modification of carboxyls (see Shani-Sekler et al. [1988] for a discussion of possible experimental differences). Inactivation of control enzyme shows the following characteristics (Shani-Sekler et al., 1988): (*a*) The rates of inactivation by DCCD of Rb and Na occlusion are identical within error, and in addition Rb ions protect against DCCD with high affinity. (*b*) As a function of DCCD concentration, inactivation of Rb and Na occlusion are also indistinguishable, and show a first-order dependence on DCCD concentration. (*c*) Inactivation of occlusion and Na/K-ATPase activity occur in parallel. (*d*) A number of hydrophobic amines (e.g., 2-naphthylamine) accelerate the rate of inactivation of occlusion. This phenomenon is possible for a carboxyl modification but is not expected if inactivation is due to cross-linking.

Based on the inactivation kinetics, we proposed that one or a small number of carboxyl residues in a hydrophobic environment are directly involved, and further that K and Na ions bind to these same residues. An alternative hypothesis, that K and Na bind to different sites, was less likely because this would imply an improbable series of coincidences to explain the identical kinetics.

The direct test of the hypothesis, however, is to demonstrate so-called "stoichiometric inactivation." The enzyme was incubated with [^{14}C]DCCD at different concentrations with and without Rb ions, and Rb-protected incorporation and occlusion were measured. There is indeed proportionality, extrapolating to about two molecules of DCCD per pump for full inactivation (Goldshleger et al., 1990). However, the accuracy of this type of experiment is not completely satisfactory due to a high nonspecific incorporation. This is also a disadvantage for localizing the labeled carboxyls in the α chain on which all specific labeling occurs.

The 19-kD membranes have a distinct advantage in this respect due to removal of much of the irrelevant protein and nonspecific labeling. In these membranes one can demonstrate a strict linear correlation of Rb-protected incorporation and inactivation of occlusion, extrapolating to exactly one DCCD per Rb site (Karlish, S. J. D., R. Goldshleger, D. Tal, S. Hoving, and W. D. Stein, manuscript in preparation).

For localization of the DCCD, 19-kD membranes were labeled with or without Rb, dissolved in SDS, and run on 10 or 16% gels. The 19-kD fragment is specifically labeled (i.e., Rb protected), as is a second smaller peptides(s) of ~9kD. The β chain or its fragments are not labeled. Above the position of the 19-kD fragment there is Rb-protected labeling throughout the gel. We have some indication that this is due to cross-linking of the 19-kD fragment, but is not the cause of inactivation (Karlish, S. J. D., R. Goldshleger, D. Tal, S. Hoving, and W. D. Stein, manuscript in preparation).

We have asked the question whether the smaller labeled peptide is a fragment of the 19-kD peptide. There are two reasons for thinking that this is not the case. First, we already know that tryptic fragmentation of the 19-kD membrane leads to loss of occlusion and therefore probably also Rb protection against DCCD. Second, by partially purifying the labeled 19-kD fragment on a TSK-3000 column a direct estimate of stoichiometry could be made. This was close to one DCCD per 19 kD, a value consistent with direct participation in Rb occlusion, and of course suggests the necessity of a second fragment. Unambiguous identification of the second fragment remains to be established.

TABLE IV
Structural Features of the Cation Sites

1. **Two carboxyl groups are involved.** This is based on labeling. In addition, observations of effects of electrical diffusion potentials on reconstituted Na/K pumps suggest that the cation transporting domain contains two negatively charged groups (Goldshleger et al., 1987); i.e., two quite different techniques point to the same conclusion.
2. **Two K and two Na ions combine with the same two carboxyl-containing sites. The third Na combines with neutral ligating groups.** This is based on inactivation kinetics.
3. **Therefore, the same transport path must serve Na and K ions, sequentially.**
4. **Cation sites are located on the α chain within *trans*-membrane segments.** This is suggested by DCCD labeling and the proteolysis data.
5. **One carboxyl is located in the COOH-terminal 19-kD fragment; another is located in a ~9-kD fragment.** Labeling.
6. **It is likely that the cation occluding structure consists of a complex of several *trans*-membrane segments.** A minimum of four α helices are required to create a cage for a K ion.
7. **Two cation sites are ~4°A distant.** This has been concluded on the basis of inhibition experiments with a series of diamines (Schuurmans-Stekhoven et al., 1988).
8. **Cation and ATP sites are separate; thus their interactions are indirect.** Proteolysis experiments.

Table IV summarizes the features of the emerging model. Clearly the challenge now is to identify the precise carboxyls and *trans*-membrane segments involved, and build a model of the occlusion cage.

Comparison with Information on Cation Binding Sites in Other Pumps

A possible approach to identification of cation binding domains is chemical modification with reactive derivatives of high affinity cation analogues. The best candidate for such a compound has been described for the gastric H/K-ATPase. This is the class of substituted pyridyl-1,2a-imidazoles (for example, SCH 28080), which act as high affinity K-competitive inhibitors at the extracellular surface (Wallmark et al., 1987). Photoreactive methylated derivatives of SCH 28080 inactivate the gastric H/K-ATPase in a K-competitive fashion with covalent incorporation (Munson and Sachs, 1988; Keeling et al., 1989) and are not released from the membrane after extensive tryptic digestion. Thus the site is close to or within the *trans*-membrane segments. Analogous high affinity inhibitors of Na/K-ATPase, suitable for labeling studies, have not yet been described. SCH 28080 itself is highly selective for H/K-ATPase.

Site-directed mutagenesis of sarcoplasmic reticulum Ca-ATPase, expressed in COS cells, has now been performed quite extensively. Of interest in the present context are mutations in residues considered to lie either within *trans*-membrane segments (Clarke et al., 1989*b*) or in the first cytoplasmic loop, the so-called "stalk" (Clarke et al., 1989*a*). Those in the "stalk" did not affect active Ca transport; i.e., these residues do not bind Ca. But several mutations in putative *trans*-membrane segments prevented active Ca transport and especially the ability of Ca to inhibit phosphorylation by P_i, but not that of P_i to phosphorylate. These residues—glu 309, glu 771, asn 796, thr 799, asp 800, and glu 908—were suggested to bind Ca ions within the putative M4, M5, M6, and M8 segments. Generally, our findings appear to be compatible with the model of a binding cage within the membrane segments from M4 toward the COOH terminus. But there are ambiguities in these assignments. For example, direct and indirect effects of the mutations on the sites might be confused because Ca binding could not be measured (the results could be explained alternatively if the protein were locked in an E_2 state). Also, the conclusion that four carboxyls are involved in transport of two Ca ions, although plausible, is not easily reconciled with observations on transient charge movements, implying that the sites contain less than four carboxyls (see Glynn and Karlish, 1990). In fact, glu 908 was recently mutated to a glutamine but Ca transport was not inhibited (Clarke et al., 1990). Another source of ambiguity is the location of the residues. The four residues analogous to glu 771, asn 796, thr 799, and asp 800 in Na/K-ATPase have been predicted to actually lie outside the membrane, flanking the M5 segment (Kawakami et al., 1985; Shull et al., 1985; Ovchinnikov et al., 1986). Only glu 309 is predicted to lie within a *trans*-membrane segment, M4, by all the models.

Important lessons for the future seem to be that the power of mutagenesis will be more fully realized if and when direct ion binding assays become possible in improved expression systems, such as the Baculovirus system (De Tomaso et al., 1990). On the other hand, the protein chemical approach is essential to prove directly the actual rather than predicted *trans*-membrane folding of the chains, and which segments or carboxyl residues are of interest for the cation binding, and hence provide a framework for rational mutation experiments.

Acknowledgments

This work was supported by a grant from the USPHS (GM-32286) and by the Weizmann Institute Renal Research Fund.

References

Clarke, D. M., K. Maruyama, T. W. Loo, E. Leberer, G. Inesi, and D. H. Maclennan. 1989*a*. Functional consequences of glutamate, aspartate, glutamine, and asparagine mutations in the stalk sector of the Ca^{2+}-ATPase of sarcoplasmic reticulum. *Journal of Biological Chemistry.* 264:11246–11251.

Clarke, D. M., T. W. Loo, and D. H. Maclennan. 1990. Functional consequences of alternations to polar amino acids located in the trans-membrane domain of the Ca^{2+}-ATPase of sarcoplasmic reticulum. *Journal of Biological Chemistry.* 265:6262-6267.

Clarke, D. M., T. W. Loo, G. Inesi, and D. H. Maclennan. 1989*b*. Location of high affinity Ca^{2+}-binding sites within the predicted transmembrane domain of the sarcoplasmic reticulum Ca^{2+}-ATPase. *Nature.* 339:476–478.

De Tomaso, T. W., P. Zdankiewicz, and R. W. Mercer. 1990. Functional expression of Na,K-ATPase in *Spondoptera frugiperda* cells. *Journal of General Physiology.* 96:19*a*. (Abstr.)

Esmann, M., and J. C. Skou. 1985. Occlusion of Na by Na,K-ATPase in the presence of oligomycin. *Biochemical and Biophysical Research Communications.* 127:857–863.

Forbush, B., III. 1987. Rapid release of ^{42}K or ^{86}Rb from two distinct transport sites on the Na,K-pump in the presence of Pi or VO_4. *Journal of Biological Chemistry.* 262:11116–11127.

Glynn, I. M., J. L. Howland, and D. E. Richards. 1985. Evidence for the ordered release of rubidium ions occluded within the Na,K-ATPase of mammalian kidney. *Journal of Physiology.* 368:453–469.

Glynn, I. M., and S. J. D. Karlish. 1990. Occluded cations in active transport. *Annual Reviews of Biochemistry.* 59:171–205.

Goldshleger, R., S. J. D. Karlish, A. Rephaeli, and W. D. Stein. 1987. The effect of membrane potential on the mammalian sodium-potassium pump reconstituted into phospholipid vesicles. *Journal of Physiology.* 387:331–355.

Goldshleger, R., S. J. D. Karlish, W. D. Stein, and D. M. Tal. 1990. Rubidium occlusion occurs in membrane-embedded tryptic fragments of renal Na,K-ATPase. *Biophysical Journal.* 57:17*a*. (Abstr.)

Jorgensen, P. L. 1975. Purification and Characterisation of Na,K-ATPase. V. Conformational changes in the enzyme: transitions between the Na form and K form studied with tryptic digestion as a tool. *Biochimica et Biophysica Acta.* 401:399–415.

Jorgensen, P. L., and J. P. Andersen. 1988. Structural basis for E_1-E_2 conformational transition in Na,K-ATPase and Ca-ATPase. *Journal of Membrane Biology.* 103:95–120.

Jorgensen, P. L., and J. H. Collins. 1986. Tryptic and chymotryptic cleavage sites in the sequence of the α-chain of Na,K-ATPase from outer medulla of mammalian kidney. *Biochimica et Biophysica Acta.* 860:570–576.

Karlish, S. J. D., R. Goldshleger, and W. D. Stein. 1990. A 19kDa C-terminal tryptic fragment of the α-chain of Na,K-ATPase is essential for occlusion and transport of cations. *Proceedings of the National Academy of Sciences, USA.* 87:4566–4570.

Karlish, S. J. D., and W. D. Stein. 1982. Passive rubidium fluxes mediated by Na,K-ATPase reconstituted into phospholipid vesicles when ATP- and phosphate-free. *Journal of Physiology.* 328:295–316.

Kawakami, K., S. Naguchi, M. Noda, H. Takahashi, T. Ohta, M. Kawamura, H. Nojima, K. Nagano, T. Hitose, S. Inayama, et al. 1985. Primary structure of the α-subunit of *Torpedo californica* Na,K-ATPase. *Nature.* 316:733–736.

Keeling, D. J., C. Fallowfield, K. M. W. Lawrie, D. Saunders, S. Richardson, and R. J. Ife. 1989. Photoaffinity labeling of the lumenal K site of the gastric H/K-ATPase. *Journal of Biological Chemistry.* 264:5552–5558.

Klemens, M. R., J. M. MacD. Stewart, J. E. Mahaney, T. A. Kuntzweiller, M. C. Sattler, and C. M. Grisham. 1988. NMR and ESR studies of active site structures and intermediate states of kidney Na,K-ATPase and SR Ca-ATPase. *In* Advances in Biotechnology of Membrane Ion Transport. P. L. Jorgensen and R. Verna, editors. Raven Press, New York. 107–124.

Kramer, T., R. Antolovic, H. J. Bruller, U. Richter, G. Scheiner-Bobis, and W. Schoner. 1990. Function and orientation of the COOH-terminal domain of the α-subunit. *Journal of General Physiology.* 96:32*a*. (Abstr.)

Lingrel, J. B., J. Orlowski, M. M. Shull, and E. M. Price. 1990. Molecular genetics of Na,K-ATPase. *Progress in Nucleic Acid Research and Molecular Biology.* 38:37–89.

Maclennan, D. H., C. J. Brandl, B. Korczak, and N. M. Green. 1985. Aminoacid sequence of a Ca-ATPase from rabbit muscle sarcoplasmic reticulum deduced from its complementary DNA sequence. *Nature.* 316:696–700.

Matsudaira, P. 1987. Sequence from picomole quantities of proteins electroblotted onto polyvinylidene difluoride membranes. *Journal of Biological Chemistry.* 262:10035–10038.

Munson, K. B., and G. Sachs. 1988. Inactivation of H/K-ATPase by a K-competitive photoaffinity inhibitor. *Biochemistry.* 27:3932–3938.

Ovchinnikov, Y. A. 1987. Probing the folding of membrane proteins. *Trends in Biochemical Science.* 12:434–438.

Ovchinnikov, Y. A., N. M. Luneva, E. A. Arystarkhova, N. M. Gevondyan, N. M. Arzamazova, A. T. Kozhich, V. A. Nesmeyanov, and N. N. Modyanov. 1988. Topology of Na,K-ATPase: identification of the extra- and intracellular hydrophilic loops of the catalytic subunit by specific antibodies. *FEBS Letters.* 227:230–234.

Ovchinnikov, Y. A., N. N. Modyanov, N. E. Broude, K. E. Petrukhin, A. V. Grishin, N. M. Arzamazova, N. A. Aldanova, G. S. Monastyrskaya, and E. D. Sverdlov. 1986. Pig kidney Na, K-ATPase primary structure and spatial organization. *FEBS Letters.* 210:237–245.

Pedemonte, C. H., and J. H. Kaplan. 1986. Carbodiimide Inactivation of Na,K-ATPase: a consequence of internal cross-linking and not carboxyl modification. *Journal of Biological Chemistry.* 261:3632–3639.

Schagger, H., G. Von Jagow. 1987. Tricine-sodium dodecyl sulfate-polyacrylamide gel electrophoresis for separation of proteins in the range from 1–100 kDa. *Analytical Biochemistry.* 166:368–379.

Schuurmans-Stekhoven, F. M. A. H., Y. S. Zou, H. G. P. Swarts, J. Leunissen, and J. J. H. H. M. De Pont. 1988. Phosphorylation of Na,K-ATPase; stimulation and inhibition by substituted and unsubstituted amines. *Biochimica et Biophysica Acta.* 937:161–176.

Shani, M., R. Goldshleger, and S. J. D. Karlish. 1987. Rb occlusion in renal (Na,K)ATPase characterized with a simple manual assay. *Biochimica et Biophysica Acta.* 904:13–21.

Shani-Sekler, M., R. Goldshleger, D. Tal, and S. J. D. Karlish. 1988. Inactivation of Rb^+ and Na^+ occlusion on Na,K-ATPase by modification of carboxyl groups. *Journal of Biological Chemistry.* 263:19331–19342.

Shull, G. E., A. Schwartz, and J. B. Lingrel. 1985. Aminoacid sequence of the catalytic subunit of Na,K-ATPase. *Nature.* 316:691–698.

Taylor, W. R., and N. M. Green. 1989. The predicted secondary structure of the nucleotide-binding sites of six cation-transporting ATPase lead to a probable tertiary fold. *European Journal of Biochemistry.* 179:241–248.

Wallmark, B., C. Briving, J. Fryklund, K. B. Munson, R. Jackson, J. Mendlein, E. Rabon, and G. Sachs. 1987. Inhibition of gastric H/K-ATPase and acid secretion by SCH 28080, a substituted pyridy (1,2a) imidazole. *Journal of Biological Chemistry.* 262:2077–2084.

Chapter 11

Structural Features of Intermediate States of Sheep Kidney Na,K-ATPase

John M. McD. Stewart, Theresa A. Kuntzweiler, Cindy Klevickis, and Charles M. Grisham

Department of Chemistry, University of Virginia, Charlottesville, Virginia 22901

Introduction

In the past five years, elucidation of the structure of the Na,K-ATPase has advanced rapidly on several fronts. In addition to the determination of primary sequences of various isoforms and mutant forms of the α- and β-subunits (MacLennan et al., 1985; Schull et al., 1985), there are also several reports of electron microscopic studies of two-dimensional crystals of the ATPase. These studies, together with the wealth of information available from chemical modification studies of the ATPase active site (Jørgensen and Andersen, 1988), provide a foundation on which future structural studies can be based. Magnetic resonance studies in our own laboratory have provided information on the conformation and arrangement of substrates and divalent and monovalent cations at the ATPase active site (Klemens et al., 1986; Grisham, 1988; Stewart et al., 1989).

In spite of these advances, little has been learned until now about the conformational changes that are presumed to modulate the catalytic and transport activities of the sodium pump. These conformation changes may involve structural changes in the ATPase protein, structural changes in the bound substrates and cations, or both. Manifestations of these conformation changes are evident in the ligand binding and kinetic behavior of the ATPases. For example, ATP and other nucleotides and analogues display biphasic binding and kinetic properties under most conditions. Thus, the K_m for phosphorylation and for ATPase activity is ~1 μM in the E_1 conformation (Jensen and Ottolenghi, 1983), whereas binding to the E_2 form requires millimolar concentrations of ATP for saturation.

Detection and characterization of these conformation changes will require a physical probe technique that is sensitive to these changes in structure, either in the protein or in its ligands, as well as a means for stabilizing the respective conformations for the times required by the probe technique. When we began in 1979 to use the substitution-inert complexes of ATP with Cr(III) and Co(III) in kinetic and spectroscopic studies of the plasma membrane Na,K-ATPase and sarcoplasmic reticulum Ca-ATPase, we were seeking merely to control the distribution and binding of divalent cations to these enzymes. We had shown (Grisham, 1979*a*) that Mg^{2+} and other divalent ions could either bind directly to the enzyme or to the ATP substrate. $Cr(H_2O)_4ATP$ and $Co(NH_3)_4ATP$ provided us with paramagnetic and diamagnetic substrate analogues, respectively, in which the divalent metal binding site on the nucleotide is already occupied. $Cr(H_2O)_4ATP$ can be used directly as a paramagnetic probe of active site geometry. On the other hand, the diamagnetic probes, $Co(NH_3)_4ATP$ and more recently $Rh(H_2O)_4ATP$, permit us to direct paramagnetic cation probes, such as Mn^{2+} and Gd^{3+}, to sites on the enzyme itself, even in the presence of nucleotide. Under conditions in which the metal–nucleotide probe is stable (0–4°C, pH 6–7.5), it has been possible to conduct a variety of kinetic and spectroscopic studies of active site structure with these enzymes (Gantzer et al., 1982; Grisham, 1988; Klemens and Grisham, 1988; Stewart and Grisham, 1988; Devlin and Grisham, 1990). In the course of these studies, however, we discovered yet another use for these metal–nucleotide analogues. As first reported by Pauls et al. (1980), under certain conditions, these complexes can slowly inactivate the transport ATPases via formation of a covalent E-P(metal) intermediate, consisting of a 1:1:1 complex of the ATPase, an aspartyl phosphate, and the trivalent metal ion. These complexes represent highly stable forms of the E-P intermediate formed by these ATPases. The notion that these stable, inhibited complexes represent true intermediate states in the catalytic cycles of the ATPases is supported by the sensitivity of the inhibition kinetics to nucleotides and cations (Mg^{2+}, Na^{+}, K^{+}, and Ca^{2+}) and by the observation that these inhibited intermediates strongly occlude Rb^{+} (in the case of Na,K-ATPase [Vilsen et al., 1987]) or Ca^{2+} (in the case of Ca-ATPase [Vilsen and Andersen, 1987]).

We have recently demonstrated (Stewart and Grisham, 1988; Stewart et al., 1989) that nuclear relaxation measurements of several different types can be used to determine the conformations of these metal–nucleotide complexes at the active sites of Na,K-ATPase and Ca-ATPase. The ability to use these complexes to form stable E-P derivatives raises the possibility of studying the binding of metal–nucleotide analogues to E-P as well. Numerous observations of ATP and ADP binding to E_1-P and E_2-P have been reported. In addition to the well-known ATP-ADP exchange reaction, which requires ADP binding to E_1-P, Plesner and Plesner (1988) observed that ADP diminishes substrate inhibition of ATP hydrolysis by Na,K-ATPase, an effect interpreted in terms of ATP binding to E_1-P. In support of this interpretation, these authors also found that free ATP competitively inhibits ATP-ADP exchange. An earlier report (Schuurmans-Stekhoven et al., 1983) describes the rapid breakdown of E-P formed from P_i and Mg^{2+} upon addition of AMPPNP to Na,K-ATPase. Karlish and Stein (1982) observed simultaneous but distinct effects of ATP and phosphate on passive rubidium fluxes catalyzed by Na,K-ATPase reconstituted in phospholipid vesicles. Similarly, we have observed simultaneous and distinct protective effects of ATP and phosphate on the inhibition of Na,K-ATPase by 2,3-butanedione, a reagent specific for arginine residues (Grisham, 1979*b*). On the basis of computer modeling of kinetic studies with the Ca-ATPase, Cable et al. (1985) concluded that ATPase activity can be stimulated by binding of ATP or AMPPCP to the phosphorylated form of the active site after departure of ADP but before the departure of inorganic phosphate. The same authors also report that ATP can bind effectively to E-P, although with an affinity fivefold less than that to the unphosphorylated enzyme.

In this paper we describe NMR spectroscopic studies comparing the conformation of M(III)ATP bound to the unphosphorylated and phosphorylated forms of Na,K-ATPase. The results are consistent with significant conformation changes for the bound nucleotide on the two forms of the ATPase.

Experimental Procedures

Materials and Enzymes

The β,γ-bidentate complex of $Co(NH_3)_4ATP$ was prepared as described by Cornelius et al. (1977). $Cr(H_2O)_4ATP$ was prepared according to Dunaway-Mariano and Cleland (1980), and $Rh(H_2O)_4ATP$ was prepared according to Lin et al. (1984). Deuterium oxide (99.8 atom %) was from the Aldrich Chemical Co. (Milwaukee, WI), Tris-d_{11} was from MSD Isotopes (St. Louis, MO), and manganese chloride was from J. T. Baker Chemical Co. (Phillipsburg, NJ). Adenosine 5′-triphosphate was from Sigma Chemical Co. (St. Louis, MO). All other reagents were of the highest purity available commercially. The Na,K-ATPase was purified from the outer medulla of sheep kidneys as previously described (O'Connor and Grisham, 1979). Ca-ATPase was prepared from rabbit muscle according to Meissner et al. (1973).

The preparation of the ATPases and metal-ATP analogues for NMR studies has been previously described (Stewart and Grisham, 1988).

Nuclear Magnetic Resonance Studies

Proton NMR spectra were obtained at three frequencies: at 500 MHz on an Omega 500 MHz NMR spectrometer (General Electric Co., Milwaukee, WI), at 361 MHz on an NT-360/Oxford spectrometer equipped with a 1200/293B data system (Nicolet Magnetics Corp., Fremont, CA), and at 300 MHz on a QN-300 spectrometer (General Electric Co.). The spectra were measured at 4°C with a 5-mm proton probe and deuterium lock. The 90° pulse width was typically 7.50 μs at 361 MHz and 10.0 μs at 300 MHz. The HDO signal was decoupled during delays before the pulse sequences. Phosphorus NMR spectra were obtained at the same frequencies, using a 10-mm phosphorus probe and a deuterium lock.

Longitudinal relaxation times were determined from an optimized inversion–recovery experiment and 16–20 τ values were used in the 180-τ-90 sequence. T_1 values and their standard deviations were calculated using a three-parameter fit to the data. This fitting routine corrects for inhomogeneities in the H_1 field which produce incomplete inversion during the 180° pulse. Each point in the titration with Mn^{2+} was measured at both 300 and 361 MHz, and 16 accumulations of 8,000 data points were collected per τ value.

Paramagnetic contributions to the longitudinal relaxation rates of the protons of $Co(NH_3)_4ATP$ were calculated using the program NONLIN (copyright M. L. Johnson, Department of Pharmacology, University of Virginia), an interactive nonlinear function minimization program that performs a weighted least-squares fit of the data points to a specified function (in this case a straight line) by a modified Gauss-Newton iteration. The theory for the use of nuclear relaxation measurements in the calculation of enzyme active site distance and geometric information has been described in detail elsewhere (Stewart and Grisham, 1988). In the limit of fast exchange, when the rate of chemical exchange of the ligand between the coordination sphere and the bulk solvent is faster than the rate of relaxation in the first coordination sphere, and assuming outer sphere contributions are negligible, the Solomon-Bloembergen equation that describes the dipolar Mn^{2+}-proton interaction is

$$r^6 = (815)^6 \{T_{1M} [f(\tau_c)\} \quad (1)$$

where r is the Mn^{2+}-proton internuclear distance, T_{1M} is the relaxation time in the vicinity of the paramagnetic probe, and $f(\tau_c)$ is the correlation function as previously described (Stewart and Grisham, 1988). A similar equation applies for relaxation of phosphorus nuclei by Mn^{2+}, but in this case the proportionality constant is 601 instead of 815.

Conformational analysis was done on a Silicon Graphics (Mountain View, CA) computer using the program MMS (copyright UCSD). The basic structure of $Co(NH_3)_4ATP$ was constructed from the x-ray data of rubidium adenosine 5′-diphosphate monohydrate and $Co(NH_3)_4H_2P_3O_{10}{\cdot}H_2O$ (Stewart and Grisham, 1988). The conformation of the enzyme-bound CoATP was then determined using the distances calculated from the NMR data.

Results and Discussion

Our approach to the problem of studying the conformation of ATP analogues bound to E and E-P depends on the following observations: (*a*) at 0–4°C, $Co(NH_3)_4ATP$, $Cr(H_2O)_4ATP$, and $Rh(H_2O)_4ATP$ do not react with the ATPases, but can form stable Michaelis-type complexes for the several hours needed for spectroscopic studies of E·(MATP), the complex of the metal–ATP analogue with the unphosphorylated ATPase; and (*b*) at 25–37°C, $Cr(H_2O)_4ATP$ and $Rh(H_2O)_4ATP$ react slowly with the ATPases to form E-P intermediates that retain the trivalent metal ion (Cr(III) or Rh(III)). In this latter case, the E-P(M) intermediate can be isolated and then incubated with fresh metal–nucleotide complex to form E-P(M)·(MATP), a Michaelis-type complex of the metal–nucleotide analogue and E-P(M):

$$\text{E} + \text{MATP} \rightarrow \text{E-P(M)} + \text{ADP}$$

$$\downarrow$$

$$\text{Centrifuge and resuspend to isolate} \rightarrow \text{E-P(M)} + \text{MATP} \rightarrow \text{E-P(M)·(MATP)}$$

This work will compare the conformation of MATP in the E·(MATP) and E-P(M)·(MATP) complexes of Na,K-ATPase. (Note: although no attempt is made here to distinguish E_1 and E_2 forms of Na,K-ATPase, it is probable that the relatively high concentrations of ATP and ATP analogues used in the NMR studies reported here have forced the enzyme into the E_1 state.)

Kinetic Studies with MATP Complexes

The ability of complexes of Cr(III), Co(III), and Rh(III) with ATP to replace MgATP in the various partial reactions of the ATPases can be demonstrated in two types of kinetic experiments: steady-state inhibition and slow inactivation. As reported previously (Gantzer et al., 1982) and as shown in Fig. 1, these complexes are effective linear competitive inhibitors with respect to MgATP and MnATP. At low MnATP, where the high affinity form of the ATP site predominates, a K_i for CoATP of 10 μM was obtained (Fig. 1*A*). From the secondary plot of the *y* intercepts, an apparent K_m for MnATP of 2.88 μM was obtained under the same experimental conditions. At high levels of MnATP, where the low affinity form of the ATP site predominates, the K_i for CoATP was found to be 1.6 mM (Fig. 1*B*) while the apparent K_m for MnATP was 0.902 mM. Similar results are obtained with $Rh(H_2O)_4ATP$ (Serpersu et al., 1990) and $Cr(H_2O)_4ATP$ (Pauls et al., 1980; Gantzer et al., 1982; Serpersu et al., 1982, 1990).

The slow inactivation of Na,K-ATPase and Ca-ATPase by $Cr(H_2O)_4ATP$ was reported first by Schoner and co-workers (Pauls et al., 1980; Serpersu et al., 1982). The inhibition in each case is accompanied by phosphorylation of the ATPase to form a stable, inhibited complex of the ATPase containing one phosphorus and one Cr(III) per phosphorylation site. Similar results are obtained with both ATPases using $Rh(H_2O)_4ATP$, as shown in Figs. 2 (inactivation) and 3 (phosphorylation). Rhodium analysis of the E-P intermediates by atomic absorption spectrophotometry yields one rhodium per phosphorylation site, in agreement with the results with $Cr(H_2O)_4ATP$. Moreover, the inhibition constants measured in these slow inactivation studies agree well with the binding constants obtained from the steady-state inhibition studies described above (Kuntzweiler, 1990; Serpersu et al., 1990).

P-31 and H-1 NMR Studies of E·(MATP)

There are several suitable NMR experiments that can provide active site conformation information, but each of these requires a judicious choice of the MATP probe to be used. The key is to select a MATP probe that binds tightly enough to the enzyme to form an observable complex, and yet is still in fast exchange. In practical terms, this normally means that the K_D for MATP from E·(MATP) is in the range 0.1–2 mM. In these studies it was found that $Rh(H_2O)_4ATP$ binds too tightly to Na,K-ATPase, so that the fast exchange condition is not satisfied for this complex. However, $Co(NH_3)_4ATP$ binds more weakly than $Rh(H_2O)_4ATP$ and does form a

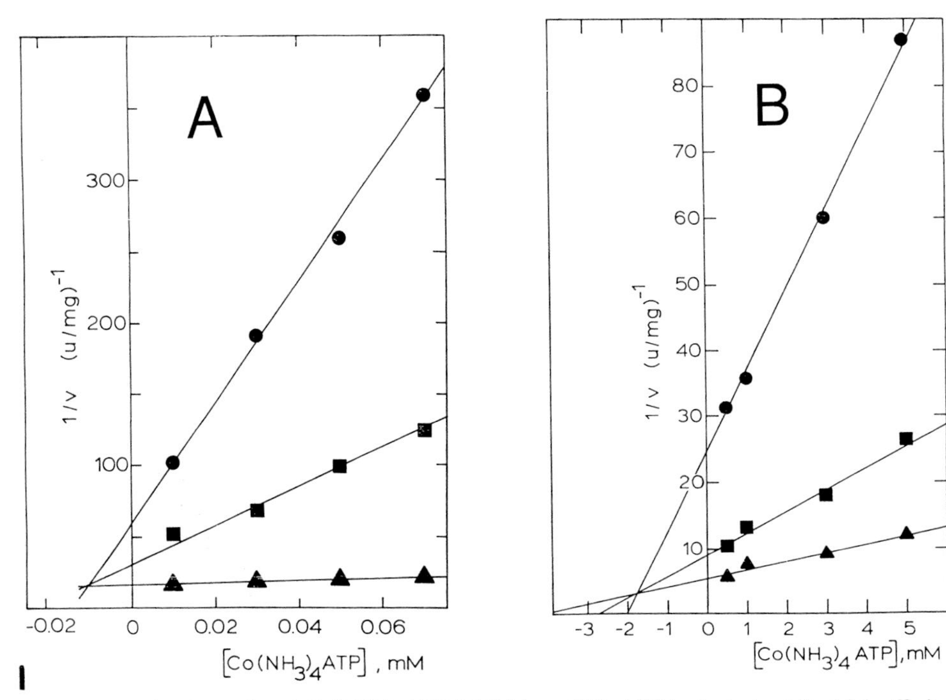

Figure 1. (*A*) Dixon plot of $Co(NH_3)_4ATP$ inhibition of MnATP hydrolysis at the high affinity substrate site of Na,K-ATPase. The solutions contained 20 mM TES-TMA, pH 7.5, 100 mM NaCl, and 10 mM KCl. The ATP concentrations were 1 μM (●), 3 μM (■), and 30 μM (▲). $T = 25°C$. (*B*) Dixon plot of $Co(NH_3)_4ATP$ inhibition of MnATP hydrolysis at the low affinity substrate site of Na,K-ATPase. Conditions as in *A*. The ATP concentrations were 0.2 mM (●), 1.0 mM (■), and 3.0 mM (▲).

fast-exchanging complex with the ATPase. In this latter case, P-31 and H-1 nuclear relaxation rates were measured in solutions of Mn^{2+} and the ATPase. As shown in Fig. 4, the presence of the ATPase causes a 6- to 10-fold enhancement of the effect of Mn^{2+} on the P-31 relaxation rates. The increases in $1/T_1$ relaxation rate for the three phosphorus nuclei of the substrate analogue are not observed in the absence of enzyme and are not observed when diamagnetic Mg^{2+} is added to the enzyme in the place of Mn^{2+}. The dissociation constant for Mn^{2+} ion from the Na,K-ATPase, determined by Mn^{2+} EPR (O'Connor and Grisham, 1979) is 0.2 μM. Using a dissociation constant for the binary Mn^{2+}-$Co(NH_3)_4ATP$ complex of 15 mM, it is

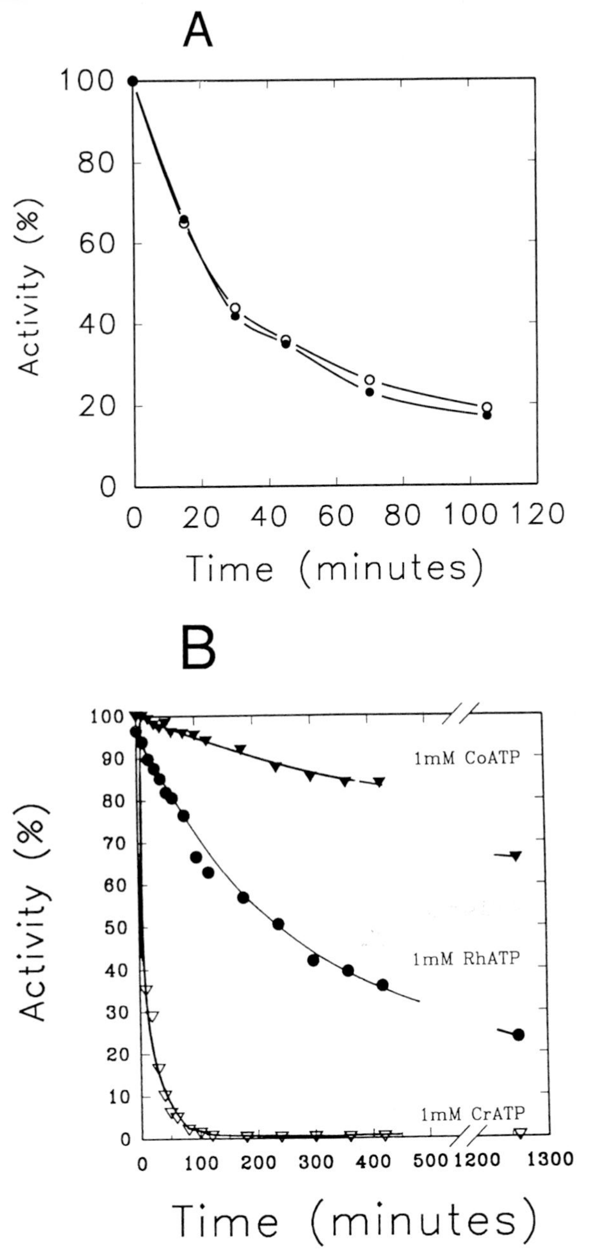

Figure 2. (*A*) Slow inactivation of Na,K-ATPase by $Rh(H_2O)_4$ ATP. Solutions contained 20 mM imidazole, pH 7.3, 10 mM NaCl, 3 mM $MgCl_2$, 0.6 mg/ml Na,K-ATPase, and 1 mM $Rh(H_2O)_4$ATP. The closed and open circles indicate two separate experiments. T = 37°C. (*B*) Slow inactivation of Ca-ATPase by MATP analogues. Solutions contained 30 mM PIPES, pH 6.8, 0.5 mM $CaCl_2$, 80 mM $MgCl_2$, 20 μM Ca-ATPase, and 1 mM MATP. T = 25°C.

easy to show that, under the conditions of these experiments, at least 97% of the added Mn^{2+} was enzyme bound. Thus, no corrections must be made for the small contribution of the binary complex to the measured paramagnetic relaxation rates (T_{1p}^{-1}). The inescapable conclusion from these data is that Mn^{2+} and $Co(NH_3)_4$ATP are bound simultaneously in a ternary complex with the Na,K-ATPase. The observed increase in $1/T_1$ could not occur in any other way. Under suitable conditions, as described previously (Klevickis and Grisham, 1982), these paramagnetic effects on

$1/T_1$ can be used to calculate the distance between the phosphorus nuclei of $Co(NH_3)_4ATP$ at the active site and enzyme-bound Mn^{2+}. The results of these calculations are shown in Table I.

As shown in Table I, addition of Mn^{2+} to solutions of ATPase and $Co(NH_3)_4ATP$ also resulted in linear increases in $1/T_1$ of all six observable proton resonances of the nucleotide analogue. The relative magnitudes of these slopes reflect the degree of

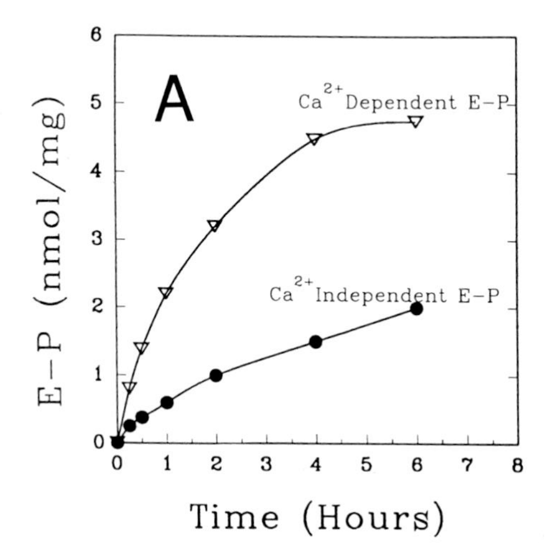

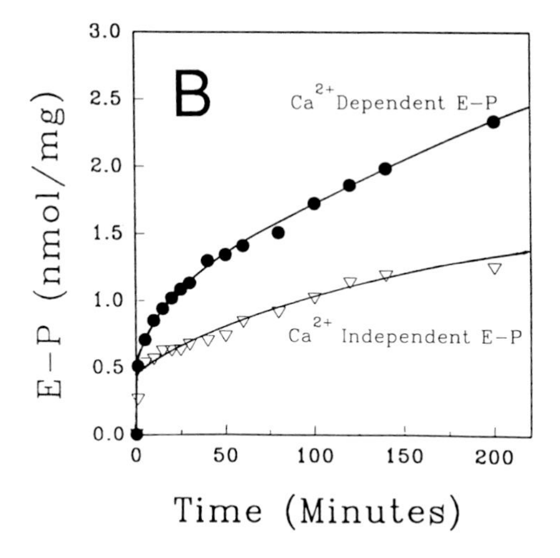

Figure 3. (*A*) Phosphorylation of Ca-ATPase by γ-^{32}P-labeled $Cr(H_2O)_4$-ATP. Solutions contained 20 mM TES, pH 7, 123 μM $CaCl_2$ or 20 mM EDTA, 10 mM $MgCl_2$, 5 μM Ca-ATPase, and 1 mM $Cr(H_2O)_4ATP$. T = 25°C. (*B*) Phosphorylation of Ca-ATPase by γ-^{32}P-labeled $Rh(H_2O)_4ATP$. Solutions contained 30 mM PIPES, pH 6.8, 1 mM $CaCl_2$ or 1 mM EGTA, 20 μM Ca-ATPase, and 0.5 mM $Rh(H_2O)_4ATP$. T = 25°C.

interaction of each of the protons of $Co(NH_3)_4ATP$ with bound Mn^{2+}. Thus, it is clear from the data in Table I that the H_8 proton of the nucleotide experiences the greatest interaction with the Mn^{2+} at its site on the enzyme. In order that these paramagnetic probe experiments might be used to determine internuclear distances at the active site, the assumption of fast exchange must be quantitatively justified. This has been done in the present case by (*a*) measurement of the paramagnetic contribution to the transverse relaxation rates of the nuclei of $Co(NH_3)_4ATP$

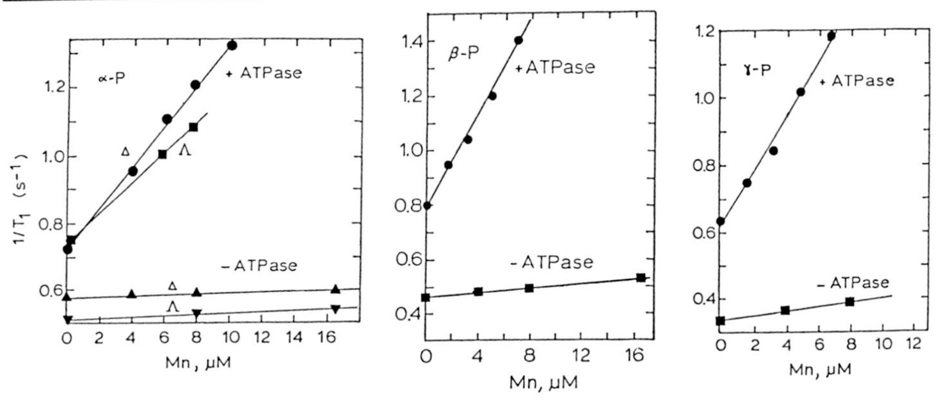

Figure 4. Effect of Mn^{2+} on the longitudinal relaxation rates of the α-, β, and γ-phosphorus nuclei of $Co(NH_3)_4ATP$ in the presence and absence of Na,K-ATPase. The solution contained 20 mM $Co(NH_3)_4ATP$, 82 mM Tris, pH 7.4, 100 mM NaCl, 10 mM KCl, and 10 μM ATPase.

(Klevickis and Grisham, 1982) and (*b*) measurement of the frequency dependence of the paramagnetic relaxation by Mn^{2+} (Stewart and Grisham, 1988). Except for the H_8 proton of $Co(NH_3)_4ATP$, both these analyses provide quantitative evidence to support the assumption of fast exchange. As shown by Stewart and Grisham (1988), the exchange rate for $Co(NH_3)_4ATP$ from the ternary complex under these conditions has been determined to be $1.4 \times 10^4\ s^{-1}$, and the dipolar correlation time τ_c has been calculated to be 5.5×10^{-10} s. These values may be used as described in Experimental Procedures to calculate the Mn^{2+}-H distances shown in Table I. Since the analysis of exchange rates indicates that the data for the H_8 proton are exchange limited, the Mn^{2+}-H_8 distance shown in Table I represents an upper limit.

The data of Table I are consistent with a model for $Co(NH_3)_4ATP$ bound at the active site in which the nucleotide adopts a bent or folded conformation and the Mn^{2+} is above and approximately planar with the adenine ring. This model, developed on a Silicon Graphics workstation using the distances of Table I, is shown in Fig. 5. The conformation of the $Co(NH_3)_4ATP$ has the glycosidic bond linking the

TABLE I
Comparison of $1/fT_{1p}$ and Mn-Nucleus Distances for $E(Co(NH_3)_4ATP)$ and $E\text{-}P(Rh)\cdot(Rh(H_2O)_4ATP$ Complexes

	Complex			
	$E(Co(NH_3)_4ATP$		$E\text{-}P(Rh)\cdot(Rh(H_2O)_4ATP)$	
Nucleus	$1/fT_{1p}$	r	$1/fT_{1p}$	*r*
	s^{-1}	Å	s^{-1}	Å
H_8	8,522	≤4.6	3,970	6.2
H_2	1,323	7.4	568	8.5
$H_{1'}$	708	8.2	388	9.1
$H_{3'}$	1,198	7.5	972	7.8
$H_{4'}$	623	8.4	561	8.6
$H_{5',5''}$	715	8.2	667	8.3

adenine base to the ribose sugar in the *anti*-configuration. The sequence of atoms defining the angle is $O_{4'}$-$C_{1'}$-N_9C_8. The torsional angle about the $C_{1'}$-N_9 bond that links base to sugar is denoted by χ. The term anti refers to the values of χ in the range of $0° \pm 90°$. Syn refers to angles of $180° \pm 90°$. The value of χ determined from the model built from the data of Table I is 35°. The anti-configuration is much more generally found in purine nucleotides and their complexes in the crystalline state as well as in solution. The conformation of the ribose ring is slightly N-type (in which $C_{2'}$ is exo and $C_{3'}$ is endo). In solution, purine ribosides normally show a small conformation preference for the S-type conformer (in which $C_{2'}$ is endo and $C_{3'}$ is exo) (Altona and Sundaralingam, 1973). The torsional angle δ is indicative of the puckering in the ribose and is defined by the atom sequence $C_{5'}$-$C_{4'}$-$C_{3'}$-$O_{3'}$ (Saenger, 1984). The value observed here is 100°. The orientation of $O_{5'}$ relative to the ribose ring is determined by the torsional angle γ, defined by the sequence $O_{5'}$-$C_{5'}$-$C_{4'}C_{3'}$

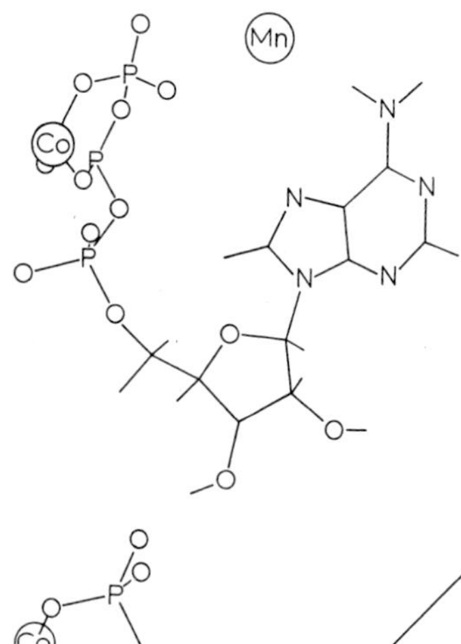

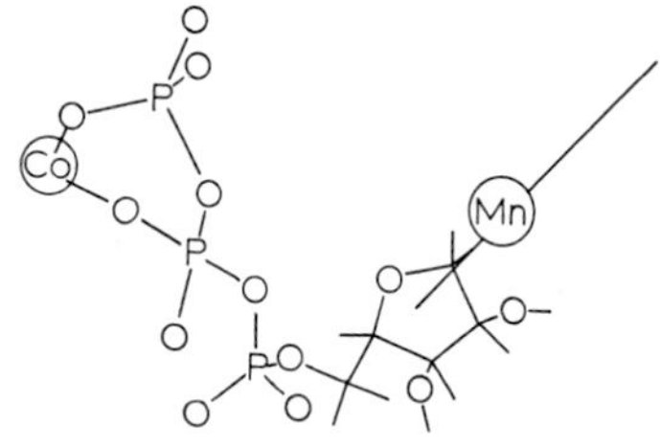

Figure 5. Conformation of $Co(NH_3)_4ATP$ and the enzyme-bound Mn^{2+} ion at the active site of kidney Na,K-ATPase. Distances were calculated from NMR relaxation data using a dipolar correlation time of 5.5×10^{-10} s.

(Saenger, 1984). The value measured in the present model is 178°, a typical value for protein-bound MgATP (Fry et al., 1985).

To corroborate these findings, we also undertook a series of transferred nuclear Overhauser effect measurements of $Co(NH_3)_4ATP$ bound to Na,K-ATPase. The initial build-up rates of the cross-peaks in two-dimensional TRNOE studies can be used to calculate interproton distances for the bound substrate analogue. The interactions measured in this way are proton–proton dipolar interactions. They are not related in any way to the paramagnetic probe interactions observed in the presence of Mn^{2+} which were described above. Thus the two-dimensional TRNOE experiment provides a totally independent means of determining the conformation of a bound nucleotide at the active site of an enzyme such as Na,K-ATPase.

H-1 and transferred nuclear Overhauser effects for the protons of $Co(NH_3)_4ATP$ were measured in solutions of Na,K-ATPase at 361 MHz. Cross-peaks were observed between H_8 and all of the ribose protons of the bound ATP analogue, and also between $H_{1'}$ and the other ribose protons. No cross-peaks were observed between the H_2 proton (on the adenine ring) and any other proton of the substrate analogue. This latter observation is by itself strong support for the anti-configuration proposed for the bound analogue from the paramagnetic probe studies. Since TRNOE cross-peaks observed after long mixing times may contain contributions from spin diffusion, initial build-up rates were measured for all the detectable cross-peaks. These initial build-up rates are inversely proportional to the sixth power of the distance between the interacting protons. Initial build-up rates were thus used to determine proton–proton distances as previously described (Stewart et al., 1989). These distances were used in a conformational analysis using the program MMS on a Silicon Graphics workstation. This analysis results in a model of $Co(NH_3)_4ATP$ bound to Na,K-ATPase which is indistinguishable from the model obtained from the paramagnetic probe measurements. The agreement between these two totally independent experiments is encouraging. This is the first time that detailed comparisons of this type could be made, even though several NMR investigations of the active site structure of Na,K-ATPase have appeared (see Stewart and Grisham, 1988, for a summary).

P-31 and H-1 NMR Studies of E-P(M)·(MATP)

As noted above, a crucial requirement of all the NMR studies described in this work is that the observed substrate analogue probe bind weakly enough to satisfy the fast exchange limitation of the spectroscopic experiment, but still bind tightly enough to form an observable complex at reasonable concentrations of the substrate analogue. These constraints mean that the dissociation constant for the ATP analogue from Na,K-ATPase under the conditions of the NMR experiment must be between 0.1 and 2 mM. In the studies of E·(MATP) described above, $Co(NH_3)_4ATP$ fell within this range, but $Rh(H_2O)_4ATP$ (the other potentially useful diamagnetic MATP complex) bound too tightly for use in the NMR studies. A different situation was found, however, for the phosphorylated forms of Na,K-ATPase. In our comparisons of unphosphorylated and phosphorylated complexes of Na,K-ATPase with the Co(III), Cr(III), and Rh(III) complexes of ATP (E·(MATP) and E-P(M)·(MATP), respectively), we have found that E-P binds MATP much more weakly than does E. (As noted above, E probably represents the E_1 state of the ATPase in the presence of the high concentrations of ATP analogue used in these spectroscopic studies.) Consequently, we found that $Co(NH_3)_4ATP$ binds too weakly to E-P(M) for use in NMR studies of E-P(M)·(MATP). $Rh(H_2O)_4ATP$, on the other hand, binds to E-P(M) (formed with $Rh(H_2O)_4ATP$) with a K_D of 0.5 ± 0.2 mM (Stewart, J., and C. Grisham, manuscript in preparation) and is thus ideally suited for NMR studies of E-P(M)·(MATP).

The preparation of E-P(M) from $Rh(H_2O)_4ATP$ and the subsequent NMR experiment are shown schematically in Fig. 6. Na,K-ATPase was incubated with $Rh(H_2O)_4ATP$ until it was 90% inactivated. The enzyme was centrifuged twice at 25,000 *g* for 30 min at 4°C and resuspended in fresh buffer to remove ADP and unreacted $Rh(H_2O)_4ATP$. Fresh $Rh(H_2O)_4ATP$ was then added to a level of 10 mM and the longitudinal relaxation rates of the P-31 and H-1 nuclei of the substrate analogue were measured as a function of added Mn^{2+} ion. As shown in Table I, substantial paramagnetic contributions to $1/T_1$ are observed for each of the phospho-

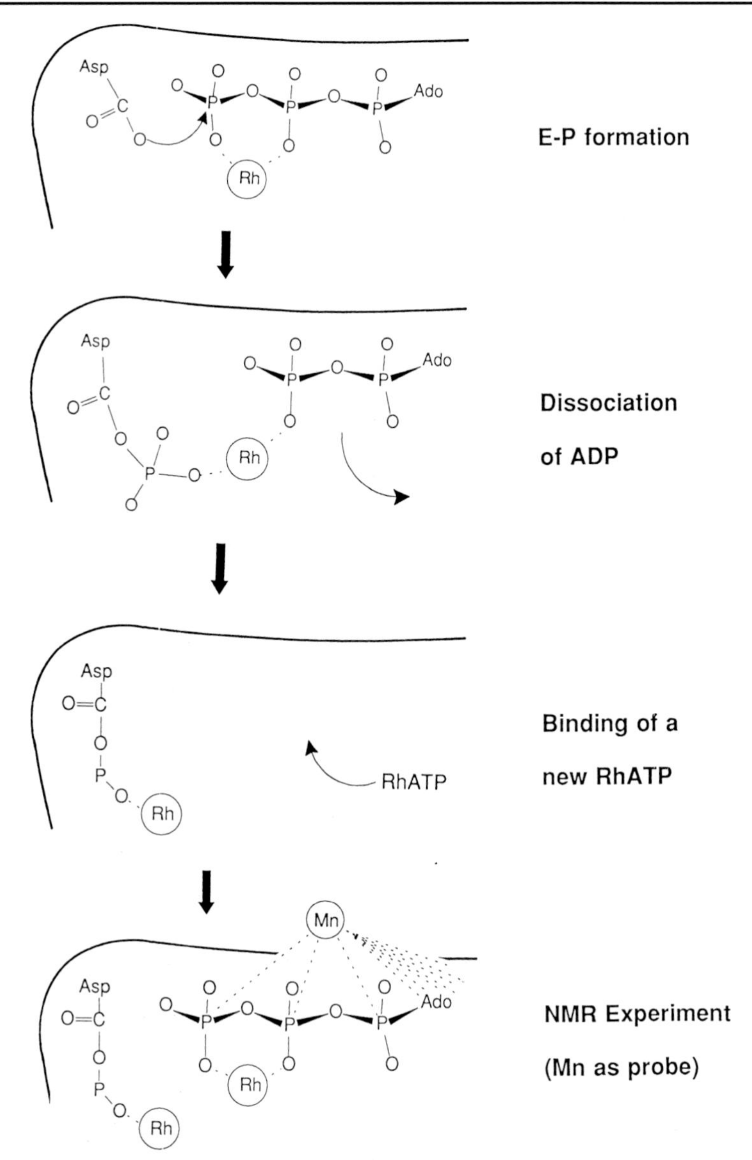

Figure 6. The protocol for the nuclear relaxation measurements of E-P(Rh)·($Rh(H_2O)_4ATP$). Incubation conditions and other details are described in the text.

rus nuclei and protons of $Rh(H_2O)_4ATP$ in the presence of E-P(M). Even more striking, however, is the observation that the observed paramagnetic effects for several of the substrate nuclei (including $H_{4'}$, $H_{5',5''}$, and to a lesser extent $H_{3'}$) are approximately the same as those observed in the unphosphorylated complex described earlier. By contrast, other nuclei (including H_8, P_β, and P_γ) display much smaller values of $1/fT_{1p}$ compared with the results with the unphosphorylated enzyme. Based on these data and the distances calculated from them, it is clear that some of the Mn^{2+}-nucleus distances change in the transition from the unphosphorylated state E·(MATP) to the phosphorylated state E-P(M)·(MATP), while other Mn^{2+}-nucleus distances remain relatively unchanged.

The distances calculated for $Rh(H_2O)_4ATP$ bound to the E-P form of the ATPase were used to model the conformation of this substrate analogue at the active site. Fig. 7 presents a comparison of the structures for MATP analogues bound to the unphosphorylated and phosphorylated states of the enzyme. Two conformational changes are apparent. The adenine ring undergoes a substantial rotation with respect to the ribose ring, and the triphosphate chain undergoes a rotation to adopt a more extended conformation in the E-P state. On the other hand, distances from Mn^{2+} to the $H_{3'}$, $H_{4'}$ and $H_{5',5''}$ protons of the ribose ring are essentially unchanged.

The measurements described in these studies shed new light on several important questions. One of these is whether the Na,K-ATPase possesses one or two ATP binding sites per "active catalytic unit." Our previous studies have clearly established the existence of one Mn^{2+} binding site per phosphorylation site. The present work

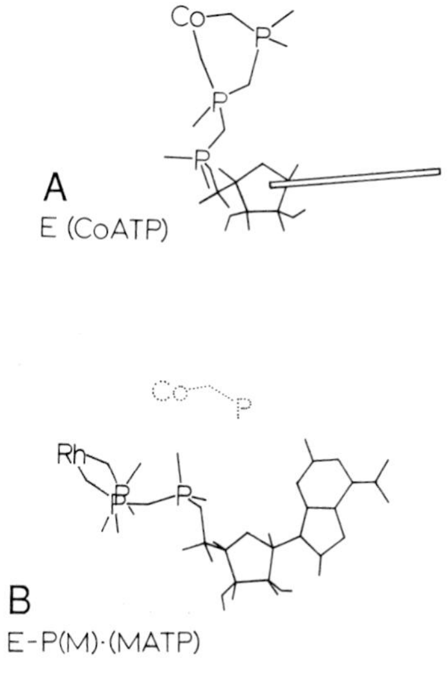

A
E (CoATP)

B
E-P(M)·(MATP)

Figure 7. A comparison of the structures determined for MATP analogues at the active site of Na,K-ATPase. The E·(CoATP) structure was obtained independently from paramagnetic relaxation and TRNOE measurements. The E-P(M)·(MATP) structure was obtained by forming E-P from $Rh(H_2O)_4ATP$, followed by isolation of the inactive intermediate and incubation with fresh $Rh(H_2O)_4ATP$ for the paramagnetic relaxation studies. In both cases the conformation was determined by fitting 10 Mn^{2+}-nucleus distances calculated from the NMR studies. For clarity, the Mn^{2+} ion is not shown, but it occupies the same position in E·(CoATP) as shown in Fig. 5. Though the structure in *B* was obtained using $Rh(H_2O)_4ATP$ to form E-P, the position of the Co(III) and γ-P from structure *A* are superimposed in dotted form in *B* for reference.

clearly shows that a second ATP molecule can bind in the vicinity of the Mn^{2+} site on the phosphorylated enzyme. The remarkably good agreement between the Mn^{2+}-H distances to the ribose ring in the unphosphorylated and phosphorylated enzymes are most easily explained by the one–ATP site model. We cannot rule out the existence of two ATP sites near the Mn^{2+} binding site, such that the Mn^{2+}-H distances to the ribose ring loci in the two sites were identical. However, the simplest model consistent with our data would require only a single ATP site.

On the other hand, if only one ATP site exists at the active site, the ability to bind a second ATP after formation of E-P requires that the covalently bound phosphate and/or the triphosphate chain of the second ATP must move somehow to accommodate each other at the single site. Two limiting cases may be imagined: (*a*) the phosphate (and metal) of the covalent E-P move to allow the second ATP to bind with the same conformation as the first, or (*b*) the phosphate (and metal) of the covalent E-P do not move, and the second ATP must adopt a different conformation at the active site. The dramatic movement of the triphosphate chain which is observed in these studies upon formation of E-P is consistent with the second limiting case. In support of this notion, the dotted letters in Fig. 7 indicate the location of the Co(III) and the phosphate of the unphosphorylated form of the enzyme. Similarly, the observation of substantially reduced affinity for nucleotides by E-P forms of the ATPases (Schuurmans-Stekhoven et al., 1983; Cable et al., 1985) is more consistent with an altered ATP conformation in the E-P form of the enzyme.

Further insights could be gained on the disposition of the phosphate and metal ion of E-P and the MATP probe in these complexes by using Cr(III) as a paramagnetic probe of the E-P complex. If E-P is formed from $Cr(H_2O)_4ATP$, leaving E-P(Cr) as a stable complex, then NMR studies of E-P(Cr)·$(Rh(H_2O)_4ATP)$ should provide information on the location of the Cr(III) probe with respect to the bound nucleotide. If, as is the case in these studies, the triphosphate chain of the second nucleotide molecule has been reoriented to accommodate the phosphate and metal of the E-P(M) complex, then it is possible that the phosphate and metal may occupy the same positions in the active site as they did during the binding and phosphorylation by the first MATP molecule. We are presently examining this question with Na,K-ATPase, but have already completed similar studies with SR Ca-ATPase. As described elsewhere (Kuntzweiler and Grisham, 1991), we have shown clearly that the Cr(III) ion of E-P(Cr) is at the same position at the active site as the Co(III) ion in the $E(Co(NH_3)_4ATP)$ complex. The implication is that the metal involved in phosphorylation (and thus perhaps also the γ-P of the ATP substrate) remains fixed at the active site during the formation of E-P. As a result, binding of the second molecule of ATP occurs with lower affinity (as noted above) and with a dramatically altered triphosphate chain orientation, as shown in this work. These observations may have important implications for the mechanisms of ATP hydrolysis and ion transport by these enzymes.

Acknowledgments

This work was supported by National Institutes of Health research grant DK-19419, a grant from the Muscular Dystrophy Association of America, and a grant from the National Science Foundation. The NMR instrumentation used in these studies was provided by grants from the National Science Foundation, University of Virginia, and the Research Corporation. C. M. Grisham is a Research Career Development Awardee of the United States Public Health Service (AM-00613).

References

Altona, C., and M. Sundaralingam. 1973. Conformational analysis of the sugar ring in nucleosides and nucleotides. *Journal of the American Chemical Society*. 95:2333–2344.

Cable, M., J. Feher, and N. Briggs. 1985. Mechanism of allosteric regulation of the Ca,Mg-ATPase of sarcoplasmic reticulum: studies with 5′-adenylyl methylenediphosphate. *Biochemistry*. 24:5612–5619.

Cornelius, R., P. Hart, and W. Cleland. 1977. Phosphorus-31 NMR studies of complexes of adenosine triphosphate, adenosine diphosphate, tripolyphosphate and pyrophosphate with cobalt(III) ammines. *Inorganic Chemistry*. 16:2799–2805.

Devlin, C., and C. Grisham. 1990. ^{1}H and ^{31}P nuclear magnetic resonance and kinetic studies of the active site structure of chloroplast CF_1 ATP synthase. *Biochemistry*. 29:6192–6203.

Dunaway-Mariano, D., and W. Cleland. 1980. Preparation and properties of chromium(III) adenosine 5′-triphosphate, chromium(III) adenosine 5′-diphosphate and related chromium (III) complexes. *Biochemistry.* 19:1496–1505.

Fry, D., S. A. Kuby, and A. S. Mildvan. 1985. NMR studies of the MgATP binding site of adenylate kinase and of a 45-residue peptide fragment of the enzyme. *Biochemistry.* 24:4680–4694.

Gantzer, M., C. Klevickis, and C. Grisham. 1982. Interactions of $Co(NH_3)_4ATP$ and $Cr(H_2O)_4ATP$ with Ca^{2+}-ATPase from sarcoplasmic reticulum and Mg^{2+}-ATPase and (Na^+ + K^+)-ATPase from kidney medulla. *Biochemistry.* 21:4083-4088.

Grisham, C. 1979*a*. The structure of the (Na^+ + K^+)-ATPase: implications for the mechanism of sodium and potassium transport. *Advances in Inorganic Biochemistry.* 1:193–218.

Grisham, C. 1979*b*. Characterization of essential arginyl residues in sheep kidney (Na^+ + K^+)-ATPase. *Biochemical and Biophysical Research Communications.* 88:229–236.

Grisham, C. M. 1988. Nuclear magnetic resonance investigations of Na,K-ATPase. *Methods in Enzymology.* 156:353–371.

Jensen, J., and P. Ottolenghi. 1983. ATP binding to solubilized (Na^+ + K^+)-ATPase: the abolition of subunit-subunit interaction and the maximum weight of the nucleotide-binding unit. *Biochimica et Biophysica Acta.* 731:282–289.

Jørgensen, P., and J. Andersen. 1988. Structural basis for E_1-E_2 conformational transitions in Na,K-pump and Ca-pump proteins. *Journal of Membrane Biology.* 103:95–120.

Karlish, S., and W. Stein. 1982. Effects of ATP or phosphate on passive rubidium fluxes mediated by Na-K-ATPase reconstituted into phospholipid vesicles. *Journal of Physiology.* 328:317–331.

Klemens, M., J. Andersen, and C. Grisham. 1986. Occluded calcium sites in soluble sarcoplasmic reticulum Ca^{2+}-ATPase. *Journal of Biological Chemistry.* 261:1495–1498.

Klemens, M., and C. Grisham. 1988. NMR studies identify four intermediate states of ATPase and the ion transport cycle of sarcoplasmic reticulum Ca^{2+}-ATPase. *FEBS Letters.* 237:4–8.

Klevickis, C., and C. Grisham. 1982. Phosphorus-31 nuclear magnetic resonance studies of the conformation of an adenosine 5′-triphosphate analogue at the active site of (Na^+ + K^+)-ATPase from kidney medulla. *Biochemistry.* 21:6979-6984.

Kuntzweiler, T. 1990. H-1 NMR studies of the conformation of a substrate analogue bound to sarcoplasmic reticulum Ca^{2+}-ATPase. Ph.D Dissertation. University of Virginia, Charlottesville, VA. 221 pp.

Kuntzweiler, T., and C. M. Grisham. 1991. NMR studies on intermediate states of SR Ca^{2+}-ATPase using Co(III), Cr(III), and Rh(III) nucleotides. *In* The Sodium Pump: Recent Developments. J. H. Kaplan and P. De Weer, editors. The Rockefeller University Press, New York. In press.

Lin, I., W. Knight, S.-J. Ting, and D. Dunaway-Mariano. 1984. Preparation and characterization of rhodium(III) polyphosphate complexes. *Inorganic Chemistry.* 23:988–991.

MacLennan, D., C. Brandl, C. Korczak, and N. Green. 1985. Amino-acid sequence of a Ca^{2+} + Mg^{2+}-dependent ATPase from rabbit muscle sarcoplasmic reticulum, deduced from its complementary DNA sequence. *Nature.* 316:696–700.

Meissner, G., E. Conner, and S. Fleischer. 1973. Isolation of sarcoplasmic reticulum by zonal centrifugation and purification of Ca^{2+}-binding proteins. *Biochimica et Biophysica Acta.* 298:246–269.

O'Connor, S., and C. Grisham. 1979. Manganese electron paramagnetic resonance studies of sheep kidney (Na^+ + K^+)-ATPase: interactions of substrates and activators at a single Mn^{2+} binding site. *Biochemistry.* 18:2315–2323.

Pauls, H., B. Bredenbrocker, and W. Schoner. 1980. Inactivation of (Na^+ + K^+)-ATPase by chromium(III) complexes of nucleotide triphosphates. *European Journal of Biochemistry.* 109:523–533.

Plesner, L., and I. Plesner. 1988. Distinction between the intermediates in Na^+-ATPase and Na^+,K^+-ATPase reactions. II. Exchange and hydrolysis kinetics at micromolar nucleotide concentrations. *Biochimica et Biophysica Acta.* 937:63–72.

Richards, D. 1988. Occlusion of cobalt ions within the phosphorylated forms of the Na^+-K^+ pump isolated from dog kidney. *Journal of Physiology.* 404:497–514.

Saenger, W. 1984. Principles of Nucleic Acid Structure. Springer-Verlag, New York. 556 pp.

Schull, G., A. Schwartz, and G. Lingrel. 1985. Amino-acid sequence of the catalytic subunit of the (Na^+ + K^+)-ATPase deduced from a complementary DNA. *Nature.* 316:691–695.

Schuurmans-Stekhoven, F., H. Swarts, J. DePont, and S. Bonting. 1983. Properties of the Mg^{2+}-induced low-affinity nucleotide binding site of (Na^+ + K^+)-activated ATPase. *Biochimica et Biophysica Acta.* 732:607–619.

Serpersu, E. H., S. Bunk, and W. Shoner. 1990. How do MgATP analogues differentially modify high-affinity and low-affinity ATP binding sites of Na^+/K^+-ATPase? *European Journal of Biochemistry.* 191:397–404.

Serpersu, E., U. Kirch, and W. Schoner. 1982. Demonstration of a stable occluded form of Ca^{2+} by the use of the chromium complex of ATP in the Ca^{2+}-ATPase of sarcoplasmic reticulum. *European Journal of Biochemistry.* 122:347–354.

Stewart, J., and C. Grisham. 1988. ^1H nuclear magnetic resonance studies of the conformation of an ATP analogue at the active site of Na,K-ATPase from kidney medulla. *Biochemistry.* 27:4840–4848.

Stewart, J., P. Jorgensen, and C. Grisham. 1989. Nuclear Overhauser effect studies of the conformation of $Co(NH_3)_4$ATP bound to kidney Na,K-ATPase. *Biochemistry.* 28:4695–4701.

Vilsen, B., and J. Andersen. 1987. Characterization of CrATP-induced calcium occlusion in membrane-bound and soluble monomeric sarcoplasmic reticulum Ca^{2+}-ATPase. *Biochimica et Biophysica Acta.* 898:313–322.

Vilsen, B., J. Andersen, J. Petersen, and P. Jørgensen. 1987. Occlusion of $^{22}Na^+$ and $^{86}Rb^+$ in membrane-bound and soluble protomeric α,β-units of Na,K-ATPase. *Journal of Biological Chemistry.* 262:10511–10517.

Chapter 12

Two-dimensional Crystals and Three-dimensional Structure of Na,K-ATPase Analyzed by Electron Microscopy

Arvid B. Maunsbach, Elisabeth Skriver, and Hans Hebert

Department of Cell Biology at the Institute of Anatomy, and the Biomembrane Research Center, University of Aarhus, DK-8000 Aarhus C, Denmark; and Structural Biochemistry Unit, CNT, Novum, Karolinska Institutet, S-14157 Huddinge, Sweden

Introduction

Analysis of regular two-dimensional crystalline arrays of membrane proteins by means of electron microscopy provides structural information that is not obtainable from studies of nonordered proteins. Therefore, to determine the structure and interactions of Na,K-ATPase in isolated, purified plasma membranes, we started about 10 years ago to develop methods to induce difference forms of two-dimensional crystal of Na,K-ATPase (Maunsbach et al., 1981; Skriver et al., 1981). At present, several different forms of these membrane crystals have been described, and for most of them a projection structure has been obtained by image processing. For three types of Na,K-ATPase crystals we have also reconstructed the three-dimensional shape of the protein units from series of electron micrographs of tilted membrane crystals. This paper emphasizes the differences between the various types of two-dimensional crystals and the derived three-dimensional models and summarizes possible pathways for the assembly of the crystals.

Materials and Methods

Membrane-bound Na,K-ATPase was purified from outer renal medulla of kidney by selective extraction of a microsomal fraction with sodium dodecyl sulfate (SDS) in the presence of ATP followed by isopycnic centrifugation (Jørgensen, 1974). The specific activity of the enzyme used for crystallization was 20–40 μmol $P_i \cdot min^{-1} \cdot mg$ $protein^{-1}$ and ouabain-insensitive ATPase activity was not detectable. Polyacrylamide gel electrophoresis in SDS showed that the preparations were 90–100% pure with respect to the content of the α- and β-subunits.

Purified Na,K-ATPase membranes were incubated with various combinations of ligands and ion concentrations under different conditions of temperature, pH, and time. For some experiments the membranes were exposed to phospholipase A_2 (Manella, 1984; Mohraz et al., 1985), and dialyzed against a calcium-containing buffer.

Assembly of crystalline arrays was monitored by negative staining and electron microscopy. Samples were taken at different times of incubation, placed on hydrophilic carbon films, and negatively stained with 1% uranyl acetate. Micrographs were taken with a JEOL 100CX electron microscope at a magnification of 54,000 and tilted projections were collected within the range of −60° to +60°. Selected areas of ordered arrays of Na,K-ATPase were digitized into 512 × 512 pixel squares. Image processing was performed with correlation averaging methods (Saxton and Baumeister, 1982; Hebert et al., 1985*a*). Calculations were made by the "EM" system (Hegerl and Altbauer, 1982). The three-dimensional analysis was performed essentially according to the methods of Henderson and Unwin (1975) and Amos et al. (1982). Fourier analysis or correlation averaging was carried out on each projection and the images were combined into a three-dimensional model.

Structure of Noncrystallized Enzyme

Purified SDS-treated plasma membranes from the thick ascending limb in the outer renal medulla appear in thin sections of conventional electron microscope preparations as small, open membrane fragments with diameters of 0.1–0.6 μm. After negative staining the Na,K-ATPase protein is observed as surface particles arranged in clusters and strands as first demonstrated by Maunsbach and Jørgensen (1974). Between the enzyme protein there are irregular areas devoid of structure which represent the enzyme-free lipid bilayer of the membranes. The density of the surface particles is 12,500/μm^2 (Deguchi et al., 1977). In noncrystallized Na,K-ATPase membranes the individual protein particles do not show a consistent substructure even at high resolution after conventional negative staining. On the basis of the particle frequency and enzyme activity of the preparation the surface particles are protomers ($\alpha\beta$-units) of the Na,K-ATPase (Deguchi et al., 1977). This conclusion is supported by the cytochemical demonstration that ouabain-sensitive, potassium-dependent phosphatase activity is associated with the particle-rich areas (Maunsbach et al., 1980). Furthermore, the particle-rich areas can be labeled with monoclonal antibodies against the α-subunit of Na,K-ATPase (Maunsbach et al., 1991).

The distribution of enzyme protein in isolated membranes can be modified by cross-linking with glutaraldehyde. If the membranes are exposed to low concentrations of glutaraldehyde the surface particles form large clusters, which demonstrates that the protein units have extensive lateral mobility in the plane of the membrane (Deguchi et al., 1977).

Two-dimensional Crystals of Na,K-ATPase Induced with Different Ligands

Since the enzyme protein shows lateral mobility in the plane of the membrane, we have systematically studied conditions required to assemble the enzyme units in regular arrays suitable for analysis with crystallographic methods. Extensive crystal formation was first observed after incubation with vanadate or phosphate and magnesium (Skriver et al., 1981). These ligand combinations favor the E_2 conformation (the potassium form) of the enzyme (Figs. 1–4). Formation of crystalline arrays is usually observed in some membrane fragments within minutes after the start of incubation. It increases gradually over the succeeding hours and days although the rate and extent of crystallization may vary from preparation to preparation. Both protomeric (monomeric) and dimeric membrane crystals can be observed at the same time in the preparations. Addition of ouabain to phosphate/magnesium medium does not change the extent or type of crystal formation. An early step in the assembly process is the formation of ladder-like rows of paired particles (Fig. 2) which then appear to assemble laterally to confluent crystals with dimers as the building blocks (Söderholm et al., 1988).

Crystals similar to those induced by vanadate/magnesium and phosphate/magnesium have subsequently been induced by the same as well as other combinations of ligands that favor the E_2 conformation (Demin et al., 1984; Mohraz and Smith, 1984; Zampighi et al., 1984). Crystalline areas also form during incubation of membranes in potassium chloride, but several crystalline centers usually appear within each membrane fragment and each lattice system is therefore small (Skriver and Maunsbach, 1983). Two different types of crystals (Figs. 5 and 6) have recently been demonstrated (Skriver et al., 1989) after incubation of Na,K-ATPase membranes with cobalt-tetrammine-ATP, which is a MgATP complex analogue that binds to the low-affinity ATP-binding site of the enzyme (Scheiner-Bobis et al., 1987). Na,K-ATPase has also been crystallized in the E_1 conformation (the sodium form of the enzyme) using oligomyocin or high concentrations of sodium (Skriver et

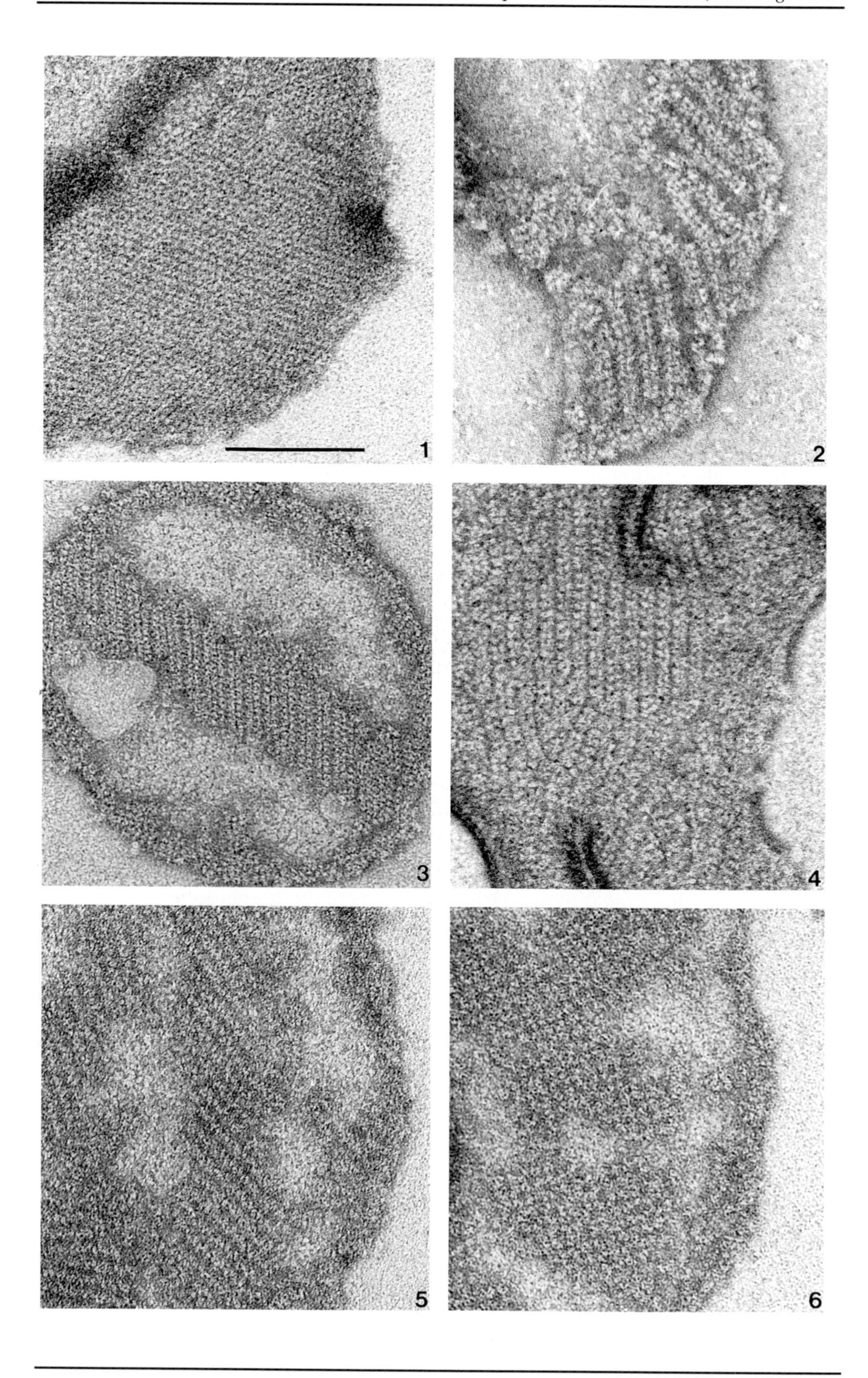
1
2
3
4
5
6

al., 1985) but these crystals are small and difficult to analyze with image processing methods.

The vanadate method for Na,K-ATPase has subsequently also been found to induce two-dimensional crystals of other membrane-bound ATPases, including Ca-ATPase from sarcoplasmic reticulum (Buhle et al., 1983; Dux and Martonosi, 1983) and H,K-ATPase from gastric mucosa (Rabon et al., 1986).

Projection Structure of Na,K-ATPase in Membrane Crystals

The different forms of crystalline arrays induced in purified Na,K-ATPase membrane have been analyzed by image processing using Fourier analysis or correlation averaging. These analyses reveal that the two-dimensional crystals induced by vanadate or phosphate and magnesium are either p1 crystals (Fig. 7) or p21 crystals (Fig. 9 and 10). The unit cell in the P1 crystals is only large enough to contain one protomer ($\sim$140,000 mol wt), while in the p21 crystals the unit cell contains one $(\alpha\beta)_2$ dimer (280,000 mol wt) (Hebert et al., 1982). Two different relationships between the interacting protomers have been observed in the p21 crystals (compare Figs. 9 and 10) and are referred to as p21 crystal types I and II (Maunsbach et al., 1988). Na,K-ATPase membranes exposed to phospholipase A_2 in combination with vanadate/magnesium show increased extents of crystallization (Manella, 1984; Mohraz et al., 1985), but the type of p1 and p21 crystalline arrays are usually the same. However, recent image analyses have shown that some of the dimeric crystals show deviations from perfect twofold symmetry in the projection structure (Fig. 11), suggesting dissimilarities between the two protomers in the unit cell (Beall et al., 1989; Hebert et al., 1990). Image processing of the linear arrays of paired protein units observed during assembly of the crystals (Fig. 8) demonstrates that such a ribbon corresponds to a row of protein pairs in the p21 crystal and that the individual protein units (protomers) shows a similar shape in both ribbons and confluent p21 crystals (Söderholm et al., 1988). Most linear arrays are built of pairs of protein units that show twofold rotational symmetry. The protein building blocks forming the two-rowed ribbons are always oriented with their large dense domains toward the

Figures 1–6. (*1*) Electron micrograph of protomeric two-dimensional crystal of Na,K-ATPase. The purified membranes were isolated from pig outer renal medulla, incubated with 0.30 μg/ml phospholipase A_2, and dialyzed against 1 mM NH_4 VO_3, 5 mM $MgCl_2$, and 5 mM $CaCl_2$ in 10 mM imidazole-HCI, pH 7.3, at 4°C for 1 d before negative staining. *Bar,* 0.1 μm. (*2*)Na,K-ATPase protein assembled in linear arrays which appear to be in the process of forming a confluent two-dimensional crystal. The enzyme was incubated in 1 mM NH_4 VO_3 and 3 mM MgCl in imidazole-HCl buffer, pH 7.5. (*3*) Dimeric two-dimensional crystal type I of Na,K-ATPase. The enzyme was incubated with phospholipase A_2 and dialyzed against 1 mM NH_4VO_3, 3 mM $MgCl_2$, and 1 mM $CaCl_2$ in 10 mM imidazole-HCl overnight, stained after 9 d. (*4*) Dimeric two-dimensional crystal type II of Na,K-ATPase. Incubation time was 14 d and the incubation medium contained 1 mM NH_4VO_3, 3 mM $MgCl_2$, 12.5 P_i, and 250 mM KCl in 10 mM imidazole-HCl buffer, pH 7.0. (*5, 6*) Electron micrograph of crystalline arrays induced in pig Na,K-ATPase membrane with cobalt-tetrammine-ATP. The Na,K-ATPase protein in Fig. 5 has assembled in linear arrays, while the protein in Fig. 6 forms square mesh patterns. Purified membranes treated with phospholipase A_2 were incubated with 0.1 mM $CO(NH_3)_4ATP$ in 10 mM Tris-HCl, pH 7.3, at 37°C for 1 h. Figs. 1–6 have the same magnification.

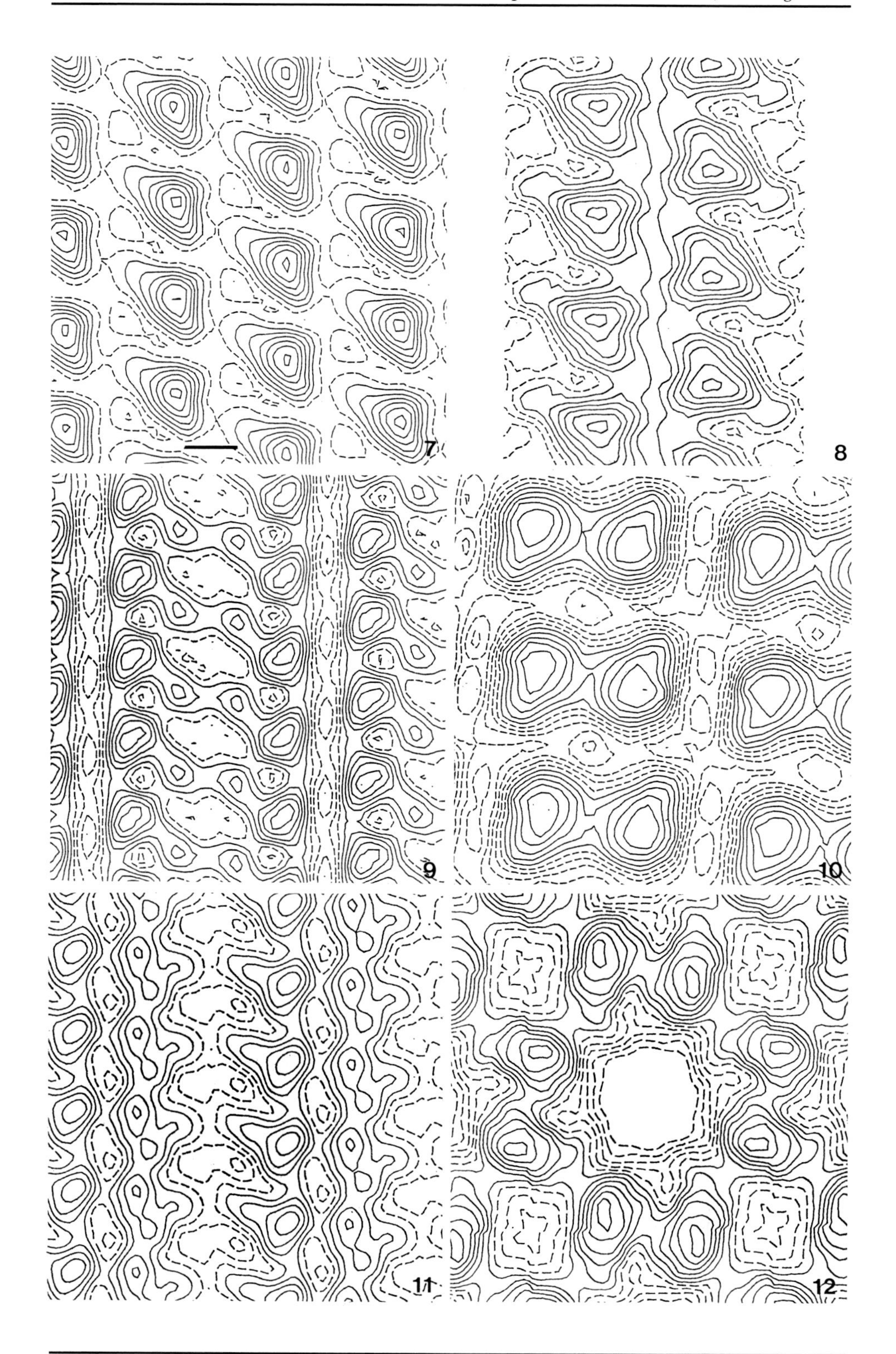
7
8
9
10
11
12

core of the ribbon (Fig. 8). Multiple-rowed arrays are built of single ribbons similar to those forming the two-rowed arrays.

In cobalt-tetrammine-ATP–induced membrane crystals the repeating building blocks consist of two symmetrically related particles forming a dimer (Figs. 11 and 12) (Skriver et al., 1989). With respect to shape and dimensions the protomeric units within each dimer resemble the protomers in the two-dimensional Na,K-ATPase crystals induced by vanadate and magnesium. This suggests that the dimers formed by $CO(NH_3)_4ATP$ also consist of two interacting αβ-units of the enzyme. In the square mesh–type membrane crystals the dimers are arranged with two-sided plane group symmetry p4 with four pairs of protein units arranged in a ring. Each unit cell contains 4 αβ-protomers and the unit cell dimensions are $a = b = 141.1$ Å (Fig. 12). Projection maps of vanadate-induced dimeric crystals (Figs. 9 and 10) illustrate that the protomers interact "base-to-base" and "tip-to-tip" (Hebert et al., 1982; Maunsbach et al., 1988; Söderholm et al., 1988). On the contrary, adjacent protomers in $CO(NH_3)_4ATP$-induced crystals are in close contact along their long sides about a twofold axis, i.e., with their "tips" in opposite directions (Fig. 12). Since the relationship between the protomers in the dimer in $CO(NH_3)_4ATP$-induced crystals is different from that in crystals formed in the presence of vanadate, the conformations may also be different. The observation that dimers and not protomers are the building blocks in cobalt-tetrammine–induced crystals is consistent with the kinetic and binding studies of Scheiner-Bobis et al. (1987, 1989) and with the possibility that two αβ-subunits of Na,K-ATPase cooperate in the catalytic cycle of the Na,K-ATPase.

Assembly of Two-dimensional Crystals

Based on the observations described in the preceding sections, the scheme in Fig. 13 suggests assembly routes for two-dimensional membrane crystals of Na,K-ATPase. Several factors determine the type of crystal(s) that assemble. In addition to the characteristics and concentrations of enzyme ligands, the $(\alpha\beta)_2/\alpha\beta$ ratio in the pool of building blocks could influence the route of the assembly. Two-rowed ribbons may be frequent if the concentration of $(\alpha\beta)_2$-units is high and may lead to the formation of large p21 crystals. The presence of αβ-protomers probably favors the formation of different multiple-rowed arrays which may cause defects in the p21 crystals. The process may be driven toward formation of p1 crystals if αβ-protomers are abundant.

Figures 7–12. Computer-averaged images from different types of Na,K-ATPase crystals. The unbroken contour lines outline repeating protein units, while dashed lines outline the uranyl acetate used for negative staining. *Bar,* 30 Å. (*7*)Vanadate/magnesium–induced p1 crystal with unit cell containing one protomer. (*8*) Reconstruction of single dimeric ribbon similar to those observed in Fig. 2. The protein units are oriented in opposite directions and their large dense regions are in close contact. The membranes were incubated with vanadate/magnesium. (*9*) Vanadate/magnesium–induced p21 crystal of type I with unit cell containing one dimer (two protomers) of the enzyme. (*10*) Vanadate/magnesium–induced p21 crystal type II containing unit cell with two protomers in a different association than in p21 crystal type I. (*11*) Vanadate/magnesium–induced crystal in phospholipase A_2–treated membrane showing unit cells with protomers of different shapes (heterodimer). (*12*) Cobalt-tetrammine-ATP–induced p4 crystal with tetrameric arrangement of dimers.

Three-dimensional Structure of Na,K-ATPase

The p1 crystals induced with vanadate and magnesium have a unit cell containing one Na,K-ATPase protomer with a height of ~120 Å (Fig. 14) (Hebert et al., 1988). This value is deduced from the location of contrast variation above the noise level and is similar to that obtained for the enzyme in p21 crystals (Hebert et al., 1985*b*). However, this height may be a minimum value since the parts of negatively stained

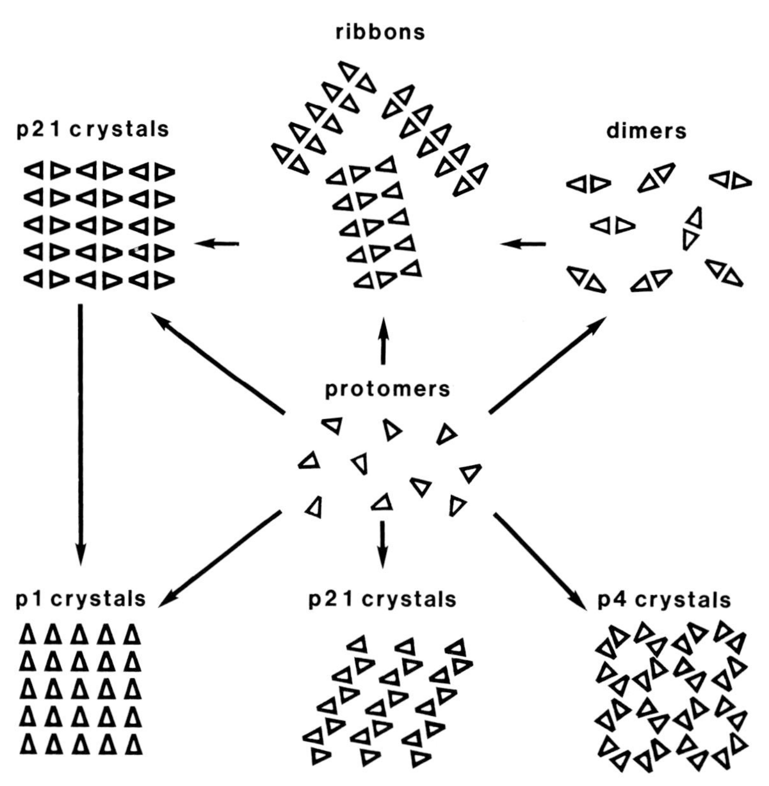

Figure 13. Diagram of proposed patterns for the assembly of two-dimensional crystals of Na,K-ATPase. Starting from randomly dispersed protomers (αβ-units) incubation of enzyme with vanadate/magnesium leads to the formation of p1 or p21 crystals, while incubation with cobalt-tetrammine-ATP leads to the formation of p4 crystals or a different type of p21 crystals. Additionally, protomers may form dimeric or multirowed ribbons that subsequently align to confluent p21 crystals. An intermediate step in the latter process may be the formation of dimers that then align to dimeric ribbons. At least some of these assembly steps appear to be reversible and, additionally, p21 crystals may change into p1 crystals upon storage.

specimens that are in contact with the carbon support film may undergo flattening (Baumeister et al., 1986). As judged from the contrast variation in the z-direction, and the location of a contrast minimum indicating the level of the lipids, the protomer is asymmetrically inserted in the membrane. It protrudes almost 60 Å on one side of the bilayer but only ~20 Å on the other side. Labeling (O'Connell, 1982) and sequence (Kawakami et al., 1985; Shull et al., 1985) studies have shown that much more protein is exposed on the cytoplasmic than on the extracellular side and therefore the 60-Å protrusion identifies the cytoplasmic side of the membrane.

The unit cell in p21 crystals induced with vanadate and magnesium contains two rod-shaped protein units representing two αβ-protomers (Fig. 15) (Hebert et al., 1985*b*). Within each unit cell the rods show a contact which is asymmetrically located in relation to the central section of the structure. The dimer protrudes ~40 Å on one side of the bilayer, probably corresponding to the cytoplasmic side, and ~20 Å on the other, probably extracellular side. The protrusion on the cytoplasmic side of the lipid bilayer molecular represents almost exclusively the α-subunit since very little of the β-subunit is exposed on the cytoplasmic side. A related three-dimensional model was presented by Ovchinnikov et al. (1985), while Mohraz et al. (1987) did not observe any protein on the extracellular side where much of the protein of the β-subunit is known to be exposed (Kawakami et al., 1986; Ovchinnikov et al., 1986).

It is noteworthy that three-dimensional reconstructions of vanadate-induced two-dimensional crystals of Ca-ATPase show that the Ca-ATPase molecule projects above the lipid bilayer into the cytoplasm in a way similar to Na,K-ATPase crystals (Castellani et al., 1985; Taylor et al., 1986). However, Ca-ATPase protein does not project on the opposite side of the membrane, in contrast to the situation in Na,K-ATPase crystals. This difference fits well with the absence of a β-subunit in Ca-ATPase and the identification of the cytoplasmic projections of both enzymes as representing α-subunits.

The protomers that represent the building blocks in cobalt-tetrammine-ATP–induced p4 crystals are related by a dyad axis. The volume of this continuous protein domain also suggests that it constitutes an $(\alpha\beta)_2$ dimer of Na,K-ATPase (Skriver et al., 1990, 1991). The two protomers are connected by a bridge (Fig. 16) as in vanadate/magnesium-induced p21 crystals (Fig. 15). However, in contrast to the latter crystals the bridge or contact area is closer to the center of the model. It corresponds to the horizontal construction in the upper part of the model in Fig. 16 and to the presumed level of the lipid bilayer. This observation is therefore also consistent with the assymmetric distribution of the Na,K-ATPase molecule relatively to the bilayer. The protomers in vanadate- and cobalt-tetrammine-ATP–induced crystals show several basic similarities with respect to size and tendency to form dimers. However, differences exist in shape and intermolecular association and may be due to differences in conformations and/or surface properties.

Some crystals induced with vanadate in membranes treated with phospholipase A_2 show the dimensions of p21 crystals, but are partially devoid of the expected twofold symmetry. Thus, sections of the model in xy-planes show that the center of the structure has a significant dyad axis but at increasing distances from the center this twofold rotational symmetry deteriorates (Hebert et al., 1990). Moreover, the mass distribution relative to the central xy-section is different in the two αβ-units. Although the individual volumes of the two protomers are equal they have different shapes and the three-dimensional reconstruction reveals that one of them protrudes more on the cytoplasmic side, while the other extends more on the extracellular side of the membrane (Figs. 18 and 19). These structural differences may be related to defined conformational states of the enzyme. It seems likely that the protomers represent two different conformational states rather than a continuous distribution since the protein units form crystals and since they show two distinct structures at the present resolution level. The different relationships between the protomers and the

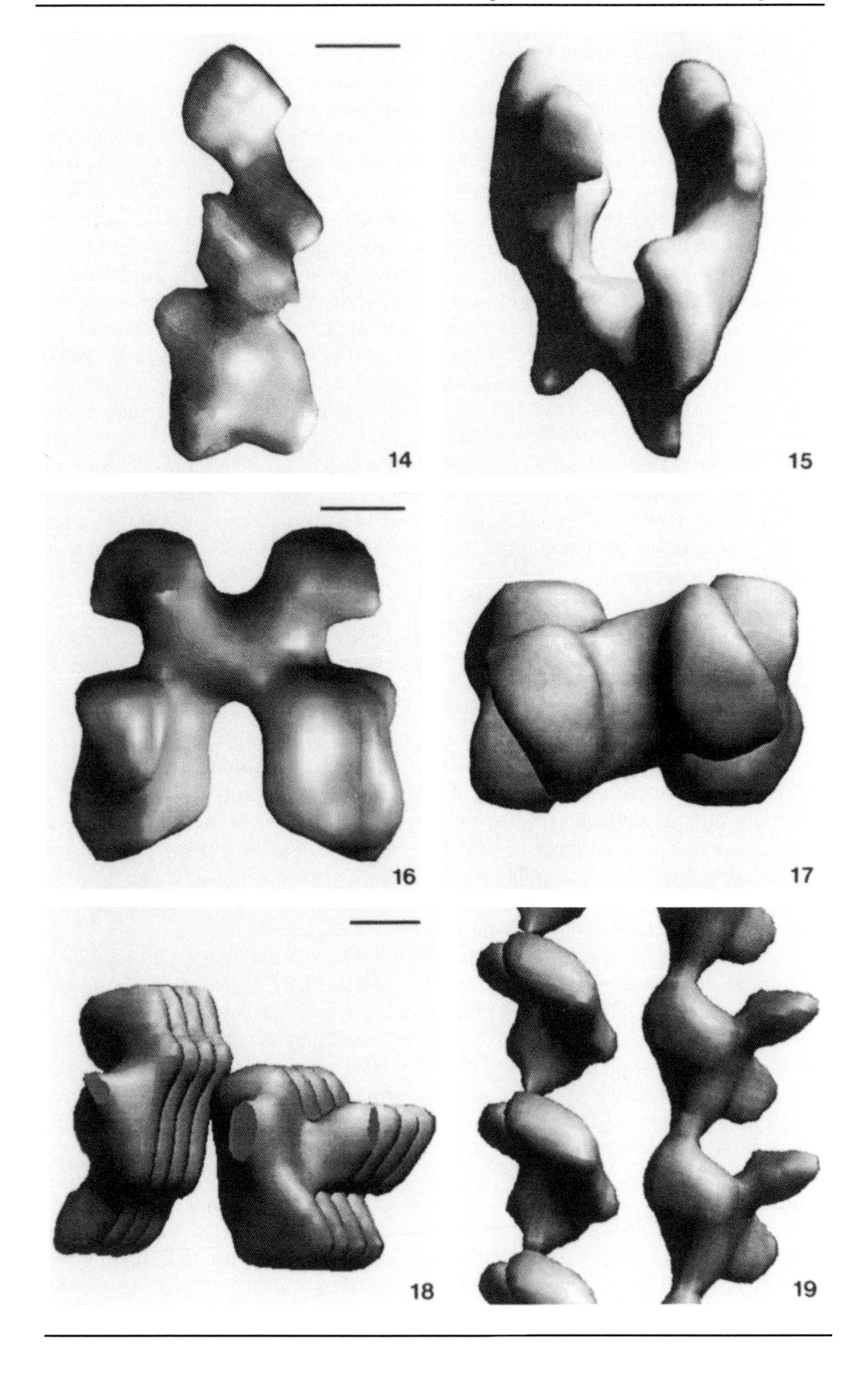
14
15
16
17
18
19

lipid bilayer may be functionally important since conformational changes during the pumping cycle of the enzyme are associated with changes in the location of the αβ-subunit relative to the lipid bilayer (Jørgensen et al., 1982).

Recently, a resolution better than 3 Å has been obtained in extremely well-ordered membrane crystals from the purple membrane on *Halobacterium halobium* (Henderson et al., 1990). At this level of resolution it starts to be feasible to identify amino acid residues in the membrane protein. In Ca-ATPase crystals a resolution of ~6 Å has recently been demonstrated (Stokes and Green, 1990). The three-dimensional reconstructions of the Na,K-ATPase protein in the p1, p21, and p4 crystals have an estimated resolution in the x and y directions of 20–25 Å. At this resolution some interactions and basic structural characteristics of the Na,K-ATPase protein can be resolved and presumably also certain conformational changes as discussed above, but ion channels or specific amino acids are far from demonstrable. The main limitation to achieving higher resolution is the internal order of the Na,K-ATPase crystals. Further progress toward high resolution studies of Na,K-ATPase membranes will therefore focus on obtaining larger Na,K-ATPase crystals with a higher degree of order as well as the application of methods for cryoelectron microscopy.

Acknowledgments

This investigation was supported by grants from the Danish Biomembrane Research Center, the Danish Medical Research Council (12-8501), the Research Foundation of the University of Aarhus, and the Swedish Medical Research Council (03x-00144).

References

Amos, L. A., R. Henderson, and P. N. T. Unwin. 1982. Three-dimensional structure determination by electron microscopy of two-dimensional crystals. *Progress in Biophysics and Molecular Biology*. 39:183–231.

Baumeister, W., M. Barth, R. Hegerl, R. Guckenberger, M. Hahn, and W. O. Saxton. 1986. Three-dimensional structure of the regular surface layer (HPI layer) of *Deinococcus radiodurans*. *Journal of Molecular Biology*. 187:241–253.

Beall, H. C., D. F. Hastings, and H. P. Ting-Beall. 1989. Digital image analysis of two-dimensional Na,K-ATPase crystals: dissimilarity between pump units. *Journal of Microscopy*. 154:71–82.

Figures 14–19. Three-dimensional models of Na,K-ATPase reconstructed from electron microscope tilt series of different types of membrane crystals of Na,K-ATPase. The suggested extracellular side is up. *Bar,* 25 Å. (*14*)Protomer in p1 crystal induced with vanadate/magnesium seen from a plane parallel to the membrane. (*15*) Dimer in p21 crystal type II induced with vanadate/magnesium seen from a plane parallel to the membrane. (*16, 17*) Dimer in tetrameric crystal induced with cobalt-tetrammine-ATP and seen from a plane parallel to the membrane (Fig. 16) and from the presumed extracellular side (Fig. 17). The level of the lipid bilayer in Fig. 16 probably corresponds to the horizontal constriction in the model. (*18, 19*) Reconstructed chain of heterodimers induced with vanadate/magnesium in membrane treated with phospholipase A_2. The two protein units within each dimer show different shapes and heights. The heterodimers are seen in a plane parallel to the membrane (Fig. 18) and from the presumed cytoplasmic side (Fig. 19).

Buhle, E. L., Jr., B. E. Knox, E. Serpersu, and U. Aebi. 1983. The structure of the Ca^{2+} ATPase as revealed by electron microscopy and image processing of ordered arrays. *Journal of Ultrastructure Research.* 85:186–203.

Castellani, L., P. M. D. Hardwicke, and P. Vibert. 1985. Dimer ribbons in the three-dimensional structure of sarcoplasmic reticulum. *Journal of Molecular Biology.* 185:579–594.

Deguchi, N., P. L. Jørgensen, and A. B. Maunsbach. 1977. Ultrastructure of the sodium pump. Comparison of thin sectioning, negative staining, and freeze-fracture of purified, membrane-bound (Na^+,K^+)-ATPase. *The Journal of Cell Biology.* 75:619–634.

Demin, V. V., A. N. Barnakov, A. V. Lunev, K. N. Dzhandzhugazyan, A. P. Kuzin, N. N. Modyanov, S. Hovmöller, and G. Farrants. 1984. Electron microscopy of membrane proteins. I. Three forms of Na,K-ATPase two-dimensional crystals. *Biology of Membranes.* 1:831–837.

Dux, L., and A. Martonosi. 1983. Two-dimensional arrays of proteins in sarcoplasmic reticulum and purified Ca^{2+}-ATPase vesicles treated with vanadate. *The Journal of Biological Chemistry.* 258:2599–2603.

Hebert, H., P. L. Jørgensen, E. Skriver, and A. B. Maunsbach. 1982. Crystallization patterns of membrane-bound Na,K-ATPase. *Biochimica et Biophysica Acta.* 689:571–574.

Hebert, H., E. Skriver, R. Hegerl, and A. B. Maunsbach. 1985*a*. Structure of two-dimensional crystals of membrane-bound Na,K-ATPase as analyzed by correlation averaging. *Journal of Ultrastructure Research.* 92:28–35.

Hebert, H., E. Skriver, U. Kavéus, and A. B. Maunsbach. 1990. Coexistence of different forms of Na,K-ATPase in two-dimensional membrane crystals. *FEBS Letters.* 268:83–87.

Hebert, H., E. Skriver, and A. B. Maunsbach. 1985*b*. Three-dimensional structure of renal Na,K-ATPase determined by electron microscopy of membrane crystals. *FEBS Letters.* 187:182–186.

Hebert, H., E. Skriver, M. Söderholm, and A. B. Maunsbach. 1988. Three-dimensional structure of renal Na,K-ATPase determined from two-dimensional membrane crystals of the p1 form. *Journal of Ultrastructure and Molecular Structure Research.* 100:86–93.

Hegerl, R., and A. Altbauer. 1982. The "EM" program system. *Ultramicroscopy.* 9:109–116.

Henderson, R., J. M. Baldwin, T. A. Ceska, F. Zemlin, E. Beckmann, and K. H. Downing. 1990. Model for the structure of bacteriorhodopsin based on high-resolution electron cryomicroscopy. *Journal of Molecular Biology.* 213:899–929.

Henderson, R., and P. N. T. Unwin. 1975. Three-dimensional model of purple membrane obtained by electron microscopy. *Nature.* 257:28–31.

Jørgensen, P. L. 1974. Purification and characterization of (Na^+ + K^+)-ATPase. III. Purification from outer medulla of mammalian kidney after selective removal of membrane components by sodium dodecyl sulphate. *Biochimica et Biophysica Acta.* 356:36–52.

Jørgensen, P. L., E. Skriver, H. Hebert, and A. B. Maunsbach. 1982. Structure of the Na,K pump: crystallization of pure membrane-bound Na,K-ATPase and identification of functional domains of the α-subunit. *Annals of the New York Academy of Sciences.* 402:207–224.

Kawakami, K., S. Noguchi, M. Noda, H. Takahashi, T. Ohta, M. Kawamura, H. Nojima, K. Nagano, T. Hiose, S. Inayama, H. Hayashida, T. Myata, and S. Numa. 1985. Primary structure of the α-subunit of *Torpedo californica* (Na^+,K^+)ATPase deduced from cDNA sequence. *Nature.* 316:733–736.

Kawakami, K., H. Nojima, T. Ohta, and K. Nagano. 1986. Molecular cloning and sequence analysis of human Na,K-ATPase β-subunit. *Nucleic Acids Research.* 14:2833–2844.

Manella, C. A. 1984. Phospholipase-induced crystallization of channels in mitochondrial outer membranes. *Science.* 224:165–166.

Maunsbach, A. B., and P. L. Jørgensen. 1974. Ultrastructure of highly purified preparations of (Na^+ + K^+)-ATPase from outer medulla of the rabbit kidney. Proceedings of the Eighth International Congress on Electron Microscopy, Canberra. 2:214–215.

Maunsbach, A. B., E. Skriver, N. Deguchi, and P. L. Jørgensen. 1980. Ultrastructure of Na,K-ATPase. *Acta Histochemica et Cytochemica.* 13:103–112.

Maunsbach, A. B., E. Skriver, and P. L. Jørgensen. 1981. Electron microscope analysis of protein distribution in purified, membrane-bound Na,K-ATPase. Proceedings of the Third International Conference on Na,K-ATPase, New Haven, CT. 3.

Maunsbach, A. B., E. Skriver, K. D. Petersen, and H. Hebert. 1991. Electron microscope analysis of the Na,K-ion pump: localization, crystallization and 3-D structure. *In* Electron Microscopy 1990. World Scientific Publishing Co. PTE. Ltd., Singapore. In press.

Maunsbach, A. B., E. Skriver, M. Söderholm, and H. Hebert. 1988. Three-dimensional structure and topography of membrane-bound Na,K-ATPase. *In.* The Na^+,K^+-pump. Part A: Molecular Aspects. J. C. Skou, J. G. Nørby, A. B. Maunsbach, and M. Esmann, editors. Alan R. Liss, Inc., New York. 39–56.

Mohraz, M., M. V. Simpson, and P. R. Smith. 1987. The three-dimensional structure of the Na,K-ATPase from electron microscopy. *The Journal of Cell Biology.* 105:1–8.

Mohraz, M., and P. R. Smith. 1984. Structure of (Na^+,K^+)-ATPase as revealed by electron microscopy and image processing. *The Journal of Cell Biology.* 98:1836–1841.

Mohraz, M., M. Yee, and P. R. Smith. 1985. Novel crystalline sheets of Na,K-ATPase induced by phospholipase A_2. *Journal of Ultrastructure Research.* 93:17–26.

O'Connell, M. A. 1982. Exclusive labeling of the extracytoplasmic surface of sodium ion and potassium ion activated adenosine-triphosphatase and a determination of the distribution of surface area across the bilayer. *Biochemistry.* 21:5984–5991.

Ovchinnikov, Yu. A., V. V. Demin, A. N. Barnakov, A. P. Kuzin, A. V. Lunev, N. N. Modyanov, and K. N. Dzhandzhugazyan. 1985. Three-dimensional structure of (Na^+ + K^+)-ATPase revealed by electron microscopy of two-dimensional crystals. *FEBS Letters.* 190:73–76.

Ovchinnikov, Yu. A., V. V. Demin, A. N. Barnakov, E. A. Svetlichni, N. N. Modyanov, and K. N. Dzandzhugazyan. 1986. Three-dimensional structure of Na^+,K^+-ATPase β-subunit revealed by electron microscopy of two-dimensional crystals. *Biology Membrane.* 3:537–541.

Rabon, E., M. Wilke, G. Sachs, and G. Zampighi. 1986. Crystallization of the gastric H,K-ATPase. *The Journal of Biological Chemistry.* 261:1434–1439.

Saxton, W. O., and W. Baumeister. 1982. The correlation averaging of a regularly arranged bacterial cell envelope protein. *Journal of Microscopy.* 127:127–138.

Scheiner-Bobis, G., M. Esmann, and W. Schoner. 1989. Shift to the Na^+ form of Na^+/K^+-transporting ATPase due to modification of the low-affinity ATP-binding site by $Co(NH_3)_4ATP$. *European Journal of Biochemistry.* 183:173–178.

Scheiner-Bobis, G., K. Fahlbusch, and W. Schoner. 1987. Demonstration of cooperating α subunits in working (Na^+ + K^+)-ATPase by the use of the MgATP complex analogue cobalt tetrammine ATP. *European Journal of Biochemistry.* 168:123–131.

Shull, G. E., A. Schwartz, and J. B. Lingrel. 1985. Amino-acid sequence of the catalytic subunit of (Na^+ + K^+)ATPase deduced from a complementary DNA. *Nature.* 316:691–695.

Skriver, E., H. Hebert, U. Kavéus, and A. B. Maunsbach. 1990. Three-dimensional structure of $CO(NH_3)_4$ATP induced membrane crystals of Na,K-ATPase. Proceedings of the XIIth International Congress for Electron Microscopy, Vol. I. Seattle, WA. 110–111.

Skriver, E., H. Hebert, and A. B. Maunsbach. 1985. Crystallization and structure of membrane-bound Na,K-ATPase. *In* The Sodium Pump. I. Glynn and C. Ellroy, editors. The Company of Biologists Ltd., Cambridge, UK. 37–44.

Skriver, E., H. Hebert, U. Kavéus, and A. B. Maunsbach. 1991. Three-dimensional structure of $CO(NH_3)_4$ATP-induced membrane crystals of Na,K-ATPase. *In* The Sodium Pump: Recent Developments. J. H. Kaplan and P. De Weer, editors. The Rockefeller University Press, New York. In press.

Skriver, E., and A. B. Maunsbach. 1983. Factors influencing the formation of two-dimensional crystals of renal Na,K-ATPase. *In* Structure and Function of Membrane Proteins. E. Quagliariello and F. Palimeri, editors. Elsevier Science Publishers, Amsterdam, New York. 211–214.

Skriver, E., A. B. Maunsbach, H. Hebert, G. Scheiner-Bobis, and W. Schoner. 1989. Two-dimensional crystalline arrays of Na,K-ATPase with new subunit interactions induced by cobalt-tetrammine-ATP. *Journal of Ultrastructure and Molecular Structure Research.* 102:189–195.

Skriver, E., A. B. Maunsbach, and P. L. Jørgensen. 1981. Formation of two-dimensional crystals in pure membrane-bound Na,K-ATPase. *FEBS Letters.* 131:219–222.

Söderholm, M., H. Hebert, E. Skriver, and A. B. Maunsbach. 1988. Assembly of two-dimensional crystals of Na,K-ATPase. *Journal of Ultrastructure and Molecular Structure Research.* 99:234–243.

Stokes, D. L., and N. M. Green. 1990. Structure of CaATPase: Electron microscopy of frozen-hydrated crystals at 6 Å resolution in projection. *Journal of Molecular Biology.* 213:529–538.

Taylor, K. A., L. Dux, and A. Martonosi. 1986. Three-dimensional reconstruction of negatively stained crystals of the Ca^{2+}-ATPase from muscle sarcoplasmic reticulum. *Journal of Molecular Biology.* 187:417–427.

Zampighi, G., J. Kyte, and W. Freytag. 1984. Structural organization of (Na^+ + K^+)-ATPase in purified membranes. *The Journal of Cell Biology.* 98:1851–1856.

Chapter 13

Functional Significance of the Oligomeric Structure of the Na,K-Pump from Radiation Inactivation and Ligand Binding

Jens G. Nørby and Jørgen Jensen

Institutes of Biophysics and Physiology, University of Aarhus, DK-8000 Aarhus C, Denmark

Abstract

The present article is concerned with the oligomeric structure and function of the Na,K-pump (Na,K-ATPase). The questions we have addressed, using radiation inactivation and target size analysis as well as ligand binding, are whether the minimal structural unit and the functional unit have more than one molecule of the catalytic subunit, α. We first discuss the fundamentals of the radiation inactivation method and emphasize the necessity for rigorous internal standardization with enzymes of known molecular mass. We then demonstrate that the radiation inactivation of Na,K-ATPase is a stepwise process which leads to intermediary fragments of the α-subunit with partial catalytic activity. From the target size analysis it is most likely that the membrane-bound Na,K-ATPase is structurally organized as a diprotomer containing two α-subunits. Determination of ADP- and ouabain-binding site stoichiometry favors a theory with one substrate site per $(\alpha\beta)_2$.

Introduction

To understand how the Na,K-pump (here also called Na,K-ATPase) effects the uphill, active transport of Na^+ and K^+ across cell membranes it is necessary to have a detailed knowledge of not only the chemical reaction mechanism of $(Na^+ + K^+)$-activated hydrolysis of ATP but also of the structural properties of the pump and its organization in the membrane. This paper is mainly concerned with the latter subject.

The building blocks of the membrane-bound Na,K-Pump are the α-subunit and the β-subunit, probably present in equimolar amounts. The protein molecular masses (m) are 112 and 35 kD for α and β, respectively (perhaps there is also a small γ-subunit with $m = \sim 10$ kD; this will be disregarded here). The α-subunit has all the ligand binding sites and is therefore called the catalytic subunit. The possible role for the β-subunit in expressing and organizing the pump is discussed in several papers in this volume.

This chapter concerns the questions of whether the functional unit of Na,K-ATPase has more than one molecule of the catalytic subunit and whether all the α-subunits really have catalytic sites. Like many other transport systems, including other transport ATPases (Nørby, 1987), the oligomeric structure and function of Na,K-ATPase is still a controversial subject that is hotly debated (Askari, 1988; Reynolds, 1988), and it appears that there are not only conflicting views but also conflicting experimental evidence.

In some of our own studies discussed below (Jensen and Nørby, 1988, 1989; Nørby and Jensen, 1989) we have used radiation inactivation to determine target sizes (TS) and radiation inactivation sizes (RIS; Beauregard et al., 1987), for the different properties of Na,K-ATPase. Furthermore we shall report and discuss some hitherto unpublished measurements of the ligand binding stoichiometry (in nanomoles per milligram protein) for ADP and ouabain binding to highly purified Na,K-ATPase preparations. The results from both types of investigation clearly point toward an oligomeric minimal functional unit containing two α-subunits, i.e., equal to $(\alpha\beta)_2$. Furthermore, if we take into account earlier experiments showing apparent negative cooperativity between high affinity ATP and ouabain binding sites (Ottolenghi and Jensen, 1983; Jensen et al., 1984), we arrive at the conclusion that the pump may be arranged in a tetrameric complex $(\alpha\beta)_4$.

Radiation Inactivation and Target Size Analysis

Fundamentals of Target Size Analysis

One of the physical methods that has been used to assess the size of the structural and functional units of enzymes and in situ membrane transport systems is radiation inactivation and target size analysis. The basic principles are as follows: when biological samples are irradiated with high energy electrons, the interaction of the electrons with the molecules studied results in loss of structure and function. The quantitative evaluation of these phenomena (target size analysis) depends on the following premises and events (Kepner and Macey, 1968; Jung, 1984; Harmon et al., 1985; Beauregard et al., 1987):

(*a*) Only hard x-rays, γ-rays, or high energy electrons from a linear accelerator, and not protons or α-particles, will fulfill the requirements for a quantitative method.

(*b*) The physical absorption of energy from fast, charged particles results in so-called primary events, i.e., ionization of atoms, where an orbital electron is stripped from the atom, leaving behind a positively charged protein.

(*c*) The ionizations occur *randomly* throughout the volume of the sample. With the type of radiation mentioned in (*a*), the distance between ionizations (or "hits") is $>1{,}000$ Å, i.e., much larger than the diameters of most proteins (50–150 Å). Under these conditions the ionization events can be considered to be *independent.*

(*d*) The primary ionization events release an enormous amount of energy (~ 60 eV per ionization; see below), enough to break several covalent bonds in a protein.

With these premises, one can use the statistical Poisson distribution formula to calculate the probability, $P(n)$, of n events taking place in a given volume, V m^3, of the sample. In our treatment here, V is a radiation-sensitive volume of the Na,K-ATPase molecule studied. If the radiation dose, D', expresses the number of inactivating events per cubic meter produced randomly throughout the material, then $V{\cdot}D'$ is the *average* number of events in V and:

$$P(n) = \frac{e^{-V \cdot D'}(V{\cdot}D')^n}{n!} \tag{1}$$

(Hutchinson and Pollard, 1961; Kempner and Schlegel, 1979; Jung, 1984; Harmon et al., 1985). If one hit (one ionization) is sufficient to destroy the structure, and thereby the function, in the sensitive volume V, then the surviving structures are those that received zero hits during the time of the experiment ($n = 0$ in Eq. 1). Expressed as the fraction of original, native molecules, this is:

$$S/S_0 = e^{-V \cdot D'} \qquad \text{or} \qquad A/A_0 = e^{-V \cdot D'} \tag{2}$$

where S and A stand for peptide structure and biological activity, respectively. To use these equations to determine target sizes or radiation inactivation sizes of sensitive domains expressed in molecular mass units, kilograms per mole, they are transformed as follows (doses are measured in Gy [joules per kilogram] and not in events per cubic meter, target size is measured in kilograms per mole = kilodaltons and not in cubic meters):

$$D(\text{Gy}) = D'(\text{events/m}^3)\cdot 1/\rho(\text{m}^3/\text{kg})\cdot Q(\text{eV/event})\cdot 1.6\cdot 10^{-19}(\text{joules/eV}) \tag{3}$$

$$m(\text{kD}) = V(\text{m}^3/\text{target})\cdot \rho(\text{kg/m}^3)\cdot 6.023\cdot 10^{23}(\text{molecules/mol}) \tag{4}$$

Combining Eqs. 3 and 4 we get:

$$V \cdot D' = \frac{m}{\rho \cdot 6.023 \cdot 10^{23}} \cdot \frac{D \cdot \rho}{Q \cdot 1.6 \cdot 10^{-19}} = \frac{m}{Q \cdot 9.64 \cdot 10^{4}} \cdot D \tag{5}$$

Insertion of Eq. 5 into Eq. 2 and logarithmic transformation lead to:

$$\ln (S/S_0) = -\gamma \cdot D \quad \text{and} \quad \ln (A/A_0) = -\gamma \cdot D \tag{6}$$

where

$$\gamma = m/(Q \cdot 9.64 \cdot 10^4).$$

If the property investigated disappears as a monoexponential function of dose, D, if D(Gy) for the sample can be measured accurately, and provided Q is known, then m can be determined from the slope ($-\gamma$, Gy^{-1}) of ln (S/S_0) or ln (A/A_0) vs. D, because:

$$m(\text{kD}) = \gamma(\text{Gy}^{-1}) \cdot Q \cdot 9.64 \cdot 10^4 \tag{7}$$

Note that the decay constant γ is the reciprocal of the dose at which the property investigated has been reduced to 37% of the original value: $\gamma = 1/D_{37}$, because $e^{-1} = 0.368$.

We shall briefly consider whether the basic requirements for the quantitative use of Eq. 7 are fulfilled. First, the dose measurements (and therefore the slope γ) are of course subject to random error, but what is more serious, there can also be systematic errors (Lo et al., 1982; Lai et al., 1987). Even if dose is measured very accurately and reproducibly by, for example, a calorimeter that accompanies the samples in the accelerator, as in our studies (Jensen and Nørby, 1988), there is generally no guarantee that the dose absorbed by the samples is the same as that absorbed by the calorimeter (but hopefully and most likely D[sample] is proportional to D[calorimeter] and/or D[sample]/D[calorimeter] is not far from 1, see below). Second, the value for Q is not known with sufficient accuracy, and it may well vary with experimental conditions (for temperature effects, see below). In early work, values of 80–110 eV, e.g., based on energy loss of charged particles in gases, was used (Hutchinson and Pollard, 1961; Rauth and Simpson, 1964). The latter authors showed, however, that for carbon, hydrocarbons, and combinations of materials of low atomic number, the average energy loss per inelastic event was lower, namely, 60 ± 10 eV. The uncertainty of the value of Q clearly can introduce an error in the calculation of m by Eq. 7. A third complicating factor, not apparent from the theoretical treatment above, is the effect of temperature on the decay constant γ. The observation is that the lower the temperature of the sample during irradiation, the lower the value of γ. The mechanism for this effect is not entirely clear, especially since the primary effects of the initial collisions, the ionizations, are not temperature dependent (Kempner et al., 1986; Beauregard et al., 1987). Kempner et al. (1986) have compared the change in radiation sensitivity with temperature of biological macromolecules and synthetic polymers and conclude that the temperature effect is related to the formation and reactions of free radicals on the affected macromolecule, subsequent to the initial collision. In some cases, where the temperature sensitivity of the dosimeter is the same as that of the biological macromolecule investigated, the temperature effect is not observed. This occurs when bleaching of

Blue Cellophane (Du Pont Co., Wilmington, DE) is used for dose measurement (Jung, 1984). The temperature factor on γ is considerable, i.e., ~ 2 per 100°C (here the dosimeter is temperature independent: Kempner and Schlegel, 1979; Kempner and Haigler, 1982; Jensen and Nørby, 1988) but there are some variations (Fluke, 1987).

Empirical Standardization of Target Size Analysis

From the above discussion it is obvious that although the linear relationship between molecular mass, m, and the decay constant, γ ($=1/D_{37}$), may reflect fundamental properties of the interaction of fast electrons with biological macromolecules, the quantitative evaluation is far from straightforward. It has therefore been necessary to determine empirically the relationship between m and γ ($=1/D_{37}$). The most commonly used equation is that of Kepner and Macey (1968), based on a collection of corresponding values for m and D_{37} published from 1951 through 1967:

$$m\,(\text{kD}) = \gamma\,(\text{Gy}^{-1})\cdot 6.4\cdot 10^{6} \tag{8}$$

All their data are from lyophilized preparations irradiated at room temperature, and m ranges from 0.5 to $5\cdot 10^{3}$ kD. The average value for Q (eV/event, see above) derived from a log/log plot of "Mol. Wt." vs. "D_{37}" is 66 eV. This value, and thus the factor 6.4, is, however, determined with a considerable degree of uncertainty, the possible range being from 50 to 100 eV. Eq. 8 has been widely used under conditions differing from those under which it was originally established. Beauregard et al. (1987) have combined this equation with the temperature dependence established empirically by Kempner and Haigler (1982):

$$\log m\,(\text{kD}) = 6.89 + \log \gamma_t\,(\text{Gy}^{-1}) - 0.0028\cdot t \tag{9}$$

(Note that this equation assumes that the factor 6.4 [Eq. 8] is valid at 30°C.)

Both Lo et al. (1982) and Lai et al. (1987) have discussed and criticized Eq. 8, and thus Eq. 9. They point out that the methods of dose measurement vary, that the original D_{37}'s often are in units that are difficult to convert to rad (or Gy) and therefore difficult to compare, and that many of the molecular weights used are incorrect and outdated. We agree with these conclusions and find the routine use of Eq. 8, combined with the temperature correction (9), too uncertain and inaccurate for our purpose. In studies of transport ATPases it is a further drawback that Kepner and Macey (1968) have no data in the m range 10^{2}–10^{3} kD.

An alternative approach to the application of Eqs. 8 and 9 is the use of one or several standard proteins and enzymes of known molecular mass which are irradiated in the same experimental setup as the samples. Subsequent measurements of structure and/or biological activity allow determination of the decay constant γ for each standard (see Eq. 6) and thereby the construction of a standard curve, which shows γ vs. m (cf. Figs. 4 and 5 in Jensen and Nørby [1988]). This might obviate the theoretical need to know the exact doses, that the samples received (the factor $= D[\text{sample}]/D[\text{dosimeter}]$), the exact value of Q, and the "temperature correction factor" (Eq. 9).

There are two types of standard curves, one based on measurement of peptide integrity and one based on activity measurements. Examples of the latter type are given by Lo et al. (1982), Venter et al. (1983), and Lai et al. (1987). We have constructed a standard curve by irradiating four different proteins with 10 MeV electrons at a dose rate of ~2 Mrad/3 min. The proteins were the albumin monomer and three enzymes: bacterial rhodopsin, glucose-6-phosphate dehydrogenase (G-6-PDH, from *Leuconostoc mesenteroides*), and β-galactosidase, the radiation-mediated breakdown was measured by SDS electrophoresis, and the molecular mass of the monomers was used. We have further demonstrated that G-6-PDH activity, which is elicited by the dimer of this enzyme, also lies on this standard curve with m corresponding to the dimer, as generally accepted (McIntyre and Churchill, 1985).

In theory, and as outlined by Kepner and Macey (1968), it should be possible from standard curves as those described above to arrive at an estimate of Q because the slope of the standard curve according to Eq. 7 is inversely proportional to Q, namely "slope" = $1/(Q \cdot 9.64 \cdot 10^4)$. It is also clear, however, that an estimate so obtained will contain a temperature factor as well as a factor characterizing the discrepancy between D(sample) and D(dosimeter). The mean value for Q from the data of Kepner and Macey (1968) was 66 eV. Our data corrected for the temperature difference (Jensen and Nørby, 1988) gave 54 eV. Recently le Maire et al. (1990) have studied the radiation-induced fragmentation of nine standard water-soluble proteins also using SDS-gel electrophoresis and developed a calibration curve of $1/D_{37}$ vs. the known monomeric molecular masses in the range of 20–120 kD. Their curve corresponds to a value for Q of ~90 eV. le Maire et al. (1990) note on p. 437 of their paper that they find a value ~35% higher than Kepner and Macey's and that our value is also higher. This is not so. Our value is ~20% lower than 66 eV and thus only ~60% of le Maire's. The fact that these values are quite compatible with those calculated or determined by purely physical principles as mentioned earlier in this paper, and that they fall within a factor of 2 of each other, suggest that the temperature effects do not differ too much between these series of experiments (see also Jensen and Nørby, 1988) and that D(sample) is not radically different from the measured doses, D(dosimeter). Nevertheless, the differences between the standard curves clearly illustrate that it is imperative to have a standard curve for each individual experimental arrangement. Considering the many factors that influence the value of Q, it is not justified to analyze the discrepancy between the Kepner and Macey standard curve ("Q" = 66 eV), which is based on activity measurements, and that of le Maire et al. ("Q" = 90 eV), based on fragmentation analysis, in the manner that le Maire et al. (1990) do. They conclude that this difference " . . . show[s] that a small but significant proportion of the molecules hit are inactivated without being fragmented, as detected by SDS/PAGE. . . . " As outlined above (even disregarding all the reservations one could have in connection with the basic data of Kepner and Macey) the difference in "Q" could just as well (and more likely?) be due to systematic deviations in the measurement of the true D(sample).

Instead of establishing a full standard curve, one may also use one internal enzyme standard. The glucose-6-phosphate dehydrogenase just mentioned has proven useful in this respect (McIntyre and Churchill, 1985; Jensen and Nørby, 1988; Nørby and Jensen, 1989).

The Oligomeric Structure of the Na,K-Pump

Results from Radiation Inactivation

The target size (TS) for the α-peptide of Na,K-ATPase and the radiation inactivation sizes (RIS) for several of its enzymatic properties, obtained by us during the last three years, are given in Table I. Without presenting a detailed review of all the published TS and RIS values for Na,K-ATPase (for some references, see Glynn, 1985; Nørby, 1987; Jensen and Nørby, 1988) we would like to emphasize that our values are generally lower than (in a few cases equal to) those published earlier. The reason for this may be that we have used our own calibration curve, and have taken great care to avoid artifacts such as secondary effects of free radicals in the medium, and to avoid aging or repeated thawing and freezing of samples after irradiation, which lead to progressive denaturation (and possibly also disintegration) of the enzyme molecules. These artifacts would result in higher decay constants and thereby higher TS and RIS values.

The values in Table I, which fall in three categories, namely, equal to, lower than, and higher than $m(\alpha)$, are described and discussed extensively in the original papers. We shall here emphasize the following points:

(*a*) TS or RIS = $m(\alpha)$. TS for the α-peptide integrity is identical to $m(\alpha) = 112$ kD. This shows that there is no transfer of radiation energy between the subunits of

TABLE I
Radiation Inactivation Sizes for Na,K-ATPase

Property	kD
Tl^+ occlusion, Tl^+ binding, total capacity	40–70
ATP, ADP, and ouabain binding, total capacity	70
K-pNPPase activity, V_{max}	106
ATP, ADP, ouabain, and vanadate binding with original affinity	110
α-peptide integrity, target size	115
Na-ATPase activity, V_{max}	135
Na,K-ATPase activity	
(10 μM ATP in assay)	140
(50 μM ATP in assay)	163
(3 mM ATP in assay), V_{max}	190
Tl^+ occlusion, Tl^+ binding with original affinity	192

Jensen and Nørby, 1988, 1989; Nørby and Jensen, 1989.

Na,K-ATPase. A similar result was obtained by Karlish and Kempner (1984) for the α-peptide, and TS close to the monomer M_r was found by Kempner and Miller (1983) for the hexameric glutamate dehydrogenase, by le Maire et al. (1990) for 12 soluble, oligomeric proteins, and by Rabon et al. (1988) for the oligomeric H,K-ATPase from gastric mucosa, although it had earlier been claimed that TS for H,K-ATPase was ~2.5 times M_r for the monomer (Saccomani et al., 1981; Beauregard et al., 1987). We consider it the rule that there is no energy transfer between noncovalently related peptides and that therefore TS is equal to the subunit size (as also concluded by le Maire et al., 1990). The unmodified, "native" binding of nucleotide, vanadate, ouabain, and the K^+-activated *p*-nitrophenylphosphatase activity all have RIS = $m(\alpha)$. This substantiates the lack of destructive energy transfer between subunits and strongly suggests that the α-monomer retains important, unmodified biological function. Destruction of a neighboring α-subunit (if there is any) or of the β-subunit seems to have no influence on these properties.

(*b*) RIS lower than $m(\alpha)$, namely for total binding capacity for ATP, ADP, ouabain, and the K^+-congener Tl^+. From our published binding experiments it is obvious that some partly damaged Na,K-ATPase molecules retain the capacity to bind the ligands mentioned, albeit with affinities that are lower than those for native binding. This is contrary to the previous general belief that one hit by a high-energy electron would completely fragment and destroy the entire target peptide (see also Introduction, Nørby and Jensen, 1989). Accordingly, radiation inactivation and fragmentation must be considered a stepwise process involving sensitive domains rather than entire peptides. In our analysis of each of the ADP, ouabain, and Tl^+-binding isotherms, we have assumed that the irradiated samples contain two populations of binding sites, one with "native" properties residing on the undamaged molecules, and another with lower affinity which is located on some of the partly fragmented α-peptides. This analysis shows that the modified binding sites are located on a "70-kD domain" in the fragmented subunits. If this domain is destroyed by a second hit the binding capacity is lost. This concept is illustrated in Fig. 1, which also shows that a RIS of 70 kD is not equal to a TS of 70 kD, i.e., the different peptides that have the intact "70-kD domain" may in principle have molecular masses in the range lower than 112 but higher than 70 kD. The model in Fig. 1

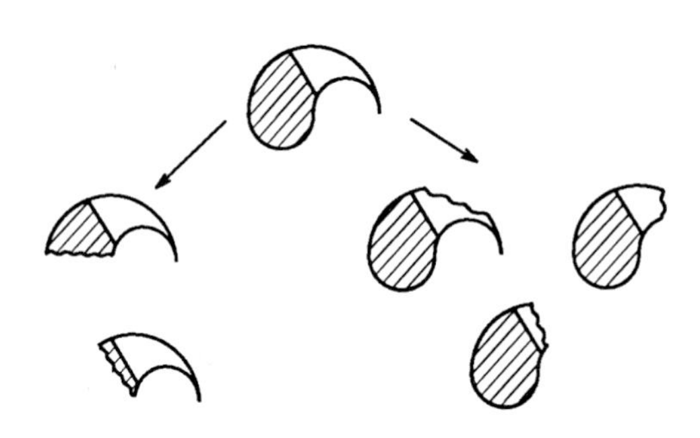

Figure 1. The stepwise destruction of an α-monomer by radiation inactivation. In this model the monomer consists of a 70-kD (*shaded*) and a 42-kD domain, and the first inactivating step is related to a hit in one of the two domains. The function of the affected domain is destroyed, but the remaining structure, the peptide, can still be larger than 70 (*right side*) or 42 kD (*left side*), so that a spectrum of peptides is produced.

provides an explanation for our inability to demonstrate sharp bands with $m < m(\alpha)$ on Coomassie blue stained SDS-PAGE gels. Furthermore, the fragments are produced and destroyed in the radiation process, and calculations show that in total they never amount to more than $\sim 18\%$ of the original amount of α in the sample (Fig. 4 in Nørby and Jensen, 1989). Recently le Maire et al. (1990) irradiated a number of soluble proteins and observed the appearance of peptides with lower M_r than the subunits. For some proteins the pattern of peptide bands produced on the SDS gels suggests that some bonds are more susceptible to radiation-induced breakage than others; for other proteins no distinct bands, but rather a smear, appears after irradiation.

In an attempt to demonstrate in a direct way the presence of functionally active fragments of the α-peptide, we have phosphorylated samples of the irradiated enzyme using [γ-^{32}P]ATP and subjected the isolated phosphoenzymes to gel electrophoresis with the result shown in Fig. 2. It appears that several peptides with a molecular mass between 100 and 40 kD, with a predominating group around 90–60 kD, can be phosphorylated, and they therefore must have a functional ATP binding site. Hints that rather small peptides may have functional capacity come from the expression studies of Fambrough and Huang (1991) in which synthesis of an ATP-binding 47-kD fragment of α was shown, and from Karlish et al. (1990), who after extensive trypsinolysis of the α-subunit could isolate a 19-kD fragment as a likely candidate for K^+ (and Na^+) occlusion. The fact that there are several fragmented, ATP-binding peptides, could justify a Scatchard plot analysis with a multitude of binding constants instead of just the two in the simple model we used. This might result in another relationship between total binding capacity and radiation dose, and thereby (an)other RIS value(s) for that property, perhaps more in accordance with the picture in Fig. 2. Unfortunately our binding data, although very accurate, cannot support a complicated analysis with more than two populations.

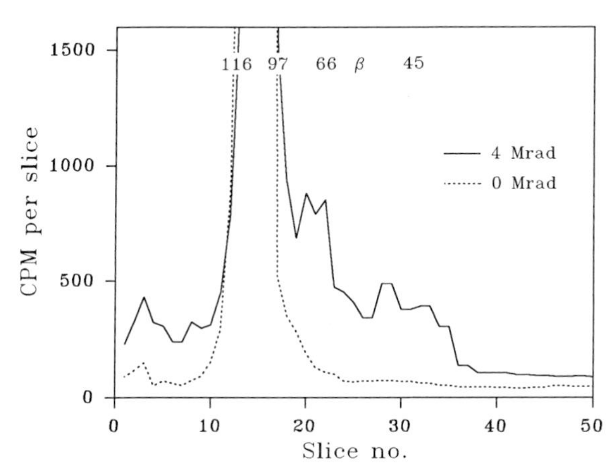

Figure 2. Demonstration of phosphorylated fragments of the α-peptide of Na,K-ATPase from pig kidney. The enzyme was irradiated (Jensen and Nørby, 1988) with 4 Mrad (40 kGy), and a sample phosphorylated with [γ-^{32}P]ATP for 60 s at 0°C (Klodos et al., 1981), the phosphoenzyme was precipitated with Na-phosphate/phosphoric acid at pH = 2.4 and spun down in an airfuge in the cold. The pellet was resuspended in ice cold trisphosphate buffer, pH 6, containing 2.5% lithium dodecylsulphate (LDS), 1.7% sucrose and 1% mercaptoethanol, and samples containing ~40 μg protein were applied to a 7.5% polyacrylamide slab-gel containing 1% LDS. Electrophoresis was performed at 4°C, 1 h at 25 V, 5 h at 100 V, with trisphosphate buffer, pH = 6, containing 1% LDS. The gel was then cut in 1.5-mm slices, which were counted for ^{32}P in the scintillation counter, full curve. A nonirradiated sample containing the same amount of unmodified α-peptide as the irradiated sample was treated likewise (*dotted curve*). Contaminating unbound ^{32}P appears at the bottom of the gel at slice 75–85 (not shown). The phosphorylation is Na^+-stimulated and the phosphoenzymes are sensitive to ADP and K^+. The positions of molecular weight standards and the β-subunit are indicated in the figure.

(*c*) RIS values larger than $m(\alpha)$. The RIS for V_{max} for Na,K-ATPase and for unmodified binding and occlusion of the K^+-congener Tl^+ is ~190 kD. The latter two phenomena are closely related and one explanation for these RIS's is that an assembly larger than an α-monomer is structurally (but not functionally) involved in performing the full Na,K-ATPase cycle and Na^+,K^+ transport. From many experiments we can say with certainty that the RIS for the full cycle functions is larger than $m(\alpha\beta) = 147$ and smaller than $m(\alpha)_2 = 224$. The comprehensive model we have developed to explain our TS and RIS data (Table I) is shown as Fig. 9 in Nørby and Jensen (1989), and in an abbreviated form in Fig. 3, *A* or *B* (see also below) of this article. In this model the α-subunits are organized as dimers, $(\alpha)_2$, in the membrane and for reasons discussed previously, there is no need for any functional assignment of the β-subunit. The radiation inactivation of the $(\alpha)_2$-dimer, also symbolized by

(112/112), follows a stepwise model involving two sensitive domains (42 and 70 kD) as described above under (*b*). A single α-subunit has full, native Na,K-ATPase and Tl^+-occluding activity provided its secondary and tertiary structure is preserved. This can be done by structural stabilization from a neighboring α-peptide in a (112/112) dimer, or from a neighboring α-peptide fragment, i.e., in a (112/70) complex. This model yields a theoretical RIS for the unmodified activity of 112 + 70 = 182 kD. The model analogously assumes that (112/112), (112/70), and (70/70) all have Na-ATPase activity, which give a calculated RIS for Na-ATPase of 140 kD. Fig. 3 of the present paper illustrates, with simplified versions of the original model (Nørby and Jensen, 1989), three hypotheses for the functional, dimeric arrangement of α-subunits and their radiation inactivation. In the native dimer, column I, the two subunits both have activity and binding sites in *A*, only one of the subunits in the dimer have activity in *B*,

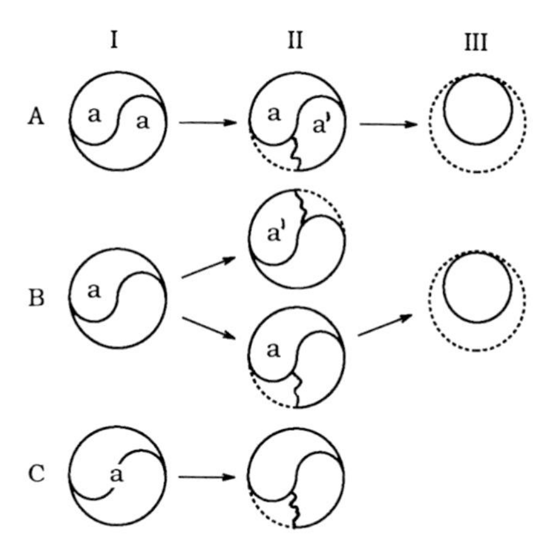

Figure 3. Three models (*A*, *B*, and *C*) for a dimeric assembly of the α-peptide of Na,K-ATPase and its radiation inactivation. Column I shows the native, undamaged dimer. The α peptides marked with *a* have full Na- and Na,K-ATPase activity, and those marked with *a′* have only Na-ATPase activity. In *C* there is a common ATP-binding site and the two peptides must cooperate to provide full activity. Column II shows the dimers after destruction of the "42-kD domain" in one of the α- subunits (cf. Fig. 1 and Nørby and Jensen, 1989). The full Na,K-ATPase activity (if there was any) of the intact α-peptide is retained (AII and BII) because it is structurally stabilized by its neighbor peptide. In column III (*A* and *B*) one α-peptide is totally destroyed (at least the "70 kD domain" has disappeared) and the intact α-peptide has lost its ATPase activity completely due to unfolding of the subunit. The RIS values for V_{max} of the Na,K-ATPase activity would be 182 (*A*), 182 (*B*), and 224 kD (*C*), and 140 (*A*), 140 (*B*), and 224 kD (*C*) for Na-ATPase activity (see Nørby and Jensen, 1989) for calculations.

and in model *C*, the two subunits share the binding site and the activity (units marked by *a* have Na- and Na,K-ATPase activity and normal Tl^+ occlusion, units marked by *a′* have Na-ATPase activity). Calculations will show that the observations in Table I do not allow us to discriminate between models *A* and *B* because they will give identical RIS values. Measurement of ligand binding stoichiometry (next section of this paper) is one way to evaluate the possibility of a dimeric α_2 structure with only one binding site like model *B*. Model *C*, however, where Na- and Na,K-ATPase is dependent upon two intact α-subunits, corresponds to a RIS of 2 × 112 = 224 kD for both Na- and Na,K-ATPase activity and must therefore be discarded.

Our conclusion from these studies is that the Na,K-ATPase is organized as a structural dimer (diprotomer $[\alpha\beta]_2$, although we cannot in fact exclude higher molecular weight oligomers) containing two structurally stabilized, but functionally independent, α-peptides. The β-subunits do not appear to have a functional, catalytic role in this complex (for some references see Nørby, 1989). Recently, the oligomeric nature of the membrane-bound Na,K-ATPase has been studied by other physical or biochemical techniques. Electron spin resonance that measures the mobility of spin labels attached to Na,K-ATPase in situ points to a minimal structural unit of $(\alpha\beta)_2$ and perhaps also higher oligomers (Esmann et al., 1989). Likewise, Hayashi et al. (1989), using a combination of techniques including low angle laser scattering photometry and high performance liquid chromatography, could demonstrate an equilibrium between $\alpha\beta$-protomers and $(\alpha\beta)_2$ diprotomers in solubilized enzyme, the equilibrium constant being such that diprotomers or higher oligomers would predominate in the membrane. All oligomeric populations, as well as the $\alpha\beta$-protomers (although they are unstable), seem to exhibit the same Na,K-ATPase activity per amount of α, as we have suggested from the radiation inactivation studies.

Minimal Functional Unit Evaluated from Ligand Binding Stoichiometry

Over the years several characteristics of the internal stoichiometry of ligand binding to Na,K-ATPase have been convincingly established: in a given nondenatured preparation of Na,K-ATPase, the number of high affinity binding sites for nucleotides, vanadate, and ouabain, and for phosphorylation by P_i and ATP are all the same (for references, see Nørby, 1983), and it has been shown that ouabain binds to the same α-subunit that is phosphorylated (Forbush and Hoffman, 1979). Furthermore, Na,K-ATPase binds and occludes two ions of K^+ or its congeners Rb^+ and Tl^+, and probably three Na^+-ions per ADP site (Jensen and Nørby, 1989; Vilsen et al., 1987).

Surprisingly enough, however, there is still considerable uncertainty as regards the minimum equivalent unit, the minimal molecular weight, as determined by the binding site concentration expressed per milligram of protein. Theoretically the minimal molecular weight of an enzyme that has one binding site per minimal unit is the reciprocal of the binding site concentration. The determination requires both accurate binding measurements corrected for unspecific binding and often involving extrapolation to infinite ligand concentration (Nørby and Jensen, 1988), and an accurate and "true" determination of the protein concentration. Generally, the binding data can be obtained on purified preparations (and due to the specificity of e.g., ADP and ATP binding, also on less pure preparations although ouabain binding has a low-affinity, unspecific component) without serious problems. The controversy as to the stoichiometry of binding is therefore concentrated on the protein determinations. Actually this controversy has been around for the last decade as is evident from inspection of the proceedings of the previous three Na,K-ATPase meetings (Nørby, 1983; Ottolenghi and Jensen, 1985; Reynolds, 1988). The routinely used methods in most laboratories are based on the Lowry method (Lowry et al., 1951) and the problem has been (or is) to what extent, if at all, these methods overestimate the protein concentration in the purified Na,K-ATPase preparations. If the minimal unit is the $\alpha\beta$-protomer with m = 147 kD, the site concentration should maximally be 6.8 nmol/mg protein, if it is the $(\alpha\beta)_2$-dimer it should be 3.4 nmol/mg protein. Laboratories that use the Lowry method with a correction factor based on amino acid analysis of either the albumin standard (Moczydlowski and Fortes, 1981, correction factor 1.07–1.18) or a pure Na,K-ATPase protein standard (e.g., Vilsen et al., 1987; Jørgensen, 1988; no correction factor published) can find values of 5–6 nmol/mg protein of TNP-ATP-, vanadate- or ouabain binding, or phosphorylation.

(Note that Hayashi et al. [1989] report that they corrected their protein determination by multiplication with 1.144, whereas in an earlier paper Hayashi et al. [1983] say that it " . . . was corrected by dividing it by a factor of 1.144 . . . "). When the Lowry values are used without correction the stoichiometry almost invariably is lower than 4 nmol/mg protein for the membrane-bound enzyme (e.g., Jensen and Ottolenghi, 1983; Serpersu et al., 1990; this paper). These differences are of course also reflected in the published specific enzyme activities (SA, U/mg protein): without correction of the Lowry protein, the purest preparations have SA around 35 U/mg protein (e.g., Serpersu et al., 1990; this paper), whereas published values based on corrected protein measurements are in the range of 40–50 (Moczydlowski and Fortes, 1981; Vilsen et al., 1987; Jørgensen, 1988; Hayashi et al., 1989).

It is not necessary for us here to repeat or elaborate on the thorough discussion given by Ottolenghi and Jensen (1985) and by Reynolds (1988) of the very important problem of site concentration. We wish, however, to report a few data obtained recently in a renewed attempt to shed some light on this controversy and also, specifically, to distinguish between models *A* and *B* in Fig. 3 (which, as noted above, will both satisfy our radiation inactivation data). During one year we prepared five batches of purified Na,K-ATPase from outer medulla of pig kidney (method of Jørgensen, 1988, Fig. 1) with SA = 33.5 ± 1.3 (mean ± SD) U/mg protein (Lowry method). The ADP binding capacity of these preparation obtained by extrapolation of straight line isotherms in Scatchard plots (Nørby and Jensen, 1988) was 3.626 ± 0.034 (mean ± SD) nmol/mg protein. The average turnover was thus 9,240 min^{-1}. This binding site stoichiometry is very close to 1 per 2 α-subunits (theoretical 3.4 nmol/mg protein). To evaluate the protein determination in our laboratory it was arranged that we and five laboratories in the Institute of Biophysics measured protein on two different purified Na,K-ATPase preparations from shark rectal gland and one albumin solution. The coefficient of variation (SD%) among laboratories was 5% both with the original Lowry method and with the modified (Peterson, 1977). The latter gave 10–15% higher protein concentrations than the original Lowry method. Our values were within ±1 SD. If we use Peterson's method our binding site concentration would be ~3.2 nmol/mg protein.

We have also investigated whether the SDS treatment during the preparation of purified Na,K-ATPase could result in denaturation and disappearance of binding sites, and we found that this was not the case. In one experiment ouabain and ADP binding sites on crude microsomes were both 0.5 nmol/mg protein, and after SDS treatment ADP- and phosphorylation sites were 0.52 nmol/mg protein. In another preparation ouabain binding sites were 0.35 and 0.36 nmol/mg before and after incubation with SDS. The turnover was increased fivefold up to 8,900 min^{-1}.

If the Lowry method measures correctly, our observations favor, but do not prove, a theory with one ATP, ADP, and ouabain binding site per $(\alpha\beta)_2$ over a theory with two binding sites. It may be argued that 3.6 is larger than the theoretical 3.4 nmol/mg, that our preparations contain significantly less than 100% active ATPase, and thus that 3.6 nmol/mg is too high for one site per active $(\alpha\beta)_2$. Alternatively, one might envisage that the α/β stoichiometry could be slightly higher than 1, allowing for binding sites on some $\alpha_2\beta$-complexes. This would raise the theoretical stoichiometry above 3.4 nmol/mg. The complexity of the problem is increased because under special conditions molecular Na,K-ATPase assemblies can be extracted that certainly seem to have a site concentration closer to two sites per $(\alpha\beta)_2$ (Jensen and Ottolenghi, 1983; Ottolenghi et al., 1986).

Conclusion

From our radiation inactivation and ligand binding measurements we consider it most likely that the membrane-bound Na,K-pump is structurally organized as a diprotomer with two α-subunits and probably two β-subunits. Although our binding-site determinations on membrane-bound enzyme favor a theory with one substrate site per $(\alpha\beta)_2$, other independent techniques will have to be used before the size of the minimal functional unit of the Na,K-pump can be definitively established.

Acknowledgments

We are grateful to Dr. Ellis S. Kempner for stimulating conversations and correspondence before the preparation of this article, to Dr. Amir Askari for suggestions on the oligomeric structure of Na,K-ATPase, and to Toke Nørby for expert assistance with the manuscript.

Our work was supported by grant 12-8174 from The Danish Medical Research Council and by The Biomembrane Research Center, University of Aarhus.

References

Askari, A. 1988. Ligand binding sites of (Na^+ + K^+)-ATPase: nucleotides and cations. *In* The Na^+, K^+-Pump, Part A: Molecular Aspects. J. C. Skou, J. G. Nørby, A. B. Maunsbach, and M. Esmann, editors. Alan R. Liss, Inc., New York. 149–165.

Beauregard, G., A. Maret, R. Salvayre, and M. Potier. 1987. The radiation inactivation method as a tool to study structure-function relationships in proteins. *In* Methods of Biochemical Analysis. D. Glick, editor. John Wiley and Sons, New York. 32:313–343.

Esmann, M., H. O. Hankovszky, K. Hideg, and D. Marsh. 1989. A novel spin-label for study of membrane rotational diffusion using saturation transfer electron spin resonance. Application to selectively labelled class I and class II-SH groups of the shark rectal gland Na^+/K^+-ATPase. *Biochimica et Biophysica Acta.* 978:209–215.

Fambrough, D. M., and R. C. C. Huang. 1991. Expression of domains of the Na,K-ATPase in a run-away plasmid replication system in bacteria. *In* The Sodium Pump: Recent Developments. J. H. Kaplan and P. De Weer, editors. The Rockefeller University Press, New York. In press.

Fluke, D. J. 1987. Effect of irradiation temperature. *In* Target-Size Analysis of Membrane Proteins. J. Venter and C. Y. Jung, editors. Alan R. Liss, Inc., New York. 21–32.

Forbush, B., III, and J. F. Hoffman. 1979. Evidence that ouabain binds to the same large polypeptide chain of dimeric Na,K-ATPase that is phosphorylated from P_i. *Biochemistry.* 18:2308–2315.

Glynn, I. M. 1985. The Na^+,K^+-transporting adenosine triphosphatase. *In* The Enzymes of Biological Membranes. 2nd ed. A. N. Martonosi, editor. Plenum Press, New York, London. 35–114.

Harmon J. T., T. B. Nielsen, and E. S. Kempner. 1985. Molecular weight determinations from radiation inactivation. *Methods in Enzymology.* 117:65–94.

Hayashi, Y., K. Mimura, H. Matsui, and T. Takagi. 1989. Minimum enzyme unit for Na^+/K^+-ATPase is the αβ-protomer. Determination by low-angle laser light scattering photometry coupled with high-performance gel chromatography for substantially simultaneous measurement of ATPase activity and molecular weight. *Biochimica et Biophysica Acta.* 983:217–229.

Hayashi, Y., T. Takagi, S. Maezawa, and H. Matsui. 1983. Molecular weights of αβ protomeric and oligomeric units of soluble (Na^+,K^+)-ATPase determined by low-angle laser light scattering after high-performance gel chromatography. *Biochimica et Biophysica Acta.* 748:153–167.

Hutchinson, F., and E. Pollard. 1961. Target theory and radiation effects on biological molecules. *In* Mechanisms in Radiobiology. M. Errera and A. G. Forsberg, editors. Academic Press, New York. 11:71–92.

Jensen, J., and J. G. Nørby. 1988. Membrane-bound Na,K-ATPase: target size and radiation inactivation size of some of its enzymatic reactions. *Journal of Biological Chemistry.* 263:18063–18070.

Jensen, J., and J. G. Nørby. 1989. Thallium binding to native and radiation-inactivated Na^+/K^+-ATPase. *Biochimica et Biophysica Acta.* 985:248–254.

Jensen, J., J. G. Nørby, and P. Ottolenghi. 1984. Sodium and potassium binding to the sodium pump of pig kidney: stoichiometry and affinities evaluated from nucleotide-binding behaviour. *Journal of Physiology.* 346:219–241.

Jensen, J., and P. Ottolenghi. 1983. ATP binding to solubilized (Na^+ + K^+)-ATPase. The abolition of subunit-subunit interaction and the maximum weight of the nucleotide-binding unit. *Biochimica et Biophysica Acta.* 731:282–289.

Jørgensen, P. L. 1988. Purification of Na^+,K^+-ATPase: enzyme sources, preparative problems, and preparation from mammalian kidney. *Methods in Enzymology.* 156:29–43.

Jung, C. Y. 1984. Molecular weight determination by radiation inactivation. *In* Molecular and Chemical Characterization of Membrane Receptors. J. C. Venter and L. C. Harrison, editors. Alan R. Liss, Inc., New York. 193–208.

Karlish, S. J. D., R. Goldshleger, and W. D. Stein. 1990. A 19-kDa C-terminal tryptic fragment of the α chain of Na/K-ATPase is essential for occlusion and transport of cations. *Proceedings of the National Academy of Sciences, USA.* 87:4566–4570.

Karlish, S. J. D., and E. S. Kempner. 1984. Minimal functional unit for transport and enzyme activities of (Na^+ + K^+)-ATPase as determined by radiation inactivation. *Biochimica et Biophysica Acta.* 776:288–298.

Kempner, E. S., and H. T. Haigler. 1982. The influence of low temperature on the radiation sensitivity of enzymes. *Journal of Biological Chemistry.* 257:13297–13299.

Kempner, E. S., and J. H. Miller. 1983. Radiation inactivation of glutamate dehydrogenase hexamer: lack of energy transfer between subunits. *Science.* 222:586–589.

Kempner, E. S., and W. Schlegel. 1979. Size determination of enzymes by radiation inactivation. *Analytical Biochemistry.* 92:2–10.

Kempner, E. S., R. Wood, and R. Salovey. 1986. The temperature dependence of radiation sensitivity of large molecules. *Journal of Polymer Science, Part D: Macromolecular Reviews* 24:2337–2343.

Kepner, G. R., and R. I. Macey. 1968. Membrane enzyme systems. Molecular size determinations by radiation inactivation. *Biochimica et Biophysica Acta.* 163:188–203.

Klodos, I., J. G. Nørby and I. W. Plesner. 1981. The steady-state mechanism of ATP hydrolysis by membrane-bound Na,K-ATPase from ox brain. II. Kinetic characterization of phospho-intermediates. *Biochimica et Biophysica Acta.* 643:463–482.

Lai, F. A., M. M. S. Lo, and E. A. Barnard. 1987. Target size determination by irradiation of enzymes as internal standards. *In* Target-Size Analysis of Membrane Proteins. J. C. Venter and C. Y. Jung, editors. Alan R. Liss, Inc., New York. 33–41.

le Maire, M., L. Thauvette, B. de Foresta, A. Viel, G. Beauregard, and M. Potier. 1990. Effects of ionizing radiations on proteins. Evidence of non-random fragmentation and a caution in the use of the method for determination of molecular mass. *Biochemical Journal.* 267:431–439.

Lo, M. M. S., E. A. Barnard, and J. O. Dolly. 1982. Size of acetylcholine receptors in the membrane. An improved version of the radiation inactivation method. *Biochemistry.* 21:2210–2217.

Lowry, O. H., N. J. Rosebrough, A. L. Farr, and R. J. Randall. 1951. Protein measurement with the folin phenol reagent. *Journal of Biological Chemistry.* 193:265–275.

McIntyre, J. O., and P. Churchill. 1985. Glucose-6-phosphate dehydrogenase from *Leuconostoc mesenteroides* is a reliable internal standard for radiation inactivation studies of membranes in the frozen state. *Analytical Biochemistry.* 147:468–477.

Moczydlowski, E. G., and P. A. G. Fortes. 1981. Characterization of a 2′,3′-O-(2,4,6-Trinitrocyclohexadienylidine)adenosine 5′-triphosphate as a fluorescent probe of the ATP site of sodium and potassium transport adenosine triphosphatase. *Journal of Biological Chemistry.* 256:2346–2356.

Nørby, J. G. 1983. Ligand interactions with the substrate site of Na,K-ATPase: nucleotides, vanadate and phosphorylation. *Current Topics in Membranes and Transport.* 19:281–314.

Nørby, J. G. 1987. Na,K-ATPase: structure and kinetics. Comparison with other transport systems. *Chemica Scripta.* 27B:119–129.

Nørby, J. G. 1989. Structural and functional organization of the Na^+,K^+-pump. *Biochemical Society Transactions.* 17:806–808.

Nørby, J. G., and J. Jensen. 1988. Measurement of binding of ATP and ADP to Na^+,K^+-ATPase. *Methods in Enzymology.* 156:191–201.

Nørby, J. G., and J. Jensen. 1989. A model for the stepwise radiation inactivation of the α_2-dimer of Na,K-ATPase. *Journal of Biological Chemistry.* 264:19548–19558.

Ottolenghi, P., and J. Jensen. 1983. The K^+-induced apparent heterogeneity of high-affinity nucleotide-binding sites in (Na^+ + K^+)-ATPase can only be due to the oligomeric structure of the enzyme. *Biochimica et Biophysica Acta.* 727:89–100.

Ottolenghi, P., and J. Jensen. 1985. Reflections on nucleotide-binding behaviour, protein determinations and the structure of membrane-bound Na,K-ATPase. *In* The Sodium Pump. I. Glynn and C. Ellory, editors. The Company of Biologists Ltd., Cambridge, UK. 219–227.

Ottolenghi, P., J. G. Nørby, and J. Jensen. 1986. Solubilization and further purification of highly purified, membrane-bound Na,K-ATPase. *Biochemical and Biophysical Research Communications.* 135:1008–1014.

Peterson, G. L. 1977. A simplification of the protein assay method of Lowry et al. which is more generally applicable. *Analytical Biochemistry.* 83:346–356.

Rabon, E. C., R. D. Gunther, S. Bassillian, and E. S. Kempner. 1988. Radiation inactivation analysis of oligomeric structure of the H,K-ATPase. *Journal of Biological Chemistry.* 263:16189–16194.

Rauth, A. M., and J. A. Simpson. 1964. The energy loss of electrons in solids. *Radiation Research.* 22:643–661.

Reynolds, J. A. 1988. The oligomeric structure of the Na,K-Pump protein. *In* The Na^+,K^+-Pump, Part A: Molecular Aspects. J. C. Skou, J. G. Nørby, A. B. Maunsbach, and M. Esmann, editors. Alan R. Liss, Inc., New York. 137–148.

Saccomani, G., G. Sachs, J. Cuppoletti, and C. Y. Jung. 1981. Target molecular weight of the gastric (H^+ + K^+)-ATPase. Functional and structural molecular size. *Journal of Biological Chemistry.* 256:7727–7729.

Serpersu, E. H., S. Bunk, and W. Schoner. 1990. How do MgATP analogues differentially modify high-affinity and low-affinity ATP binding sites of Na^+/K^+-ATPase. *European Journal of Biochemistry.* 191:397–404.

Venter, J. C., C. M. Fraser, J. S. Schaber, C. Y. Jung, G. Bolger, and D. J. Triggle. 1983. Molecular properties of the slow inward calcium channel. Molecular weight determinations by radiation inactivation and covalent affinity labeling. *Journal of Biological Chemistry.* 258:9344–9348.

Vilsen, B., J. P. Andersen, J. Petersen, and P. L. Jørgensen. 1987. Occlusion of ^{22}Na and ^{86}Rb in membrane-bound and soluble protomeric αβ-Units of Na,K-ATPase. *Journal of Biological Chemistry.* 262:10511–10517.

Reaction Kinetics and Transport Modes

Chapter 14

Conformational Transitions in the α-Subunit and Ion Occlusion

Peter Leth Jørgensen

Biomembrane Research Center, August Krogh Institute, Copenhagen University, 2100 Copenhagen OE, Denmark

Introduction

The αβ-unit is the minimum asymmetric unit in crystals of the purified membrane-bound Na,K-ATPase (Jørgensen et al., 1982). It forms cation binding sites and occlusion cavities for Na^+ or K^+ (Rb^+) within its structure in both the membrane and soluble states (Vilsen et al., 1987). ATP binding and phosphorylation take place in the large cytoplasmic protrusion of the α-subunit, while cation sites may be located in intramembrane domains (Clarke et al., 1989; Karlish et al., 1990). Transduction of the energy from ATP to movement of the cations may therefore involve long-range structural transitions in the protein. The purpose of the present work is to establish relationships between ion binding or occlusion and structural changes in the α-subunit that can be detected by specific proteolytic cleavage (Jørgensen, 1975, 1977). The protein conformations of different complexes of ^{22}Na or ^{86}Rb with Na,K-ATPase are determined to see if the different exposure of bonds to proteolysis reflects the orientation and specificity of the cation sites.

Na-K-induced Conformational Changes

Mixing purified Na,K-ATPase with Na^+ or K^+ in the absence of other ligands stabilizes two conformations of the α-subunit, the Na form (E_1) or the K form (E_2), as monitored by tryptic or chymotryptic digestion (Fig. 1*A*). Definition of E_1 and E_2 conformations of the α-subunit of Na,K-ATPase involves identification of cleavage points in the protein as well as association of cleavage with different rates of inactivation of Na,K-ATPase and K-phosphatase activities (Jørgensen, 1975, 1977). Chymotrypsin cleaves at Leu^{266} (C_3) and both Na,K-ATPase and K-phosphatase are inactivated with a monoexponential time course. Trypsin cleaves the E_1 form rapidly at Lys^{30} (T_2) and more slowly at Arg^{262} (T_3) to produce the characteristic biphasic pattern of inactivation. Localization of these splits was achieved by sequencing NH_2 termini of fragments after isolation on high resolution gel filtration columns (Jørgensen and Collins, 1986).

The E_2 form is not cleaved by chymotrypsin, but trypsin cleaves at Arg^{438} (T_1) and subsequently at Lys^{30} (T_2) and tryptic inactivation of E_2K or E_2P forms is linear and associated with cleavage at Arg^{438} (T_1). Inactivation of K-phosphatase is delayed because cleavage of T_1 and T_2 in sequence is required for inactivation of K-phosphatase activity. Thus, transition from E_1 to E_2 consists of an integrated structural change involving protection of bond C_3 or T_3 in the second cytoplasmic domain and exposure of T_1 in the central domain, while the position of T_2 in the NH_2 terminus is altered relative to the central domain (T_1) so that cleavage of T_2 becomes secondary to cleavage of T_1 within the same α-subunit in the E_2 form.

In medium containing K^+ (E_2[2K]), the addition of ATP or ADP causes a transition to the E_1 pattern of cleavage (ATP-$E_1$2K). In Na^+ medium, the addition of MgATP causes transition to the E_2 pattern of cleavage since the equilibrium between the phosphoforms (E_1P-E_2P) is poised in favor of E_2P in Na,K-ATPase from kidney. In addition to bonds exposed to proteolysis, the structural changes accompanying exchange of Na^+ and K^+ or ATP binding and phosphorylation are reflected in changes in the intensity of fluorescence from intrinsic tryptophan or from extrinsic probes such as fluorescein covalently attached to Lys^{501} (Karlish, 1980) or methio-

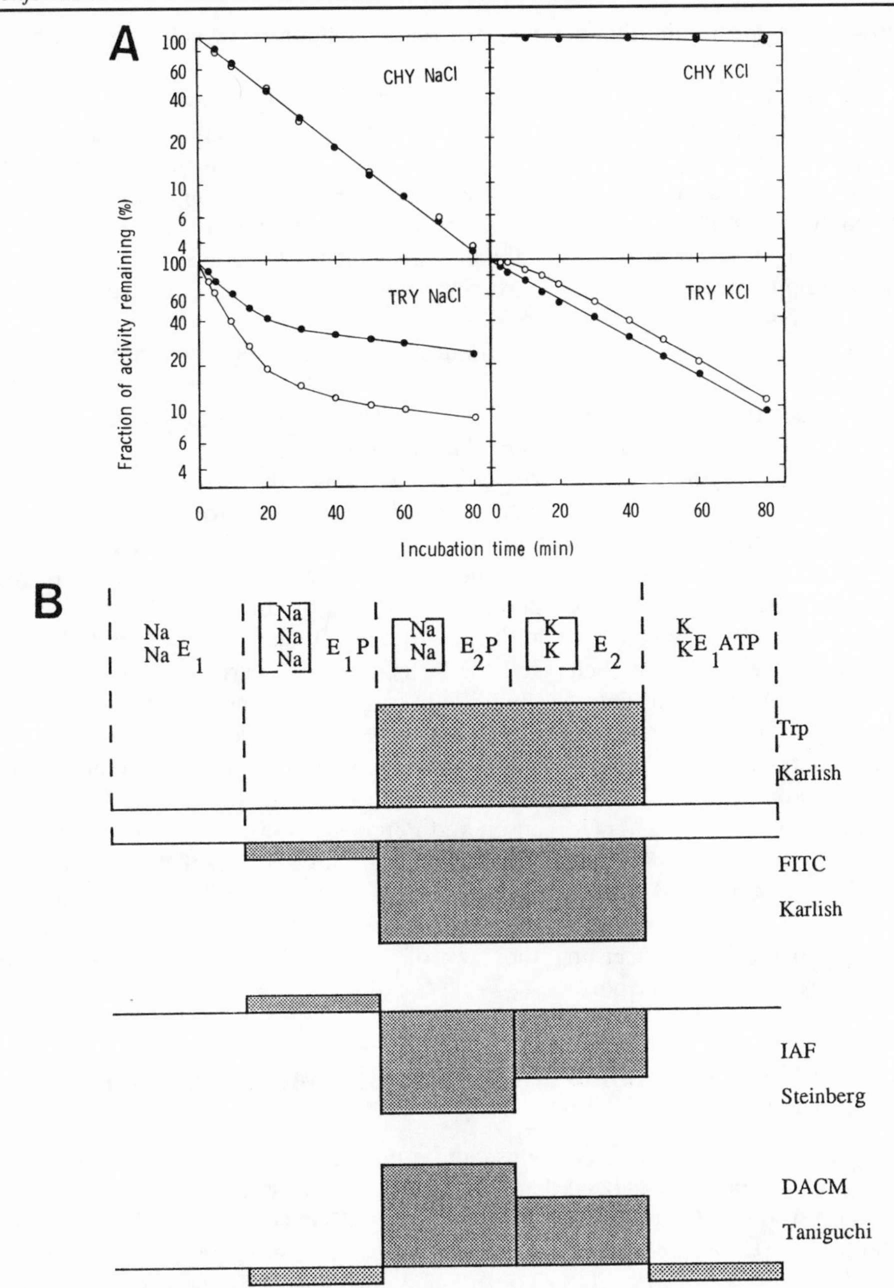

Figure 1. (*A*) Time course of inactivation of Na, K-ATPase (●) or phosphatase (○) from outer renal medulla by cleavage with chymotrypsin or trypsin in NaCl or KCl media. See text for explanations. Redrawn from Jørgensen (1977) and Jørgensen et al. (1982). (*B*) Relative fluorescence intensities from tryptophan (*Trp*), fluorescein at Lys^{501} (*FITC*), fluorescein at Cys^{457} (*IAF*), or dimethyl-amino-coumarinyl-maleimide (*DACM*) attached to sulfhydryls of Na,K-ATPase from the outer renal medulla. The horizontal line corresponds to the level of fluorescence in NaCl medium. Compiled from Karlish (1980), Steinberg et al. (1988), and Sakuraya et al. (1987).

nines of the α-subunit (Steinberg et al., 1988) or fluorescent sulfhydryl reagents (Sakuraya et al., 1987) (Fig. 1 *B*).

As seen in Fig. 1, *A* and *B*, these conformational transitions occur together in the native Na,K-pump and they are correlated with changes in affinity for ATP and phosphoenzyme properties. A single split at Leu^{266} in the α-subunit blocks the conformational transition as well as ion exchange without affecting the ability of the protein to bind ATP, phosphorylate, or occlude Na^+ (Jørgensen et al., 1982). These observations form the basis for a mechanochemical hypothesis for the Na,K-translocation process in which the structural change in the protein is involved in the coupling of ATP hydrolysis to changes in capacity, affinity, and accessibility of the cation binding sites. In E_1 forms the cation sites are accessible from the cell interior with a capacity for binding and occluding three Na^+ ions, while the cation sites of the E_2 forms face the extracellular phase with a capacity for the binding and occlusion of two Na^+ or two K^+ ions (Jørgensen and Andersen, 1988).

Arguments against this hypothesis have been that the structural changes observed by proteolysis or fluorescence represent only a few among a multitude of conformational transitions in the protein and that some of the other structural changes may be more important for the ion transport process. Reference is also made to observations that in K^+ medium only the E_1 pattern of tryptic digestion is observed in FITC-labeled Na,K-ATPase although it occludes Rb^+ and changes fluorescence when Na^+ is exchanged for K^+ (Glynn and Karlish, 1990). However, cleavage of the α-subunit into 58- and 47-kD fragments (bond T_2, Arg^{438}) of FITC-enzyme has been discussed earlier (Karlish, 1980). It is also important that the covalent attachment of FITC at Lys^{501} has an ATP-like effect in reducing the apparent affinity for K^+ ($K_d \sim 34$ μM) about fivefold relative to that of native Na,K-ATPase ($K_d \sim 7$ μM) (Jørgensen and Petersen, 1985). Fig. 2 shows a distinct shift between the two patterns of tryptic inactivation of K-phosphatase activity for both FITC-enzyme and control enzyme. Therefore, when the difference in affinity for K^+ is taken into consideration, the previous observations on FITC-enzyme do not support conclusions concerning the lack of relevance of these conformational transitions for cation transport.

Protein Conformation of Na-occluded E_1 and E_2 Forms of Na,K-ATPase

It is generally accepted that Na^+ ions can be occluded in E_1P forms. Occlusion of three Na^+ ions per EP has been demonstrated in chymotrypsin cleaved enzyme and in the Cr-ADP-E_1P[3Na] complex (Vilsen et al., 1987). Three Na^+ ions can also be occluded per EP in a complex stabilized by oligomycin in the absence of Mg^{2+} or phosphate (Shani-Sekler et al., 1988). In Fig. 3, the capacity for occlusion of ^{22}Na in the oligomycin complex is compared with that of a complex with ouabain, Mg^{2+}, and phosphate. It is seen that a maximum of two Na^+ ions are occluded per α-subunit in the ouabain complex, while occlusion in the oligomycin complex is three Na^+ per α-subunit.

The cleavage patterns of these two complexes were determined using trypsin or chymotrypsin (data not shown). As expected, the oligomycin complex displays patterns identical to those of the E_1Na form, while the ouabain complex cleaves like the E_2K form. This demonstrates that Na^+ can be occluded in both E_1 and E_2 forms of the protein. The important problem is how the observation of Na occlusion in the Mg-E_2P[2Na]-ouabain complex is relevant for the transport mechanism. In the scheme in Fig. 4 (Jørgensen and Andersen, 1988), the E_1P-E_2P transition releases a single Na^+ ion at the extracellular surface and E_2P[2Na] represents an occluded state in transition to E_2P-2Na with Na^+ leaving the sites making them accessible for

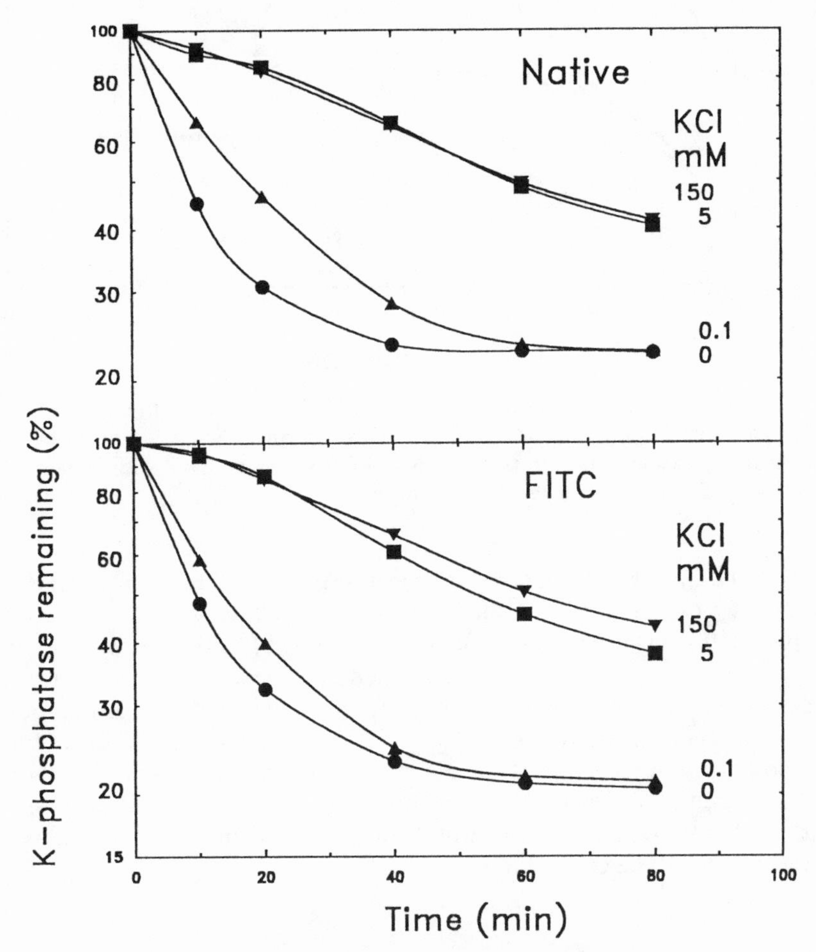

Figure 2. Time course of inactivation of K-phosphatase activity during tryptic cleavage at 0 (●), 0.1 (▲), 5 (■), and 150 (▼) mM KCl of native and FITC-labeled Na,K-ATPase from outer renal medulla. Labeling with FITC (Karlish, 1980) reduced Na,K-ATPase activity to 3%, while K-phosphatase activity remained at 87%. Procedure for tryptic cleavage as in Jørgensen (1977).

binding of K^+ from the extracellular phase. In a scheme involving two cycles, Yoda and Yoda (1987) used the term E*P for this intermediate and it was observed that ouabain reacts with E*P (Lee and Fortes, 1985). Our observations show that the correct notation for E*P is E_2P[2Na], since E_2P can occlude either two Na^+ or two K^+ without altering protein conformation as detected by proteolysis. In Na^+ medium, the E_2P[2Na] intermediate is sensitive to ADP because binding of one Na^+ allows it

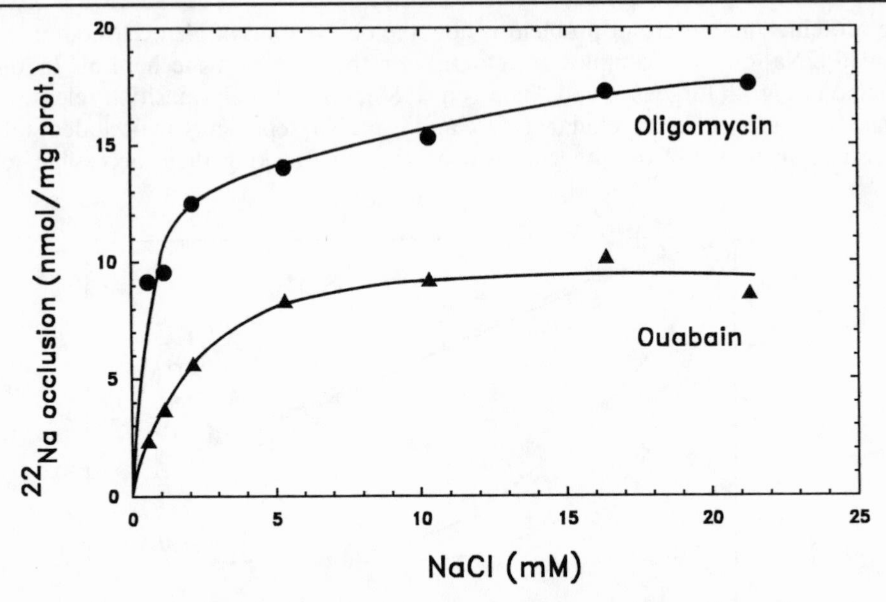

Figure 3. ^{22}Na occlusion in oligomycin (●) or ouabain (▲) complexes with membrane-bound Na,K-ATPase from kidney. Procedure as described by Shani-Sekler et al. (1988) using incubation at increasing concentrations of Na^+ for 15 min at 20°C with either 30 μM oligomycin or 1 mM ouabain, 1 mM $MgCl_2$, and 1 mM P_i-Tris.

to return to the $E_1P[3Na]$ form for reaction with ADP and formation of ATP. After addition of K^+, exchange of Na^+ for K^+ at the extracellular surface would lead to dephosphorylation. The apparent ambiguity of the $E_2P[2Na]$ form with respect to reactivity to ADP and K^+ is therefore explained by the cation site occupancy, while the protein conformation of the $E_2P[2Na]$ intermediate is the same as that of other E_2 forms. With these properties of the $E_2P[2Na]$ complex, the ADP-sensitive fraction of the phosphoenzyme comprises the Na-occluded E_1 and E_2 forms, $E_1P[3Na]$ and $E_2P[2Na]$. Therefore the sum of the amounts of ADP- and K-sensitive phosphoenzyme is equal to $E_1P[2Na]$ plus $2 \times E_2P[2Na]$ plus $E_2P[0]$ and thus exceeds the total EP by the amount of the $E_2P[2Na]$ intermediate.

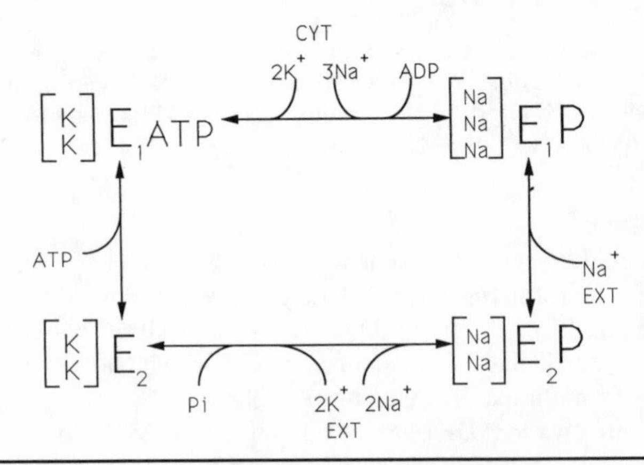

Figure 4. Minimal E_1-E_2 reaction cycle of the Na,K-pump with four major occluded conformations and ping-pong sequential cation translocation. Na or K within brackets are occluded, with phosphoforms occluding Na^+ and dephosphoforms occluding K^+ or Rb^+.

Comparison of ^{86}Rb- and ^{22}Na-occluded Forms of Na,K-ATPase

Fig. 5 shows the capacity and apparent affinity for the occlusion of ^{22}Na or ^{86}Rb in the ouabain-MgE$_2$P complex in comparison with occlusion of ^{86}Rb in the unliganded Na,K-ATPase. The apparent affinity of these E$_2$ forms for the cations varies over a wide range from $K_{1/2}$(Rb) ~ 9 μM to $K_{1/2}$(Na) ~ 1.7 mM, but without changes in proteolytic cleavage patterns. The apparent affinity of ^{86}Rb for formation of the Mg-ouabain-E$_2$P[Rb] complex is >100-fold higher than for Na. The data show an equal capacity for occlusion of either 2Rb$^+$ or 2Na$^+$ per αβ-unit. The difference in apparent affinity for the complexes in Fig. 5 is sufficient for allowing exchange of Na for Rb(K) at the extracellular surface, but the transition from E$_2$P[2Na] to E$_2$[2Rb] has no influence on the patterns of proteolytic digestion or on the fluorescence levels referred to in Fig. 1, *A* and *B*.

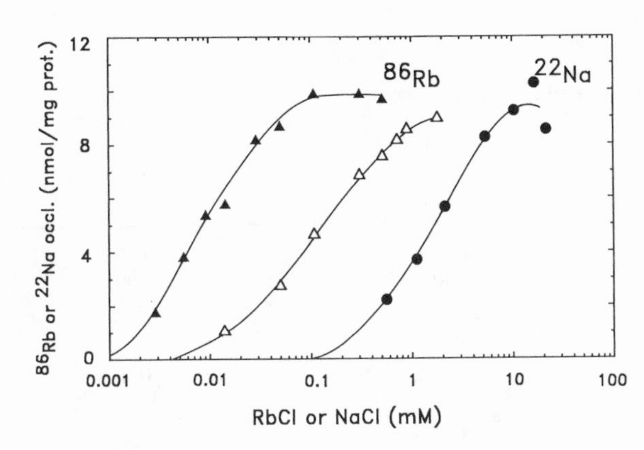

Figure 5. Occlusion of ouabain complexes with ^{22}Na (●) or ^{86}Rb (△) and unliganded Na,K-ATPase with ^{86}Rb (▲). Procedure as in Fig. 3.

Effect of C$_3$ Cleavage on E$_1$P-E$_2$P Transition and Cation Exchange

The alternating exposure of C$_3$ (Leu^{266}) or T$_3$ (Arg^{262}) in the E$_1$ form and T$_1$ (Arg^{438}) in the E$_2$ form reflects that motion within the segment (M_r 18,170) between these bonds including the phosphorylated residue (Asp^{369}) is an important element in E$_1$-E$_2$-transition. This is illustrated by the widely different consequences of selective cleavage of C$_3$ and T$_1$ for E$_1$-E$_2$ transition and cation exchange.

C$_3$ cleavage is a selective and particularly efficient tool for examining structure–function relationships of the second cytoplasmic domain. Binding affinities for ADP and ATP are reduced four- to fivefold, while TNP-ATP binds with the same affinity as in native Na,K-ATPase. Nucleotide binding is not affected by K$^+$ or Rb$^+$ although cation sites are undamaged. Conversely, the cleaved enzyme also binds ^{86}Rb with high affinity and occludes the cations, but cation binding and occlusion are unaffected by nucleotide (Table I).

Transport studies in reconstituted vesicles show that C$_3$ cleavage blocks the relatively fast Na-Na or K-K exchange (20–40 s^{-1}) and Na-K exchange (500 s^{-1}), but the slow passive ouabain-sensitive Rb-Rb-exchange (1 s^{-1}) and occlusion of K$^+$ or Na$^+$ are only partially affected. C$_3$ cleavage allows formation of E$_1$P[3Na], but prevents charge transfer coupled to Na$^+$ translocation in purified Na,K-ATPase after adsorption to a planar lipid bilayer (Apell et al., 1987). In combination with the observation of the occlusion of $2Na^+$ in the $E_2P[2Na]$ these data show that the E_1P-E_2P transition represents a charge-translocating step and that the single Na^+ ion released at the extracellular surface represents this transfer of charge (cf. Gadsby et al., 1991). C_3 cleavage thus interferes with structural changes that alter the capacity of the cation sites for occlusion of Na^+ and presumably their orientation.

Mutagenesis in Yeast H-ATPase

The notion that the segment containing C_3 and T_3 is important for conformational adaptability of the protein is supported by mutations in yeast. Portillo and Serrano (1988) induced mutations in the genes of the H-ATPase of *Saccharomyces cerevisiae* and found a temperature-sensitive mutant after mutation of Gly^{254}→Ser. Ghislain et al. (1987) showed that the Gly^{268}→Asp substitution is responsible for a mutant

TABLE I
Properties of the C_3 Cleaved Derivative of NA,K-ATPase

Enzymatic activity, ligand binding, or transport capacity	Cleavage C_3 Leu^{266}	Control
Na,K-ATPase	0%	100%
ATP-ADP exchange	400–500%	100%
ADP binding		
Capacity	100%	100%
Affinity, $K_d/\mu M$	0.075	0.045
Phosphorylation		
Capacity	100%	100%
E_1P/E_2P ratio	100/0	14/84
Vanadate binding		
Capacity	0%	100%
Rb-binding		
Capacity	100%	100%
Affinity, $K_d/\mu M$	9–12	9–12

Compiled from Jørgensen et al. (1982) and Jørgensen and Andersen (1988).

phenotype in *Schizosaccharomyces pombe.* The mutant H-ATPase has reduced ATPase and proton pumping activity and it is vanadate resistant. The mutation may thus produce a higher concentration of E_1 forms and less of the vanadate-binding E_2 form during steady-state ATP hydrolysis. In interpreting the work on mutations in H-ATPase, the view was advanced that mutations in the second cytoplasmic loop disrupt an endogenous phosphatase activity (Portillo and Serrano, 1988). Inactivation of phosphatase activity after the Glu^{233}→Gln mutation (Glu^{233} in H-ATPase is homologous to Glu^{183} in Ca-ATPase) suggested that the defect was in the hydrolytic step of the catalytic cycle. However, the measurements of dephosphorylation rates in Ca-ATPase indicate that the conformational change and the E_1P-E_2P interconversion are blocked in the mutants rather than the hydrolysis of E_2P (Clarke et al., 1990).

Selective Proteolytic Cleavage and Mutagenesis of Ca-ATPase from Sarcoplasmic Reticulum

In the second cytoplasmic domain between M_2 and M_3 in Ca-ATPase of the sarcoplasmic reticulum, cleavage of bonds at Lys^{234} and Arg^{236} (Imamura and Kawakita, 1989) or cleavage between Glu^{231} and Ile^{232} with V8 protease (le Maire et al., 1990) inactivate the enzyme (Fig. 6). The cleavage product receives the γ-phosphate from ATP, but it is unable to undergo the interconversion between E_1P and E_2P forms of the phosphoenzyme. Phosphorylation is activated by Ca^{2+}, but occlusion data are not reported. Site-directed mutagenesis of Gly^{233} (for Val, Gln, or Arg) of Ca-ATPase located between the two proteolytic splits gives mutants that are inactive with respect to Ca-ATPase or Ca transport. Again, ATP and Ca^{2+} can bind and form an ADP-sensitive phosphoenzyme that is deficient in the transition to the ADP-insensitive form. In this mutant the apparent affinity in the backward reaction with P_i is reduced ($K_{1/2}$ 250 μM) relative to the wild type ($K_{1/2}$ 50 μM), suggesting that the E_2P

```
Ca-ATPase.S.R.   272 L R V D Q S I L T G E S V S
                         . : : . :   : : : : :
Na,K-ATPase α1   204 C K V D N S S L T G E S E P
```

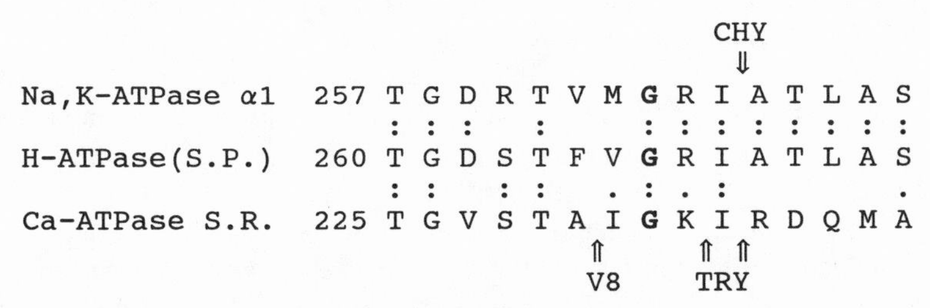

Figure 6. Amino acid sequences around the C_3 cleavage site in the second cytoplasmic loop of the α-subunit of Na,K-ATPase and homologous segments in H-ATPase (Ghislain et al., 1987) and Ca-ATPase of sarcoplasmic reticulum (Clarke et al., 1990). See text for explanations.

intermediate is destabilized (Andersen et al., 1989). It is proposed that the region around Gly^{233} is involved in energy transduction between the phosphate site and the cation binding sites in the transmembrane region.

Mutations in another region, the second cytoplasmic loop between M_2 and M_3 in Ca-ATPase of sarcoplasmic reticulum (Thr^{181}→Ala, Gly^{182}→Ala, and Glu^{183}→Gln), also result in a complete loss of Ca transport and Ca-ATPase activity associated with a dramatic reduction of the rate of phosphoenzyme turnover (Clarke et al., 1990). These mutations do not affect the affinity of the enzyme for P_i and therefore resemble the Pro^{312} mutants (Vilsen et al., 1987) in that they affect only the E_1P-E_2P conformational change and not the affinities for ATP, Ca^{2+}, or P_i.

These data show interesting similarities between conformational transitions in Ca-ATPase in sarcoplasmic reticulum and α1-subunit of Na,K-ATPase, but caution is necessary in drawing parallels between the Ca transport reaction and Na,K-pumping. A Ca-occluded E_2 form of Ca-ATPase has not been demonstrated and there is no exchange of Ca^{2+} with another cation at the extracellular surface. It has also been difficult to produce evidence for two-state models with different orientations of the Ca binding sites relative to the membrane. In the Ca-ATPase, dephosphorylation is followed by reaction with cytoplasmic Ca^{2+}, while in the normal cycle of Na,K-ATPase dephosphorylation is followed by binding and occlusion of two extracellular K^+ ions to form the relatively stable E_2[2K] form that must react with ATP for transition to the E_1 form.

Discussion and Conclusions

The data discussed in this article provide evidence for the existence of two conformations with different capacities and orientation of cation sites. In the E_1 form of Na,K-ATPase, the exposure of $C_3(T_3)$ to cleavage reflects that the cation sites of the phosphoprotein are in an inwardly oriented conformation with a capacity for the occlusion of three Na^+ ions. The E_2 form with exposed T_1 and protected $C_3(T_3)$ occludes either $2Na^+$ or $2Rb^+(K^+)$ in the phosphoform or $2Rb^+(K^+)$ in the unliganded enzyme.

The sequence around C_3 in the α-subunit of Na,K-ATPase is homologous with segments of Ca-ATPase from sarcoplasmic reticulum and H-ATPase from yeast and plants. Selective proteolytic cleavage of Ca-ATPase and site-directed mutagenesis of both pumps stabilizes E_1P and interferes with structural transitions and cation translocation in the same manner as observed after C_3 cleavage in Na,K-ATPase. The observations show that it is a general feature that this region (Fig. 6) of the cation pump proteins is involved in structural transitions accompanying energy transduction between the phosphate site and the cation sites.

The selective C_3 cleavage experiments show that the E_1P-E_2P isomerization is coupled to the major charge-carrying step in the reaction cycle, the deocclusion of one Na^+ ion during transition from E_1P[3Na] to E_2P[2Na] with release of the Na^+ ion to the extracellular side. The implication of two Na-occluded forms does not require more than one set of sites. The change in conformation of the protein that opens and closes barriers may also alter the position of a given set of cation binding sites. Mutagenesis experiments have shown that carboxyl groups in transmembrane segment M_4, in Na,K-ATPase (Glu^{329}; Lingrel, 1991) and in Ca-ATPase (Glu^{309}; Clarke et al., 1989) contribute to formation of Na^+ and Ca^{2+} binding sites. It is possible that the conformational transition causes a rotation or tilting of the M_4 transmembrane helix, altering the number of cation coordinating groups and their orientation.

Another attraction of the demonstration of two Na-occluded intermediates (E_1P[3Na] and E_2P[2Na]) is that it provides explanations for apparent discrepancies with respect to the kinetics and quantities of the phosphoenzyme forms. The definition of the two phosphoenzyme intermediates explains why, in Na^+ medium, the ADP-sensitive phosphoenzyme comprises both Na^+-occluded forms. Since the E_2P[2Na] form will appear in both the ADP- and K-sensitive fractions of the phosphoenzyme, the amount of the E_2P[2Na] form can be determined as the sum of the ADP- and K-sensitive phosphoenzymes minus the total phosphoenzyme.

The loss of ATP-K antagonism with unaffected occlusion of Na^+ or Rb^+ after chymotryptic cleavage (Jørgensen and Andersen, 1988) has been used as an argument against the E_1-E_2 nomenclature for the conformational transitions (Glynn and Karlish, 1990). Obviously, the cleavage patterns in Fig. 1 *A* and the model in Fig. 4, including the E_1-E_2 notation, are only valid for native Na,K-ATPase. Occlusion of $3Na^+$ or $2Rb^+$ ions is still possible after C_3 cleavage and even after removal of the entire cytoplasmic protrusion by extensive cleavage (Karlish et al., 1990). The operational definitions of the E_1E_2 conformations by proteolysis or fluorescence are no longer valid for these truncated enzymes. The observation that the machinery can be taken apart for examination of the individual components and their interactions should not prevent the development of concepts for the function of its intact structure.

References

Andersen, J. P., B. Vilsen, E. Leberer, and D. H. MacLennan. 1989. Functional consequences of mutations in the β-strand sector of the Ca^{2+}-ATPase of sarcoplasmic reticulum. *Journal of Biological Chemistry*. 264:21018–21023.

Apell, H.-J., R. Borlinghaus, and P. Lauger. 1987. Fast charge translocations associated with partial reactions of the Na,K-pump. II. Microscopic analysis of transient currents. *Journal of Membrane Biology*. 97:179–191.

Clarke, D. M., T. W. Loo, G. Inesi, and D. H. Maclennan. 1989. Location of high affinity Ca^{2+}-binding sites within the predicted transmembrane domain of the sarcoplasmic reticulum Ca^2-ATPase. *Nature*. 339:476–478.

Clarke, D. M., T. W. Loo, and D. H. MacLennan. 1990. Functional consequences of mutations of conserved amino acids in the β-strand domain of the Ca-ATPase of sarcoplasmic reticulum. *Journal of Biological Chemistry*. 265:14088–14092.

Gadsby, D., M. Nakao, and A. Bahinski. 1991. Voltage-induced Na/K pump charge movements in dialyzed heart cells. *In* The Sodium Pump: Structure, Mechanism, and Regulation. J. H. Kaplan and P. De Weer, editors. The Rockefeller University Press, New York. 355–371.

Ghislain, M., A. Schlesser, and A. Goffeau. 1987. Mutation of a conserved glycine residue modifies the vanadate sensitivity of the plasma membrane H-ATPase from *Schizosaccharomyces pombe. Journal of Biological Chemistry*. 262:17549–17555.

Glynn, I. M., and S. J. D. Karlish. 1990. Occluded cations in active transport. *Annual Review of Biochemistry*. 59:171–205.

Imamura, Y. and M. Kawakita. 1989. Purification of limited tryptic fragments of Ca^{2+}, Mg^{2+}-adenosine triphosphatase of the sarcoplasmic reticulum and identification of conformation-sensitive cleavage sites. *Journal of Biochemistry*. 105:775-781.

Jørgensen, P. L. 1975. Purification and characterization of Na,K-ATPase. V. Conformational changes in the enzyme: transitions between the Na-form and the K-form studied with tryptic digestion as a tool. *Biochimica et Biophysica Acta*. 401:399–415.

Jørgensen, P. L. 1977. Purification and characterization of Na,K-ATPase. VI. Differential tryptic modification of catalytic functions of the purified enzyme in presence of NaCl and KCl. *Biochimica et Biophysica Acta*. 466:97–108.

Jørgensen, P. L., and J. P. Andersen. 1988. Structural basis for E_1-E_2 conformational transitions in Na,K-pump and Ca-pump proteins. *Journal of Membrane Biology*. 103:95–120.

Jørgensen, P. L., and J. H. Collins. 1986. Tryptic and chymotryptic cleavage sites in the sequence of α-subunit of Na,K-ATPase from outer medulla of mammalian kidney. *Biochimica et Biophysica Acta*. 860:570–576.

Jørgensen, P. L., and J. Petersen. 1985. Chymotryptic cleavage of α-subunit in E_1-forms of renal Na,K-ATPase: effects on enzymatic properties, ligand binding and cation exchange. *Biochimica et Biophysica Acta.* 821:319–333.

Jørgensen, P. L., E. Skriver, H. Hebert, and A. B. Maunsbach. 1982. Structure of the Na,K-pump: Crystallization of pure membrane-bound Na,K-ATPase and identification of functional domains of the α-subunit. *Annals of the New York Academy of Sciences.* 402:203–219.

Karlish, S. J. D. 1980. Characterization of conformational changes in Na,K-ATPase labeled with fluorescin at the active site. *Journal of Bioenergetics and Biomembranes.* 12:11–136.

Karlish, S. J. D., R. Goldshleger, and W. D. Stein. 1990. A 19-kDa C-terminal tryptic fragment of the α-chain of Na,K-ATPase is essential for occlusion and transport of cations. *Proceedings of the National Academy of Sciences, USA.* 87:4566–4570.

Lee, J. A., and P. A. G. Fortes. 1985. Anthroylouabain binding to different phosphoenzyme forms of Na, K-ATPase. *In* The Sodium Pump. I. M. Glynn and C. Ellory, editors. Company of Biologists, Ltd., Cambridge, UK. 277–282.

le Maire, M., S. Lund, A. Viel, P. Champeil, and J. V. Møller. 1990. Ca^{2+}-induced conformational change and location of Ca^{2+} transport sites in sarcoplasmic reticulum Ca^{2+}-ATPase as detected by the use of proteolytic enzyme (V8). *Jounal of Biological Chemistry.* 265:111–123.

Lingrel, J., J. Orlowski, E. M. Price, and B. G. Pathak. 1991. Regulation of the α-subunit genes of the Na,K-ATPase and determinant of cardiac glycoside sensitivity. *In* The Sodium Pump: Structure, Mechanism, and Regulation. J. H. Kaplan and P. De Weer, editors. The Rockefeller University Press, New York. 1–16.

Portillo, F., and R. Serrano. 1988. Dissection of functional domains of the yeast proton-pumping ATPase by directed mutagenesis. *EMBO Journal.* 7:1793–1798.

Sakuraya, M., K. Taniguchi, K. Suzuki, A. Kudo, S. Nakamura, and S. Iida. 1987. Changes in fluorescence energy transfer between sulfhydryl fluorescent residues during ouabain sensitive Na^+,K^+-ATP hydrolysis. *Japanese Journal of Pharmacology.* 44:311–321.

Shani-Sekler, M., R. Goldshleger, D. M. Tal, and S. J. D. Karlish. 1988. Inactivation of Rb^+ and Na^+ occlusion on Na,K-ATPase by modification of carboxyl groups. *Journal of Biological Chemistry.* 263:19331–19341.

Steinberg, M., P. A. Tyson, E. T. Wallick, and T. L. Kirley. 1988. Identification and localization of the 5-iodoacetamido-fluorescein reporter site on dog kidney Na,K-ATPase. *Progress in Clinical Biological Research.* 273:39–44.

Vilsen, B., J. P. Andersen, J. Petersen, and P. L. Jørgensen. 1987. Occlusion of ^{22}Na and ^{86}Rb in membrane bound and soluble protomeric αβ-units of Na,K-ATPase. *Journal of Biological Chemistry.* 262:10511–10517.

Yoda, A., and S. Yoda. 1987. Two different phosphorylation-dephosphorylation cycles of Na,K-ATPase proteoliposomes accompanying Na^+ transport in the absence of K^+. *Journal of Biological Chemistry.* 262:110–115.

Chapter 15

A Hofmeister Effect on the Phosphoenzyme of Na,K-ATPase

Robert L. Post and Kuniaki Suzuki

Department of Molecular Physiology and Biophysics, Vanderbilt University Medical School, Nashville, Tennessee 37232-0615

Introduction

Hofmeister (1888) added various salts to a solution of egg albumin and recorded the lowest concentration in each case that just produced turbidity. He arranged ions in a series according to their potency. Such a series is now called a Hofmeister or lyotropic series. In 1985, Collins and Washabaugh summarized characteristics of the Hofmeister effect somewhat as follows: (*a*) The effects appear at concentrations between 10 mM and 1 M. (*b*) The effects of anions predominate over those of cations. (*c*) Different measures of the series rank ions similarly; e.g., SO_4^{2-} = HPO_4^{2-} > F^- > Cl^- > Br^- > I^- = ClO_4^- > SCN^-. Ions to the right of Cl^- destabilize the structure of water and are called "chaotropes;" ions to the left stabilize it and are called "kosmotropes." (*d*) The effects are approximately additive over all species. Hofmeister effects appear in conformational equilibria of proteins including unfolding and formation of crystals (von Hippel and Schleich, 1969; Ries-Kautt and Ducruix, 1989). Dani et al. (1983) investigated the effects of lyotropic anions on Na^+ channel gating. The and Hasselbach (1975) studied the effects of chaotropic anions on the Ca-ATPase of sarcoplasmic reticulum.

In the reaction sequence of Na,K-ATPase there are two principal reactive states, E_1 and E_2 (Glynn, 1988; Nørby and Klodos, 1988; Glynn and Karlish, 1990). These are defined variously in the literature according to (*a*) the kind of transported cation bound, Na^+ or K^+; (*b*) the side of the membrane to which the bound cations have access, cytoplasmic or extracellular; and (*c*) the reactivity of the phosphate group of the phosphorylated intermediate, E_1P or E_2P. The phosphate group of E_1P equilibrates with the terminal phosphate group of ATP, whereas the phosphate group of E_2P equilibrates with inorganic phosphate; the phosphate group is attached covalently to the β-carboxyl group of the same aspartyl residue in each case. In this report we use the third definition. Each definition suggests a different kind of conformational change.

Taniguchi and Post (1975) converted E_2P to E_1P by adding a concentrated sodium chloride solution. They supposed that binding of sodium ion to a low affinity extracellular transport site on E_2P converted its low energy phosphate group to a high energy phosphate group on E_1P. Hara and Nakao (1981) reported that high concentrations of NaCl favored E_1P over E_2P in phosphoenzyme formed from ATP. Nørby et al. (1983) studied these kinetics further and introduced an intermediate state, their "pool B," between E_1P and E_2P. Yoda and Yoda (1986, 1987*a, b,* 1988) reported further evidence in favor of this intermediate state, which they named "E*P." They found more factors to influence the distribution between E_1P and E_2P. Lee and Fortes (1985) supported the participation of a third reactive state of the phosphoenzyme in the reaction sequence. In all these experiments the anion was chloride.

We report here an effect of anions on the distribution of the phosphoenzyme between E_1P and E_2P. The potency of anions ranks them in a Hofmeister series.

Materials and Methods

Membrane-bound Na,K-ATPase from dog kidney outer medulla was prepared by Jørgensen's methods (Jørgensen, 1974). The enzyme was stored at 4°C in 40 mM histidine, 1 mM H_4EDTA, and 3.26 M glycerol with <5% loss of activity in 50 d. The specific activity and the phosphoenzyme level were 10–35 U/mg at 37°C and 1–3 nmol/mg protein, respectively. Phosphorylation and dephosphorylation of the enzyme were done at 2°C as follows unless otherwise stated: In 0.9 ml the reaction system contained 65–113 μg membrane protein, 40 μmol imidazole/MOPS (3-[*N*-morpholino]propanesulfonic acid) at pH 7.1, 0.2–10 μmol $MgCl_2$, and a salt. The enzyme was phosphorylated for 20–70 s from 10 or 20 nmol [γ-^{32}P]ATP added in 0.1 ml. Phosphorylation was stopped either by an acid quench or by 0.1 ml of a chase solution containing 1–2 μmol unlabeled MgATP without or with 8 μmol MgADP or 50 μmol KCl. The ATP chase solution was neutralized with $NaHCO_3$ or Tris. The concentrations shown on the abscissa of the figures are within 10% of that during the chase. After various times as indicated, the chase was terminated with 10 ml of 0.35 M trichloroacetic acid containing 0.6 mM ATP and 13 mM H_3PO_4. The samples were filtered, washed, and counted. NaBr was "Suprapur" from Merck, Darmstadt, Germany.

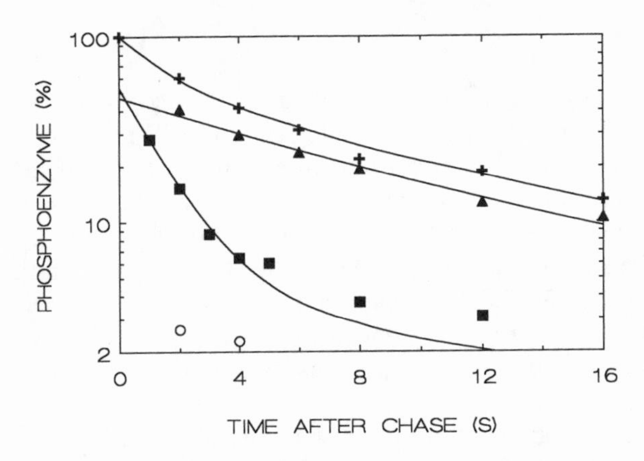

Figure 1. Kinetics of dephosphorylation in 160 mM NaCl after chases without or with ADP or K^+. The phosphorylation system is described under Materials and Methods. It contained 0.3 μmol $MgCl_2$ and 160 μmol NaCl in 1 ml of the labeling mixture. The chase solution contained 3 μmol of Na_4CDTA without (+) or with 1.1 μmol ADP (▲) or 6 μmol KCl (■) or both (○). The solid lines fit a model consisting of independent rapid and slow components in each of which E_1P is the only precursor to E_2P and both forms undergo spontaneous hydrolysis. See an expanded abstract by Suzuki and Post (1991) for evidence for rapid and slow components.

Results

Reaction System

To distinguish between E_1P, phosphoenzyme sensitive to ADP, and E_2P, phosphoenzyme sensitive to K^+, the enzyme was first phosphorylated from [^{32}P]ATP and then further formation of the radioactive phosphoenzyme was prevented by a "chase" with an excess of unlabeled ATP or with a chelator of Mg^{2+}, CDTA (cyclohexylenediamine tetraacetic acid), which is required for phosphorylation. To test sensitivity to ADP or to K^+ these ligands were added to the chase solution. Fig. 1 illustrates the system. In most experiments phosphoenzyme was estimated only at 4 s after the chase. If the level after K^+ was higher than that after ADP, then E_1P was favored; if the level after ADP was higher than that after K^+, then E_2P was favored. Thus a preponderance of one phosphoform over the other could be produced by effects on the initial distribution (partition) between the phosphoforms or on their rates of disappearance. In most experiments a chase with a combination of ADP and K^+ left $<3\%$ of the initial level of phosphoenzyme after 1 s or longer.

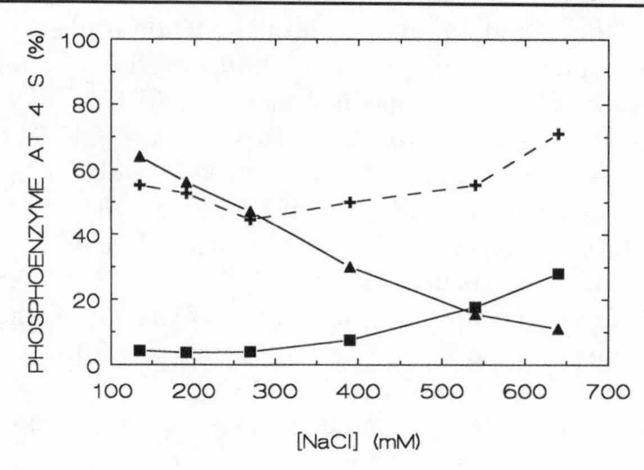

Figure 2. Effect of NaCl concentration on sensitivity of phosphoenzyme to ADP or K^+. Data are from three enzyme preparations. In two experiments the buffer was Tris·Cl at pH 7.5. $MgCl_2$ was 10 μmol. In one experiment $MgSO_4$ replaced $MgCl_2$. The chase solution contained 2 μmol of unlabeled ATP without (+) or with 8 μmol ADP (▲) or 50 μmol KCl (■). The duration of the chase was 4 s.

Lyotropic Anions

Effects of changing the concentration of NaCl on 4-s chases are shown in Fig. 2. At low concentrations the phosphoenzyme was insensitive to ADP; thus E_2P predominated. At high concentrations the phosphoenzyme was insensitive to K^+; thus E_1P predominated. A concentration of NaCl of 540 mM gave equal levels of each kind of phosphoenzyme. We call this concentration a "crossover" concentration. Replacement of Cl^- with I^- produced a similar pattern except that the crossover concentration was lower, ~100–170 mM (Fig. 3). Further experiments with Br^-, NO_3^-, CNS^-, or ClO_4^- showed similar patterns with crossover concentrations progressively lower than that of Cl^-. In contrast to these ions, carboxylate ions showed higher crossover concentrations as in Fig. 4.

Different enzyme preparations gave different results (Fig. 3). Specific single experiments (not shown) indicated two factors that favored E_2P; (*a*) use of a light fraction from the isopycnic zonal centrifugation rather than a heavy fraction, and (*b*) storage of the enzyme preparation at 4°C for 55 d (5% loss of activity). Storage at −70°C for 44 d preserved the initial response. Addition of 0.01% bovine serum albumin had no effect on the response.

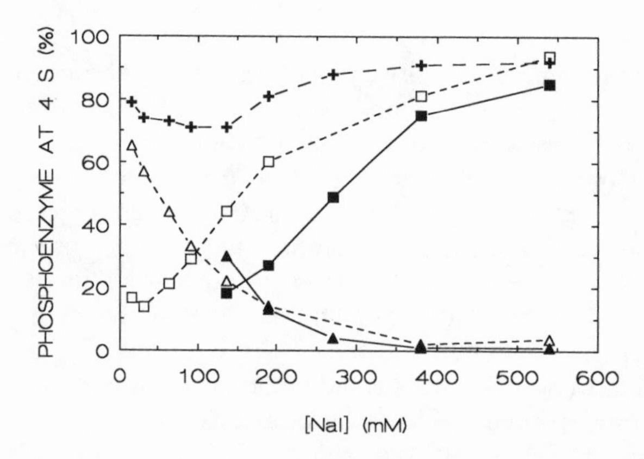

Figure 3. Effect of I^- concentration and enzyme age on phosphoenzyme sensitivity to ADP or K^+. Data are from five experiments on four enzyme preparations. The reaction system contained 10 μmol $MgCl_2$, 16–540 μmol NaI, and 2% (mol/mol) as much cysteine as NaI. Chase solutions, duration, and symbols are the same as those in Fig. 2. Open symbols show data from enzyme preparations stored at 4°C for 8–32 d; closed symbols show storage for 91 d. The age of the enzyme in the blank chase was 30 d. In other experiments there was no effect of storage on phosphoenzyme level after a blank chase.

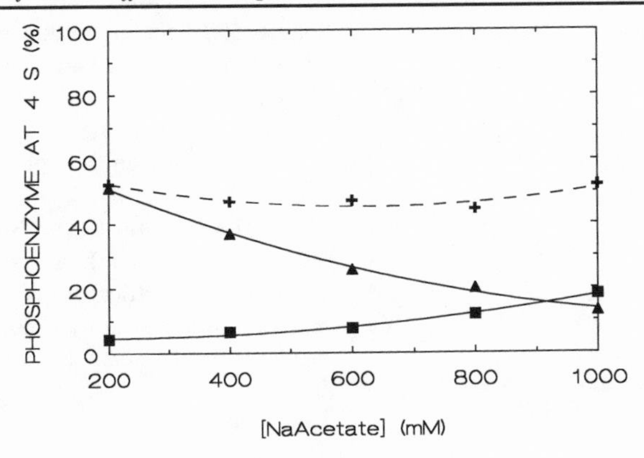

Figure 4. Effect of acetate concentration on phosphoenzyme sensitivity to ADP or K^+. The reaction system contained 6 μmol of $MgCl_2$ and 200, 400, 600, 800, or 1,000 μmol of sodium acetate. Chase solutions, duration, and symbols are the same as those in Fig. 2. Data are from one experiment.

Rank of Anions

To rank anions at a constant Na^+ concentration we used a single enzyme preparation. The rank of the chaotropic ions was $Cl^- < Br^- < NO_3^- < SCN^- = I^- < ClO_4^-$ (Fig. 5). The rank of kosmotropic anions was citrate^{3-} > aspartate$^-$ > succinate^{2-} > SO_4^{2-} = acetate$^-$ > Cl^- (Fig. 6 and text). Polybasic anions were compared with monobasic ions at normal, not molar, concentrations. Other experiments comparing pairs of anions using the same membrane preparation showed succinate^{2-} > SO_4^{2-}. Isethionate, monochloro- and dichloroacetate$^-$ were between acetate$^-$ and Cl^-. Iso- and *n*-butyrate$^-$ were like succinate^{2-}; pivalate$^-$ was like aspartate$^-$. Thus the potency of the various anions ranks them in a sequence that qualifies as a Hofmeister series (Collins and Washabaugh, 1985).

Additivity of the Actions of Anions

To test if the effect of one anion adds to that of another, we compared Cl^- with a mixture of NO_3^-, a more chaotropic anion, and acetate, a more kosmotropic anion. In preliminary experiments a mixture of 160 mM NO_3^- and 240 mM acetate gave almost the same responses as 400 mM Cl^- (not shown).

Lyotropic Anions on Spontaneous Dephosphorylation

Anions affected not only the relative sensitivity of the phosphoenzyme to ADP or K^+, but also the rate of spontaneous hydrolysis as shown by the phosphoenzyme level 4 s after a blank chase with ATP alone. The more chaotropic anions decreased the rate and the more kosmotropic anions increased the rate (Figs. 5 and 6). We compared

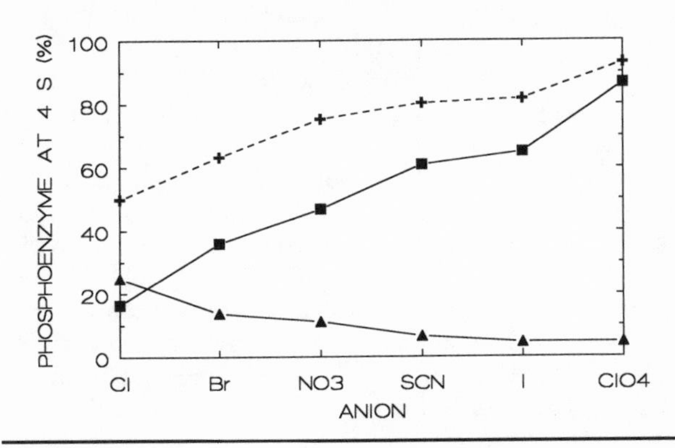

Figure 5. Effect of chaotropic anions on sensitivity of phosphoenzyme to ADP or K^+. The reaction system contained 300 μmol of the sodium salts of Cl, Br, NO_3, SCN, I, or ClO_4. Together with I^- the system contained 2% (mol/mol) cysteine as in Fig. 3. Chase solutions, duration, and symbols are the same as those in Fig. 2. Data are from one experiment.

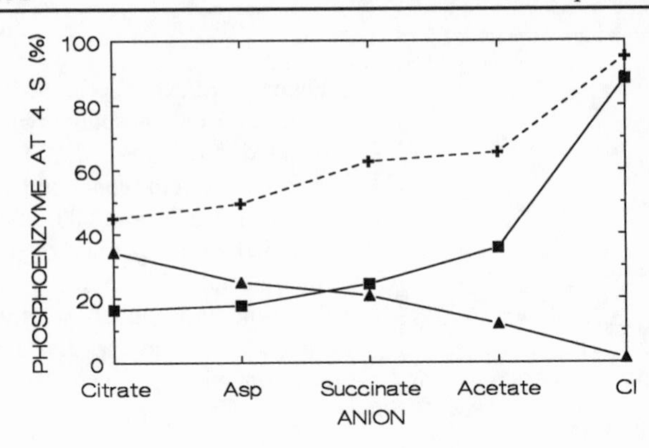

Figure 6. Effect of kosmotropic anions on the sensitivity of the phosphoenzyme to ADP or K^+. The reaction system contained one milliequivalent of the sodium salts of Na_3citrate (*Citrate*), aspartate (*Asp*), Na_2 succinate (*Succinate*), acetate (*Acetate*), or chloride (*Cl*). Chase solutions, duration, and symbols are the same as those in Fig. 2. Data are from one experiment.

sodium salts of Br^-, Cl^-, formate, and citrate at various concentrations. In all cases at concentrations between 24 and 100 mM the rate increased as the [Na^+] increased (Fig. 7). At higher concentrations the rate tended to decrease and the magnitude of the decrease was greater the more chaotropic the anion (Fig. 7). In 160 mM $NaNO_3$ replacement of 100 mM of [Na^+] with tetramethylammonium, *N*-methylglucamine, choline, lysine, guanidine, or arginine decreased the rate of dephosphorylation by at least a factor of 2 in preliminary experiments (not shown). Thus the stimulation of dephosphorylation was more prominent with Na^+.

Cations

When organic cations were varied in a medium with a constant anion composition, the ratio of E_1P to E_2P varied. The cations formed a series. They showed the following relative potency in favoring E_1P: arginine > guanidine > lysine > choline ≥ *N*-methylglucamine > tetramethyl ammonium > triethylamine > tripropylamine (Fig. 8). In another experiment $Tris^+$ = $choline^+$ (not shown). We do not know if this is a Hofmeister series.

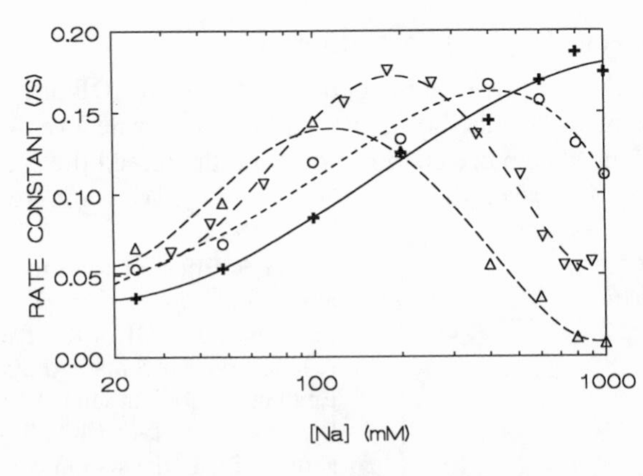

Figure 7. Effect of concentration of sodium salts on spontaneous hydrolysis of phosphoenzyme. Data are from 13 experiments on 6 enzyme preparations. The reaction system contained 0.225–10 μmol $MgCl_2$, 32–900 μmol NaCl (▽), 8–333 μmol Na_3–citrate (+), 24–1,000 μmol Na-formate (○), or NaBr (△). The chase solution contained 0.8–2 μmol unlabeled ATP (80:1 to 20:1 molar ratio of cold ATP to hot ATP) or 4.5 μmol of CDTA (CDTA:Mg = 20:1) plus 0.1 μmol unlabeled ATP (cold ATP:hot ATP = 10:1). After various times the chase was terminated with acid. The rate constant was calculated assuming single exponential decay. The values for each anion were normalized by reference to rate constants that were obtained from a single enzyme preparation in the presence of 0.1 N NaCl, Na_3-citrate, Na-formate, or NaBr.

Na^+ was different from these cations and compared with them Na^+ stimulated dephosphorylation. With respect to the partition of the phosphoenzyme between E_1P and E_2P compared with other cations, the action of Na^+ depended on the accompanying anion. In the following experiments (not shown) the amounts of E_1P and E_2P were estimated by extrapolation back to the onset of the chase. In 0.4 M NaCl E_1P was 67%, whereas replacement of 0.36 M Na^+ with $Tris^+$ decreased E_1P to 45% (not shown). In Cl^-, Na^+ favored E_1P more than $Tris^+$ did. Yet in 0.6 N Na_3–citrate E_1P was ~37%, whereas replacement of 585 mM of the Na^+ with $Tris^+$ increased E_1P to 62% (not shown). In citrate, Na^+ favored E_1P less than $Tris^+$ did. Thus chaotropic anions made Na^+ more effective in favoring E_1P. A preliminary experiment suggested that Hofmeister effects could be observed when the cations were 20 mM Na^+ plus 500 mM Lys^+. Thus Na^+ did not appear to be required for Hofmeister effects.

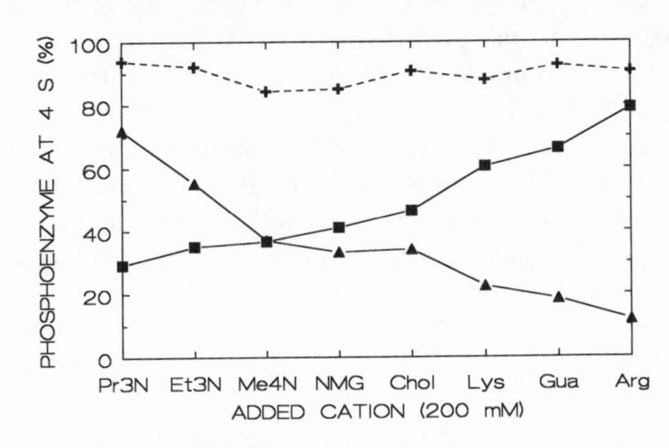

Figure 8. Effect of organic cations on sensitivity of phosphoenzyme to ADP or K^+. The reaction system contained 4 μmol $Mg(NO_3)_2$, 50 μmol $NaNO_3$, and 200 μmol of the nitrate salts of arginine (*Arg*), guanidine (*Gua*), lysine (*Lys*), choline (*Chol*), *N*-methylglucamine (*NMG*), tetramethylammonium (*Me4N*), triethylamine (*Et3N*), and tripropylamine (*Pr3N*). The chase solution contained 1.1 μmol unlabeled ATP without (+) or with 8.8 μmol ADP (▲) or 6.6 μmol KNO_3 (■). The chase was terminated after 4 s.

Effects of Anions on Activity

(Na^+)-ATPase and (Na^+ + K^+)-ATPase activities were estimated at 40 and 400 mM [Na^+]. Replacement of Cl^- with other anions had no effect at 40 mM. At 400 mM replacement with carboxylate anions had no effect but replacement with NO_3^-, SCN^-, or ClO_4^- inhibited both activities; NO_3^- was less effective than the other two anions (not shown). Thus anions that favored E_1P inhibited activity.

Miscellaneous Results

Preliminary results indicated that dimethylformamide (20%, vol/vol) favored E_2P; urea (1 M) had little effect (not shown). Our experiments did not show a significant amount of E∗P. We have seen effects of anions on the amount of phosphoenzyme before the chase; we have not studied them. In some experiments low initial phosphorylation and anomalous dephosphorylation were seen, particularly with trichloroacetate as anion and with guanidinium thiocyanate; we think that the enzyme may have been partially denatured.

Discussion

A Hofmeister effect on the reactive state of an enzyme is an unusual finding. A Hofmeister effect often relates to the folding or association of macromolecules. The change in reactive state between E_1P and E_2P implies a conformational change since it involves the reactivity of a single phosphate group. It is possible that the Hofmeister effect is acting by modifying the folding of the protein rather than by participation of anions directly in the active center, or by modification of the structure of the specific water molecule that attacks the bond of the active site phosphate group, particularly in the E_2P state.

Although chaotropic ions usually favor dissociation or unfolding of macromolecules, in at least one case they do not (Ries-Kautt and Ducruix, 1989). Thus it is not possible to be sure whether E_1P or E_2P is the more unfolded state. The usual expectation would be that E_1P is more unfolded. See also Blasie et al. (1985) for the movement of protein toward the membrane in E_2P of the Ca-ATPase of sarcoplasmic reticulum.

We hope that this Hofmeister effect will provide a new means for control and investigation of the reactive states of the phosphoenzyme.

These results leave many questions unanswered. When a high concentration of Na^+ favors E_1P, does it act at a low affinity transport site in communication with the extracellular medium or nonspecifically as a lyotropic ion? To what extent does the partition between E_1P and E_2P represent an equilibrium and to what extent a steady state? What is the mechanism of the unexpected sensitivity of the enzyme to aging or to the fraction collected (Fig. 3)? Is it changes in membrane lipids (Yoda and Yoda, 1987*a*, *b*)?

Acknowledgments

This work was supported by grant HL-01974 from the National Heart, Lung, and Blood Institute of the National Institutes of Health.

References

Blasie, J. K., L. G. Herbette, D. Pascolini, V. Skita, D. H. Pierce, and A. Scarpa. 1985. Time-resolved x-ray diffraction studies of the sarcoplasmic reticulum membrane during active transport. *Biophysical Journal.* 48:9–18.

Collins, K. D., and M. W. Washabaugh. 1985. The Hofmeister effect and the behaviour of water at interfaces. *Quarterly Review of Biophysics.* 18:323–422.

Dani, J. A., J. A. Sanchez, and B. Hille. 1983. Lyotropic anions. Na channel gating and Ca electrode response. *Journal of General Physiology.* 81:255–281.

Glynn, I. M. 1988. Overview: the coupling of enzymatic steps to the translocation of sodium and potassium. *In* The Na^+,K^+-Pump. Part A: Molecular Aspects. J. C. Skou, J. G. Nørby, A. B. Maunsbach, and M. Esmann, editors. Alan R. Liss, Inc., New York. 435–460.

Glynn, I. M., and S. J. D. Karlish. 1990. Occluded cations in active transport. *Annual Review of Biochemistry.* 59:171–205.

Hara, Y., and M. Nakao. 1981. Sodium ion discharge from pig kidney Na^+,K^+-ATPase: Na^+-dependency of the $E_1P \Leftrightarrow E_2P$ equilibrium in the absence of KCl. *Journal of Biochemistry.* 90:923–931.

Hofmeister, F. 1888. II. Zur Lehre von der Wirkung der Salze. Zweite Mittheilung. *Archiv für Experimentelle Pathologie und Pharmakologie.* 24:247–260.

Jørgensen, P. L. 1974. Purification and characterization of (Na^+ + K^+)-ATPase. III. Purification from the outer medulla of mammalian kidney after selective removal of membrane components by sodium dodecylsulphate. *Biochimica et Biophysica Acta.* 356:36–52.

Lee, J. A., and P. A. G. Fortes. 1985. Anthroylouabain binding to different phosphoenzyme forms of Na,K-ATPase. *In* The Sodium Pump. 4th International Conference on Na,K-ATPase. I. M. Glynn and C. Ellory, editors. The Company of Biologists, Ltd., Cambridge, UK. 277–282.

Nørby, J. G., and I. Klodos. 1988. Overview: the phosphointermediates of Na,K-ATPase. *In* The Na^+,K^+-Pump. Part A: Molecular Aspects. J. C. Skou, J. G. Nørby, A. B. Maunsbach, and M. Esmann, editors. Alan R. Liss, Inc., New York. 240–270.

Nørby, J. G., I. Klodos, and N. O. Christiansen. 1983. Kinetics of Na-ATPase activity by the Na,K pump. Interactions of the phosphorylated intermediates with Na^+, $Tris^+$, and K^+. *Journal of General Physiology.* 82:725–759.

Ries-Kautt, M. M., and A. F. Ducruix. 1989. Relative effectiveness of various ions on the solubility and crystal growth of lysozyme. *Journal of Biological Chemistry.* 264:745–748.

Suzuki, K., and R. L. Post. 1991. Slow and rapid components of phosphorylation kinetics of Na,K-ATPase. *In* The Sodium Pump: Recent Developments. J. H. Kaplan and P. De Weer, editors. The Rockefeller University Press, New York. In press.

Taniguchi, K., and R. L. Post. 1975. Synthesis of adenosine triphosphate and exchange between inorganic phosphate and adenosine triphosphate in sodium and potassium ion transport adenosine triphosphatase. *Journal of Biological Chemistry.* 250:3010–3018.

The, R., and W. Hasselbach. 1975. The action of chaotropic anions on the sarcoplasmic calcium pump. *European Journal of Biochemistry.* 53:105–113.

von Hippel, P. H., and T. Schleich. 1969. Ion effects on the solution structure of biological macromolecules. *Accounts of Chemical Research.* 2:257–265.

Yoda, A., and S. Yoda. 1987*a*. Two different phosphorylation–dephosphorylation cycles of Na,K-ATPase proteoliposomes accompanying Na^+ transport in the absence of K^+. *Journal of Biological Chemistry.* 262:110–115.

Yoda, A., and S. Yoda. 1988. Cytoplasmic K^+ effects on phosphoenzyme of Na,K-ATPase proteoliposomes and on the Na^+-pump activity. *Journal of Biological Chemistry.* 263:10320–10325.

Yoda, S., and A. Yoda. 1986. ADP- and K^+-sensitive phosphorylated intermediate of Na,K-ATPase. *Journal of Biological Chemistry.* 261:1147–1152.

Yoda, S., and A. Yoda. 1987*b*. Phosphorylated intermediates of Na,K-ATPase proteoliposomes controlled by bilayer cholesterol: interaction with cardiac steroid. *Journal of Biological Chemistry.* 262:103–109.

Chapter 16

Rate-limiting Steps in Na Translocation by the Na/K Pump

Bliss Forbush III and Irena Klodos

Department of Cellular and Molecular Physiology, Yale University School of Medicine, New Haven, Connecticut 06510; Mount Desert Island Biological Laboratory, Salsbury Cove, Maine 04672; and Institute of Biophysics, University of Aarhus, DK-8000 Aarhus, Denmark

Introduction

Na and K movements catalyzed by the Na/K pump have been adequately described by modified versions of the Post-Albers model (Post et al., 1965; Albers et al., 1968). For the purposes of this paper, the essential features of the model are shown in Scheme 1. Note that only steps 2 and 3 are presumed to be composed of single reactions; steps 1′, 4′, 5′, and 6′ each involve multiple reactions lumped together for ease of discussion.

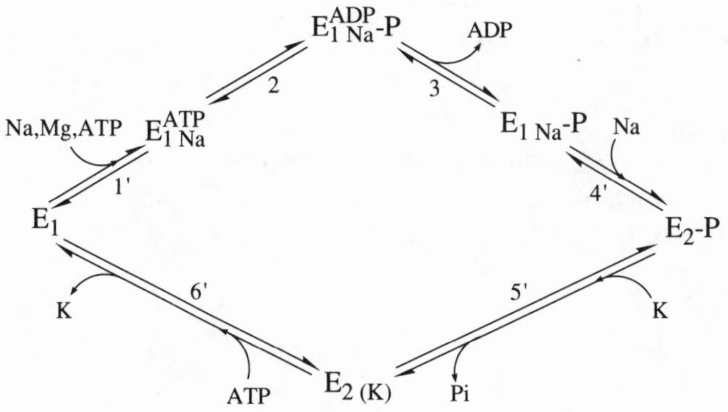

$E_{1\,Na}^{ADP}$-P ADP

2 3

Na,Mg,ATP $E_{1\,Na}^{ATP}$ $E_{1\,Na}$-P Na

1' 4'

E_1 E_2-P

6' 5'

K K

ATP $E_{2\,(K)}$ Pi

Scheme 1

Among the questions to be asked of such a model is, "What are the slow steps?" There are several reasons to ask. First, the slow steps are the reactions that are easiest to study experimentally, and they are the reactions that dominate kinetic transients involving multiple reactions. Second, the slow steps are physiologically important since they determine the performance limits of the enzyme. Third, an understanding of these steps seems particularly valuable in understanding structure–function relationships and the mechanism of transport: the structure of the enzyme must have been optimized through evolution to increase the rate of these steps.

The steady-state performance of Na,K-ATPase is maximal near pH 7.5 and declines at low pH and high pH (Skou, 1957). The rate limitation at low pH is accounted for by the ATP-stimulated deocclusion of K ions as part of the process of K translocation (Forbush, 1987). The rate of K deocclusion increases with increasing pH, so that at high pH this reaction is much faster than the overall turnover rate. We therefore proposed that there must be steps in Na translocation which become rate limiting (Forbush, 1987). This report summarizes our attempts to determine the rates of Na translocation steps, and to pinpoint the slow reactions at high pH.

Methods

Experiments with the Fluorescent Aminostyrylpyridinium Dyes RH-160 and RH-421

Experiments were performed in a simple filter fluorometer described elsewhere (Klodos and Forbush, 1988; Forbush, B., and I. Klodos, manuscript in preparation; 540 nm excitation, >610 nm emission). RH-160 or RH-421 (6 μM) was bound to SDS-washed membranes (~0.2 mg/ml) in a solution that typically contained 100 mM

NaCl, 5–40 mM TrisHEPES, and 0.1 mM EGTA. In experiments involving caged ATP, a 20-μl sample contained 2 mM Mg, 0.5 mM caged ATP, and 10 mM glutathione; ~60% of the caged ATP was photolyzed in <0.1 ms with the light of a short arc flashlamp focused with an elliptical mirror; data could be recorded after a 10-ms delay for dye phosphorescence to decay. In experiments involving "caged Mg" (Mg/DM-nitrophen), the solution contained 0.25 mM ATP, 0.4 mM DM-nitrophen, and 0.2–0.4 mM Mg. For stop-flow experiments a 1.5 mm (i.d.) × 8 mm plexiglass chamber was mounted in the filter fluorometer—two microsyringes each delivered 8-μl aliquots via a simple "T" mixer, displacing a teflon plunger. The observational dead time was estimated to be ~3 ms. One of the two solutions contained 4 mM MgATP and 20 mM HEPES/Tris buffer; the other contained Na,K-ATPase, RH-dye, Na, and 0.1 mM Mg as above.

Dephosphorylation

Dephosphorylation experiments were performed using the rapid filtration apparatus previously described (Forbush, 1984*b,* 1987, 1988). Phosphorylation from 10 μM [γ^{32}P]ATP was performed for 5 s at 0°C in the presence of 1 M NaCl and 2 mM Mg to form E_1-^{32}P. The sample was diluted in 1.5 M NaCl, 0.3 mM EDTA, and 8 mM imidazole (pH 7.2), filtered, and rinsed with the same solution. After transfer to the rapid filtration apparatus, the sample was further rinsed with the same solution or with a different solution as noted in the text, and dephosphorylation was observed on changing to a solution typically containing 100 mM Na, with or without 10 mM K. In most experiments the rinse solution in the rapid filtration apparatus was at 0°C, but essentially the same results were obtained when this solution was at 20°C.

In some experiments it was determined whether dephosphorylation occurred via release of $^{32}P_i$ or [^{32}P]ATP. To each filtrate sample was added 2 ml of isobutanol/hexane and 1 ml of 0.1 M molybdate in 10% H_2SO_4. After 10 s of vigorous vortexing, 1 ml of each phase was counted to determine the two chemical species.

Results

Rate of Na Translocation Steps Revealed by Fluorescent Dyes

We have found that aminostyrylpyridinium dyes such as RH-160 and RH-421 are very useful indicators of pump cycle changes in Na,K-ATPase from various sources (Klodos and Forbush, 1988): neurospora plasma membrane H-ATPase (Nagel et al., 1989), sarcoplasmic reticulum Ca-ATPase, and gastric H,K-ATPase (Klodos, I., and B. Forbush, manuscript in preparation). The dyes partition reversibly into the lipid bilayer and exhibit large changes in fluorescence emission during different states in the pump cycle without affecting enzyme activity. The mechanism of conformation sensing is probably related to the electrochromic nature of the dye molecules: it is likely that at least part of the dye signal is due to changes in the electrostatic field surrounding the pump molecule as ions are bound and released (Klodos and Forbush, 1988; Post and Suzuki, 1991).

The largest change in the fluorescence of RH-160 (or RH-421) bound to Na,K-ATPase is the increase seen on addition of ATP to the NaE$_1$ form of the enzyme, resulting in the $E_1 \rightarrow \rightarrow E_2$-P reactions (steps 1–4 in Scheme 1) and Na translocation. Using caged ATP to initiate the reaction, we have found that the rate constant of the fluorescence change is ~45 s^{-1} at pH 7.2 and 20°C (Klodos and

Forbush, 1988). The rate is in good agreement with other measures of the Na translocation steps, including ^{22}Na movement (Forbush, 1984*a,* 1985), charge movement (Nakao and Gadsby, 1986; Borlinghaus et al., 1987; Gadsby et al., 1991), and the conformational change in iodoacetamidofluorescein-modified enzyme (Steinberg and Karlish, 1989; Stürmer et al., 1989).

Caged ATP and Measurements of Na,K-ATPase Transients

The use of caged ATP to initiate rapid reactions of Na,K-ATPase has always been clouded by the fact that caged ATP itself binds to Na,K-ATPase in competition with ATP, raising the possibility that the observed transients contain components that are affected by the interaction (Forbush, 1984*a;* Borlinghaus et al., 1987; Fendler et al., 1987, 1991). The situation is complicated by the fact that the various components of a transient curve do not carry signatures, and by the ambiguities in assigning derived rate constants to individual steps in multistep processes.

To circumvent these difficulties, we used two other means to initiate pump turnover. In the first, DM-nitrophen, a photosensitive derivative of EDTA (Kaplan and Ellis-Davies, 1988), was used to produce a rapid increase in magnesium concentration. Due to a rather low affinity of DM-nitrophen for Mg, an initial finite level of free Mg was unavoidable, so some of the Na,K-ATPase was not in the desired E_1 state at the outset. Thus the observed transient was somewhat smaller than seen with caged ATP. The rate constant of the fluorescence transient was, however, identical to that measured with caged ATP, $\sim$45–50 s^{-1}, demonstrating that this rate is intrinsic to the Na/K pump and is not an artifact of interaction with caged ATP (Klodos and Forbush, 1988; Forbush, B., and I. Klodos, manuscript in preparation).

We have also taken the straightforward approach and initiated the catalytic cycle by mixing Na,K-ATPase with ATP in a stop-flow fluorometer. The results of one such experiment are illustrated in Fig. 1 (upper panel). Again the results confirmed those from the earlier experiments with caged ATP, in that the rate constant of the fluorescence transient was found to be 45–60 s^{-1} in dog and pig kidney Na,K-ATPase.

pH Profile of the RH-160,RH-421 Transient

The turnover rate of kidney Na,K-ATPase is $\sim$20 s^{-1} at 20°C (cf. Forbush, 1987); therefore, the Na translocation steps account for about half the rate limitation of steady-state pump activity under normal pump conditions, while K deocclusion accounts for the other half (Karlish and Yates, 1978; Forbush, 1987). In view of the expectation that slowing of Na translocation would account for the decrease in steady-state Na,K-ATPase activity at high pH, we were especially interested in the pH dependence of the RH-dye transient. Several curves from a representative experiment are illustrated in Fig. 1. They illustrate that, in agreement with the prediction, the rate of fluorescence increase was found to decrease from 45–60 s^{-1} at pH 7.1 to 8–16 s^{-1} at pH 8.7. Essentially identical results were obtained whether the reaction was started by photolysis of caged ATP to release ATP, photolysis of DM-nitrophen to release Mg (Klodos and Forbush, 1988), or by mixing ATP with the enzyme in the stop-flow apparatus (Forbush, B., and I. Klodos, manuscript in preparation). Thus, as summarized in Fig. 2, the pH dependence of Na,K-ATPase activity is adequately described as limited by the rate of K deocclusion at low pH, and by the rate of Na translocation at high pH.

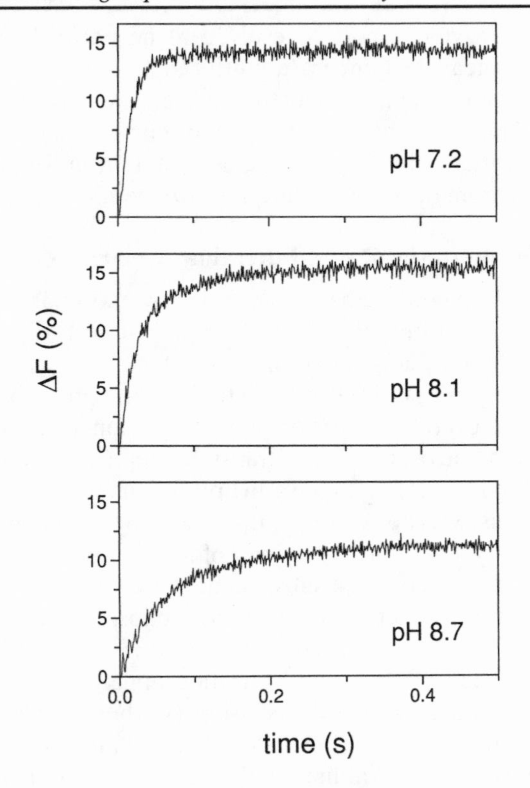

Figure 1. Transient fluorescence change on mixing RH-421 bound to pig kidney Na,K-ATPase with a solution containing 4 mM MgATP. Both solutions contained 100 mM Na and additional 0.1 mM Mg. The pH was 7.2, 8.1, or 8.7, as indicated. Rate constants for the fit of single exponentials to these curves were 60, 31, and 16 s^{-1}, respectively.

Species Difference in the Rate of the RH-421 Transient

It has often proven to be difficult to reconcile related transient kinetic results obtained in different laboratories. A complicating factor has been the use of Na,K-ATPase from different species and tissues. In this regard, we have noted that the rate of ATP-stimulated ^{86}Rb deocclusion at 20°C is substantially faster with eel Na,K-ATPase (Forbush, 1985) than with pig or dog kidney Na,K-ATPase (Forbush, 1987) or bovine brain Na,K-ATPase (Forbush, B., unpublished results). In experiments similar to those shown above, we compared the rate of the increase in RH-421 fluorescence on mixing ATP with Na,K-ATPase prepared from different sources. We found that for Na,K-ATPase from pig kidney, eel electroplax, and shark rectal gland the rate constant was ~50, ~145, and ~65 s^{-1}, respectively, at 20°C and pH 7.3 (data not shown). The high reaction rate in eel Na,K-ATPase is particularly

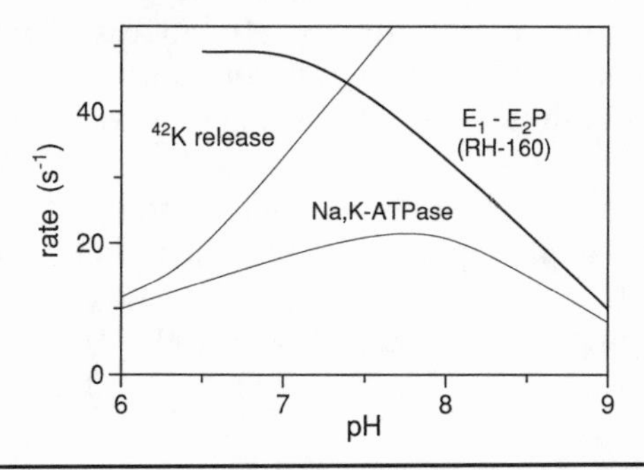

Figure 2. Schematic diagram comparing reaction rates in kidney Na,K-ATPase. Shown are the turnover rate of Na,K-ATPase activity (Forbush, 1987), the rate constant of ATP-stimulated ^{42}K release, and the RH-dye transient indicative of the E_1-E_2-P conformational change (Klodos and Forbush, 1988, and Fig. 1).

interesting; it suggests that the eel enzyme may have evolved to be particularly effective at a lower body temperature, near the temperature utilized in these assays. The large difference between the rate constant in eel electroplax Na,K-ATPase and that in mammalian kidney Na,K-ATPase may help to explain earlier discrepancies among results from different laboratories, and should serve as a caution in future attempts to construct a unified model from results with differing enzyme sources.

Measurements of Dephosphorylation in the Rapid Filtration Apparatus

Na binding, translocation, and release is known to be accompanied by phosphorylation of Na,K-ATPase from ATP and a conformational change from E_1-P to E_2-P; in conventional schemes, K-dependent dephosphorylation can proceed only after Na is released, completing the translocation process. If this is so, rate-limiting steps in Na translocation must also be part of the overall process of phosphorylation–dephosphorylation. Thus we sought to measure the overall rate of transition/dephosphorylation from E_1-P to E_2 as a function of pH, expecting to see a pH profile similar to that determined with RH dyes. Na,K-ATPase was poised in the E_1-^{32}P state by incubation with [^{32}P]ATP in the presence of very high concentrations of NaCl, and in the absence of K. K was added and Na was diluted in the subsequent chase, allowing an evaluation of the rate of the E_1-P $\rightarrow$ E_2-P transition followed by dephosphorylation (steps 4′ and 5′ in Scheme 1).

These experiments were performed with the rapid filtration apparatus used previously to measure ^{22}Na translocation and ^{86}Rb deocclusion (Forbush, 1984*a,* 1987). The apparatus is ideally suited for determination of "off" rates of membrane-bound ligands; however, experiments are somewhat limited by the requirement that after filtration/rinsing the sample is held on the filter for 10–20 s at 0°C before the observation is performed. We used a highly lyotropic rinse solution (1–2 M NaBr, NaI, or NaCl) which stabilizes E_1-P (Post, 1988; Post and Suzuki, 1991) and greatly slows the E_1-P to E_2-P conversion (Klodos and Forbush, 1991). Then in the rapid filtration apparatus, the E_1-P $\rightarrow$ E_2-P transition was initiated by switching to a nonlyotropic medium with or without K to promote dephosphorylation of E_2-P.

As illustrated in Fig. 3*A*, a complete time course of dephosphorylation was readily obtained with one sample. As in all prior studies, rapid dephosphorylation required K because E_2-P breaks down slowly in its absence. In this and numerous similar experiments the measured rate constant of dephosphorylation from E_1-^{32}P in the presence of K was 60–90 s^{-1} at pH 7.0–7.5 and 20°C. This is a lower estimate of the actual rate constant because of limitation of chamber washout in the apparatus (120–140 s^{-1}; Forbush, 1988). It may also be noted that the maximal rate of ^{32}P release is not attained immediately upon changing the solution (cf. Fig. 3*A*); this is indicative of an additional step in the process ($k > 100\ s^{-1}$). In any case, for the purposes of discussion in this paper, even the lower of the rate constants obtained in Fig. 3*A* is significantly faster than the overall $E_1 \rightarrow E_2$-P transition seen with the RH dyes (Fig. 1).

In other experiments, when we allowed the phosphorylated intermediate to relax from E_1-P to E_2-P by rinsing with a nonlytropic medium for 30 ms in the rapid filtration apparatus, addition of K resulted in an extremely rapid release of ^{32}P, too fast to resolve adequately with this apparatus (Fig. 3*B*, $k > 100\ s^{-1}$). This is consistent with previous findings that K-stimulated dephosphorylation of E_2-^{32}P is very fast.

ADP was also effective in promoting dephosphorylation (Fig. 3 *C*). The rate constant of ^{32}P release seen here is 21 s^{-1}. It was confirmed that the released ^{32}P was indeed [^{32}P]ATP, not $^{32}P_i$ (not shown). When a lower concentration of ADP was used, the *amount* of rapidly released ^{32}P was less, but the rate constant of release remained the same (lower curves, Fig. 3 *C*). This is consistent with models in which binding of ADP competes with the E_1-P → E_2-P transition, but in which a step other than ADP binding limits the rate of [^{32}P]ATP release. This slow step has been shown to be dissociation of [^{32}P]ATP (Mardh and Post, 1977; step 1 in Scheme 1).

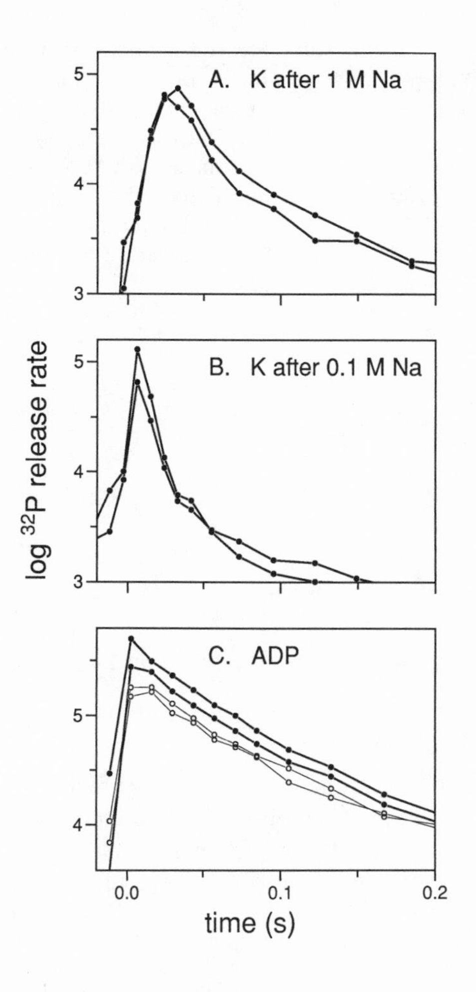

Figure 3. Release of ^{32}P from Na,K-ATPase monitored in the rapid filtration apparatus. The E_1^{32}P intermediate was formed and collected on a filter as described in Methods. The base 10 logarithm of the rate of release (arbitrary units) is plotted. In the rapid filtration apparatus, the first solution contained 2 M NaCl (*A*), 0.1 M NaCl (*B*), or 1 M NaI (*C*); the second solution contained 100 mM NaCl and either 10 mM KCl (*A* and *B*), 6.4 mM ADP (*C*, *heavy lines*), or 0.2 mM ADP (*C*, *light lines*). *C* was from a different experiment than *A* and *B*.

pH Profile of ^{32}P Release from E_1-^{32}P

Time courses of transition/dephosphorylation at various pHs are illustrated in Fig. 4 (center column); also shown are control curves obtained in the absence of K (left column) or when the E_1-P → E_2-P transition was blocked by high salt (right column). The rate constants of ^{32}P release are replotted in Fig. 5 (heavy line). To our surprise, transition/dephosphorylation was found to be much higher than both the RH-dye transient and the turnover rate of Na,K-ATPase at high pH (dashed lines, replotted from Fig. 3). Thus, above pH 7.5 the net rate of the reactions encompassing E_1-P →

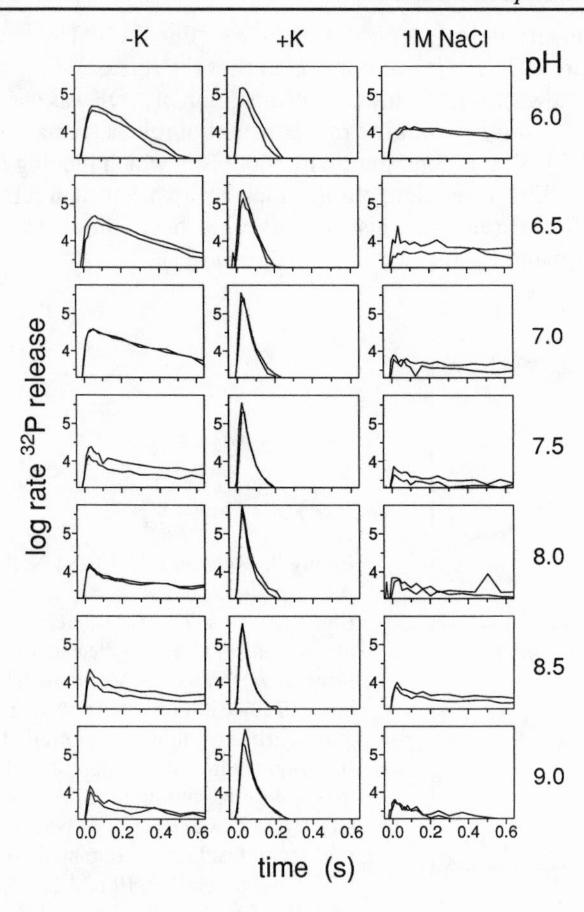

Figure 4. ^{32}P release from $E_1{}^{32}P$. In the rapid filtration apparatus the first solution contained 2 M NaCl. The second solution contained 100 mM Na (*left column*), 100 mM Na/10 mM K (*middle column*), or 1 M NaCl (*right column*) at the indicated pH.

$E_2 + P_i$ is about two to four times faster than the $E_1 \rightarrow E_2$-P reactions manifested by the styryl dyes.

ADP Effect on Dephosphorylation Kinetics

Of many possible explanations for the above discrepancy in rates are models in which the limiting step at high pH involves formation of the phosphorylated intermediate; these steps (steps 2 and 3 in Scheme 1) are part of the sequence of events seen by the styryl dyes, but are not a part of the dephosphorylation process starting from E_1-P. As

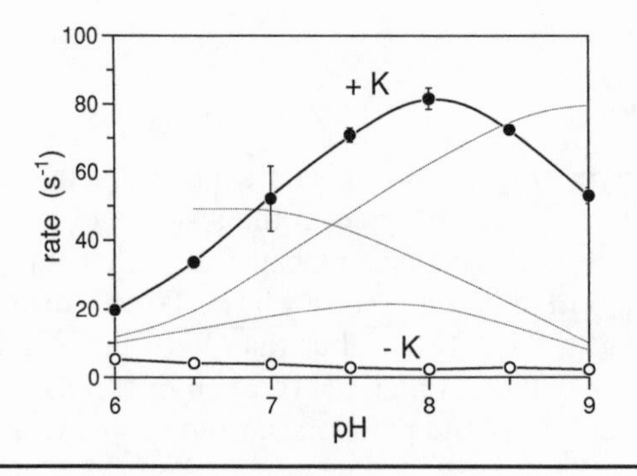

Figure 5. The rate constants of ^{32}P release, obtained from fits of single exponentials to the data in Fig. 4 (from left and center columns). Error bars denote the range between duplicates. For release in the presence of K, the data between 40 and 100 ms were used. Light dashed lines reproduce the curves from Fig. 2.

a test of such models we examined the effect of a brief preexposure to ADP on the rate of dephosphorylation. By including ADP with high salt in the rinse solution, the events of phosphorylation should be reversed (in 20–30 ms, <50% dissociates as [^{32}P]ATP; cf. Fig. 3 *C*). As a result of this procedure, it was anticipated that much of the enzyme would be in the E_1(ATP) and E_1-P(ADP) forms. If the forward rates from these states to E_1-P are very fast (steps 2 and 3), then the subsequently observed dephosphorylation rate would be unaffected by the ADP prepulse. However, if either of the rates is low, a lower rate of $^{32}P_i$ production is expected following ADP; furthermore, a continued dissociation of [^{32}P]ATP should be observed during the observation period.

The results of one such experiment are illustrated in Fig. 6. In each sample we assayed the released ^{32}P by organic extraction of phosphomolybdate in order to

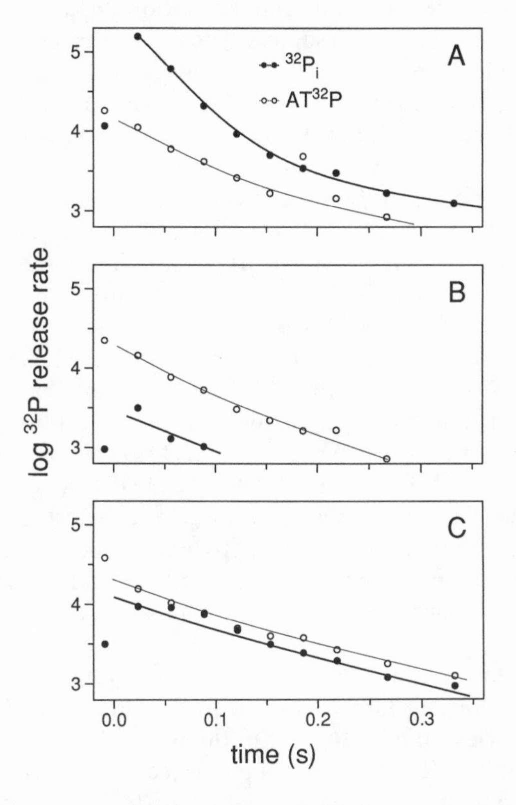

Figure 6. ^{32}P release after exposure to ADP. In the rapid filtration apparatus, the first solution contained 1.5 M NaCl (*A*) or 1.5 M NaCl and either 10 mM ADP/10 mM EDTA (*B*) or 10 mM ADP/5 mM Mg (*C*); the second solution contained 100 mM Na/10 mM K and either 10 mM EDTA (*A* and *B*) or 5 mM Mg (*C*). $^{32}P_i$ (●) and [^{32}P]ATP (○) were separated from one another by extraction of P_i as phosphomolybdate.

resolve $^{32}P_i$ and [^{32}P]ATP. As expected, most of the ^{32}P was released as $^{32}P_i$ from the control sample (panel *A*) under conditions similar to those in the experiments of Figs. 3 *A* and 4. When the sample was prerinsed with 10 mM ADP in the absence of Mg there was almost no subsequent release of $^{32}P_i$ (panel *B*). This could be attributed to the dissociation of Mg from the E_1([^{32}P]ATP) complex, which would prevent subsequent rephosphorylation. However, even when Mg was included in both the ADP-containing rinse and the second solution, only about half of the remaining ^{32}P was released as $^{32}P_i$ and half as [^{32}P]ATP, and the rate of release was low (panel *C*). According to the arguments presented above, these results strongly favor models in which there is a slow step *preceding* the E_1-P → E_2-P transition.

^{22}Na Occlusion by the Uninhibited Na,K Pump

It would be extremely informative to examine the fate of Na ions as they are translocated across the membrane, coupled to the individual phosphorylation–dephosphorylation events. Glynn et al. (1984) have shown that Na ions become occluded on the NEM-inhibited or chymotrypsin-nicked Na/K pump when it is phosphorylated from ATP, and that these occluded ions dissociate upon reversal of phosphorylation. However, there has been no direct demonstration of occluded Na on the functional uninhibited enzyme. Our previous studies showed that the overall Na translocation process had a rate constant near 30–40 s^{-1} (Forbush, 1984*a*, 1985), similar to the overall $E_1 \rightarrow E_2$-P rate constant determined by the styryl dyes. It seemed that it should be possible to occlude ^{22}Na ions in E_1-P and measure the rate of release of the ^{22}Na upon relaxation from this state. We have recently made numerous attempts to study the uninhibited enzyme, using the rapid filtration apparatus to examine the release of ^{22}Na. These have not met with success. The limiting experimental factor is probably that our procedure requires a ^{22}Na-free rinse period of at least 5–10 s at 0°C, and during this period all of the putative occluded ^{22}Na may be lost.

In typical experiments, 1–3 mM ^{22}Na, 1 μM–1 mM ATP, and 0.1–1 mM Mg were incubated with dog kidney (or pig kidney or bovine brain) Na,K-ATPase for 0.5–5 s at <1 s at 20°C. In view of the strong effect of lytropic ions in stabilizing E_1-P, we tested the effect of 0.5–2 M *N*-methylglucamine iodide or bromide in the initial incubation medium, and have routinely used 1.5–2 M *N*-methylglucamine or sodium chloride, bromide, or iodide in all rinse media. These conditions are known to stabilize E_1-P for minutes at 0°C and for several seconds at 20°C (Klodos and Forbush, 1991; Klodos, I., and B. Forbush, manuscript in preparation). Media of various pH values were also examined. We looked for ^{22}Na release at 20°C in the rapid filtration apparatus upon addition of ADP or upon lowering the concentration of lytropic ions. In none of these experiments were we able to detect ^{22}Na release that could be ascribed to release of occluded ^{22}Na, presumably because the occluded ions exchange with the medium before the observation period begins. One conclusion can be drawn from this otherwise disappointing result: if indeed ^{22}Na ions are occluded in the E_1-P state of the uninhibited Na/K pump as generally presumed, the occluded Na ions are able to exchange with the medium at a significant rate while the pump remains in the E_1-P state.

This negative result also suggests that the E_1-P state stabilized by lyotropic ions is different than that stabilized in the various inhibited forms of the pump, since we previously had little difficulty demonstrating ^{22}Na occlusion by NEM-inhibited (Forbush, 1985) or oligomycin-inhibited (Forbush, B., unpublished results) Na,K-ATPase. In those experiments the rate of ^{22}Na release (~ 60 s^{-1} at 20°C) was much more rapid than the release of [^{32}P]ATP (~ 15 s^{-1}), showing that Na dissociates rapidly from E_1(ATP) or possibly from E_1-P(ADP) (Forbush, 1985).

Discussion

Using styryl dyes to report events in the Na/K pump cycle we have found that steps in the Na translocation process are partially rate limiting under normal conditions, and that the overall rate of Na translocation decreases as the pH is raised (Klodos and Forbush, 1988). By mixing ATP with Na,K-ATPase in a stop-flow mixing apparatus we have confirmed previous results obtained with caged ATP and with Mg/DM-nitrophen. The close agreement demonstrates that interaction of caged ATP with Na,K-ATPase has not been a limiting factor in resolving the slow step in translocation, at least under the conditions used in our experiments with kidney membranes.

We expected to find that within the overall Na translocation process, steps immediately preceeding the formation of E_2-P would be rate limiting (either the E_1-P → E_2-P transition or associated Na dissociation events). We therefore anticipated that the rate of dephosphorylation starting from E_1-^{32}P would be essentially the same as the overall set of reactions E_1 → → E_2-P monitored by the styryl dyes. However, as shown above, the reactions leading to dephosphorylation were found to be substantially faster than the reactions measured in the experiments with RH-421. There are a number of possible explanations for this discrepancy:

(*1*) It is possible that the dephosphorylation rates observed here are unrelated to the normal steps in the reaction cycle. It can be argued that exposure to high salt might cause the Na,K-pump to assume a nonphysiological state (E_1-P!), and that from this state the transition E_1-P! → E_2P is different than the normal transition E_1P → E_2P. For instance, a two-step process involving a "relaxation" from the high salt condition (E_1-P! → E_1-P → E_2-P) would be slower than the single step E_1-P → E_2-P, and indeed a "lag" phase seen in Fig. 3*A* would be consistent with a "relaxation" step; however, this is in the wrong direction to explain our discrepancy. It is also possible that a reaction E_1-P! → E_2-P might be faster than the normal conformational change. If this is the case, dephosphorylation experiments after exposure to high salt are of little use in interpreting the normal cycle. We have no evidence against this ad hoc explanation for the dephosphorylation results, which if accepted would leave open the possibility that the slow step in Na translocation is the E_1P → E_2P transition.

(*2*) It has been proposed previously that steps involved in Na deocclusion could be the slowest steps in the Na translocation process (Forbush, 1985), and recent results pointing to a "potential well" on the extracellular face of the membrane (Gadsby et al., 1991; Post and Suzuki, 1991) have been interpreted in terms of a rate-limiting step in Na release. Froehlich and Fendler (1991) have suggested that the Na deocclusion process may occur *after* the E_1-P → E_2-P conformational change and after the K-stimulated dephosphorylation of E_2-P. This proposal would explain the apparent discrepancy discussed here, since it predicts rapid ^{32}P release and a slower overall change involving Na release. It does not, however, predict the results in Fig. 6. In addition, since a maximal rate of Na release should require that the preceding steps be rapid enough, one of the consequences of this model is that the rate of Na release should depend on K for stimulation of dephosphorylation. On the contrary, the transition seen in Na transport, fluorescence, and voltage transients is very little affected by external K (Forbush, 1984*a*; Borlinghaus et al., 1987; Fendler et al., 1987; Steinberg and Karlish, 1989).

(*3*) The slow step in the E_1 → E_2P process may precede the E_1P → E_2P conformational change, and thus not be measured in Fig. 3. Two possibilities for the slow reaction include the phosphorylation step itself, E_1(ATP) → E_1-P(ADP) (Scheme 1, step 2), and the release of ADP, E_1-P(ADP) → E_1-P (Scheme 1, step 3). This explanation is strongly supported by the result illustrated in Fig. 6, that exposure to ADP results in slow subsequent release of the remaining ^{32}P, and in approximately equal rates of release as [^{32}P]ATP and $^{32}P_i$. If the forward rates involved in formation of E_1-P were very fast (steps 2 and 3), all of the ^{32}P would dissociate as $^{32}P_i$ after the pulse of ADP. The data do not allow a discrimination between the two steps (phosphorylation and ADP dissociation) as to which is the slow process. However, previous workers have concluded that the ADP release process (step 3) is very rapid (cf. Mardh and Post, 1977; Hobbs et al., 1985).

Earlier phosphorylation studies are usually summarized in terms of a very rapid series of steps leading to phosphorylation. In eel electroplax Na,K-ATPase the E_1(ATP) → E_1-P(ADP) reaction (Scheme 1, step 2) is ~175 s^{-1} at 21°C (Hobbs et al., 1985), and in bovine brain it is ~200 s^{-1} at 25°C (Mardh and Zetterqvist, 1974). Mardh and Post (1977) found a rate of ~50 s^{-1} on addition of 5 μM ATP to guinea pig kidney Na,K-ATPase at 25°C; however, this concentration of ATP may not have been saturating. Given the uncertainties involved in comparing rates at different temperatures and among enzymes from different tissue sources, these determinations may be compatible with a rate of 60 s^{-1} for step 2 in mammalian kidney Na,K-ATPase at pH 7.5. We know of no evidence against the proposition that this step becomes much slower as the pH is raised.

Several previous experiments can be interpreted in support of rate limitation at step 2 in Scheme 1, in that they provide evidence for the accumulation of E(ATP) during steady-state Na,K-ATPase activity. Kanazawa et al. (1970) and Klodos and Norby (1979) have pointed out that in the presence of K the level of E-P is too low to account for the steady-state turnover rate. Kanazawa et al. (1970) suggested that the discrepancy could be explained by a scheme in which bound ATP accounts for the difference. While that particular scheme has been shown to be incorrect in other respects, the possibility that E_1(ATP) accumulates in the steady state has not been disproven. Lowe and Reeve (1983) have also presented preliminary evidence supporting this idea.

The rate of loss of ^{32}P from E_1-^{32}P directly demonstrated in Figs. 3 and 4 is a lower limit to the rate of the E_1-P → E_2-P transition, since it includes the rate of the dephosphorylation step, and may also include some time for the enzyme to "adjust" to the lower concentration of lyotropic ions. The rate (>80 s^{-1} at 20°C) is fast enough for E_1-P to be an intermediate in the turnover of the pump (~20 s^{-1} for the entire cycle), providing that the steady-state level of E_1-P is ~25% of the enzyme. Although measurements of this level have not been reported for kidney Na,K-ATPase at saturating [^{32}P]ATP and 20°C, this is similar to the level in bovine brain at 25°C (Mardh, 1975; 100 μM ATP). An earlier conclusion that E_1-P was *incompetent* at 0°C (Norby et al., 1983) was based on measurements in the presence of high concentrations of lytropic ions. It is now clear that the E_1-P → E_2-P reaction is greatly retarded in lyotropic media (Post, 1988; Klodos and Forbush, 1991; Klodos and Plesner, 1991; Post and Suzuki, 1991).

In summary, our data demonstrate that Na translocation steps are slow enough to partially limit the rate of pump turnover at neutral pH, and that these steps become the major rate limitation in pump activity at high pH. Dephosphorylation experiments indicate that while the conformational transition E_1-P → E_2-P (step 3 in Scheme 1) is slow enough to contribute some of the rate limitation at neutral pH, it is not the slow step at high pH; the data support the hypothesis that the rate of phosphorylation (step 2 in Scheme 1) becomes rate limiting above pH 8.0.

Acknowledgments

We thank Grace Jones and John T. Barberia for excellent technical assistance.

This work was supported by the NIH (GM-31782), the Danish Medical Research Council (M-12-7961, 12-9089), and the Aarhus University Foundation (1987-7131/01-5-LF 3-7).

References

Albers, R. W., G. J. Koval, and Siegel. 1968. Studies on the interaction of ouabain and other cardio-active steroids with sodium-potassium-activated adenosine triphosphatase. *Molecular Pharmacology.* 4:324–336.

Borlinghaus, R., H. J. Apell, and P. Lauger. 1987. Fast charge translocations associated with partial reactions of the Na,K-pump. 1. Current and voltage transients after photochemical release of ATP. *Journal of Membrane Biology.* 97:161–178.

Fendler, K., J. Froehlich, S. Jaruschewski, A. Hobbs, W. Albers, E. Bamberg, and E. Grell. 1991. Correlation of charge translocation with the reaction cycle of the Na,K-ATPase. *In* The Sodium Pump: Recent Developments. J. H. Kaplan and P. De Weer, editors. The Rockefeller University Press, New York. In press.

Fendler, K., E. Grell, and E. Bamberg. 1987. Kinetics of pump currents generated by the Na^+,K^+-ATPase. *FEBS letters.* 224:83–88.

Forbush, B., III. 1984*a*. Na^+ movement in a single turnover of the Na pump. *Proceedings of the National Academy of Sciences, USA.* 81:5310–5314.

Forbush, B., III. 1984*b*. An apparatus for rapid kinetic analysis of isotopic efflux from membrane vesicles and of ligand dissociation from membrane proteins. *Analytical Biochemistry.* 140:495–505.

Forbush, B., III. 1985. Rapid ion movements in a single turnover of the Na^+ pump. *In* The Sodium Pump. I. Glynn and C. Ellory, editors. The Company of Biologists Ltd., Cambridge, UK. 599–611.

Forbush, B., III. 1987. Rapid release of ^{42}K and ^{86}Rb from an occluded state of the Na,K-pump in the presence of ATP or ADP. *Journal of Biological Chemistry.* 262:11104–11115.

Forbush, B., III. 1988. Rapid ^{86}Rb release from an occluded state of the Na,K-pump reflects the rate of dearsenylation or dephosphorylation. *Journal of Biological Chemistry.* 263:7961–7969.

Froehlich, J. P., and K. Fendler. 1991. The partial reactions of the Na^+- and $Na^+ + K^+$–activated adenosine triphosphatases. *In* The Sodium Pump: Structure, Mechanism, and Regulation. J. H. Kaplan and P. De Weer, editors. The Rockefeller University Press, New York. 227–247.

Gadsby, D. C., M. Nakao, and A. Bahinski. 1991. Voltage-induced Na/K pump charge movements in dialyzed heart cells. *In* The Sodium Pump: Structure, Mechanism, and Regulation. J. H. Kaplan and P. De Weer, editors. The Rockefeller University Press, New York. 355–371.

Glynn, I. M., Y. Hara, and D. E. Richards. 1984. The occlusion of sodium ions within the mammalian sodium-potassium pump: its role in sodium transport. *Journal of Physiology.* 351:531–547.

Hobbs, A. S., R. W. Albers, and J. P. Froehlich. 1985. Quenched-flow determination of the E_1P to E_2P transition rate constant in electric organ Na,K-ATPase. *In* The Sodium Pump. I. Glynn and C. Ellory, editors. The Company of Biologists Ltd., Cambridge, UK. 355–361.

Kanazawa, T., M. Saito, and Y. Tonomura. 1970. Formation and decomposition of a phosphorylated intermediate in the reaction of Na^+-K^+ dependent ATPase. *Journal of Biochemistry.* 67:693–711.

Kaplan, J. H., and G. C. Ellis-Davies. 1988. Photolabile chelators for the rapid photorelease of divalent cations. *Proceedings of the National Academy of Sciences, USA.* 85:6571–6575.

Karlish, S. J. D., and D. W. Yates. 1978. Tryptophan fluorescence of ($Na^+ + K^+$)-ATPase as a tool for study of the enzyme mechanism. *Biochimica et Biophysica Acta.* 527:115–130.

Klodos, I., and B. Forbush III. 1988. Rapid conformational changes of the Na/K pump revealed by a fluorescent dye. *Journal of General Physiology.* 92:46*a.* (Abstr.)

Klodos, I., and B. Forbush III. 1991. Transient kinetics of dephosphorylation of Na,K-ATPase after dilution of NaCl. *In* The Sodium Pump: Recent Developments. J. H. Kaplan and P. De Weer, editors. The Rockefeller University Press, New York. In press.

Klodos, I., and J. G. Norby. 1979. Effect of K^+ and Li^+ on intermediary steps in the Na,K-ATPase reaction. *In* Na,K-ATPase Structure and Kinetics. J. C. Skou and J. G. Norby, editors. Academic Press, London. 331–342.

Klodos, I., and L. Plesner. 1991. Anion effects on the steady-state ratio of the phosphoenzymes of Na,K-ATPase as measured by dephosphorylation and oligomycin inhibition. *In* The Sodium Pump: Recent Developments. J. H. Kaplan and P. De Weer, editors. The Rockefeller University Press, New York. In press.

Lowe, A. G., and Reeve. 1983. Aspects of the presteady state hydrolysis of ATP by Na,K-ATPase. *Current Topics in Membranes and Transport.* 19:577–580.

Mardh, S. 1975. Bovine brain Na^+, K^+-stimulated ATP phosphohydrolase studied by a rapid-mixing technique: K^+-stimulated liberation of [^{32}P]-phosphoenzyme and resolution of the dephosphorylation into two phases. *Biochimica et Biophysica Acta.* 391:448–463.

Mardh, S., and R. L. Post. 1977. Phosphorylation from adenosine triphosphate of sodium- and potassium-activated adenosine triphosphatase. Comparison of enzyme-ligand complexes as precursors to the phosphoenzyme. *Journal of Biological Chemistry.* 252:633–638.

Mardh, S., and O. Zetterqvist. 1974. Phosphorylation and dephosphorylation reactions of bovine brain ($Na^+ + K^+$)-stimulated ATP phosphohydrolase studied by a rapid-mixing technique. *Biochimica et Biophysica Acta.* 350:473–483.

Nagel, G., C. L. Slayman, and I. Klodos. 1989. Fluorescence probing of a major conformation change in the plasma membrane H^+-ATPase of Neurospora. *Biophysical Journal.* 55:338*a.* (Abstr.)

Nakao, M., and D. C. Gadsby. 1986. Voltage dependence of Na translocation by the Na/K pump. *Nature.* 323:628–630.

Norby, J. G., I. Klodos, and N. O. Christiansen. 1983. Kinetics of Na-ATPase activity by the Na,K pump. Interactions of the phosphorylated intermediates with Na^+, $Tris^+$, and K^+. *Journal of General Physiology.* 82:725–759.

Post, R. L. 1988. Lyotropic ions on reactivity of phosphoenzyme of sodium, potassium-ATPase. *Biophysical Journal.* 53:138*a.* (Abstr.)

Post, R. L., A. K. Sen, and A. S. Rosenthal. 1965. A phosphorylated intermediate in adenosine triphosphate-dependent sodium and potassium transport across kidney membranes. *Journal of Biological Chemistry.* 240:1437–1445.

Post, R. L., and K. Suzuki. 1991. A Hofmeister effect on the phosphoenzyme of Na,K-ATPase. *In* The Sodium Pump: Structure, Mechanism, and Regulation. J. H. Kaplan and P. De Weer, editors. The Rockefeller University Press, New York. 201–209.

Skou, J. C. 1957. The influence of some cations on an adenosine triphosphatase from peripheral nerves. *Biochimica et Biophysica Acta.* 23:394–401.

Steinberg, M., and S. J. D. Karlish. 1989. Studies on conformational changes in Na,K-ATPase labeled with 5-iodoacetamidofluorescein. *Journal of Biological Chemistry.* 264:2726–2734.

Stürmer, W., H.-J. Apell, I. Wuddel, and P. Läuger. 1989. Conformational transitions and charge translocation by the Na,K pump: comparison of optical and electrical transients elicited by ATP-concentration jumps. *Journal of Membrane Biology.* 110:67–86.

Stürmer, W., R. Bühler, H. J. Apell, and P. Läuger. 1991. Charge translocation by the sodium pump: ion binding and release studied by time-resolved fluorescence measurements. *In* The Sodium Pump: Recent Developments. J. H. Kaplan and P. De Weer, editors. The Rockefeller University Press, New York. In press.

Chapter 17

The Partial Reactions of the Na^+- and $Na^+ + K^+$–activated Adenosine Triphosphatases

Jeffrey P. Froehlich and Klaus Fendler

Laboratory of Cardiovascular Science, National Institute on Aging, National Institutes of Health, Baltimore, Maryland 21224; and Max-Planck-Institut für Biophysik, D-6000, Frankfurt am Main, Germany

In the Proceedings of the Vth International Conference on the Na^+ Pump, Ian Glynn contributed an overview on the pump mechanism in which the Albers-Post scheme and modifications thereto were discussed in relation to current views and "certain awkward problems" (Glynn, 1988). The primary purpose of this paper is to extend that discussion of awkward problems and to offer new alternatives in light of current evidence derived largely from rapid mixing experiments. A separate section deals with the kinetic behavior of the phosphoenzyme conformational transition which couples ATP hydrolysis to Na^+ translocation in the transport cycle.

Historical Background

The Albers-Post mechanism has become the frame of reference for studies on the Na,K-ATPase because of its ability to accommodate a wide variety of biochemical and physiological observations relating to the operation of the Na,K pump. In the early 1980s papers began to appear that questioned the adequacy of this scheme in explaining the kinetic behavior of the enzyme activity associated with the Na,K pump, and modifications were proposed that attempted to correct these shortcomings (Plesner et al., 1981; Nørby et al., 1983). More recently, these modifications have themselves been challenged (Hobbs et al., 1985*a*; Nørby and Klodos, 1988) as new evidence has accumulated on the phosphoenzyme intermediates and pathways involved in ATP hydrolysis.

A logical point at which to begin examining this question is a series of rapid mixing experiments carried out by Mårdh (1975*a*) in which he measured the time course of phosphoenzyme decomposition produced by the addition of K^+ CDTA or K^+ + unlabeled ATP to the phosphorylated bovine brain Na,K-ATPase. In these experiments, the enzyme was incubated with ATP in the presence of Na^+ and Mg^{2+} to achieve a constant (steady-state) level of phosphorylation, and then a chase containing K^+ plus an agent to prevent rephosphorylation was added to initiate dephosphorylation. Decomposition of the phosphoenzyme exhibited a triphasic decay pattern in which the fast component was attributed to the hydrolysis of E_2P, while the intermediate phase was interpreted as the dephosphorylation of E_1P taking place via E_2P. The slow component, representing $<10\%$ of the total EP, was identified as a K^+-insensitive phosphoenzyme (Post et al., 1972; Mårdh, 1975*b*) which turns over too slowly to be in the main catalytic pathway. The intermediate phase (referred to as the slow phase by Mårdh) decomposed with a rate constant of 28 s^{-1} when CDTA was added to prevent rephosphorylation of the enzyme. This rate constant increased to 46 s^{-1} when unlabeled ATP was substituted for CDTA in the chase solution. If K^+ (10 mM) was included in the phophorylation medium, only the intermediate and slow decay fractions were observed; however, the steady-state level of the intermediate component was lower (by $\sim 25\%$) and its rate of disappearance faster (77 s^{-1}) than observed when K^+ was absent during phosphorylation. As proof that the more rapidly disappearing phosphoenzyme is a kinetically competent intermediate, Mårdh (1975*a*) showed that the amplitude of this component multiplied by its rate of disappearance was in good agreement with the overall Na,K-ATPase activity measured under identical conditions.

Although the results of Mårdh's dephosphorylation experiments were compatible with the qualitative predictions of the Albers-Post mechanism, there were certain quantitative aspects of his analysis that raised concern among other investigators in the field. One problem was that the rate of decay of the intermediate fraction depended on the choice of reagent (CDTA or unlabeled ATP) used to prevent rephosphorylation and that only the faster of these rates gave a calculated overall velocity of ATP hydrolysis that agreed with the measured value (Plesner et al., 1981). In making this calculation the back reaction was assumed to be negligible, so that the overall ATPase velocity was determined as the product of the concentration of the intermediate fraction times its rate of decay. A potential source of error in this calculation is that the measured rate of phosphoenzyme decomposition may underestimate the actual rate at which the phosphoenzyme is converted to its product state because of continued ^{32}P-labeling of the enzyme by bound ATP after the addition of the chase. This behavior has been observed in electric organ Na,K-ATPase (Hobbs et al., 1980*a*) where the EDTA-induced rate of dephosphorylation after labeling in the presence of K^+ was found to be two to three times slower than the rate of E_2P hydrolysis activated by the addition of K^+. An additional problem noted by Plesner et al. (1981) was that the rate of conversion of E_1P to E_2P calculated from kinetic results obtained in the absence of K^+ was 10 times smaller than the experimentally determined value in the presence of K^+ (77 s^{-1}). As pointed out by Glynn (1988), this discrepancy could mean that K^+ accelerates the rate of the phosphoenzyme transition, although there is other evidence (Nørby et al., 1983; Klodos, 1988) indicating that it has just the opposite effect. The magnitude of the rate of E_1P to E_2P conversion in the absence of K^+ is critical because if it is as slow as 7 s^{-1} and the spontaneous rate of E_2P breakdown is only 3–4 s^{-1} (Mårdh, 1975*a*; Hobbs et al., 1980*a*), then there must be a significant back reaction at the conformational step to account for the 25-fold increase in the overall ATPase activity that results from the addition of K^+. A direct measurement of the reverse rate constant (Klodos et al., 1981), however, failed to provide experimental support for this assumption, leading Plesner et al. (1981) to conclude that the enzyme catalyzes ATP hydrolysis by parallel Na^+- and $Na^+ + K^+$–activated pathways. This notion of separate reaction pathways was incorporated into a catalytic mechanism referred to as the "bicycle scheme" (Plesner et al., 1981) in which one of the cycles is activated exclusively by Na^+ and includes the acid-stable phosphoenzymes E_1P and E_2P. The alternate cycle, which operates when both monovalent cations are present, initially contained no acid-stable phosphoenzymes but was subsequently modified to include these as intermediates (Plesner and Plesner, 1988). It should be noted that the phosphoenzymes in the Na^+-activated cycle were found to turn over rapidly enough to account for the overall velocity of the Na-ATPase reaction. While this is a necessary criterion for kinetic competence, it is not sufficient, as demonstrated in subsequent studies involving the electric organ Na,K-ATPase (see below).

Further proof of the inadequacy of the Albers-Post mechanism came from experiments carried out by Nørby et al. (1983) in which they examined the kinetics of the ox brain Na,K-ATPase at 0°C. Using various combinations of the ligands ADP, ATP, and K^+ to dephosphorylate the phosphoenzyme formed in the absence of K^+, they showed that the sum of the ADP- and K^+-sensitive phosphoenzymes was $>100\%$ of the total EP, confirming an earlier observation by Kuriki and Racker (1976). The decay pattern obtained with ADP was triphasic with a fast phase that increased as a function of the Na^+ concentration; dephosphorylation by K^+ + unlabeled ATP produced a biphasic decay pattern with a secondary phase that also increased with increasing [Na^+]. In analyzing this behavior it was found that a minimum of three phosphoenzyme pools or intermediates (E_aP, E_bP, and E_cP) were necessary to explain the decay pattern induced by ADP, and that one of these pools (E_b) had to be sensitive to both ADP and K^+ to account for the discrepancy between the sum of the phosphoenzyme intermediates and the total EP. The model assumed that these phosphoenzymes were consecutive intermediates in a single catalytic pathway and that the second state, E_bP, dephosphorylated in the presence of ADP to ATP via the first state, E_aP, at elevated Na^+ concentrations. The rapid phase produced by K^+ + unlabeled ATP was assumed to result from the parallel decomposition of E_bP and E_cP to P_i, while the subsequent slower phase represented the decomposition of E_aP to intermediates situated downstream in the reaction sequence. The rate constant for the slower phase of K^+-induced dephosphorylation was found to be lower than the computer-simulated rate of E_aP breakdown in the absence of K^+ and too slow to account for the overall velocity of ATP hydrolysis measured in the presence of Na^+ and K^+.[1] To account for these observations Nørby et al. (1983) proposed that K^+, in addition to activating the hydrolysis of E_bP and E_cP, also blocks the conversion of E_aP to E_bP, excluding the latter as a kinetically competent intermediate in the pathway activated by Na^+ and K^+. In the absence of K^+, the interconversion of phosphorylated states occurred rapidly enough to satisfy the criterion for intermediates in the Na-ATPase reaction. Additional support for the three-pool consecutive model came from a similar set of dephosphorylation experiments carried out by Yoda and Yoda using the Na,K-ATPase prepared from the eel electric organ (1986). They found that by changing the temperature or the concentration of cholesterol in the membrane they were able to control the proportions of the phosphorylated intermediates (1987*b*), mimicking the effects produced by $[Na^+]$. These effects were accompanied by changes in the transport coupling ratios indicating that stabilization of the intermediate ($E{*}P$ in their terminology) pool leads to the deocclusion and release of Na^+ from the transport protein before translocation (1987*a*). A lucid discussion of the three-pool model together with some of the conceptual problems that have arisen in light of recently acquired experimental evidence can be found in Nørby and Klodos (1988).

Kinetic Behavior of the Na-ATPase Reaction in Electric Organ Microsomes: Dephosphorylation Reaction

A third challenge to Mårdh's interpretation of the dephosphorylation experiments came from work that was carried out in this laboratory which focused on the pre–steady state kinetic behavior of the acid-stable phosphorylated intermediates in electric organ Na,K-ATPase (Hobbs et al., 1985*a*, 1988). The rationale behind these experiments is depicted in Fig. 1, which shows the theoretical time courses for the ADP- and K^+-sensitive phosphoenzymes in relation to the total acid-stable EP. Because E_1P and E_2P are consecutive intermediates in the transport cycle, the proportions of these species present at different phosphorylation times will reflect the rate at which the ADP-sensitive state is converted to the K^+-sensitive state. Fig. 2*A* shows two double mixing experiments in which phosphorylation by ATP was carried out for either 10 (pre–steady state) or 116 ms (steady state) and then a chase

[1] The slow phase of dephosphorylation in Nørby et al. (1983) corresponds to the intermediate decay phase in the acid quench experiments of Hobbs et al. (1985*a*).

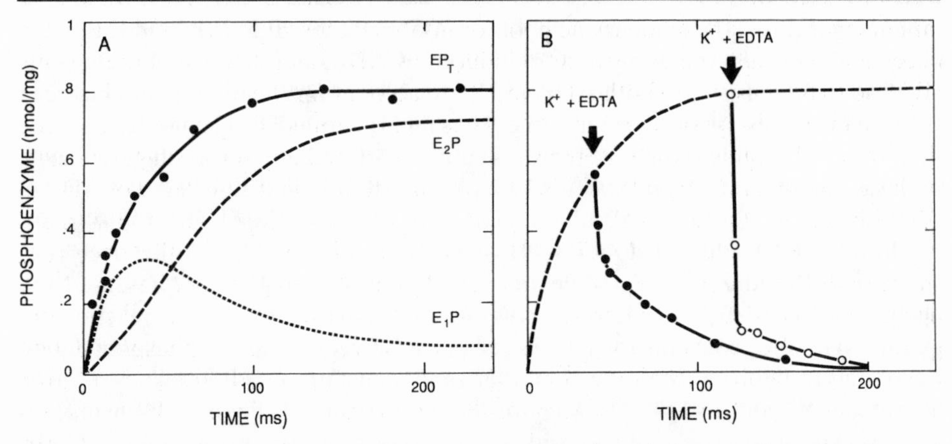

Figure 1. Theoretical time courses of the E_1P and E_2P acid-stable intermediates and the total phosphoenzyme, EP_T. (*A*) The solid circles illustrate the time course of phosphorylation by ATP in the absence of K^+. E_1P (*dotted line*) and E_2P (*dashed line*) represent consecutive intermediates in the Albers–Post reaction mechanism. (*B*) Theoretical time course of K^+ + EDTA–induced dephosphorylation in the pre–steady state (*solid circles*) and steady state (*open circles*) of phosphorylation.

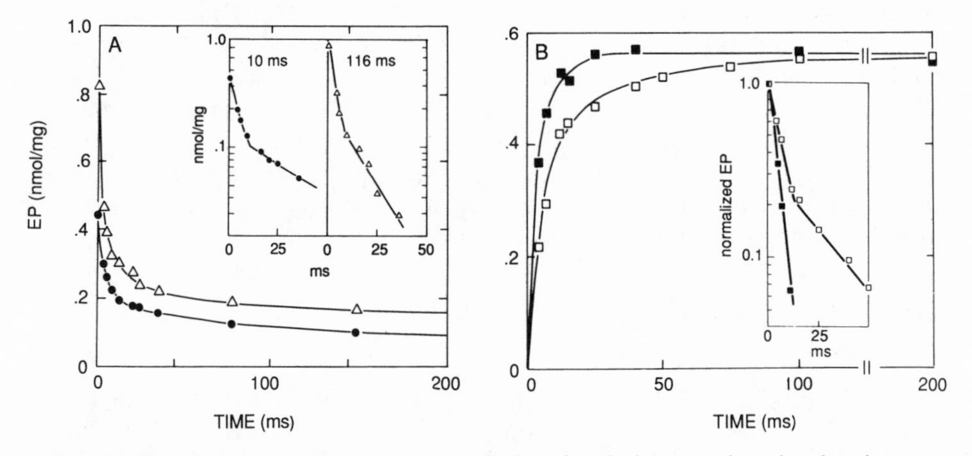

Figure 2. Multiphasic phosphorylation and dephosphorylation reactions in electric organ Na,K-ATPase. (*A*) Electric organ microsomes (1 mg protein/mg) were phosphorylated in a medium containing 10 μM [^{32}P]ATP, 25 mM NaCl, 1 mM $MgCl_2$, and 60 mM Tris–HCl (pH 7.5) at 21°C. Phosphorylation was interrupted at 10 (*closed circles*) or 116 ms (*open triangles*) by the addition of a chase containing 6.66 mM KCl and 10 mM EDTA and the reaction was allowed to proceed for the indicated times before the addition of acid. Inset shows the data replotted on a semilogarithmic scale after subtraction of the slowly decaying component. (*B*) Electric organ microsomes equilibrated with 60 mM Tris (pH 7.4) buffer (*open squares*) or buffer plus 25 mM NaCl (*solid squares*) were mixed with an equal volume of 200 μM [^{32}P]ATP, 2 mM $MgCl_2$ and either 50 or 25 mM NaCl to initiate phosphorylation. At the indicated times acid was added to terminate the reaction. Inset shows a semilogarithmic plot of EP formation obtained by taking the difference between the normalized maximum level of phosphorylation and each of the preceding data points and plotting these values as a function of time.

containing K^+ + EDTA was added to induce dephosphorylation. At 116 ms phosphoenzyme decomposition obeyed triphasic kinetics, similar to the patterns obtained in the bovine (Mårdh, 1975*a*) and ox brain enzymes (Norby et al., 1983) under steady-state labeling conditions. The fast phase, which accounts for ~50% of the total EP (Table I), corresponds to the hydrolysis of E_2P, while the slow component, representing 25% of the total ^{32}P incorporation, is due in part to tightly sealed vesicles with sequestered K^+ sites (Hobbs et al., 1980*a*). This interpretation derives additional support from the experimental results in Fig. 3, which show that preincubation with FCCP and monensin (ionophores that increase the permeability of the membrane to H^+ and K^+) accelerate the disappearance of the ^{32}P-protein–bound label in a major fraction of the slowly decaying sites. The remaining stable fraction represents a K^+-insensitive phosphoenzyme in which the K^+-activated dephosphorylation sites are presumably blocked (Post et al., 1972; Mårdh, 1975*a*).

TABLE I
Effect of Phosphorylation Time and [Na^+] on K^+- and ADP-induced Dephosphorylation of the Na-ATPase Phosphoenzyme

Phosphorylation time	[K^+]	[ADP]	[Na^+]	Fast	Intermediate	Slow
ms	*mM*	*mM*	*mM*		%	
10	6.66	—	25	49 (235)	24 (17)	27 (.6)
116	6.66	—	25	54 (288)	22 (28)	24 (.9)
116	6.66	—	200	52 (132)	17 (16)	24 (1.2)
116	53.3	—	200	53 (303)	23 (23)	24 (1.0)
116	—	1	25	4 (*)	80 (6.4)	16 (.3)
116	—	1	200	24 (283)	69 (33)	7 (*)

With the exception of the changes in the ion concentrations noted in the table, the conditions used to phosphorylate the electric organ microsomes were identical to those described in the legend to Fig. 2. After 10 or 116 ms phosphorylation was interrupted by the addition of a chase containing either 6.66 mM KCl + 10 mM EDTA or 1 mM ADP + 10 mM EDTA. The components are expressed as percentages normalized to the EP level present at the time the chase was added. Numbers in parentheses are the first-order decay constants in s^{-1}.
*Rates too rapid or too slow to be meaningfully determined.

The marked stability of these components to dephosphorylation by K^+ justifies their exclusion from the main catalytic pathway and their subtraction from the dephosphorylation reaction before kinetic analysis (Fig. 2*A*, *inset*).

If the phosphoenzyme responsible for the intermediate phase of dephosphorylation is the precursor to the K^+-sensitive EP, then its rate of decay should approximate the rate of conversion of E_1P to E_2P, assuming that the kinetics of this reaction are not altered by the chase. Judging from the relatively slow decay of the intermediate component at 116 ms (28 s^{-1}), E_2P is expected to represent a much smaller fraction of the total phosphoenzyme at 10 ms of phosphorylation than at 116 ms. However, as seen in Table I, the proportions of the fast and intermediate components were very similar in the two experiments, implying that these phosphointermediates accumulate very rapidly. Because P_i release and rephosphorylation of the enzyme in the absence of K^+ take place slowly (Froehlich et al., 1976), the phosphoenzymes corresponding to these phases are presumably in a state of rapid equilibrium to account for the relatively high levels of the intermediate decay fraction present in the steady state. However, in that case both species should have disappeared rapidly when phosphorylation was interrupted by the addition of K^+ + EDTA. One could argue that K^+, in addition to activating the hydrolysis of E_2P, also blocks the conversion of E_1P to E_2P, which is equivalent to the proposal of Nørby et al. (1983). The difficulty with this explanation (apart from having to introduce a new effect of K^+ in the transport cycle) is that it does not account for the high steady-state levels of the intermediate component that occur in the absence of K^+. In these experiments the

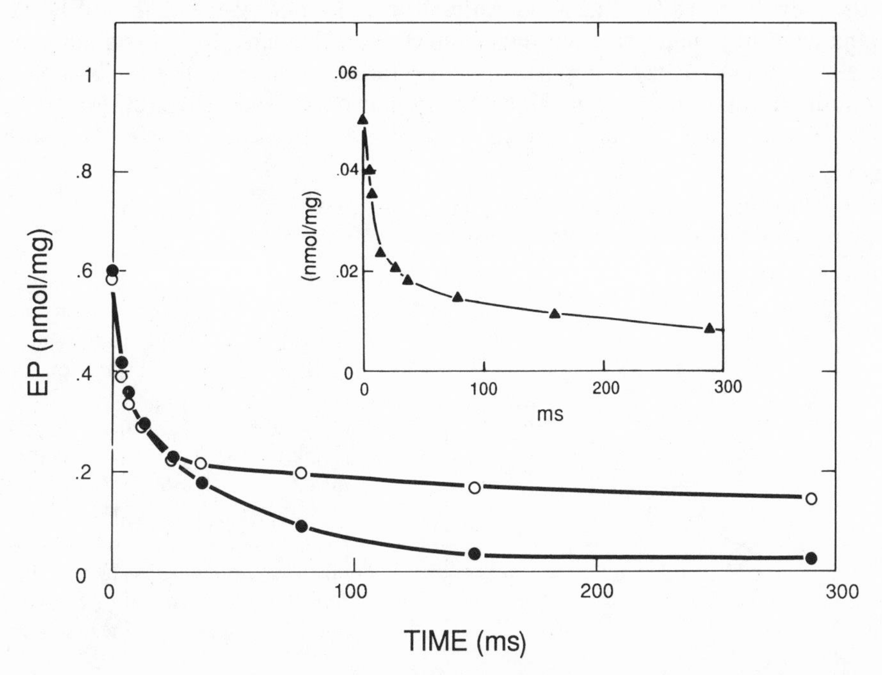

Figure 3. Dephosphorylation of the Na-ATPase and Na,K-ATPase phosphoenzymes. Electric organ microsomes were pre-equilibrated with (*closed circles*) or without (*open circles*) monensin and FCCP and phosphorylated in the absence of K^+ before measuring dephosphorylation as described in the legend to Fig. 2. Inset shows the time course of dephosphorylation of electric organ microsomes phosphorylated in the presence of Na^+ and K^+. The labeling conditions were identical to those in Fig. 2*A* except that the incubation medium also contained 10 mM KCl. After 116 ms phosphorylation was interrupted by the addition of 0.5 mM unlabeled ATP and dephosphorylation was allowed to proceed for the indicated times before the addition of acid.

Na^+ concentration in the phosphorylation medium was only 25 mM, which is 10–20 times less than the amount needed to stabilize the formation of E_1P (Yoda and Yoda, 1986).

The intermediate component of dephosphorylation accounts for ~25% of the phosphorylated sites at 25 mM Na^+. If the phosphoenzyme responsible for this phase is E_1P or an ADP-sensitive state, then the addition of ADP should elicit a rapid phase of dephosphorylation accompanied by ATP formation. Fig. 4*A* shows that when a chase containing ADP + EDTA was added to the phosphoenzyme formed in the absence of K^+, the resulting pattern of EP decay was biphasic and inorganic phosphate release was stoichiometrically related to phosphoenzyme decomposition in both phases. These results strongly suggest that the fast and intermediate decay fractions represent two different forms of the K^+-sensitive phosphoenzyme with different rates of hydrolysis or sensitivities toward K^+ (Hobbs et al., 1985*b*). The absence of the fast phase in the control experiment with EDTA alone (Fig. 4 *B*) implies that one of these intermediates is able to bind ADP, resulting in activation of its rate of hydrolysis. At 200 mM $[Na^+]$, addition of 1 mM ADP caused ~25% of the labeled sites to disappear rapidly (~300 s^{-1}; Table I), which is characteristic of ADP sensitivity (Hobbs et al., 1983*a*, *b*). An identical experiment using K^+ + EDTA to dephosphorylate the enzyme showed no effect of this increase in $[Na^+]$ on the level of the intermediate decay component, although the rate was slowed because of competition between Na^+ and K^+ at the dephosphorylation sites (cf. $[K^+] = 53.3$ mM; Table I). Thus, the appearance of ADP sensitivity does not correlate with a

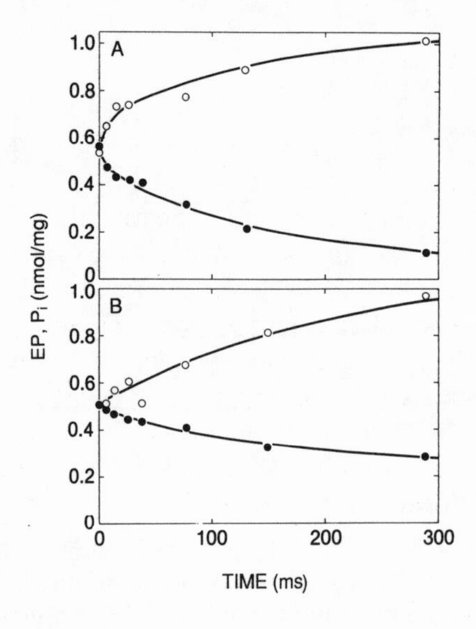

Figure 4. Dephosphorylation of the Na-ATPase phosphoenzyme by ADP + EDTA or EDTA alone. The conditions used to phosphorylate the enzyme were identical to those described in Fig. 2 *A*. At 116 ms either 1 mM ADP + 10 mM EDTA (*A*) or 10 mM EDTA (*B*) was added to dephosphorylate the enzyme and EP decomposition (*closed circles*) and P_i release (*open circles*) were measured by quenching the reaction with acid.

change in the level of the intermediate component, providing further evidence that the latter is not an ADP-sensitive species.

The implication that the intermediate decay component is not an ADP-sensitive phosphoenzyme requires modification of the Albers-Post mechanism. The minimal alternative is a consecutive scheme in which the intermediate and fast phases result from the dephosphorylation of consecutive ADP-insensitive phosphoenzyme intermediates, E_2P' and E_2P:

$$E + ATP \underset{1}{\rightleftharpoons} E_1ATP \underset{2}{\overset{ADP}{\rightleftharpoons}} E_1P \underset{3}{\rightleftharpoons} E_2P' \underset{4}{\rightleftharpoons} E_2P \underset{5}{\overset{P_i}{\rightleftharpoons}} E_2 \underset{6}{\rightleftharpoons} E_1$$

$$E_2P \xrightarrow[7]{K^+} E_2 + P_i$$

Spontaneous (step 5) and K^+-activated P_i release (step 7) occur only at the second of these intermediates, requiring that E_1P and E_2P' both dephosphorylate via E_2P. Because essentially all of the phosphoenzyme is ADP-insensitive by 116 ms, and since the K^+-activated rate of E_2P hydrolysis (k_7) is much faster than the spontaneous decay rate (k_5), we were able to simplify the analysis of the dephosphorylation reaction at 116 ms by using a reduced version of the above scheme surrounded by the dashed line. The solutions to the time-dependent equations for this mechanism are (Klodos et al., 1981):

$$[E_2P'] = C_1 \cdot e^{-At} + C_2 \cdot e^{-Bt}$$

$$[E_2P] = C_3 \cdot e^{-At} + C_4 \cdot e^{-Bt}$$

$$[EP] = H \cdot e^{-At} + G \cdot e^{-Bt}$$

where $[EP] = [E_2P'] + [E_2P], A = [p + q]/2, B = [p - q]/2$, and

$$p = (k_4 + k_{-4} + k_7)$$

$$q = (p^2 - 4 \cdot k_4 \cdot k_7)^{1/2}$$

The model was tested by assigning specific values to the rate constants k_4, k_{-4}, and k_7 (K^+-activated P_i release was assumed to be irreversible) and comparing the behavior of the simulated dephosphorylation reaction to the experimental time course. The rate coefficients A and B correspond to the fast (288 s^{-1}) and intermediate (28 s^{-1}) decay rates in Table I so that $B = 10.3A$. This relationship can be used to solve for k_{-4} in terms of k_4 and k_7, reducing the number of independent variables from three to two. To insure that the survey was comprehensive, k_4 and k_7 were assigned values from 10 to 1,010 s^{-1} in increments of 20 s^{-1}, holding one parameter constant while varying the other. The requirement that the values for H and G be positive (negative values of H or G yield phosphoenzyme decay curves that bend downward from the origin) was used to limit the range of values assigned to the k_1. Once a set of k_i was found that produced agreement between the calculated and actual decay rates, it was used to compute the initial concentrations of E_2P' and E_2P in the dephosphorylation experiments using the following relationships:

$$v = k_4 \cdot E_2P' - k_{-4} \cdot E_2P$$

$$[EP] = [E_2P'] + [E_2P]$$

In the electric organ Na,K-ATPase, v = 4–5 nmol/mg per s and [EP] = 0.6–0.7 nmol/mg at 21°C (Froehlich et al., 1976). Having the values for k_4, k_{-4}, k_7, $[E_2P']$, and $[E_2P]$, the amplitude coefficients H and G were determined and compared with the actual amount of phosphoenzyme present in the fast and intermediate decay phases. Using this procedure we were able to identify a range of values for k_4 (30–290 s^{-1}), k_{-4} (16–135 s^{-1}), and k_7 (30–290 s^{-1}) that yielded calculated decay rates close to the measured values (28 and 288 s^{-1}). By further adjustment of these parameters a set of k_1 was found that produced agreement between the calculated ratio of amplitude coefficients, G:H, and the observed 2.5:1 ratio of fast to intermediate components; viz., k_4 = 28 s^{-1}, k_{-4} = 2 s^{-1}, and k_7 = 284 s^{-1}. Because the calculated G:H ratio was very sensitive to small (5%) changes in the k_i, the kinetic constants yielding the proper rate and amplitude coefficients all fall within a very narrow range.

With a rate of transformation of E_2P' to E_2P of only 28 s^{-1}, the fraction of fast component in the pre–steady state of phosphorylation will be significantly smaller than the corresponding fraction at 116 ms. We can estimate approximately how much of the phosphoenzyme will be present as E_2P at 10 ms by assuming that the formation of E_2P from E_2P' is a simple first-order process that begins at $t = 0$ and has a rate constant of 28 s^{-1} (for a complete model, the actual proportion of E_2P at 10 ms will be less than this estimated value because of the delays caused by ATP binding, phosphorylation, and the conversion of E_2P' to E_2P). This approximation yields a 10-ms fast fraction of 0.25, which is only 50% of the measured value (0.49; Table I). It is therefore apparent that the rate of conversion of E_2P' to E_2P must be greater than 28 $^{-1}$ in order to simulate the behavior of the dephosphorylation reaction at 10 ms, and that this model does not have a set of rate constants that will satisfy the quantitative features of dephosphorylation in both the pre–steady-state and steady-state phases of EP formation.

An important distinction between the above scheme and the modification proposed by Nørby et al. (1983) is that activation of phosphoenzyme hydrolysis by K^+ in their model involves both ADP-insensitive species. The failure of the above consecutive scheme to account for the dephosphorylation results justifies the consideration of a separate mechanism that would allow E_2P' and E_2P to dephosphorylate in parallel when K^+ is present:

$$\begin{array}{ccc} E_2P' & \overset{4}{\longleftrightarrow} & E_2P \\ K^+ \Big\downarrow 8 & & 7 \Big\downarrow K+ \\ E_2' + P_1 & & E_2 + P_i \end{array}$$

This scheme has three independent variables (the fourth rate constant is automatically determined by the relationship between the rate coefficients A and B) making the search for an appropriate set of k_i relatively laborious. The range of values assigned to k_4 and k_{-4} can be reduced, however, by the appropriate use of kinetic information from the pre–steady state. According to Table I, the fraction of phosphoenzyme with the fast decay rate at 10 ms of phosphorylation (0.49) is 93% of the corresponding fraction at 116 ms (0.54), which corresponds to an apparent rate of conversion of E_2P' to E_2P of 265 s^{-1}. Here we have assumed that the time-dependent shift in these components can be approximated by a reversible first-order transition that begins at $t = 0$ and has a rate equal to the sum of k_4 and k_{-4}. In modeling dephosphorylation, we allowed $k_4 + k_{-4}$ to have values of 200, 250, or 300 s^{-1} and then varied k_7 and k_8 between 10 and 1,010 s^{-1} in increments of 20 s^{-1}. As in the previous case, we found that while some selections gave a reasonably close approximation to the measured decay rates (e.g., $A = 329\ s^{-1}$; $B = 30\ s^{-1}$), they were unable to reproduce the amplitude coefficients for the fast and intermediate phases of dephosphorylation ($H = 0.68$; $G = 0.02$). Conversely, assignments that gave a good fit to the measured amplitudes (e.g., $H = 0.47$; $G = 0.23$) failed to yield the observed first-order decomposition rates ($A = 919\ s^{-1}$; $B = 85\ s^{-1}$). These results cast doubt on the adequacy of three-state consecutive models to explain the multiphasic dephosphorylation kinetics of the phosphoenzyme formed in the absence of K^+. Our disagreement with other investigators regarding the adequacy of these schemes stems from the inclusion of pre–steady-state data in our analysis, which places certain constraints on the values assigned to the rate constants for the partial reactions. It should be noted that the apparent failure of these models to explain the dephosphorylation results does not rule out the possibility of a third phosphorylated intermediate in the Na-ATPase pathway. Although our results indicate that the phosphoenzymes responsible for the fast and intermediate decay phases are not confined to a single catalytic pathway, the actual number of phosphointermediates and their reactivity toward ADP and K^+ are unknown.

Kinetic Behavior of the Na-ATPase in Electric Organ Microsomes: Phosphorylation Reaction

Additional kinetic evidence favoring modification of the Albers-Post mechanism derives from rapid mixing experiments measuring the pre–steady-state time dependence of the phosphorylation reaction in the absence of K^+. Mårdh and Post (1977) first reported biphasic phosphorylation of the mammalian Na-ATPase in a rapid acid quench study in which they examined the effects of the early addition of ligands on the initial rate of phosphoenzyme formation. In particular, they found that preexposure to ATP in either the presence or absence of Na^+ gave a biexponential pattern of phosphorylation upon completion of the reaction mixture. A subsequent report by Hobbs et al. (1988) demonstrated similar biphasic behavior in the electric organ Na-ATPase and showed that it could not be explained by a consecutive arrangement of phosphorylated intermediates under the conditions in which the experiments were conducted. Those experiments also ruled out explanations involving a slow conversion of E_2 and E_1 because the biphasicity was present even at relatively high (100 μM) ATP concentrations where the conformational equilibrium would have strongly favored E_1.

Fig. 2 *B* shows an acid quench experiment in which EP formation was measured after the simultaneous addition of ATP, Na^+, and Mg^{2+} to an enzyme suspended in a buffer without ligands. The time control of phosphoenzyme formation obeyed biphasic kinetics (Fig. 2 *B*, *inset*), demonstrating that preincubation with ATP is not obligatory for biphasic phosphorylation. The fast phase, which accounts for about two-thirds of the phosphoenzyme, displayed a rate constant of 300 s^{-1} or greater, while the subsequent phase was 10 times slower. The ratio of the fast to slow phases as well as their rates remained fairly constant over a wide range of ATP concentrations (Hobbs et al., 1988). Preincubation of the enzyme with Na^+ before adding ATP and Mg^{2+} eliminated the biphasicity (Fig. 2 *B*, *inset*) without changing the total EP. Similar results were obtained if both Na^+ and ATP were present before initiating phosphorylation (not shown). Together with the previous results, these experiments suggest that there are two distinct pathways for phosphorylation and that including Na^+ in the preincubation medium accelerates EP formation in the slower pathway. Because Na^+ is expected to rapidly equilibrate with its binding sites, the effects produced by preincubation are likely to reflect a slow process such as a Na^+-induced conformational change or diffusion through a narrow space (e.g., an access channel) in the transport protein. It is interesting that the fast and slow phases of phosphorylation have roughly the same proportions (2:1) as the intermediate and fast phases of K^+ EDTA-induced dephosphorylation. Although the significance of this is unclear, it raises the possibility that the slow component of phosphorylation and the intermediate component of dephosphorylation are separate manifestations of the same catalytic activity.

Kinetics of the Na,K-ATPase Reaction in Electric Organ Microsomes: Dephosphorylation Reaction

In the absence of K^+ the electric organ enzyme displays kinetic behavior that does not conform to a simple consecutive reaction mechanism. Is there evidence for kinetic heterogeneity when Na^+ and K^+ are both present during phosphoenzyme formation? We addressed this question by examining the pattern of dephosphorylation of the phosphoenzyme formed in the presence of Na^+ and K^+ using unlabeled ATP to prevent rephosphorylation (Fig. 3, *inset*). Analysis of these data by curve fitting showed that the phosphoenzyme decays biexponentially with an initial rate

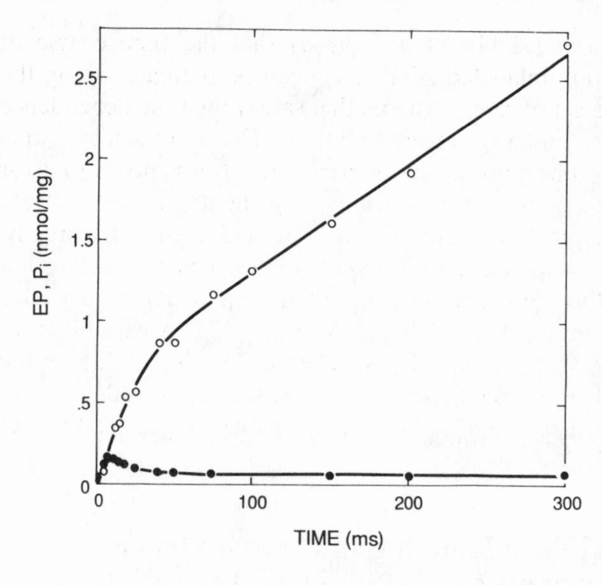

Figure 5. Transient state phosphorylation and P_i release by electric organ Na,K-ATPase after pre-equilibration with Na^+ and K^+. Electric organ microsomes (1 mg protein/ml) suspended in a medium containing 130 mM NaCl, 10 mM KCl, 3 mM $MgCl_2$, and 25 mM imidazole, pH 6.2, were mixed with an equal volume of the same medium containing 20 μM [^{32}P]ATP at 24°C. The reaction was allowed to proceed for the indicated times before quenching with acid. *Closed circles,* EP formation; *open circles,* P_i release.

that is six times faster (135 s^{-1}) and a final rate that is seven times slower (4 s^{-1}) than the rate of the intermediate phase of dephosphorylation. The latter component, which represents 25% of the phosphorylated sites in the absence of K^+, is virtually undetectable when both ions are present. Comparison of the experiments in Figs. 2*A* and 3 (*inset*) shows that addition of K^+ before phosphorylation also greatly reduces the steady-state level of EP formation. These results demonstrate that K^+, in addition to activating the hydrolysis of E_2P, markedly reduces the level of the intermediate component by changing either its rate of synthesis or its rate of decomposition. In the former situation K^+ would have to bind to the enzyme before phosphorylation and block the formation of the intermediate component without affecting the K^+-sensitive EP or the total number of active sites. Neither of these possibilities appears likely because the maximum level of K^+-activated P_i production during the first turnover of the Na,K-ATPase (Fig. 5) closely approximates the steady-state level of EP formation in the absence of K^+ (cf. Fig. 2*A*). If K^+ blocked the formation of the intermediate component, then the maximum level of EP formation would have exceeded P_i production in the pre–steady state. Alternatively, the disappearance of the intermediate component could reflect an increased rate of turnover secondary to accumulation of K^+ at the transport sites. In this situation one could imagine that the K^+ dephosphorylation sites are separated from the extracellular space by an access channel or an intermittent gate (Forbush, 1988), either of which would prevent added K^+ from immediately occupying those sites during mixing. To activate dephosphorylation K^+ would have to pass through the channel or gate introducing a delay. Preincubation with K^+ would allow the ion sufficient time to reach these sites and induce rapid dephosphorylation of E_2P. In the mechanism proposed by Nørby et al. (1983), the slow turnover of the intermediate component results from an inhibitory effect of K^+ on the conversion of the ADP-sensitive EP to a K^+-sensitive form. An advantage of the restricted access model is that there is no need to introduce an entirely new site or role for K^+ in the transport mechanism to explain these effects. Instead, one is left with having to explain why some sites are not immediately accessible to K^+.

We have previously shown (Hobbs et al., 1980*a*) that the time course of dephosphorylation following an unlabeled ATP chase can be simulated using the Albers-Post mechanism and a set of rate constants that satisfy the time dependence of EP formation and P_i release under turnover conditions. This approach to testing the kinetic competency of the phosphoenzyme intermediates is less prone to error because it avoids the complications that arise from using the amplitude and rate coefficients of dephosphorylation to compute the rate of ATP hydrolysis directly. Glynn (1988) has argued that an apparent weakness in the design of the dephosphorylation experiments is that the agent used to chase the phosphoenzyme may alter the kinetics of the reaction, noting that unlabeled ATP and CDTA gave different rates of dephosphorylation in the ox kidney Na,K-ATPase (Mårdh, 1975*a*). In the electric organ Na,K-ATPase it appears unlikely that the chase influences the kinetics of this reaction since unlabeled ATP and EDTA gave indistinguishable patterns of dephosphorylation (Froehlich et al., 1983).

Kinetics of the Na,K-ATPase Reaction in Electric Organ Microsomes: Phosphoenzyme Conformational Transition

The complex behavior of the K^+ EDTA-induced dephosphorylation reaction (Fig. 2*A*) makes it difficult to evaluate the rate of the phosphoenzyme conformational transition from these data. An alternative approach involves measuring the pre–steady-state time dependence of EP formation and P_i release in the presence of Na^+ and K^+ and then simulating this behavior using a model in which the rate of conversion of E_1P to E_2P represents the only unknown parameter (Froehlich et al., 1983; Hobbs et al., 1985*a*). An important consequence of including both ions in the reaction mixture is that the intermediate decay component vanishes so that the data can be analyzed using the Albers-Post mechanism. An added advantage is that E_2P breaks down immediately, generating a "burst" of P_i production and a transient overshoot in EP formation. Because the K^+-activated hydrolytic reaction is very fast ($> 1{,}000\ s^{-1}$ at 24°C), the kinetics of the phosphoenzyme transition preceding E_2P hydrolysis are influential in controlling the initial rate of P_i release and the decay of the phosphoenzyme overshoot during the first turnover. Using the Albers-Post mechanism to model this behavior, we estimated the rate of conversion of E_1P to E_2P to be 500–1,000 s^{-1} at 21°C (Froehlich et al., 1983). This method of evaluation requires that P_i production during the burst phase originate exclusively from the turnover of E_2P. In agreement with this we found that oligomycin, which blocks the conversion of E_1P to E_2P (Fahn et al., 1966), completely abolished the P_i burst (Hobbs et al., 1983*a*, 1985*a*). An additional technical problem involves the slowly dephosphorylating, K^+-insensitive phosphoenzyme which must be subtracted from the data before the simulation of EP formation. This was resolved by selectively inhibiting the formation of the rapidly decaying phosphoenzyme with vanadate and directly measuring the time course of formation of the slow component of dephosphorylation in the pre–steady state (Hobbs et al., 1980*b*).

Attempts to resolve the kinetics of Na^+ translocation by other methods have led to lower estimates for the rate of the phosphoenzyme conformational transition. Using a rapid filtration technique, Forbush (1984) measured $^{22}Na^+$ efflux during the initial turnover of the pump activated by the photolytic release of ATP from caged ATP (Kaplan et al., 1978). At 21°C he obtained a value of 50 s^{-1} for the initial rate of efflux which was similar to the rate of the E_1P to E_2P transition reported by Taniguchi and co-workers (Taniguchi et al., 1983, 1985) and Steinberg and Karlish (1989) using stopped-flow fluorescence and covalent labeling of the α-subunit with conformation-sensitive probes. Another approach used to investigate the kinetics of the translocation step involves determination of the transient pump currents generated in response to a voltage perturbation (Gadsby et al., 1985) or a light-driven ATP concentration jump (Fendler et al., 1985). In the latter technique membrane fragments containing Na,K-ATPase are adsorbed to a planar lipid bilayer and the ATP-dependent pump currents are measured via capacitive coupling to the underlying bilayer. Using a laser flash to rapidly generate ATP, Fendler et al. (1985) reported that the current transient consisted of a rapidly (100 s^{-1}) rising phase followed by a slower biexponential decay (12 s^{-1} and 3 s^{-1}). The larger of the decay constants was dependent on [ATP], supporting its identification with the ATP binding reaction, while the rising phase was correlated with the E_1P to E_2P transition, which was assumed to be electrogenic. In cardiac myocytes the current signals produced by a voltage step at 20°C have a relaxation rate of 40 s^{-1} (Gadsby et al., 1991), similar to the rate of $^{22}Na^+$ efflux measured by rapid filtration. The suggestion from these studies that the charge translocation step is partially rate limiting is supported by other evidence showing that the velocity of the pump is sensitive to changes in the membrane potential under voltage clamp conditions (De Weer et al., 1988; Rakowski et al., 1989).

One of the difficulties in comparing kinetic results obtained by different methods is that the experimental conditions used are often different. To avoid this problem we recently made a comparison of the enzymatic behavior of the Na,K-ATPase partial reactions in electric organ microsomes with their electrical behavior measured under conditions that were identical with respect to preparation, ionic composition, temperature, and pH. These experiments were conducted at pH 6.2 to achieve a rapid release of ATP from caged ATP (McCray et al., 1980) while minimizing the pH-dependent loss of enzymatic activity. Fig. 5 shows an example of an acid quench experiment in which the time course of EP formation and P_i release were measured by pre-equilibrating the enzyme with Na^+ and K^+ before mixing with ATP. Compared with the data at alkaline (7.4) pH (Hobbs et al., 1980*a*; Froehlich et al., 1983), the results are quite similar and indicate that E_2P hydrolysis is rapid. After subtraction of the stable phosphoenzyme, these data were simulated using the Albers–Post mechanism by allowing the phosphoenzyme transition rate constant to vary while holding the rates of the remaining steps at constant, predetermined values.[2] As seen in Fig. 6, a close fit to both the EP and P_i data was obtained by assigning the conformational transition a rate of 600 s^{-1}. In repeat experiments

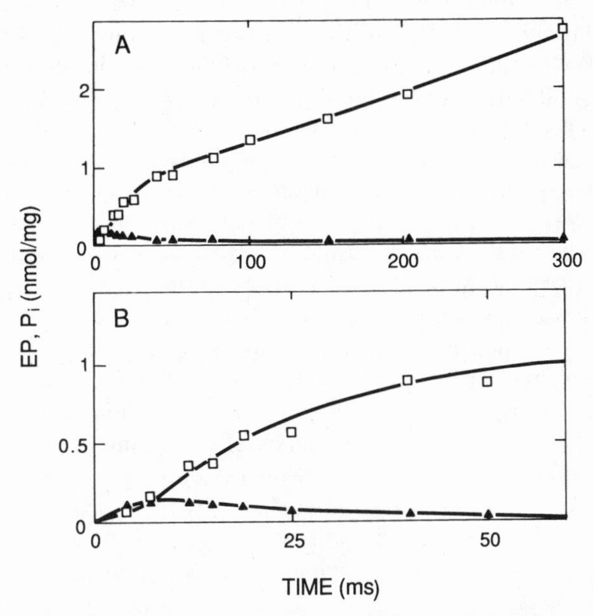

Figure 6. Simulations of transient state phosphorylation and P_i release in electric organ Na,K-ATPase. The experimental data are taken from Fig. 5. The data were simulated (*solid lines*) using the Albers–Post mechanism:

$$E_1 + ATP \overset{1}{\rightleftharpoons} E_1ATP \overset{2}{\rightleftharpoons} E_1P + ADP \overset{3}{\rightleftharpoons} E_2P \overset{4}{\rightleftharpoons} E_2 + P_i \overset{5}{\rightleftharpoons} E_1$$

and the following set of separately determined rate constants (in s^{-1}): $k_1[ATP]/k_{-1}$ =60/50; k_2/k_{-2} = 550/0; k_3/k_{-3} = 600/0; k_4/k_{-4} = 1,000/0; k_5/k_{-5} = 8/0. The slow component of dephosphorylation, determined in the presence of 2 μm vanadate (Hobbs et al., 1980*b*), was subtracted from the phosphoenzyme before plotting the data. *Solid triangles,* EP formation; *open squares,* P_i release. (*A*) Complete time course; (*B*) expanded time scale.

values of 400 and 1,000 s^{-1} were obtained, which overlap with previous estimates at higher pH and lower temperatures.

In parallel experiments using the bilayer apparatus, transient pump currents were measured which consisted of a lag phase, a monoexponential rise, and a biexponential decay (not shown). The pump current was fit with a five-state linear consecutive model in which the final step is electrogenic. The corresponding rate

[2] Unpublished acid quench experiments conducted at pH 6.2 and 24°C.

constants were (in s^{-1}): $k_1 = 750, k_2 = 460, k_3 = 360, k_4 = 69$ in the presence of K^+ (10 mM), and $k_1 = 750, k_2 = 500, k_3 = 370, k_4 = 69$ in the absence of K^+. In addition, the characteristic reciprocal time constant of the measuring circuit (Fendler et al., 1987) had to be included (3.8 and 9 s^{-1} in the presence and absence of K^+, respectively). For an optimal fit the lag phase had to be represented by two exponentials, k_1 and k_2, with the faster of these being determined by the photolytic release of ATP from caged ATP (1,400 s^{-1}) and the rise time of the amplifier ($1/t$ = 1,300 s^{-1}). The rate constant k_4 is in reasonable agreement with the apparent rate of phosphorylation in the presence of caged ATP (Fendler et al., 1991) supporting its correlation with ATP binding. The two remaining constants, k_2 and k_3, are similar to the rate constants for phosphorylation and the conformational transition evaluated from the rapid mixing studies, implicating the latter reaction in charge translocation. This is consistent with an earlier proposal by Karlish et al. (1985) but conflicts with other evidence linking charge translocation to a slower step in the transport cycle (Gadsby et al., 1991; Stürmer et al., 1991). There may, of course, be more than one charge-carrying step in the reaction mechanism, in which case Na^+ translocation can be envisioned as a two-step process consisting of a rapid conformational transition (400–1,000 s^{-1}) followed by a slower Na^+ deocclusion step (40–50 s^{-1}). It should be noted that K^+-activated E_2P hydrolysis is 10–20 times faster than the presumptive Na^+ deocclusion reaction, implying that K^+ binds to the dephosphorylation sites before Na^+ is released from the transport protein. These results are difficult to explain but are compatible with the notion of an access channel separating the extracellular space from the transport sites for Na^+ and K^+ (Stürmer et al., 1991). The main difficulty with this interpretation is that it fails to explain why two-thirds of the labeled sites dephosphorylate immediately upon addition of K^+ (Table I). Such behavior would seem to require a channel that selectivity retards Na^+ release without imposing a kinetic restriction on K^+ entry.

Summary and Conclusions

Kinetic investigations carried out in a number of laboratories have accumulated evidence favoring modification of the Albers–Post mechanism. The results of the rapid mixing studies involving the eel enzyme indicate that the complex kinetic behavior is confined to the Na^+-activated reaction pathway (Na-ATPase). The main conceptual problem in interpreting the dephosphorylation experiments involves the intermediate component, which turns over too slowly to account for the overall velocity of P_i production in the presence of Na^+ and K^+ and exhibits behavior compatible with an ADP-insensitive phosphoenzyme. Attempts to simulate the dephosphorylation reaction using schemes in which the intermediate component represents a precursor to the K^+-sensitive phosphoenzyme, E_2P, were unsuccessful in reproducing both the pre–steady-state and steady-state time dependence. When Na^+ and K^+ were both present during phosphorylation, the time course of dephosphorylation showed no evidence of an intermediate decay component, implying that K^+ either prevents its formation or accelerates its turnover. Complex kinetic behavior was also observed in the phosphorylation reaction under conditions where the reaction was initiated by the simultaneous addition of ATP, Na^+, and Mg^{2+}. Preincubation with Na^+ eliminated the biexponential pattern of accumulation so that only the fast phase was seen. The proportion of EP in the slow phase of phosphorylation was approximately equal to the fraction of EP in the intermediate phase of dephosphorylation (roughly one-third of the sites), suggesting that the two may be related to the same catalytic activity.

To try to explain these observations using recent modifications to the Albers–Post mechanism is difficult without invoking additional complex effects of the transported ions. We propose that a series model for phosphorylation is inadequate and that further modification of the mechanism is required. The alternative to a consecutive mechanism is a parallel pathway scheme:

$$E_1 \xrightleftharpoons{ATP} E_1{\cdot}ATP \xrightleftharpoons[fast]{ADP} E_1P \longrightarrow E_2P \xrightleftharpoons[fast]{P_i} E_2 \longrightarrow E_1$$

$$E_1' \xrightleftharpoons{ATP} E_1{\cdot}ATP' \xrightleftharpoons[slow]{ADP} E_1P' \longrightarrow E_2P' \xrightleftharpoons[slow]{P_i} E_2' \longrightarrow E_1'$$

In this model the enzyme exists in two distinct forms which are distributed in the upper and lower pathways in a ratio of 2:1. In the lower pathway the rates of phosphorylation and E_2P hydrolysis are controlled by the kinetics of ligand binding because of a structural constraint (ion channel?) imposed by the transport protein. When phosphorylation is carried out in the presence of Na^+ alone, E_2P and E_2P' accumulates rapidly and give rise to the fast and intermediate components of dephosphorylation, respectively. Preincubation with Na^+ and K^+ eliminates the functional differences between these pathways by removing the kinetic dependence of ligand binding, resulting in behavior that conforms to the Albers–Post mechanism. An important advantage of this scheme over previous proposals is that it uses the established binding sites for the monovalent cations, eliminating the need to introduce new sites or additional ion-mediated effects. Additional experimental support for the parallel pathway model was recently obtained by Suzuki and Post (1991), who found that phosphorylation of the dog kidney Na-ATPase leads to a kinetically heterogeneous species.

A parallel pathway mechanism places fewer constraints on the mechanism of formation of the phosphoenzymes corresponding to the fast (E_2P) and intermediate (E_2P') decay fractions because they occur in separate pathways. The principal requirement is that the rate constants governing their formation be very large to account for the rapid rate at which they reach their final distribution (Table I). Because the phosphointermediates in the upper and lower pathways are formed independently of each other, it is easy to find a set of rate constants that yield a dephosphorylation pattern in which the sum of the ADP- and K^+-sensitive phosphoenzymes exceeds the total EP. However, it is difficult to explain why high levels of $[Na^+]$ increase the intermediate component at the expense of the fast component because, as we have seen, the former does not represent an ADP-sensitive phosphoenzyme. Post and Suzuki (1991) have reported that high concentrations of sodium salts can produce changes in the levels of K^+- and ADP-reactive phosphorylated states by a mechanism (protein unfolding?) that is distinct from the conformational effects induced by ions at the low affinity transport sites. NaCl may therefore have two conformational effects at elevated concentrations, one of which is to stabilize

E_1P, while the other might be to convert enzyme in the upper to that in the lower pathway, thereby decreasing the access of K^+ to the dephosphorylation sites in a larger fraction of the enzyme population. The latter effect cannot be attributed to direct competition between Na^+ and K^+ because rapid dilution of the $[Na^+]$ at the time of addition of K^+ did not lead to an increase in the proportion of rapidly-decaying K^+-sensitive EP (Forbush and Klodos, 1991).

The possibility that kinetic heterogeneity may be responsible for the complex patterns of phosphorylation and dephosphorylation raises questions concerning the metabolic relationship of these pathways and their functional significance. Are these patterns due to post-translationally modified enzymes or do they reflect differences in the membrane microenvironment that become manifest only under specific conditions (e.g., absence of K^+)? With regard to the first of these possibilities, Rowe et al. (1988) observed that the NH_3 terminus of the α-subunit lacks a segment consisting of 15 amino acid residues which has been implicated in the regulation of ion binding because of its high lysine content. Whether this reflects a specific post-translational modification or is due to proteolysis during isolation of the enzyme has not been resolved. Among the other possible sources of functional heterogeneity, isozymes can be eliminated on the grounds that there is no evidence for more than one isoform of the Na,K-ATPase in the electric organ preparation. Another consideration is that the enzymes catalyzing these pathways may not operate independently but may, in fact, be coupled through a conformational interaction involving adjacent subunits in an oligomer (Askari, 1988). In this case the multiphasic behavior of the Na-ATPase phosphorylation and dephosphorylation reactions are manifestations of the subunit coupling interaction and reflect the degree of oligomerization. In the presence of oligomycin, the eel Na,K-ATPase exhibits kinetic behavior that is strongly suggestive of a dimeric interaction between α-subunits (Hobbs et al., 1983*a*). The primary difficulty of applying this model to the Na-ATPase data is that the observed 2 to 1 ratio of fast to slow components of EP formation suggests a higher degree of oligomerization, namely a trimer, for which there is little supporting evidence. The question of physiologic significance is even more perplexing since these complex kinetic features appear to be limited to the Na-ATPase reaction and would not be present (as far as we can discern with our methods) under physiological conditions.

The principal conclusions from the studies on the phosphoenzyme conformational transition are (*a*) that a simple model (i.e., the Albers–Post mechanism) will suffice to explain the enzymatic and electrical behavior in the presence of Na^+ and K^+, and (*b*) that the E_1P to E_2P transition is fast ($600\ s^{-1}$ at 24°C). A corollary to (*a*) is that E_1P and E_2P are kinetically competent intermediates in the Na,K-ATPase reaction pathway. Whether this scheme needs to be modified to allow K^+ to bind to the enzyme before Na^+ is released is an interesting mechanistic question requiring further investigation by fast kinetic methods.

References

Askari, A. 1988. Ligand binding sites of (Na^+,K^+)-ATPase: nucleotides and cations. *In* The Na^+,K^+ Pump, Part A: Molecular Aspects. J. C. Skou, J. G. Nørby, A. B. Maunsbach, and M. Esmann, editors. Alan R. Liss, Inc., New York. 149–165.

De Weer, P., D. C. Gadsby, and R. F. Rakowski. 1988. Voltage dependence of the Na,K pump. *Annual Review of Physiology.* 50:225–241.

Fahn, S., G. J. Koval, and R. W. Albers. 1966. Sodium–potassium-activated adenosine triphosphatase of Electrophorus electric organ. I. An associated sodium-activated transphosphorylation. *Journal of Biological Chemistry.* 241:1882–1889.

Fendler, K., J. Froehlich, S. Jaruschewski, A. Hobbs, W. Albers, E. Bamberg, and E. Grell. 1991. Correlation of charge to translocation with the reaction cycle of the Na,K-ATPase. *In* The Sodium Pump: Recent Developments. J. H. Kaplan and P. De Weer, editors. The Rockefeller University Press, New York. In press.

Fendler, K., E. Grell, and E. Bamberg. 1987. Kinetics of pump currents generated by the Na^+,K^+-ATPase. *FEBS Letters.* 224:83–88.

Fendler, K., E. Grell, M. Haubs, and E. Bamberg. 1985. Pump currents generated by the purified Na^+,K^+-ATPase from kidney on black lipid membranes. *EMBO Journal.* 4:3079–3085.

Forbush, B., III. 1984. Na^+ movement in a single turnover of the Na^+ pump. *Proceedings of the National Academy of Sciences, USA.* 81:5310–5314.

Forbush, B., III. 1988. Overview: occluded ions and the Na,K-ATPase. *In* The Na^+,K^+ Pump, Part A: Molecular Aspects. J. C. Skou, J. G. Nørby, A. B. Maunsbach, and M. Esmann, editors. Alan R. Liss, Inc., New York. 229–248.

Forbush, B., III, and I. Klodos. 1991. Rate-limiting steps in Na translocation by the Na/K pump. *In* The Sodium Pump: Structure, Mechanism, and Regulation. J. H. Kaplan and P. De Weer, editors. The Rockefeller University Press, New York. 211–225.

Froehlich, J. P., R. W. Albers, G. J. Koval, R. Goebel, and M. Berman. 1976. Evidence for a new intermediate state in the mechanism of the (Na^+,K^+)-adenosine triphosphatase. *Journal of Biological Chemistry.* 251:2186–2188.

Froehlich, J. P., A. S. Hobbs, and R. W. Albers. 1983. Evidence for parallel pathways of phosphoenzyme formation in the mechanism of ATP hydrolysis by Electrophorus Na,K-ATPase. *Current Topics in Membrane Transport.* 19:513–535.

Gadsby, D. C., J. Kimura, and A. Noma. 1985. Voltage dependence of Na/K pump current in isolated heart cells. *Nature.* 315:63–65.

Gadsby, D. C., M. Nakao, and A. Bahinski. 1991. Voltage-induced Na/K pump charge movements in dialyzed heart cells. *In* The Sodium Pump: Structure, Mechanism, and Regulation. J. H. Kaplan and P. De Weer, editors. The Rockefeller University Press, New York. 355–371.

Glynn, I. M. 1988. The coupling of enzymatic steps to the translocation of sodium and potassium. *In* The Na^+,K^+ Pump, Part A: Molecular Aspects. J. C. Skou, J. G. Nørby, A. B. Maunsbach, and M. Esmann, editors. Alan R. Liss, Inc., New York. 435–460.

Hobbs, A. S., R. W. Albers, and J. P. Froehlich. 1980*a*. Potassium-induced changes in phosphorylation and dephosphorylation of ($Na^+ + K^+$)-ATPase observed in the transient state. *Journal of Biological Chemistry.* 255:3395–3402.

Hobbs, A. S., R. W. Albers, and J. P. Froehlich. 1983*a*. Effects of oligomycin on the partial reactions of the sodium plus potassium-stimulated adenosine triphosphatase. *Journal of Biological Chemistry.* 258:8163–8168.

Hobbs, A. S., R. W. Albers, and J. P. Froehlich. 1983*b*. ADP sensitivity of the oligomycin-treated Na,K-ATPase. *Current Topics in Membrane Transport.* 19:569–572.

Hobbs, A. S., R. W. Albers, and J. P. Froehlich. 1985*a*. Quenched-flow determination of the E_1P to E_2P transition rate constant in electric organ Na,K-ATPase. *In* The Sodium Pump. I. M. Glynn, and C. Ellory, editors. The Company of Biologists, Ltd., Cambridge, UK. 355–361.

Hobbs, A. S., R. W. Albers, and J. R. Froehlich. 1988. Complex time dependence of phosphoenzyme formation and decomposition in electroplax Na,K-ATPase. *In* The Na^+,K^+ Pump, Part A: Molecular Aspects. J. C. Skou, J. G. Nørby, A. B. Maunsbach, and M. Esmann, editors. Alan R. Liss, Inc., New York. 307–314.

Hobbs, A. S., R. W. Albers, J. P. Froehlich, and P. F. Heller. 1985*b*. ADP stimulates hydrolysis of the "ADP-insensitive" phosphoenzyme in Na^+,K^+-ATPase and Ca^{2+}-ATPase. *Journal of Biological Chemistry.* 260:2035–2037.

Hobbs, A. S., J. P. Froehlich, and R. W. Albers. 1980*b*. Inhibition by vanadate of the reactions catalyzed by the (Na^+,K^+)-stimulated ATPase. *Journal of Biological Chemistry.* 255:5724–5727.

Kaplan, J. H., B. Forbush, III, and J. F. Hoffman. 1978. Rapid photolytic release of adenosine 5′-triphosphate from a protected analogue: Utilization by the Na:K pump of human red blood cell ghosts. *Biochemistry.* 17:1929–1935.

Karlish, S. J. D., A. Rephaeli, and W. D. Stein. 1985. Transmembrane modulation of cation transport by the Na,K pump. *In* The Sodium Pump. I. M. Glynn and C. Ellory, editors. The Company of Biologists, Ltd., Cambridge, UK. 487–499.

Klodos, I., J. G. Nørby, and I. W. Plesner. 1981. The steady state kinetic mechanism of ATP hydrolysis catalyzed by membrane-bound (Na^+ + K^+)-ATPase from ox brain. II. Kinetic characterization of phosphointermediates. *Biochimica et Biophysica Acta.* 643:463–482.

Klodos, I. 1988. ATP synthesis by Na,K-ATPase. Effect of K^+ on the ADP-dependent dephosphorylation of phosphoenzyme. *In* The Na^+,K^+ Pump, Part A: Molecular Aspects. J. C. Skou, J. G. Nørby, A. B. Maunsbach, and M. Esmann, editors. Alan R. Liss, Inc., New York. 149–165.

Kuriki, Y., and E. Racker. 1976. Inhibition of (Na^+,K^+) adenosine triphosphatase and its partial reactions by quercetin. *Biochemistry.* 16:4951–4956.

Mårdh, S. 1975*a*. Bovine brain Na^+,K^+-stimulated ATP phosphohydrolase studied by a rapid-mixing technique: K^+-stimulated liberation of [^{32}P] orthophosphate from [^{32}P] phosphoenzyme and resolution of the dephosphorylation into two phases. *Biochimica et Biophysica Acta.* 391:448–463.

Mårdh, S. 1975*b*. Bovine brain Na^+,K^+-stimulated ATP phosphohydrolase studied by a rapid-mixing technique. Detection of a transient [^{32}P] phosphoenzyme formed in the presence of potassium ions. *Biochimica et Biophysica Acta.* 391:464–473.

Mårdh, S., and R. L. Post. 1977. Phosphorylation from adenosine triphosphate of sodium- and potassium-activated adenosine triphosphatase. *Journal of Biological Chemistry.* 252:633–638.

McCray, J. A., L. Herbette, T. Kihara, and D. R. Trentham. 1980. A new approach to time-resolved studies of ATP-requiring biological systems: laser flash photolysis of caged ATP. *Proceedings of the National Academy of Science, USA.* 77:7237–7241.

Nørby, J. G., I. Klodos, and N. O. Christiansen. 1983. Kinetics of Na-ATPase activity of the Na,K pump. Interactions of the phosphorylated intermediates with Na^+, $Tris^+$, and K^+. *Journal of General Physiology.* 82:725–759.

Nørby, J. G., and I. Klodos. 1988. The phosphointermediates of Na,K-ATPase. *In* The Na^+, K^+ Pump, Part A: Molecular Aspects. J. C. Skou, J. G. Nørby, A. B. Maunsbach, and M. Esmann, editors. Alan R. Liss, Inc., New York. 249–270.

Plesner, I. W., L. Plesner, J. G. Nørby, and I. Klodos. 1981. The steady state kinetic mechanism of ATP hydrolysis catalyzed by membrane-bound (Na$^+$ + K$^+$)-ATPase from ox brain. III. A minimal model. *Biochimica et Biophysica Acta.* 643:483–494.

Plesner, L., and I. W. Plesner. 1988. Distinction between the intermediates in Na$^+$-ATPase and Na$^+$,K$^+$-ATPase reactions. II. Exchange and hydrolysis kinetics at micromolar nucleotide concentrations. *Biochimica et Biophysica Acta.* 937:63–72.

Post, R. L., C. Hegyvary, and S. Kume. 1972. Activation by adenosine triphosphate in the phosphorylation kinetics of sodium and potassium ion transport adenosine triphosphatase. *Journal of Biological Chemistry.* 247:6530–6540.

Post, R. L., and K. Suzuki. 1991. A Hofmeister effect on the phosphoenzyme of Na,K-ATPase. *In* The Sodium Pump: Structure, Mechanism, and Regulation. J. H. Kaplan and P. DeWeer, editors. The Rockefeller University Press, New York. 201–209.

Rakowski, R. F., D. C. Gadsby, and P. De Weer. 1989. Stoichiometry and voltage dependence of the sodium pump in voltage-clamped, internally dialyzed squid axon. *Journal of General Physiology.* 93:903–941.

Rowe, P. M., W. T. Link, A. K. Hazra, P. G. Pearson, and R. W. Albers. 1988. Antibodies to synthetic peptides as probes of the structure of the α subunit of the Na,K-ATPase. *In* The Na$^+$,K$^+$ Pump, Part A: Molecular Aspects. J. C. Skou, J. G. Nørby, A. B. Maunsbach, and M. Esmann, editors. Alan R. Liss, Inc., New York. 115–120.

Steinberg, M., and S. J. D. Karlish. 1989. Studies on conformational changes in Na,K-ATPase labeled with 5-iodoacetamidofluorescein. *Journal of Biological Chemistry.* 264:2726–2734.

Stürmer, W., R. Bühler, H.-J. Apell, and P. Läuger. 1991. Charge translocation by the sodium pump. Ion binding and release studied by time-resolved fluorescence measurements. *In* The Sodium Pump: Recent Developments. J. H. Kaplan and P. De Weer, editors. The Rockefeller University Press, New York. In press.

Suzuki, K., and R. L. Post. 1991. Slow and rapid components of dephosphorylation kinetics of Na,K-ATPase. *In* The Sodium Pump: Recent Developments. J. H. Kaplan and P. De Weer, editors. The Rockefeller University Press, New York. In press.

Taniguchi, K., K. Suzuki, and S. Iida. 1983. Stopped flow measurement of conformational change induced by phosphorylation in (Na$^+$,K$^+$)-ATPase modified with N-[p-(2-benzimidazolyl)phenyl] maleimide. *Journal of Biological Chemistry.* 258:6927–6931.

Taniguchi, K., K. Suzuki, D. Kai, M. Kudo, K. Tomita, and S. Iida. 1985. ATP hydrolysis in the presence of Na$^+$, with or without K$^+$, occurs accompanying conformational changes via $E_1 \cdot$ ATP, $E_1 \cdot$ ATP, E_1P, E_2P, and E_2 in sequence. *In* The Sodium Pump. I. M. Glynn, and C. Ellory, editors. The Company of Biologists, Ltd., Cambridge, UK. 363–367.

Yoda, A., and S. Yoda. 1987*a*. Two different phosphorylation–dephosphorylation cycles of Na,K-ATPase proteoliposomes accompanying Na$^+$ transport in the absence of K$^+$. *Journal of Biological Chemistry.* 262:110–115.

Yoda, S., and A. Yoda. 1986. ADP- and K$^+$-sensitive phosphorylated intermediate of Na,K-ATPase. *Journal of Biological Chemistry.* 261:1147–1152.

Yoda, S., and A. Yoda. 1987*b*. Phosphorylated intermediates of Na,K-ATPase proteoliposomes controlled with bilayer cholesterol. *Journal of Biological Chemistry.* 262:103–109.

Chapter 18

Successes and Failures of the Albers-Post Model in Predicting Ion Flux Kinetics

John R. Sachs

Department of Medicine, State University of New York at Stony Brook, Stony Brook, New York 11794-8151

The Albers-Post model of the reaction mechanism of the Na,K pump was developed more than twenty years ago to account for the effects of Na, K, ATP, and ADP on transphosphorylation reactions carried out by Na,K-ATPase preparations. In its current expanded form the scheme has been remarkably effective in accounting for the results of ion transport and transphosphorylation measurements, of physical and biochemical studies of the changes of pump conformation which occur during a transport cycle, of measurements of electrical events which accompany ion transport, and of measurements of occlusion of Rb and Na ions by Na,K-ATPase preparations (Glynn, 1985; Glynn, 1988; Glynn and Karlish, 1990). Nevertheless, many observations have been reported which do not seem to be consistent, or even compatible, with the Albers-Post scheme as it is currently formulated. In view of the success of the scheme in accounting for a broad range of diverse observations, it certainly does not seem reasonable to discard it because of observations which do not seem compatible, but on the other hand, many of the incompatible observations are too compelling to be ignored. It seems likely that some modifications of the Albers-Post scheme will be necessary to accommodate all the findings, but it is not yet clear what those modifications will be. In this article, I will review some experiments which seem to provide support for the scheme, and then discuss studies which suggest that modification is necessary and explore some modifications which may help to resolve the discrepancies.

Cleland has written that "because it looks at the reaction while it is taking place, a kinetic study is the ultimate arbiter of mechanism" (Cleland, 1989). The studies described here involved measurements of ion transport or ATPase activity using human red blood cells, red cell ghosts, or broken membranes prepared from red cells. Red cells were used because they are readily available, the concentration of ions on the two sides of the red cell membrane is easily controlled, and the enzyme can be studied without changing its relationship with the cell membrane. The studies were designed to determine the order in which the ions combine with and are released from the two sides of the membrane during a transport cycle. According to the Albers-Post scheme, Na must be able to leave the enzyme before K is added from the outside, and K must be able to bind to Na-free enzyme, and K must be able to leave the enzyme before Na reacts at the inside, and Na must be able to combine with K-free enzyme; i.e., the reaction mechanism must be ping-pong with respect to Na and K. Many, but not all, of the experiments utilize the methods of steady-state kinetics which have the added advantage that it is possible to investigate the reaction mechanism while the enzyme is turning over, and it is possible to carry out experiments at 37°C where the turnover rate is maximal.

Fig. 1 shows a contemporary version of the Albers-Post mechanism. In addition to Na-K exchange, the pump is capable of carrying out Na–Na exchange (Garrahan and Glynn, 1967*a*, *c*) through the sequence:

$$E_1ATP \underset{Na_c}{\rightleftharpoons} E_1ATP3Na \underset{ADP}{\rightleftharpoons} E_1P\,3(Na) \longleftrightarrow E_1P3Na \overset{Na_o}{\rightleftharpoons}$$

$$E_1P2N_a \overset{Na_o}{\rightleftharpoons} E_1PN_a \longleftrightarrow E_2PNa \overset{Na_o}{\rightleftharpoons} E_2P$$

and K–K exchange (Glynn et al., 1970; Simons, 1974) through the sequence:

$$E_1ATP \underset{K_c}{\rightleftharpoons} E_1ATP2K \underset{ATP}{\rightleftharpoons} E(2K) \longrightarrow E_2P2K \overset{K_o}{\rightleftharpoons} E_2P$$

In addition, it has long been known that the pump is capable of carrying out a Na efflux uncoupled to the entry of K or any other cation (Garrahan and Glynn, 1967*a*, *b*); this is accounted for in the Albers-Post scheme by the pathway from E_2P to E_1ATP characterized by the rate constant k_{12}. Binding and release of ions at both sides of the membrane is assumed to be rapid equilibrium.

Interaction of Na and K with Cation Binding Sites

At the inside surface, Na and K ions are assumed to bind randomly and independently to three equivalent sites on the same enzyme form; the model was first described by Garay and Garrahan (1973). The affinity of the sites for Na are assumed to be equivalent and characterized by the equilibrium constant K_N, and the affinity of

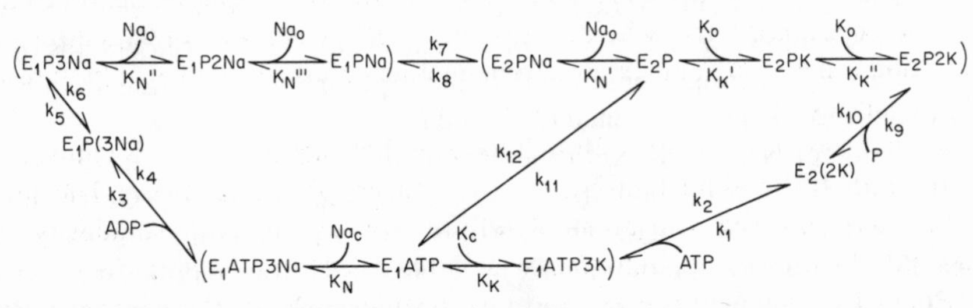

Figure 1. A version of the Albers-Post reaction mechanism discussed in the text. The portions of the reaction mechanism within parentheses are assumed to be in rapid equilibrium. Na_c is intracellular Na, Na_o extracellular Na, K_c intracellular K, and K_o extracellular K. K_N, K_K, etc. are dissociation constants and k_1, k_2, etc. are rate constants

the sites for K are also assumed to be equivalent and characterized by the equilibrium constant K_K. Binding of Na or K to one of the sites does not prevent binding of the other ion to one or both of the other sites. Na efflux occurs when all three sites are combined with Na, and K efflux occurs when all three are combined with K. From this model one can calculate the fraction of the total pool E_1ATP (E_1ATP_T) which is bound to three Na ions as a function of cell Na and K concentration as:

$$E_1ATP3Na/E_1ATP_T = 1\bigg/\left[1 + \frac{K_N}{Na_c}\left(1 + \frac{K_c}{K_K}\right)\right]^3 \tag{1}$$

and the fraction which is bound to three K ions as:

$$E_1ATP3K/E_1ATP_T = 1\bigg/\left[1 + \frac{K_K}{K_c}\left(1 + \frac{Na_c}{K_N}\right)\right]^3 \tag{2}$$

Observations which support this formulation include:

(*a*) Pump activity (Na-K exchange) is competitively inhibited by K, and K_N is a linear function of K over a wide range of K concentrations, as predicted by Eq. 1 (Garay and Garrahan, 1973; Sachs, 1986*a*).

(*b*) At fixed Na concentration, the reciprocal of pump activity (Na-K exchange) is a parabolic function of K concentration, as predicted by Eq. 1 (Sachs, 1986*a*).

(*c*) K–K exchange is competitively inhibited by Na_c (Simons, 1974; Sachs 1986*a*) as predicted by Eq. 2 and, as required for a competitive inhibitor, sufficiently high concentrations of Na_c drive the K–K exchange to zero (Sachs, 1986*a*).

(*d*) At fixed K concentration, the reciprocal of the K–K exchange is a parabolic function of cell Na concentration, as predicted by Eq. 2 (Sachs, 1986*a*).

Rate equations based on this model describe the effect of cell Na and K concentration on the pump rate of red cell preparations surprisingly well despite the unlikely assumption that the dissociation constants of the three sites for Na and for K are the same and not affected by the state of occupancy of the other sites, and despite the fact that the rate equations derived from Fig. 1 are much more complex than the simple equations used in evaluating this model (see below). In other pump preparations more complex relations between pump rate and cell Na concentration have been observed (Karlish and Stein, 1985; Rossi and Garrahan, 1989*a*). In red cells, rate equations derived from Eqs. 1 and 2 describe successfully the variation in Na-K and K–K exchange rate determined simultaneously as a function of cell Na concentration using the same values of K_N and K_K (Sachs, 1986*a*). Moreover, the conclusion that Na and K compete at a single set of sites at the inside surface depends on the shapes of the curves arising from the inhibition experiments, and these are not very sensitive to the exact values assigned to K_N and K_K.

Interaction of Na_o and K_o with the cation binding sites at the outside is more complex. Na_o competitively inhibits activation of pump rate by K_o (Garrahan and Glynn, 1967*b*; Sachs, 1967). In the absence of Na_o, the curve relating pump activity to external K concentration is either nearly hyperbolic or antisigmoid; i.e., it rises more rapidly than a rectangular hyperbola at low concentrations of K_o (Lew et al., 1973; Sachs, 1977*a*). As Na_o increases, $K_{1/2}$ for K_o (the concentration of K_o at which pump rate is half maximal) increases and the activation curve becomes progressively more sigmoid. These findings indicate that Na_o and K_o combine with the same enzyme form (E_2P) and that, in contrast with the situation for the binding sites at the inside surface, binding of Na and K is mutually exclusive. Despite the fact that there must be multiple binding sites for Na_o, Dixon plots of the reciprocal of the K influx measured at a fixed concentration of K_o as a function of Na_o concentration are linear up to at least 150 mM Na_o (Sachs, 1974). To account for the linear Dixon plot, it is necessary to assume either that the affinity of the first Na binding site on E_2P for Na_o is much greater than the affinity of the remaining two sites, or, more likely, that the binding of Na results in a conformational change $E_2P \rightarrow E_1P$ (or E*P):

$$E_1PNa) \longrightarrow (E_2PNa \xrightarrow{Na_o} E_2P \xrightarrow{K_o} E_2PK$$

The release of Na at the outside surface occurs in multiple steps, probably separated by conformational changes (Nørby et al., 1983; Taniguchi et al., 1986; Yoda and Yoda, 1986); little information is available about the association constants for Na at the sequential steps. The model presented in Fig. 1 does not distinguish between the individual conformational changes, but assumes that the major change occurs in the transition $E_1PNa \longleftrightarrow E_2PNa$. Also missing from the model is a pathway for the uncoupled efflux of Na observed in proteoliposomes in which a single Na ion is released to the outside for each ATP hydrolyzed (Yoda and Yoda, 1987), nor does the model permit direct competition between Na and K at the K binding sites, although such competition must occur (probably with low affinity for Na_o) because it is known that Na_o can replace K_o in activating Na efflux (Lee and Blostein, 1980).

Evaluation of the Reaction Mechanism by Steady-State Kinetics

Using standard methods steady-state rate equations can be derived from the scheme shown in Fig. 1; they have the form $v = N/D$, where v is the reaction rate, N is a numerator, and D is a denominator. The denominator is common to each of the rate equations, but the numerator varies depending on how the pump rate is measured. The rate equations are very complicated with many terms, many rate constants, and many dissociation constants; they are not very useful and will not be given here. All useful modifications of the complete rate equation require elimination of some of the terms. One way in which this may be done is by making a judgment about the relative magnitudes of some of the rate constants or dissociation constants (e.g., one of the rate constants is so small or one of the dissociation constants is so large that the term in which it appears vanishes) based on experimental findings from rapid reaction experiments or equilibrium binding experiments. In this case care must be taken that the judgment about the magnitude of the rate constant or the dissociation constant does not implicitly assume the mechanism which is to be tested. A preferable way to simplify the rate equations is to set the concentration of one or more of the substrates so high during the measurement of the reaction rate that the terms in which they appear as a denominator disappear, or to set the concentration of one or more of the products at zero so that the terms in which they appear as numerators disappear. By setting the concentration of Na_o, K_c, ADP, and phosphate equal to zero, the rate equation from Fig. 1 becomes

$$v = N/D$$

where the common denominator is

$$D = \left(1 + \frac{K_N}{\mathrm{Na_c}}\right)^3 \left[k_1 k_5 k_7 \left(k_9 + k_{12}\frac{K'_K K''_K}{\mathrm{K}_0^2}\right)\right] + \left(\frac{K'_K K''_K}{\mathrm{K}_0^2} + 2\frac{K'_K}{\mathrm{K}_0} + 1\right) k_1 k_3 k_5 k_7$$
$$+ \frac{K'_K K''_K}{\mathrm{K}_0^2}(k_1 k_3 k_5 k_{12}) + k_1 k_3 k_9 (k_5 + k_7) + k_3 k_7 (k_5 k_9 + k_1 k_{12}).$$

When the pump rate is measured by estimating unidirectional efflux of Na, net efflux of Na, or rate of production of ADP

$$N = k_1 k_3 k_5 k_7 E_T \left(k_9 + k_{12}\frac{K'_K K''_K}{\mathrm{K}_0^2}\right)$$

and when it is measured by estimating unidirectional influx of K, net movement of K, or rate of production of phosphate

$$N = k_1 k_3 k_5 k_7 k_9 E_T$$

where E_T is total enzyme concentration. It is obvious that Na_o and K_c can be kept at zero only in preparations (intact cells, proteoliposomes) in which the inside and outside can be separated and not in solubilized or particulate enzyme preparations.

Inspection of the rate equation shows that the relation between pump rate and Na concentration will be described by $v = V_M/(1 + K_N/\text{Na}_c)^3$ only if the values of the other terms in the denominator are insignificant relative to the value of the term containing Na_c. As discussed above, the simple relation is satisfactory for experiments performed with red cell preparations, but has been found to be unsatisfactory with some other preparations. It would not be surprising if the value of one or more of the rate constants differ from preparation to preparation. Steady-state kinetic evaluations of a reaction measurement usually involve the determination of the variation in V_M (the maximal velocity at saturating substrate concentration) and $K_{1/2}$ (the concentration of substrate at which saturation is half maximal) in response to some experimental manipulation. These are directly measured experimental values and make no assumptions about the values of kinetic constants or dissociation constants. At any rate, even with the rate equation derived from the relatively simple mechanism shown in Fig. 1, it hardly seems likely that unambiguous values for the rate constants and dissociation constants could be extracted from any realistic number and variety of experiments. In what follows, the rationale for a kinetic approach will, where possible, be developed intuitively from mass action considerations, but all of the intuitive interpretations are always supportable by quantitative deductions from the rate equations.

According to the Albers-Post scheme, the reaction mechanism of the Na,K pump should be ping-pong with respect to Na_c and K_o; i.e., it should be possible for Na to be released to the outside before K is bound and for K to be released to the inside before Na is bound:

$$\begin{array}{cccccc} & \text{Na}_c & \text{Na}_o & & \text{K}_o & \text{K}_c \\ & \downarrow & \uparrow & & \downarrow & \uparrow \\ \hline \text{E}_1 & & & \text{E}_2 & & \end{array}$$

For ping-pong mechanisms, measurement of V_M and $K_{1/2}$ for one substrate at several fixed concentrations of the other substrate in the absence of products shows that V_M and $K_{1/2}$ for the variable substrate increase at a fixed ratio as the concentration of the fixed substrate increases. In the scheme above, increasing concentrations of Na_c in the absence of Na_o increases the amount of E_2, and therefore V_M increases at saturating K_o. Because the amount of E_2 increases with increasing Na_c, the concentration of K_o necessary to half saturate also increases. Unfortunately, the Albers-Post scheme is not strictly ping-pong and is better represented as:

$$\begin{array}{cccccc} & \text{Na}_c & \text{Na}_o & & \text{K}_o & \text{K}_c \\ & \downarrow & \uparrow & & \downarrow & \uparrow \\ \hline \text{E}_1 & \longleftarrow & \longleftarrow & \text{E}_2 & & \end{array}$$

in which the direct pathway between E_2 and E_1 results from the uncoupled Na efflux. Although increasing the concentration of Na_c increases the pump rate at saturating K_o, the amount of E_2 will be less than in the pure ping-pong mechanism because it can convert to E_1 through the uncoupled pathway. As a result, $K_{1/2}$ increases less than V_M does at low concentrations of Na_c, and the ratio $V_M/K_{1/2}$ is not constant.

When K influx was measured in human red cells at a number of fixed concentrations of Na_c in the absence of K_c or Na_o, $K_{1/2}$ for K_o and V_M increased as the concentration of Na_c increased (Sachs, 1977*b*). The ratio $V_M/K_{1/2}$ increased as a sigmoidal function of Na_c concentration, and the $K_{1/2}$ for Na_c was about the same as the $K_{1/2}$ usually found when pump rate is measured as a function of Na_c concentration. The results were described well by a model similar to that shown in Fig. 1 which included an uncoupled Na efflux; the value of the uncoupled efflux predicted from the experiments, in which Na-K exchange was measured, was about the same as the value usually found when Na efflux is measured in solutions free of Na and K (Sachs, 1979). Several other studies have been reported in which measurements were made in the absence of Na_o and K_c; in squid axons (Baker et al., 1969) and in human red cells (Chipperfield and Whittam, 1974, 1976) $K_{1/2}$ for K_o and V_M both increased as Na_c increased, $K_{1/2}$ more rapidly than V_M. Different results were obtained when measurements were made in the presence of products. Two studies have been reported in which measurements were made in red cells which contained K (Hoffman and Tosteson, 1971; Garay and Garrahan, 1973); in one of the studies external Na was also present. In a third study, pump rate was measured during ATP-dependent Na–Na exchange (in which Na_o replaces K_o) using Na,K-ATPase incorporated into proteoliposomes (Cornelius and Skou, 1988); K_c was absent, but one of the products of the reaction, Na_o, was necessarily present but varied in concentration. In the red cell experiments, V_M increased but $K_{1/2}$ for K_o remained unchanged when Na_c was varied, and V_M increased but $K_{1/2}$ for Na_c was unchanged when K_o was increased. Similarly, in the proteoliposome experiments, $K_{1/2}$ for Na_c did not vary with Na_o, nor did $K_{1/2}$ for Na_o vary with Na_c, although V_M increased in both cases. The lack of change in $K_{1/2}$ in these experiments has led to the suggestion that binding sites for Na_c and for K_o exist simultaneously, that the sites do not interact, and that sites for Na must be occupied by Na_c and sites for K by K_o at the same time for a transport cycle to occur; this is a special case of a sequential mechanism and is clearly inconsistent with the Albers-Post scheme. The difference between the results of these experiments and those performed in the complete absence of products has not been explained. The complex rate equation obtained from Fig. 1 when Na_o and K_c are not set at zero predicts that the concentration of these ions will alter the effect of Na_c or K_o on the $K_{1/2}$ for the other substrate cation. Perhaps Na_o and K_c can render the $K_{1/2}$ for one substrate independent of the concentration of the other by altering the distribution of enzyme between E_1 and E_2.

Characteristics of the K–K Exchange

The Na–Na and K–K exchanges carried out by the Na,K pump provide strong support for the ping-pong mechanism provided that they are true exchanges and that they occur as parts of the overall Na-K exchange pathway rather than as manifestations of alternative pathways. If the pump reaction mechanism were sequential rather than ping-pong, e.g.,

$$\begin{array}{cccccc} K_o & Na_c & K_c & Na_o & K_o & Na_c \\ \downarrow & \downarrow & \uparrow & \uparrow & \downarrow & \downarrow \\ \hline \end{array}$$

it is clear that the presence of Na_c will be necessary for K–K exchange to occur, and that the presence of K_o will be necessary for Na–Na exchange to occur. Na_c not only is not required for K–K exchange, but, in fact, inhibits it (Simons, 1974), and, similarly, K_o is not required for Na–Na exchange, but inhibits it (Garrahan and Glynn, 1967*c*).

Models have been suggested which suppose that the exchanges occur when coexisting and noninteracting internal and external sites are simultaneously filled with internal and external K for K–K exchange or with internal and external Na for Na–Na exchange (Garrahan and Garay, 1976; Levitt, 1980). It is possible to distinguish kinetically between this mechanism and a true exchange:

$$E_i \overset{K_c}{\rightleftharpoons} E_iK \longrightarrow E_oK \overset{K_o}{\rightleftharpoons} E_o$$

For a true exchange, V_M and the $K_{1/2}$ for internal K increase with increasing concentration of K_o, and V_M and $K_{1/2}$ for K_o increase with increasing concentration of K_c; in either case, the ratio $V_M/K_{1/2}$ should be constant. The reason for this can be seen intuitively. As the concentration of K_c increases, the fraction of the enzyme which is in the form E_o increases; as a result, the exchange rate at saturating K_o concentration (V_M) increases. Increasing concentration of E_o increases the concentration of K_o necessary for half saturation; $K_{1/2}$ for K_o also increases. It has been shown that, when K–K exchange is measured in human red cells, V_M and $K_{1/2}$ for K_o increase with increasing K_c concentration, as predicted (Sachs, 1981). It is likely, then, that the K–K exchange is a true exchange. There are many reasons for believing that the K–K exchange is part of the overall Na-K exchange which come from a comparison of the properties of the ion fluxes with the biochemical characteristics of purified Na,K-ATPase preparations (Glynn and Karlish, 1982). There are two reasons for accepting the identity of the two exchanges which do not rely on the biochemical data (Sachs, 1986*a*):

(*a*) From Fig. 1, one would expect that, if the K–K exchange is a partial reaction of the Na-K exchange, the $K_{1/2}$ for K_o of the Na-K exchange and the $K_{1/2}$ for K_o of the K–K exchange should be the same when measured under identical conditions. In fact, the two values differ markedly (Sachs, 1980). The discrepancy is attributable to the uncoupled Na efflux. The difference in the $K_{1/2}$ for K_o of the two exchanges means that the ratio of the value of the Na-K exchange to the value of the K–K exchange at each K_o concentration is not constant, but varies with K_o concentration: Na-K exchange/K–K exchange $= f(K_o)$. When rate equations for the Na-K exchange and the K–K exchange are derived from the reaction mechanism of Fig. 1, the denominators in each case are the same, but the numerators differ. The ratio of the numerator of the Na-K exchange rate equation to the numerator of the K–K exchange equation turns out to be a function of K_o concentration, but only if the value of k_{12} is not zero, i.e., only if there is an uncoupled Na efflux. Since it is known that low concentrations of Na_o suppress the uncoupled Na efflux (Garrahan and Glynn, 1967*a*), it is possible to experimentally test the conclusion derived from the rate equation. When the $K_{1/2}$ for K_o of the Na-K exchange and the $K_{1/2}$ for K_o of the K–K exchange were measured at a low concentration of Na_o sufficient to partially suppress the uncoupled Na efflux, the two values approached each other. The rate equations derived from Fig. 1 satisfactorily explain a puzzling aspect of the activation of Na-K exchange and K–K exchange by K_o provided the two exchanges share the same reaction mechanism.

(*b*) As discussed above, intracellular Na competitively inhibits the K–K exchange (at sufficiently high concentrations of Na_c the magnitude of the exchange goes to zero) and intracellular K competitively inhibits the Na-K exchange. All aspects of the competitive interaction between the two ions have been investigated in detail (Sachs, 1986*a*), and it can be concluded that mutual inhibition results from competition for the same set of binding sites.

Since the K–K exchange is part of the overall Na-K exchange, and since Na_c is not required for the exchange to occur, Na_c must bind to the pump after K is released.

Inspection of Fig. 1 leads to the conclusion that, if there is an uncoupled Na efflux, and if all the steps in the reaction mechanism are reversible so that K–K exchange is possible, then there should be an uncoupled K efflux which, like the uncoupled Na efflux, is inhibited by low concentrations of Na_o. Rate equations derived from Fig. 1 also predict the presence of an uncoupled K efflux provided the value of the rate constant k_{12} is not zero. Such an uncoupled K efflux can be demonstrated in red blood cells; it is inhibited by the same concentrations of Na_o that inhibit the uncoupled Na efflux, and its magnitude is about what one would predict from measurements of the magnitude of the uncoupled Na efflux (Sachs, 1986*a*).

Evaluation of the Reaction Mechanism Using Reversible Inhibitors

An independent approach to the evaluation of pump reaction mechanism depends on the determination of the characteristics of pump inhibition by reversible inhibitors which combine with specific enzyme forms. The rationale for the procedure can be understood intuitively. In the reaction mechanism

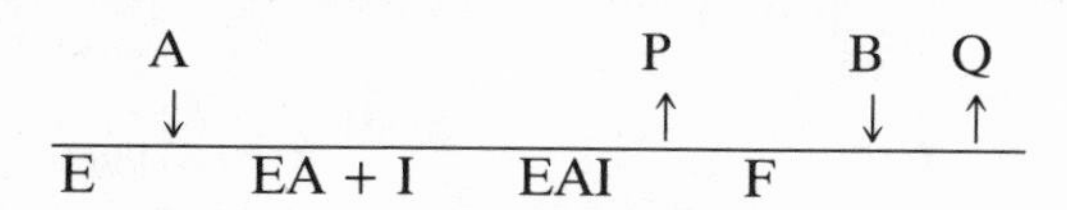

in which the inhibitor binds reversibly to the enzyme from EA, the effect of the inhibitor on the curve relating velocity to the concentration of substrate B at a fixed concentration of substrate A depends on whether or not the step between the addition of I to EA and the addition of B to F is reversible or irreversible. The step is reversible if the product P is present and irreversible if it is absent. If P is present, EA can be formed from F either in the forward direction after addition of B or in the backward direction after addition of P. The steady-state concentration of EA, and therefore fractional inhibition at any fixed concentration of I, will be independent of the concentration of B (Fig. 2, right panel). Plots of velocity as a function of concentration of substrate B in the presence and absence of inhibitor differ by a constant factor; $K_{1/2}$ for B does not change, but V_M is reduced; inhibition is noncompetitive with respect to B. On the other hand, if P is absent, EA can be formed from F only in the forward direction after addition of substrate B. The steady-state concentration of EA, and therefore fractional inhibition at any concentration of I, increases as the concentration of B increases. When velocity is plotted as a function of the concentration of substrate B, the difference between the two curves increases as the concentration of B increases to saturating levels. As a result, $K_{1/2}$'s for substrate B as well as V_M decrease in the presence of inhibitor; inhibition is uncompetitive with respect to substrate B. If, therefore, removal of a product converts noncompetitive inhibition with respect to a particular substrate to uncompetitive inhibition, release of the product must occur between the point at which the inhibitor binds and the point at which the substrate is added.

Two reversible inhibitors of the Na,K pump of red blood cells have proven useful in evaluating the pump reaction mechanism (Sachs, 1980, 1986*b*). There is a great deal of evidence from other preparations that oligomycin reacts preferentially with E_1 forms of the enzyme which have bound Na. In red cells, oligomycin inhibits Na-K exchange and Na–Na exchange, but does not inhibit K–K exchange unless the cells contain a small amount of Na. Fractional inhibition of Na–Na exchange increases with increasing concentration of Na_c and with increasing concentration of Na_o; inhibition of the exchange is uncompetitive with both Na_c and Na_o. Oligomycin inhibits, therefore, by combining with an enzyme form which occurs between the addition of Na at the inside and its release at the outside. Inhibition of the Na-K exchange by oligomycin is noncompetitive with respect to K_o if Na_o is present, but uncompetitive if the solution is Na-free:

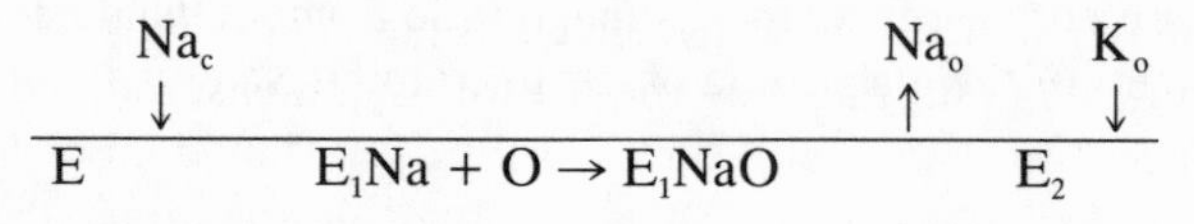

where O is oligomycin. From the discussion above it is clear that since removal of a product converts noncompetitive inhibition to uncompetitive inhibition, the product

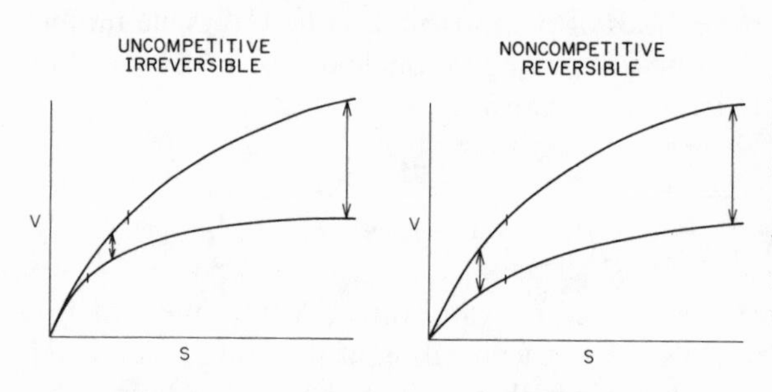

Figure 2. A plot of reaction rate vs. substrate concentration in the absence (*upper curve*) and presence (*lower curves*) of an inhibitor. The vertical lines intersecting the curves indicate $K_{1/2}$. For discussion, see text.

must be released between the addition of the inhibitor and the addition of the substrate; Na must be released before K is added.

A second inhibitor, vanadate, inhibits the K–K exchange at low concentrations, but does not inhibit the Na–Na exchange. Fractional inhibition of the Na-K exchange by vanadate increases with increasing K_o concentration, and fractional inhibition of K–K exchange increases with increasing K_c concentration. The results indicate that vanadate inhibits by combining with an enzyme form which occurs between the addition of K_o at the outside and the addition of K_c at the inside; in other preparations, vanadate is known to bind to E_2 forms. At low ATP concentrations, vanadate inhibition is noncompetitive with Na_c at high concentrations of K_c, but uncompetitive in the absence of K_c:

$$\begin{array}{cccc} & K_o \downarrow & & K_c \uparrow \quad Na_c \downarrow \\ \hline E_2 & E_2K + V \rightarrow E_2KV & & E_1 \end{array}$$

where V is vanadate. Since removal of the product, K_c, converts inhibition which is noncompetitive with Na_c to uncompetitive inhibition, K must be released to the inside before Na binds.

Taken together, the two studies show that the reaction mechanism is ping-pong with respect to Na_c and K_o.

Alternative Reaction Pathways

According to Fig. 1, phosphate acts as a product inhibitor of Na,K pump rate. Rate equations derived from Fig. 1 predict that phosphate inhibition should be strictly competitive with ATP, phosphate should increase the $K_{1/2}$ for K_o, and phosphate inhibition should be uncompetitive with Na_c in K-free cells. None of these predictions are fully verified.

Phosphate inhibition has been found to be noncompetitive or only partially competitive with ATP (Hexum et al., 1970; Apell et al., 1986; Robinson et al., 1986; Sachs, 1988) and not competitive as predicted by the Albers-Post scheme.

When the effect of phosphate on the $K_{1/2}$ for K_o is measured, the results depend on the ATP concentration at which the measurement is made. Beaugé and Di Polo (1979) and Eisner and Richards (1982) found that phosphate increased the $K_{1/2}$ for K_o as predicted by the model when the measurement was made at low ATP. When the measurement was made at high ATP, phosphate inhibition was uncompetitive with K_o; $K_{1/2}$ for K_o was reduced by phosphate (Sachs, 1988).

Inhibition by phosphate was found to be noncompetitive or partially competitive with Na_c in the absence of K_c, and not uncompetitive as predicted by the Albers-Post scheme (Sachs, 1988).

The characteristics of phosphate inhibition cannot, therefore, be explained by the Albers-Post scheme. An explanation for the discrepancy may be available from a consideration of the characteristics of the K–K exchange. When phosphate and ATP are low, K–K exchange proceeds through the Albers-Post pathway (Karlish et al., 1982; Eisner and Richards, 1983):

$$E_2P \xrightleftharpoons{K_o} E_2PK \xrightleftharpoons{P} E_2K \xrightleftharpoons{ATP} E_1ATPK \xrightleftharpoons{K_c} E_1ATP$$

but when either phosphate or ATP concentration is high, the exchange proceeds through an alternative pathway in which an enzyme form appears which simultaneously binds ATP and phosphate (Sachs, 1981; Karlish et al., 1982; Eisner and Richards, 1983):

$$E_2P \xrightleftharpoons{K_o} E_2PK \xrightleftharpoons{ATP} EATPPK \xrightleftharpoons{K_c} E_1ATPP \xrightleftharpoons{P} E_1ATP$$

The discrepancies between the characteristics of phosphate inhibition and the predictions of the Albers-Post scheme can be resolved if, in the presence of phosphate, an alternative Na-K exchange pathway appears in which ATP binds to E_2PK:

$$E_2P \xrightleftharpoons{K_o} E_2PK \xrightleftharpoons{ATP} E\,PATPK \xrightleftharpoons{K_c} E_1ATPP \xrightleftharpoons{P} E_1ATP \xrightleftharpoons{Na_c} E_1ATPNa$$

A rate equation derived from the alternative pathway predicts phosphate inhibition noncompetitive with ATP, uncompetitive with K_o, and partially competitive with

Na_c. In the presence of phosphate, part of the Na-K exchange takes place through the Albers-Post pathway and part through the alternative pathway. Since phosphate inhibition is competitive with ATP in the Albers-Post pathway and noncompetitive in the alternative pathway, combined operation should result in partially competitive inhibition (Robinson et al., 1986); inhibition uncompetitive with K_o at high ATP can result from flow through the alternative pathway, and increase of $K_{1/2}$ for K_o by phosphate at low ATP can result from flow through the Albers-Post pathway; and the combination of inhibition uncompetitive with Na_c through the Albers-Post pathway with inhibition competitive with Na_c through the alternative pathway accounts for the observed noncompetitive inhibition. Kinetic evaluation of phosphate inhibition of H,K-ATPase has also led to the suggestion of the presence of a pathway in which ATP adds to E_2P (Reenstra et al., 1988).

There is a second phenomenon which suggests that alternative transport pathways may occur. When pump rate is measured as a function of K_o concentration in solutions containing Na_o and vanadate, the curve is biphasic; pump rate in the presence of vanadate at first increases almost as rapidly as in the absence of vanadate, reaches a peak, and then decreases (Bond and Hudgins, 1975, 1979, 1982; Beaugé, 1979; Beaugé et al., 1980). Na_o protects against vanadate inhibition since inhibition markedly increases if Na_o is removed while vanadate remains constant. K_o competes with Na_o at the protective site, and binding of K_o promtoes vanadate inhibition. The K congeners Tl, Rb, Cs, and NH_4 also promote vanadate inhibition in the presence of Na_o, and the relative affinity of the protective site for these ions is the same as the relative affinity of the transport sites (Bond and Hudgins, 1979). The transport sites nevertheless are not the protective sites (Sachs, 1987). If Na_o protects against inhibition and K_o promotes inhibition by competing at the two transport sites, inhibition would be uncompetitive with K_o in the presence of Na_o as it is in Na-free solutions; biphasic activation curves would not occur. Biphasic K activation curves could result, however, if pumps with only one of the two transport sites occupied by K_o are capable of transport, if Na_o bound to the second transport site protects such cycles against vanadate inhibition, and if pumps with both transport sites occupied by K_o are capable of both transport and inhibition by vanadate. If protection occurs when Na_o occupies one transport site and K the other, the Na either will or will not be transported inward with the K ion. If Na is transported inward with the K ion, the Na:K coupling ratio for Na-K exchange during protected cycles will be 2:1 and there will be a measurable Na influx along with the K influx; if Na is not transported inward, the coupling ratio will be 3:1 for protected cycles. The Na:K coupling ratio did not differ from 3:2 when measured while Na_o was protecting against vanadate inhibition, and there was no evidence of an Na influx accompanying the K influx (Sachs, 1987). The site at which Na_o protects against vanadate inhibition is not one of the two transport sites which move K inward. Under circumstances in which Na_o protects against vanadate inhibition, the reaction mechanism must be different from when it does not, in that the steady-state level of the species which binds vanadate is reduced. The alternative pathway must not be very different from the usual pathway since Na_o does not much change the pump rate, and Na_o does not protect against inhibition by phosphate or arsenate.

By determining the concentration of Na_o necessary for half maximal protection at fixed vanadate and several fixed K_o concentrations, and the concentration of K_o which maximally promotes vanadate inhibition at fixed vanadate and several fixed

Na_o concentrations, it was possible to estimate the dissociation constant of the protective sites for Na_o and K_o; it was well under 1 mM for both ions (Sachs, 1987). As discussed above, a high affinity site for Na_o at which it inhibits uncoupled Na efflux has long been known, but whether K_o also interacts with this site, and whether it is the same as the protective site is not known. The protective site has many of the characteristics of a transport site; perhaps the site which transports the third Na ion is critically different from the sites which transport the other two Na ions and the K ions, and perhaps the third site can operate independently of them during transport in such a way that it can serve as the protective site and account for other deviations from ping-pong kinetics. Such a situation would markedly complicate the kinetic behavior of the pump.

Modification of the Reaction Mechanism by Dimer Interaction?

Promotion of vanadate inhibition by K_o in the presence of Na_o appears to occur with sigmoid dependence on K_o concentration (Beaugé, 1979), and Na_o protection against vanadate inhibition shows sigmoid dependence on Na_o concentration (Sachs, 1987). The situation is very complex, and multiple binding sites are not the only explanation for sigmoid curves; nevertheless, it has been suggested that sigmoid curves may indicate that more than one site is involved in modulating vanadate inhibition (Beaugé, 1979). In fact, if multiple sites are involved, the behavior of the system is remarkably similar to the behavior of the cation transport sites at the inside of the pump described above (Sachs, 1987).

Multiple high affinity sites for Na_o and K_o in addition to the transport sites and on the same α chain would certainly be surprising. It occurred to us that Na_o protection at multiple sites could conceivably arise from subunit interaction. We assume that each $\alpha\beta$ protomer can carry out a complete Na-K exchange cycle, that dimers of $\alpha\beta$ protomers exist, and that each protomer of a dimer can modify the reaction mechanism of its neighbor. Interaction of Na_o and K_o with transport sites of a nontransporting $\alpha\beta$ protomer might modify an Na-K exchange cycle carried out by a neighboring protomer in such a way that vanadate inhibition does or does not occur.

A method of testing this hypothesis is outlined in Fig. 3. Fig. 3 *A* shows $\alpha\beta$ protomers interacting in a dimer. When the K activation curve is measured in solutions containing Na, the activation curve in the absence of vanadate is approximately hyperbolic, but in its presence, the curve is biphasic. Fig. 3 *B* shows the results expected after reacting pumps with an inhibitor which combines randomly and irreversibly with $\alpha\beta$ protomers; binding of the inhibitor to one protomer of a pair does not affect binding of the inhibitor to the other, nor does inhibition of one subunit affect the function of the second. Fig. 3 *B* assumes 50% inhibition. In a quarter of the dimers neither protomer has bound inhibitor; the K activation curves in the presence and absence of vanadate are the same as those shown in Fig. 3 *A*. In half of the pumps, one protomer has combined with inhibitor. For these pumps, the K activation curve in the absence of vanadate is qualitatively the same as that shown in Fig. 3 *A*, but Na_o does not protect against vanadate inhibition since dimer interaction is not possible. In the remaining quarter of dimers, both subunits are inhibited. The expected behavior of the total enzyme preparation is obtained by

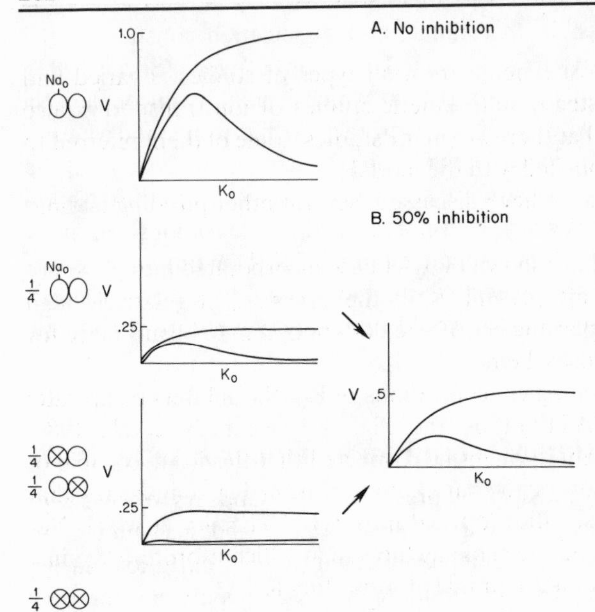

Figure 3. Effect of a random and irreversible inhibitor in a system exhibiting dimer interaction. For discussion, see text.

summing the contribution of each subgroup. Na_o protection against vanadate inhibition is much lower in the inhibited preparation than in the uninhibited preparation.

Inhibition of the Na,K pump by ouabain has the characteristics described above, and we have used it to evaluate the dimer interaction model in the experiment summarized in Table I. Cells were preincubated with ouabain at the concentration listed in the table, and the cells were then removed from contact with the inhibitor. The fraction of the pump rate which remained uninhibited at each ouabain concentration is listed in the table. The cells were then used to measure K influx in high Na solutions with and without 20 μM vanadate; measurements were made at a concentration of K_o low enough (2.26 mM) to permit significant protection by Na_o against vanadate inhibition and at a high K_o concentration (22.6 mM) where there is little protection. Table I lists the fraction of the pump rate which was not inhibited by vanadate at the two K_o concentrations. According to the dimer-interaction model, protection by Na_o against vanadate inhibition at 2.26 mM K_o should have decreased in the cells in which pumps were inhibited by ouabain; in fact, there was little change and the experiment does not support the model.

TABLE I

Ouabain Preincubation	Fraction Uninhibited	Fraction Uninhibited by Vanadate	
		K_o 2.24 *mM*	K_o 22.6 *mM*
0	—	0.72	0.15
0.5×10^{-8}M	0.75	0.75	0.18
1.0×10^{-8}M	0.59	0.74	0.19
4.0×10^{-8}M	0.24	0.75	0.20

Conclusion

The evidence for the Albers-Post scheme from all types of studies is varied and extensive. The evidence from steady-state kinetic studies of ion transport is also strong, but it must be admitted that there are many studies, some of them referred to above, which have not been reconciled with the model.

In addition to the examples we have discussed, several other puzzling findings should be mentioned.

(*a*) Yoda and Yoda (1988) have shown that ATPase incorporated into phospholipid vesicles is stimulated by cytoplasmic K in the presence of relatively high concentrations of ATP and cytoplasmic Na. There does not seem to be any place for such stimulation in the Albers-Post scheme.

(*b*) In the presence of saturating concentrations of K_o, Na_o inhibits pump rate; inhibition requires high ADP/ATP ratios and high K_c (Kennedy et al., 1986; Hoffman et al., 1985). At first sight, it might be suspected that this is an example of product inhibition, but evaluation of the rate equation from Fig. 1 indicates that product inhibition should be insignificant at saturating K_o. The site at which Na_o inhibits pump rate is unlikely to be the same as the site at which it protects against vanadate inhibition since protection is minimal at saturating K_o.

(*c*) Steady-state kinetic studies of the rate of ATP hydrolysis in the presence of Na and K suggest that ATP binds to an enzyme form which has both Na and K bound (Plesner and Plesner, 1985). Rossi and Garrahan (1989*a*, *b*) have also measured steady-state hydrolysis rate carried out by purified enzyme preparations and find serious discrepancies with the Albers-Post scheme.

When steady-state kinetic methods are used to evaluate the reaction mechanism of the Na,K pump, the results are decidedly mixed. Some findings are strongly confirmative, while some cannot be accommodated by the scheme. It may be that the discrepancies can be explained by alternative reaction sequences or by dimer interaction (Plesner [1987] has proposed a model involving dimer interaction to explain aspects of steady-state kinetic studies.) But it is hard to escape the suspicion that some significant aspect of the exchange mechanism is being missed.

Acknowledgments

This work was supported by grant AM 19185 from the United States Public Health Service.

References

Apell, H.-J., M. T. Nelson, M. M. Marcus, and P. Laüger. 1986. Effect of ATP, ADP and inorganic phosphate on the transport rate of the $Na,^{+}K^{+}$-pump. *Biochimica et Biophysica Acta.* 857:105–115.

Baker, P. F., M. P. Blaustein, R. D. Keynes, J. Manil, T. Shaw, and R. A. Steinhardt. 1969. The ouabain-sensitive fluxes of sodium and potassium in squid giant axons. *Journal of Physiology.* 200:459–496.

Beaugé, L. A. 1979. Vanadate-potassium interactions in the inhibition of the Na,K-ATPase. *In* Na,K-ATPase: Structure and Kinetics. J. C. Skou and J. G. Nørby, editors. Academic Press, Inc., London. 373-378.

Beaugé, L. A., J. D. Cavieres, I. M. Glynn, and J. J. Grantham. 1980. The effects of vanadate on the fluxes of sodium and potassium ions through the sodium pump. *Journal of Physiology.* 301:7–23.

Beaugé, L. A. and DiPolo. 1979. Sidedness of the ATP-Na^+-K^+ interactions with the Na^+ pump in squid axons. *Biochimica et Biophysica Acta.* 553:495–500.

Bond, G. H., and P. M. Hodgkins. 1982. Low affinity Na^+ sites on (Na^+ and K^+)-ATPase modulate inhibition of Na^+-ATPase activity by vanadate. *Biochimica et Biophysica Acta.* 687:310—314.

Bond, G. H., and P. M. Hudgins. 1975. Inhibition of human red cell Na,K-ATPase by magnesium and potassium. *Biochimica et Biophysica Acta.* 66:645–650.

Bond, G. H., and P. M. Hudgins. 1979. Kinetics of inhibition of Na,K-ATPase by Mg^{2+}, K^+ and vanadate. *Biochemical and Biophysical Research Communications.* 18:325–331.

Chipperfield, A. R., and R. Whittam. 1974. Evidence that ATP is hydrolysed in a one-step reaction of the sodium pump. *Proceedings of the Royal Society of London, Series A.* 187:269–280.

Chipperfield, A. R., and R. Whittam. 1976. The connection between the ion binding sites of the sodium pump. *Journal of Physiology.* 260:371–385.

Cleland, W. W. 1989. Enzyme kinetics revisited: a commentary. *Biochimica et Biophysica Acta.* 1000:209–212.

Cornelius, F., and J. C. Skou. 1988. The sided action of Na^+ on reconstituted shark Na^+/K^+-ATPase engaged in Na^+-Na^+ exchange accompanied by ATP hydrolysis. II. Transmembrane allosteric effects on Na affinity. *Biochimica et Biophysica Acta.* 944:223–232.

Eisner, D. A., and D. E. Richards. 1982. Inhibition of the sodium pump by inorganic phosphate in resealed red cell ghosts. *Journal of Physiology.* 326:1–10.

Eisner, D. A., and D. E. Richards. 1983. Stimulation and inhibition by ATP and orthophosphate of the potassium-potassium exchange in resealed red cell ghosts. *Journal of Physiology.* 335:495–506.

Garay, R. P., and P. J. Garrahan. 1973. The interaction of sodium and potassium with the sodium pump in red cells. *Journal of Physiology.* 231:297–325.

Garrahan, P. J., and R. P. Garay. 1976. The distinction between sequential and simultaneous models for sodium and potassium transport. *Current Topics in Membranes and Transport.* 8:29–97.

Garrahan, P. J., and I. M. Glynn. 1967*a*. The behavior of the sodium pump in red cells in the absence of external potassium. *Journal of Physiology.* 192:159–174.

Garrahan, P. J., and I. M. Glynn. 1967*b*. The sensitivity of the sodium pump to external sodium. *Journal of Physiology.* 192:175–188.

Garrahan, P. J., and I. M. Glynn. 1967*c*. Factors affecting the relative magnitudes of the sodium:potassium and sodium:sodium exchanges catalysed by the sodium pump. *Journal of Physiology.* 192:189–216.

Glynn, I. M. 1985. The Na^+,K^+-transporting adenosine triphosphatase. *In* The Enzymes of Biological Membranes. Vol. 3. 2nd ed. A. N. Martonosi, editor. Plenum Publishing Corp., New York. 35–114.

Glynn, I. M. 1988. The coupling of enzymatic steps to the translocation of sodium and potassium. *In* the Na^+,K^+-pump. Part A: Molecular Aspects. J. C. Skou, J. G. Nørby, A. B. Maunsbach, and M. Esmann, editors. Alan R. Liss, Inc., New York. 435–460.

Glynn, I. M., and S. J. D. Karlish. 1982. Conformation changes associated with K^+ transport by the Na-K^+-ATPase. *In* Membranes and Transport. Vol. 1. A. N. Martonosi, editor. Plenum Publishing Corp., New York. 529–536.

Glynn, I. M., and S. J. D. Karlish. 1990. Occluded cations in active transport. *Annual Review of Biochemistry.* 59:171–205.

Glynn, I. M., V. L. Lew, and U. Lüthi. 1970. Reversal of the potassium entry mechanism in red cells, with and without reversal of the entire pump cycle. *Journal of Physiology.* 207:371–391.

Hexum, T., F. E. Samson, Jr., and R. H. Himes. 1970. Kinetic studies of (Na^+-K^+-Mg^{24})-ATPase. *Biochimica et Biophysica Acta.* 212:322–331.

Hoffman, J. F., S. Dissing, B. G. Kennedy, G. Lunn, R. Marín, and M. Milanik. 1985. Ligands and the control of the red cell Na,K-pump. *In* The Sodium Pump. I. M. Glynn and J. C. Ellory, editors. The Company of Biologists, Ltd., Cambridge, England. 551–553.

Hoffman, P. G., and D. C. Tosteson. 1971. Active sodium and potassium transport in high potassium and low potassium sheep red cells. *Journal of General Physiology.* 58:438–466.

Karlish, S. J. D., W. R. Lieb, and W. D. Stein. 1982. Combined effects of ATP and phosphate on rubidium exchange mediated by Na-K-ATPase reconstituted into phospholipid vesicles. *Journal of Physiology.* 328:333–350.

Karlish, S. J. D., and W. D. Stein. 1985. Cation activation of the pig kidney sodium pump: transmembrane allosteric effects of sodium. *Journal of Physiology.* 359:119–149.

Kennedy, B. G., G. Lunn, and J. F. Hoffman. 1986. The effects of altering the ATP/ADP ratio on pump-mediated Na/K and Na/Na exchanges in resealed human red blood cell ghosts. *Journal of General Physiology.* 87:47–72.

Lee, K. H., and R. Blostein. 1980. Red cell sodium fluxes catalysed by the sodium pump in the absence of K^+ and ADP. *Nature.* 285:338–339.

Levitt, D. G. 1980. The mechanism of the sodium pump. *Biochimica et Biophysica Acta.* 604:321–345.

Lew, V. L., M. A. Hardy, and J. C. Ellory. 1973. The uncoupled extrusion of Na^+ through the Na^+ pump. *Biochimica et Biophysica Acta.* 323:251–266.

Nørby, J. G., I. Klodos, and N. O. Christiansen. 1983. Kinetics of Na-ATPase activity by the Na,K pump. *Journal of General Physiology.* 82:725-759.

Plesner, I. W. 1987. Application of the theory of enzyme subunit interactions to ATP-hydrolyzing enzymes. The case of Na,K ATPase. *Biophysical Journal.* 51:69–78.

Plesner, I. W., and L. Plesner. 1985. Kinetics of (Na^+ + K^+)-ATPase: analysis of the influence of Na^+ and K^+ by steady-state kinetics. *Biochimica et Biophysica Acta.* 818:235–250.

Reenstra, W. W., J. D. Bettencourt, and J. G. Forte. 1988. Kinetic studies of the gastric H,K-ATPase. Evidence for simultaneous binding of ATP and phosphate. *Journal of Biological Chemistry.* 263:19618–19625.

Robinson, J. D., C. A. Leach, R. L. Davis, and L. J. Robinson. 1986. Reaction sequences for (Na^+-K^+)-dependent hydrolytic activities: new quantitative kinetic models. *Biochimica et Biophysica Acta.* 872:204–304.

Rossi, R. C., and P. J. Garrahan. 1988*a*. Steady-state kinetic analysis of the Na^+/K^+-ATPase. The activation of ATP hydrolysis by cations. *Biochimica et Biophysica Acta.* 981:95–104.

Rossi, R. C., and P. J. Garrahan. 1989*b*. Steady-state kinetic analysis of the Na,K-ATPase. The inhibition by potassium and magnesium. *Biochimica et Biophysica Acta.* 981:105–114.

Sachs, J. R. 1967. Competitive effects of some cations on active potassium transport in the human red blood cell. *Journal of Clinical Investigation.* 46:1433–1441.

Sachs, J. R. 1974. Interaction of external K, Na and cardiotonic steroids with the Na-K pump of the human red blood cell. *Journal of General Physiology.* 63:123–143.

Sachs, J. R. 1977*a*. Kinetics of the inhibition of the Na-K pump by external sodium. *Journal of Physiology.* 264:449–470.

Sachs, J. R. 1977*b*. Kinetic evaluation of the Na,K pump reaction mechanism. *Journal of Physiology.* 273:489–514.

Sachs, J. R. 1979. A modified consecutive model for the Na,K-pump. *In* Na,K-ATPase: Structure and Kinetics. J. C. Skou and J. G. Nørby, editors. Academic Press, New York. 463–473.

Sachs, J. R. 1980. The order of release of sodium and addition of potassium in the sodium pump reaction mechanism. *Journal of Physiology.* 302:219–240.

Sachs, J. R. 1981. Mechanistic implications of the potassium-potassium exchange carried out by the sodium-potassium pump. *Journal of Physiology.* 316:263–273.

Sachs, J. R. 1986*a*. Potassium-potassium exchange as part of the overall reaction mechanism of the sodium pump of the human red blood cell. *Journal of Physiology.* 374:221–244.

Sachs, J. R. 1986*b*. The order of addition of sodium and release of potassium at the inside of the sodium pump of the human red cell. *Journal of Physiology.* 381:149–168.

Sachs, J. R. 1987. Inhibition of the Na,K pump by vanadate in high Na solutions. Modification of the reaction mechanism by external Na acting at a high-affinity sites. *Journal of General Physiology.* 90:291–320.

Sachs, J. R. 1988. Phosphate inhibition of the human red cell sodium pump: simultaneous binding of adenosine triphosphate and phosphate. *Journal of Physiology.* 400:545–574.

Simons, T. J. B. 1974. Potassium: potassium exchange catalysed by the sodium pump in human red cells. *Journal of Physiology.* 237:123–155.

Taniguchi, K., K. Suzuki, T. Sasaki, H. Shinokobe, and S. Iida. 1986. The change in light scattering following formation of ADP-sensitive phosphoenzyme in Na^+,K^+ -ATPase modified with N-[p-(2-benzimidazolyl)phenyl]maleimide. *Journal of Biological Chemistry.* 261:3272–3281.

Yoda, A., and S. Yoda. 1987. Two different phosphorylation-dephosphorylation cycles of Na,K-ATPase proteoliposomes accompanying Na^+ transport in the absence of K^+. *Journal of Biological Chemistry.* 262:110–115.

Yoda, A., and S. Yoda. 1988. Cytoplasmic K^+ effects on phosphoenzyme of Na,K-ATPase proteoliposomes and on the Na^+-pump activity. *Journal of Biological Chemistry.* 263:10320–10325.

Yoda, S., and A. Yoda. 1986. ADP- and K^+-sensitive phosphorylated intermediate of Na,K-ATPase. *Journal of Biological Chemistry.* 261:1147–1152.

Chapter 19

The Kinetics of Uncoupled Fluxes in Reconstituted Vesicles

Flemming Cornelius

Institute of Biophysics, University of Aarhus, Aarhus, Denmark

Introduction

When we speak of uncoupled Na^+-efflux it is the flux mode the Na pump performs in the absence of extracellular Na^+ and K^+. It is accompanied by ATP hydrolysis and can be inhibited by ouabain (Garrahan and Glynn, 1967). It can be rationalized in the context of the Albers-Post reaction scheme (Fig. 1) and comprises kinetically a subset of states in the overall Na^+/K^+-exchange reaction. A version of the Albers-Post scheme is shown in Fig. 1 (modified from Karlish et al., 1978). In the absence of both K^+ and Na^+ at the extracellular surface, the three cytoplasmic Na^+ ions taken up, occluded, and released to the extracellular side are followed by a return to an E_1A state by ATP binding, but without occluded cations (Fig. 2). In this truly uncoupled mode the normal 3 Na_{cyt}: $2K_{ext}$: 1 ATP stoichiometry is changed to 3 Na_{cyt}: 1 ATP; i.e., both modes are electrogenic and would contribute to the cell membrane potential.

In this article three questions will be addressed in studies on reconstituted Na,K-ATPase: the first concerns the stoichiometry of the uncoupled flux-mode; is the coupling ratio 3 Na_{cyt}: 1 ATP as expected, and is it fixed or variable? The second question relates to the electrogenicity of the uncoupled Na^+-efflux, especially whether this flux mode is truly uncoupled, or accompanied by co- or counterion transport. The third question that is considered is the interaction with this flux mode of extracellular cations.

To clarify these questions the hydrolytic activity of reconstituted shark Na,K-ATPase has been investigated under conditions of uncoupled Na^+-efflux and compared with its capacity for charge translocation at varied cytoplasmic Na^+ concentrations under different conditions of pH from $5.5 \leq pH \leq 8.0$.

Reconstitution has been applied because it has several advantages over studies in whole cells or vesicles. It particularly avoids interference from other membrane constituents. However, reconstitution has many pitfalls too; therefore, a brief characterization of the reconstituted system employed is given.

Reconstitution

The reconstitution procedure uses the nonionic detergent $C_{12}E_8$ to solubilize and purify shark rectal gland Na,K-ATPase (Cornelius and Skou, 1984; Cornelius, 1988). The specific hydrolytic activity of the solubilized preparation is 600–900 μmol/mg·h at 23°C, corresponding to a turnover number of 45–70 s^{-1}. By cosolubilization with pure phospholipids of the appropriate composition, followed by detergent elimination by adsorption to bio-beads, unilamellar vesicles with incorporated Na,K-ATPase, which retain the full hydrolytic activity and transport capacity are formed, so-called proteoliposomes.

In Fig. 3 a sketch of a shark liposome is shown with incorporated Na,K-ATPase drawn approximately to scale. For quantitative investigations several properties of the proteoliposome have to be established:

(*a*) To account for the sidedness of the Na,K-ATPase, the symmetry of incorporation must be known, in our case ~65% of the Na pumps are right-side out (r:o), 10% have the opposite orientation, inside out (i:o), and the remaining 25% have both sides exposed, nonoriented (n-o), as determined by functional assay (Cornelius, 1988).

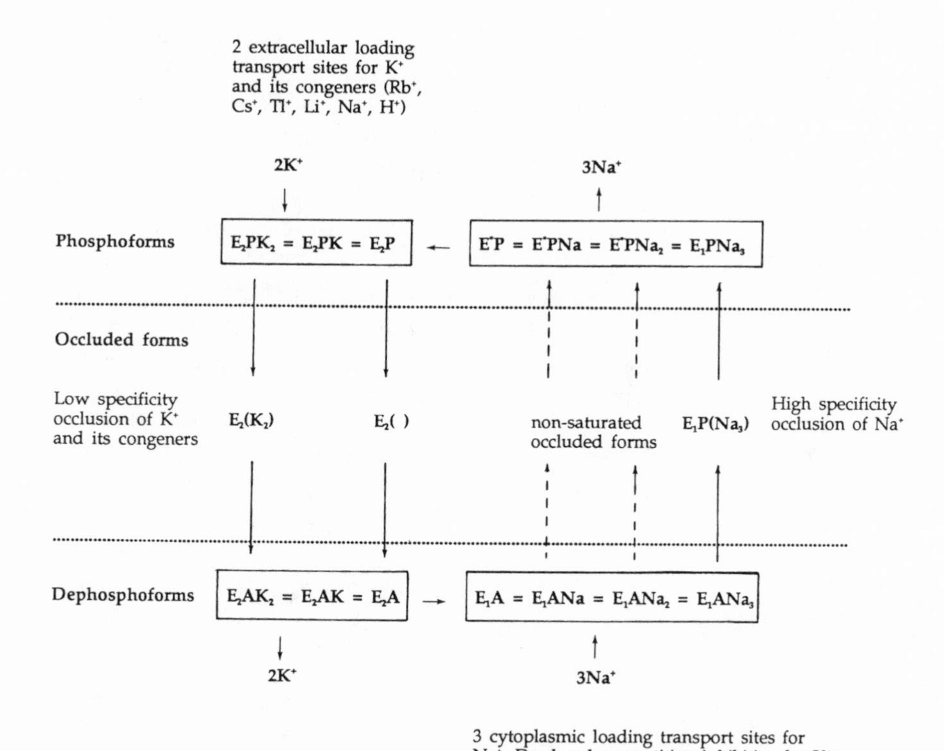

Figure 1. The Albers-Post scheme as modified by Karlish et al. (1978) and Cornelius and Skou (1985). The symbols E, EP, and EA indicate the unphosphorylated form, the phosphorylated form, and the form with ATP bound, respectively. Three major enzyme conformations are indicated by subscript 1 (the "sodium form"), by subscript 2 (the "potassium form"), and by asterisks. Enzyme species enclosed in boxes are assumed to be in rapid equilibrium. In this consecutive kinetic scheme 3 Na^+ ions are taken up at the cytoplasmic side and bound to an E_1A form of the enzyme. The enzyme is phosphorylated and the Na^+ ions are transferred from sites accessible from the cytoplasmic aspect into the membranous phase, where they are occluded without access from either side of the membrane. It is possible that enzyme species with less than three Na^+ bound can be phosphorylated, and occlude Na^+ (*broken arrows*). Few congeners for cytoplasmic Na^+ are known in the sense that they are occluded and translocated instead of Na_{ext}, possibly only H^+ and Li^+, whereas dead-end competition by cytoplasmic K^+ takes place to the E_1A form of the enzyme. By a spontaneous deocclusion the Na^+ ions are transferred to the extracellular side, where they are released in a stepwise reaction, the first of which is accompanied by a change in conformation from an E_1P form to an E^*P form (Yoda and Yoda, 1986). After another conformational change to the E_2P form, with high K^+ affinity, two extracellular K^+ ions are taken up and after dephosphorylation are occluded. A number of alkali cations, including Na^+, are known to be able to substitute for extracellular K^+ in this reaction. After low-affinity ATP binding the K^+ ions are deoccluded to the cytoplasmic side and the cycle is completed by a conformational change to the high-affinity sodium form E_1A. The dephosphorylation of E_2P without bound cations is believed to represent the pathway leading to the so-called uncoupled Na^+ efflux.

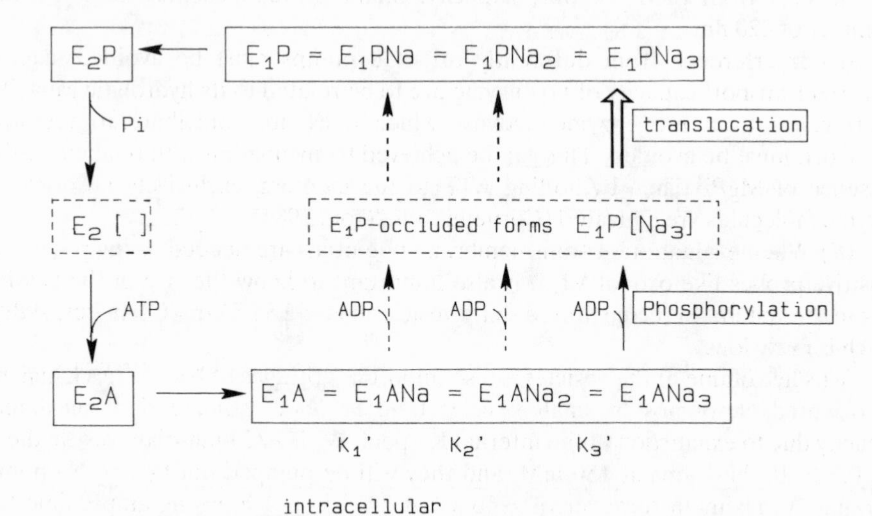

Figure 2. The subset of kinetic states in the over all Albers-Post scheme, which comprises the uncoupled Na^+ efflux.

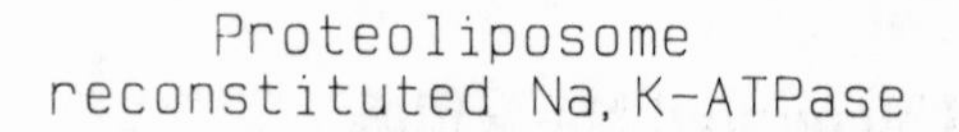

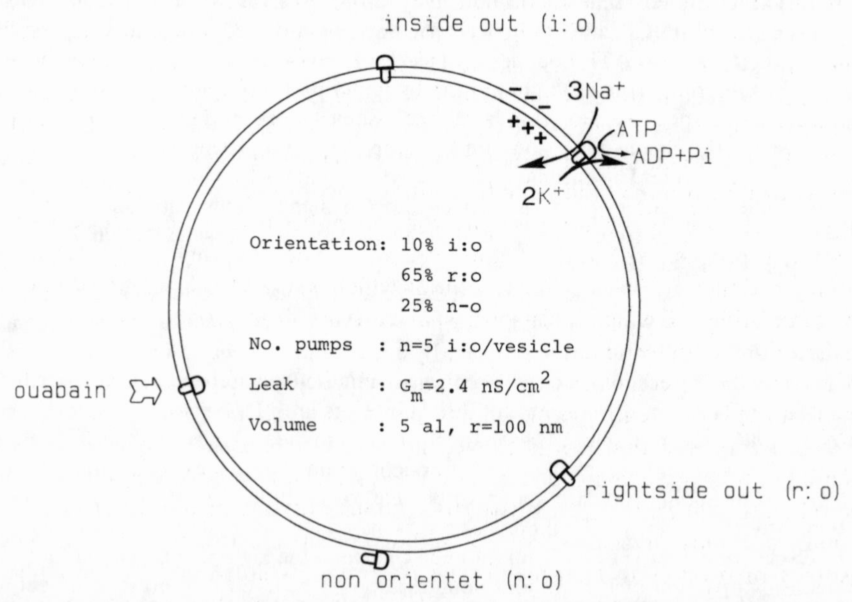

Figure 3. Sketch of shark proteoliposome, showing the three possible orientations, right-side out, inside-out, and nonoriented. The inside-out but not right-side out enzyme can be activated by ATP in the medium. The right-side out and nonoriented enzymes, but not inside-out enzyme, can be inhibited by ouabain in the medium.

(*b*) The average number of pumps per vesicle can be calculated from the amount of protein per mass of phospholipid, if the average size of the vesicles are known, e.g., from their trapping capacity. Shark proteoliposomes have a mean diameter of 220 nm.

(*c*) Interference from differently oriented pumps must be avoided, e.g., if vectorial transport capacity of i:o enzyme are to be related to its hydrolytic capacity, interference from n-o enzyme activity, which does not contribute to vectorial transport, must be avoided. This can be achieved by incubation with ouabain in the presence of MgPi. Then by adding ATP to the medium, exclusively i:o oriented enzyme molecules are activated (Cornelius and Skou, 1984).

(*d*) When evaluation of transmembrane potentials are needed, using potential-sensitive probes like oxonol VI, it is also important to know the size of the passive leak in the system. In reconstituted shark vesicles it is ~ 2.5 nS/cm^2 (Cornelius, 1989), which is fairly low.

(*e*) The volume of the vesicles is also important. Studies of Na^+/K^+ exchange by i:o oriented Na-pumps in small vesicles tend to underestimate their maximum capacity due to exhaustion of the internal K^+-pool. With a 220-nm-diam vesicle there are 0.5×10^6 Na^+ ions at 150 mM, and they will be pumped out by one Na pump working at maximum turnover of ~ 50 s^{-1} at 23°C in ~ 1 h, giving ample time for initial measurements. However, a vesicle of o.d. 50 nm contains only 3,500 ions at the same concentration, giving only ~ 1 min to empty. In such a vesicle only 10 turnovers by a pump changes the internal concentration by more than 1 mM.

Stoichiometry of Uncoupled Na^+ efflux

As mentioned above, the uncoupled Na^+ efflux should be electrogenic with a coupling ratio of 3 Na_{cyt}:ATP (Fig. 2). However, in red cell ghosts this flux mode is apparently electroneutral (Dissing and Hoffman, 1983, 1990) and this also seems to be the case in reconstituted kidney Na,K-ATPase, at least at pH < 7.5, whereas at higher pH it becomes electrogenic (Karlish et al., 1988; Goldshleger et al., 1990).

To measure coupling stoichiometry of exchange reactions one can measure hydrolytic capacity and at the same time either tracer fluxes, or, alternatively, rate of charge translocation. The latter is more convenient because it can be achieved using fluorescent probes, which does not necessitate the separation of vesicles and medium, and therefore has a time resolution which is superior. The method does not, however, distinguish which ionic species are carrying the current; only its magnitude and direction are determined.

For reconstituted i:o oriented Na pumps, which upon activation take up external (cytoplasmic) Na^+, creating a membrane potential inside positive, it is most convenient to use probes that are negative and accumulate inside the vesicles as the potential develops. Such a probe is oxonol VI, which has a large potential-sensitive spectral shift giving great sensitivity. Its response time is of the order of 300 ms, adequate for most measurements of sodium pump electrogenicity. The specific sensitivity of oxonols to transmembrane potentials is illustrated by the ability of lipid-permeable anions like SCN^- or TPB^- to dissipate inside positive potentials, whereas permeable cations like TBA^+ and TPP^+ are less effective or ineffective due to the opposing gradient of electrical potential (cf. Grinius et al., 1970).

In Fig. 4 the fluorescence of oxonol VI increases upon activation of uncoupled

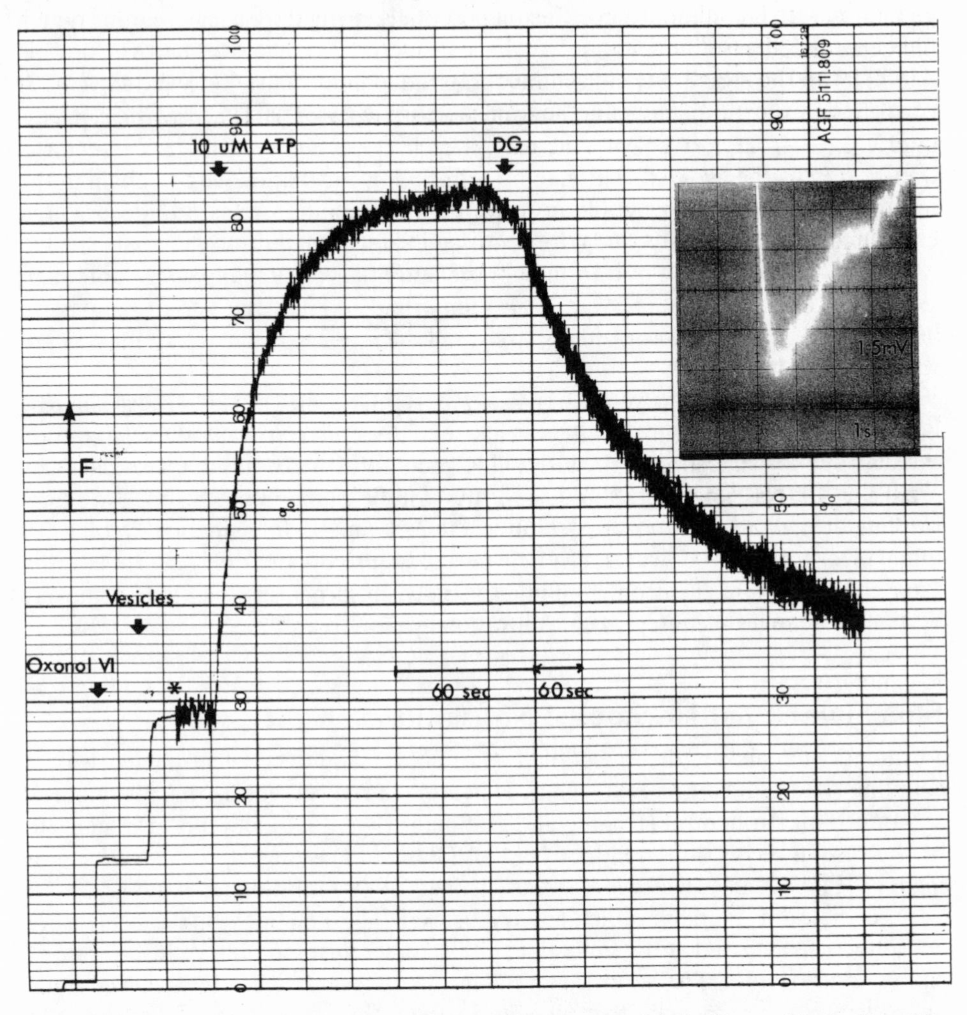

Figure 4. The oxonol VI response to transmembrane potential (inside positive) elicited by activation of uncoupled Na^+ efflux (uptake in proteoliposomes) on inside-out oriented enzyme with ATP (10 μM). 100 μl proteoliposomes containing sucrose (260 mM), Mg^{2+} (2 mM), and histidine (30 mM, pH 7.0) are added to a cuvette with 2.8 ml of a medium containing 130 mM Na^+ instead of sucrose, and 540 nM oxonol VI. The temperature is 23°C. Excitation and emission wavelengths are 580 and 660 nm, respectively. The steady state membrane potential is 240 mV. Addition of digitoxigenin (DG) arrests the pump and the potential declines with a time constant of 420 s, due to the passive equilibration of ions. From the initial rate of rise in membrane potential (dV/dt) the turnover rate of charge transfer (v_o) can be calculated as described in the text. The inset shows a stopped-flow experiment demonstrating early events in fluorescence change after rapid mixing (the downstroke in trace) of proteoliposomes (syringe I) with ATP (syringe II). The oscilloscope trace shows the initial 2 s after mixing, with the oscilloscope sensitivity on 1.5 mV/div. (Taken from Cornelius, 1989.)

Na^+ efflux with ATP in reconstituted Na,K-ATPase. The vesicles contain sucrose and the medium contained 130 mM Na^+ and Mg^{2+}. The fluorescence signal can be calibrated with K^+-valinomycin diffusion potentials giving the transmembrane potential. Because the sodium pump can be considered as an electromotive force in parallel with the membrane capacitance (C_m) and conductance (G_m), the current (I) generated by the pump inducing the transmembrane potential (V_m) is:

$$I = AC_m(dV_m/dt)$$

where A is the surface area of the proteoliposomes.

Initially this current is carried exclusively by the pump because the leak current is zero; therefore, the turnover of net charges, ν_o, for one of n pumps is:

$$\nu_o = AC_m(dV_m/dt)/ne$$

where e is the elementary charge.

Because n is proportional to the surface area of the vesicles, ν_o becomes independent of vesicle size. The membrane conductance, G_m, can be calculated from the time constant (τ) of voltage decay when the pump is inhibited because $\tau = C_m/G_m$. With shark proteoliposomes the time constant is 420 s, giving a specific membrane conductance of only 2.4 nS/cm^2, assuming a bilayer capacitance of 1 $\mu F/cm^2$. Due to the very low G_m, very high transmembrane potentials of up to 250 mV can be developed by the reconstituted pump.

In the inset to Fig. 4, the voltage response is shown for a stopped-flow experiment performed to investigate whether the electrogenic response is elicited by incoming Na^+, turning on electrogenic Na^+/Na^+ exchange (Cornelius and Skou, 1985). However, on a time scale within one turnover of the sodium pump, where build up of internal Na^+ is negligible, there is still no indication of a delay in the fluorescence response, strongly indicating that the uncoupled Na^+ flux mode is electrogenic. From the initial slope of the fluorescence response, an initial turnover of net charges can be calculated to be $8.8 \pm 0.2\ s^{-1}$ (mean ± SE, $n = 10$).

Measurements in parallel experiments of the rate of Pi-liberation in MgPi-ouabain preincubated proteoliposomes reveal a hydrolytic activity of 42.0 ± 4.1 μmol/mg·h (mean ± SE, $n = 10$). With 3.7 nmol sites/mg protein this corresponds to a catalytic turnover of $3.2 \pm 0.3\ s^{-1}$, ~5% of turnover during Na^+/K^+ exchange. By comparing the measured turnover of net charge and catalysis a stoichiometry of 2.8 ± 0.1 is found, in agreement with 3 Na^+ translocated per ATP split (Cornelius, 1989, 1990).

The Electrogenicity of Uncoupled Na^+ Efflux

There are, therefore, apparent discrepancies between the characteristics of uncoupled efflux in red blood cells, reconstituted kidney, and shark Na,K-ATPase. In red cells the efflux is electroneutral, apparently coupled to cotransport of anions. In kidney it is also electroneutral at pH < 7.5, apparently by countertransport of H^+, and in shark it is electrogenic and probably truly uncoupled. Do these differences reflect the existence of three different kinds of uncoupled efflux modes, or could the difference be only apparent? A crucial question to be answered in this respect is whether the accompanying ions in red cells and kidney are transported via binding sites on the Na pump or whether they are passively distributing according to the membrane potential via another pathway. If the first possibility is correct, then we should expect occlusion and activation of hydrolytic activity by cytoplasmic anions in red cells, and by extracellular H^+ in kidney. In a recent paper Dissing and Hoffman (1990) did not find anion occlusion accompanying uncoupled Na^+ efflux. The possible occlusion of H^+ has not been investigated, but in inside-out vesicles from red cells (IOV) there seems to be rather firm evidence that H^+ is transported on the pump in exchange for Na^+ and activates hydrolysis (Polvani and Blostein, 1988).

Very tight preparations are needed to detect the electrogenicity of the pump to decide whether the coupling is only electrical and the co- and counterions move down the electrical gradient created by the pump or their transport is primarily on the pump. It is questionable whether the red cell ghosts and kidney proteoliposomes are tight enough. In SO_4^{2-} buffer the red cell ghosts of Dissing and Hoffman (1990) equilibrate upon ouabain addition with a time constant of ~40 s, giving a conductance of ~17 nS/cm^2; in IOV it is between 50 and 150 nS/cm^2 (Polvani and Blostein, 1989), and in reconstituted kidney vesicles it is up to 50 nS/cm^2 (Goldshleger et al., 1990), compared with shark proteoliposomes where it is 2.4 nS/cm^2. The high passive leaks in the first two preparations are reflected in the very low steady state potentials they can support, 1–5 mV for Na^+/K^+ exchange in red blood cells and IOV, and 1–4 mV for Na^+/Na^+ exchange and uncoupled Na^+ efflux (high pH) in kidney proteoliposomes compared with the 200–250 mV in shark proteoliposomes during Na^+/K^+ exchange, Na^+/Na^+ exchange, or uncoupled Na^+ efflux. It is therefore possible that the electroneutral responses found in red cells and kidney proteoliposomes during uncoupled Na^+ efflux are the result of electrical short circuiting by ions that equilibrate passively fast enough to keep pace with the very low rate of pumping during uncoupled Na^+ efflux. Another problem with red cells is the existence of the very powerful anion pathway of capnophorin (band 3), which can still be significant even after treatment with DIDS.

If we take a closer look at the Albers-Post scheme for uncoupled Na^+ efflux, it is normally assumed that only enzyme species saturated with 3 Na^+ are phosphorylated and occluded, resulting in the translocation of 3 Na^+ from the cytoplasmic to the extracellular aspect of the pump. The pump then reverses to the E_1A state without cations bound. Two questions which will influence the degree of electrogenicity arise. Can enzyme species with fewer than 3 Na_{cyt} turn over and hydrolyze ATP; and can other ions such as H^+ participate in the reaction, acting at the extracellular side by binding to E_2P as a K^+ congener (cf. Fig. 2)?

Variable Stoichiometry or Variable Exchange Partners

The 3 Na_{cyt}:2 K_{ext}:1 ATP coupling ratio during Na^+/K^+ exchange seems to be remarkably constant, although indications that H^+ may be a counterion in uncoupled Na^+ efflux are present from studies in both reconstituted kidney vesicles and red cells (Hara and Nakao, 1986; Polvani and Blostein, 1988). Recent observations on inside-out vesicles from red cells indicate that at low Na_{cyt} the coupling ratio decreases (Blostein, 1983; Polvani and Blostein, 1989). The evidence for a reduced coupling at low Na_{cyt} deduced from measurements of membrane potentials is, however, made uncertain by the fact that H^+ conceivably can substitute for both Na_{cyt} and K_{ext} (Polvani and Blostein, 1988). Because, as mentioned above, the measurements in red cells are difficult to interpret due to a large leak conductance (up to 150

nS/cm^2), and resulting small membrane potential (~ 1 mV) and possible interference from residual anion-exchange activity, similar studies with reconstituted Na,K-ATPase would be useful for comparison.

If the stoichiometry changes or co- or countertransport take place in uncoupled Na^+ efflux during low Na_{cyt} or low pH it should result in a changed net charge:ATP stoichiometry. This has been investigated using shark proteoliposomes during uncoupled Na^+ efflux at saturating cytoplasmic Na^+, by comparing the rate of net charge translocation using oxonol VI with the initial rate of ATP hydrolysis at different pH in the range of $5.5 \leq pH \leq 8.0$ (Cornelius, 1990). In Fig. 5 the fluorescence response at pH 7.0 is compared with the two experimental extremes, pH 5.5 and pH 8.0, and in Fig. 6 the calculated stoichiometries of net charge:ATP are depicted. The net charge:ATP stoichiometry is 3:1 for pH $\geq$ 7, whereas it decreases to 1.5:1 for pH $<$ 7. This latter effect could be due to: (*a*) E_1A enzyme species with <3 Na_{cyt} participate in the overall reaction, (*b*) some anions are cotransported or cations countertransported,

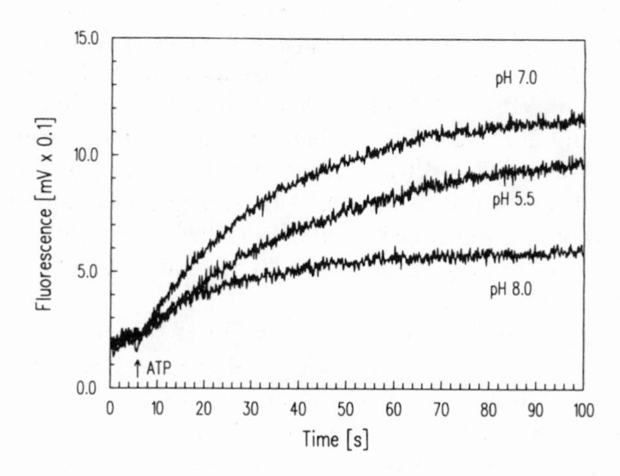

Figure 5. Development of transmembrane potentials as detected with the fluorescence of oxonol VI of sodium pump engaged in uncoupled Na^+ efflux (uptake in proteoliposomes) at three different pH values ($pH_{in} = pH_{out}$). 100 μM ATP is added as indicated; otherwise the conditions are described in the legend to Fig. 4. The initial rates of development of transmembrane potentials as calculated from the slopes of the three recordings are 1.95 mV/s (pH 5.5), 4.7 mV/s (pH 7.0), and 1.54 mV/s (pH 8.0). (Taken from Cornelius, 1990.)

or (*c*) a dissociation in the coupling of Na_{cyt} translocation and ATP hydrolysis occurs. Because in these experiments the Na_{cyt} concentration is saturating and constant, the most likely explanation is that extracellular H^+ at low pH are countertransported in exchange for cytoplasmic Na^+. Circumstantial support for this is that a slight reactivation in hydrolytic activity associated with uncoupled Na^+ efflux is observed at the lowest pH tested (Fig. 6); this is an effect of extracellular H^+. Furthermore, H^+ countertransport has previously been demonstrated in reconstituted kidney Na,K-ATPase (Forgac and Chin, 1982; Hara and Nakao, 1986) and in inside-out red-cell vesicles (Polvani and Blostein, 1988), where hydrolytic activation by H^+ in the absence of alkali cations was also demonstrated. H^+ activation of hydrolytic activity in the absence of Na^+ as well as K^+ (H^+/H^+ exchange), which we have recently measured to be ~ 2.5 μmol/mg·h at pH 5.5, corresponding to $\sim 5\%$ of uncoupled Na^+ efflux (unpublished results) provides more direct evidence. This very low activity presumably indicates the low efficiency of H^+ as a Na_{cyt} congener.

The absence of H^+ countertransport at pH ≥ 7 is probably the result of the very low H^+ concentration inside the proteoliposomes, at pH 7 there is less than one free proton present per proteoliposome, increasing to ~8 at pH 5.5, the majority being bound to buffer anions. A similar explanation for a shift from electroneutral uncoupled Na^+ efflux at pH < 7.5 to electrogenic at pH > 7.5 was given by Goldshleger et al. (1990) in reconstituted kidney Na,K-ATPase. The reported difference in electrogenicity between shark and kidney enzyme at different pH values could be due to differences in the pK's of groups close to Na^+ binding sites.

To investigate if the concentration of Na_{cyt} influenced the stoichiometries, activation curves for cytoplasmic Na^+ for either hydrolytic activity or net charge translocation were measured at pH values in the same range of 5.5 ≤ pH ≤ 8.0. If a fixed stoichiometry exists at all Na_{cyt}, then the two kinds of activation curves should superimpose at all Na_{cyt} concentrations. Results from pH 5.5, 7.0, and 8.0 are shown in Fig. 7. At pH 7.0 (*B*) a steeply increasing *S*-shaped activation curve with $K_{0.5}$ for Na_{cyt} of 3–4 mM is obtained for both ATP hydrolysis and net charge translocation.

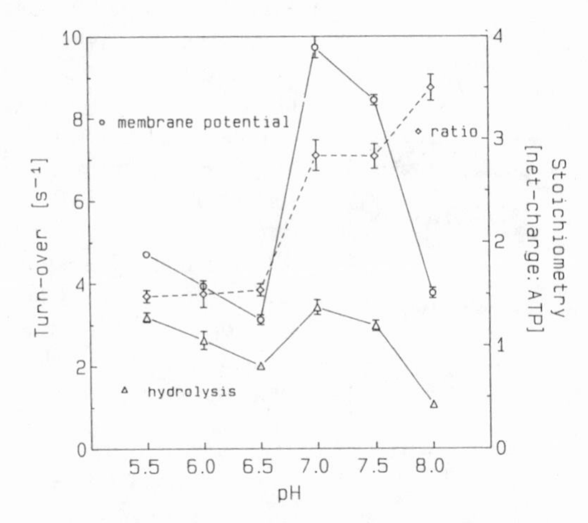

Figure 6. Comparison at different pH values of turnover of charge translocation (*upper full curve*) and hydrolysis (*lower full curve*) during conditions of uncoupled Na^+ efflux. The Na^+ concentration in the medium is in all cases saturating (130 mM). The dashed curve indicates the calculated stoichiometry of net charge translocated per ATP molecule split.

Calculation of the net charge/ATP stoichiometry gives 3:1 for all Na_{cyt} studied, in accordance with a 3 Na_{cyt}/ATP stoichiometry, i.e., the stoichiometry is independent of Na_{cyt} concentration at pH 7.0. This also holds at pH 7.5 (not shown). However, outside 7.0 ≤ pH ≤ 7.5 the net charge/ATP stoichiometries found at *saturating* Na_{cyt} (Fig. 6) are decreased at decreasing Na_{cyt}. In Fig. 7 results are shown in this pH range. At both pH 5.5 and 8.0 a shift in the curves of rate of net charge translocation relative to rate of hydrolysis vs. Na_{cyt} is observed (*A* and *C*). At pH 5.5 the stoichiometry decreases with decreasing Na_{cyt} from the 1.5:1 obtained at saturating Na_{cyt} toward 1:1 net charges/ATP, whereas at pH 8.0 it decreases from the 3:1 found at saturating Na_{cyt} toward 2:1, when the Na_{cyt} concentration decreases (Cornelius, 1990).

Because the indicated shifts in the activation curves comparing net charge translocation with hydrolysis are taking place at fixed pH, the most straightforward explanation is that at nonoptimal pH (outside 7.0–7.5) the capability for enzyme species with <3 Na_{cyt} to be phosphorylated and translocate Na^+ is enhanced, perhaps

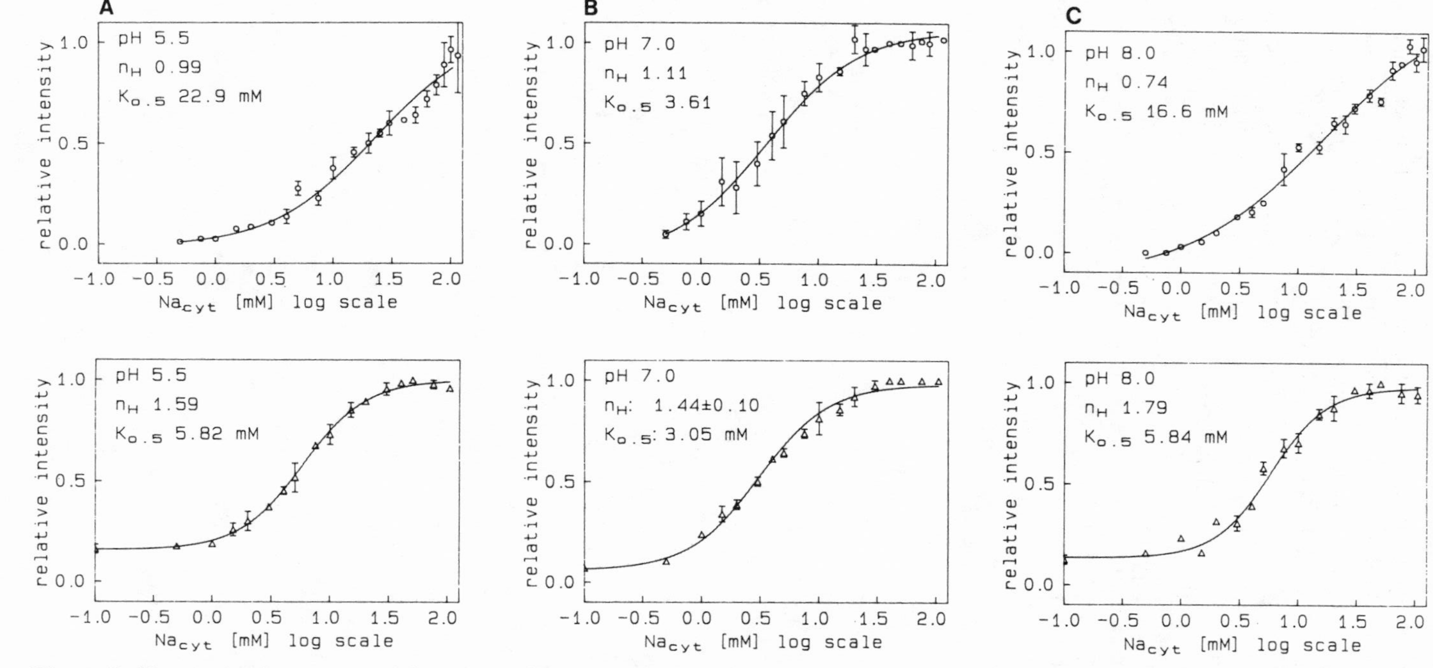

Figure 7. Na_{cyt} activation curves of (*upper panels*) turnover of charge translocation and (*lower panels*) hydrolysis during conditions of uncoupled Na^+ efflux at pH 7.0 and at the two pH extremes tested (pH 5.5 and 8.0). Note that the abscissa is logarithmic. Each curve is the regression line calculated according to the Hill equation and the fitted parameters (n_H, the Hill coefficient and $K_{0.5}$, the half-maximum activation constant) are given in each panel. Each point is the mean of four experiments with SE indicated. (Taken from Cornelius, 1990.)

TABLE I
Characterization of Uncoupled Na^+-Flux

Preparation	Electrogenicity	Stoichiometry Na^+/ATP	Co/counter ions	Max rate V_{max}	Leak conductance	Na_{ext} activation
					nS/cm²	
Shark PL	Electrogenic $5.5 \leq pH \leq 8.0$	3:1 variable	H^+?	5%	2	Sigmoid
Kidney PL	Electrogenic, pH > 7 Neutral, pH ≤ 7	3:1	H^+?	1%	50	Sigmoid
RBC	Neutral, pH 7.4	3:1	Anions	15%	15	Biphasic
IOV	?	1.6:1 Variable	H^+	10%	50–150	Biphasic

through ionization of some groups with pK values in regions near the Na_{cyt} binding sites causing a shift in the Na^+ affinity (Cornelius, 1990). In IOV of red cells, a decreased stoichiometry at low Na_{cyt} concentration has also been observed (Blostein, 1983; Polvani and Blostein, 1989). The apparent reversal of polarity of membrane potential observed with IOV at $Na_{cyt} < 0.4$ mM has not been confirmed by measurements of oxonol VI signals in shark proteoliposomes.

Table I shows a collection of data comparing the uncoupled Na^+ fluxes characterized in various preparations. Some differences exist. However, they appear to be variations of the same theme, rather than results of different categories of uncoupled fluxes, the major differences being between reconstituted preparations and RBC in their sensitivity to extracellular Na^+.

The final issue addressed in this article involves the interaction of extracellular cations with the uncoupled Na^+ efflux. In red blood cells there is a dip in the activation curve for extracellular Na^+ at concentrations near 5 mM (Glynn and Karlish, 1976), which might mean that at low extracellular Na^+ concentrations

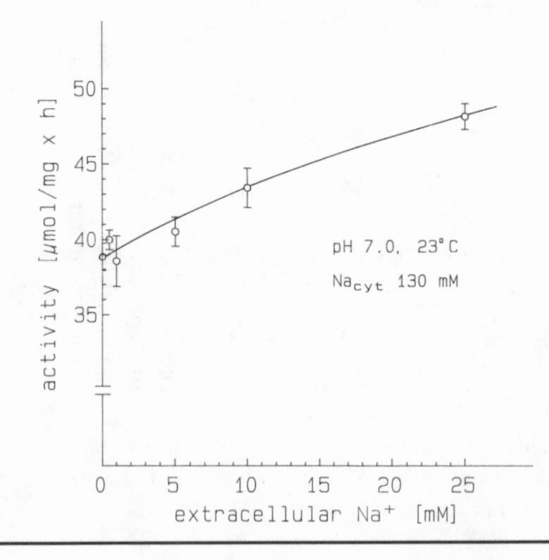

Figure 8. Activation by extracellular Na^+ of ATPase activity of reconstituted inside-out Na,K-ATPase. The proteoliposomes made in sucrose (260 mM), Mg^{2+} (2 mM), and histidine (30 mM, pH 7.0) are equilibrated overnight at 4°C to obtain indicated intraliposomal (extracellular) concentrations of Na^+. Points are means of three experiments with SE indicated.

uncoupled Na^+ efflux proceeds faster than Na^+/Na^+ exchange accompanied by ATP hydrolysis. Such biphasic activation by extracellular Na^+ is not found in reconstituted systems, where Na^+/Na^+ exchange at all Na_{ext} concentrations exceeds uncoupled Na^+ efflux (see Fig. 8). Sigmoid activation rather than biphasic activation by extracellular Na^+ has been found both for reconstituted kidney enzyme (Karlish and Stein, 1985) and for reconstituted shark enzyme (Cornelius and Skou, 1988). The absence of biphasic activation curves for extracellular Na^+ suggest a lower probability in reconstituted preparations for E_2P to dephosphorylate without Na^+, giving uncoupled efflux, than when Na^+ is bound extracellularly, giving Na^+/Na^+ exchange, whereas in red cells the rate constant for dephosphorylation of "empty" E_2P giving uncoupled Na^+ efflux is presumably larger. As seen from Table I, this is also reflected in a higher maximal rate of uncoupled Na^+ efflux in RBC than is found in reconstituted Na,K-ATPase.

The maximum catalytic activities given in Table I are the ones found during supposedly extracellular alkali cation-free conditions, whereas in the presence of extracellular alkali cations, especially K^+, any contribution from uncoupled Na^+ efflux will be almost eliminated due to the high affinity of E_2P for extracellular Na^+ and K^+, which will lead to a redistribution within the E_2P pool with very little "empty" E_2P.

Acknowledgments

T. Feddema and H. Zakarias are gratefully acknowledged for technical assistance and Professor J. C. Skou is gratefully acknowledged for helpful discussions.

The Danish Medical Research Council, Novo Foundation, and The Danish Biotechnology Centre for Biomembranes are gratefully acknowledged for financial support.

References

Blostein, R. 1983. The influence of cytoplasmic sodium concentration on the stoichiometry of the sodium pump. *Journal of Biological Chemistry.* 258:12228–12232.

Cornelius, F. 1988. Incorporation of $C_{12}E_8$-solubilized Na^+,K^+-ATPase into liposomes: determination of sidedness and orientation. *Methods in Enzymology.* 156:156–167.

Cornelius, F. 1989. Uncoupled Na^+-efflux on reconstituted shark Na,K-ATPase is electrogenic. *Biochemical and Biophysical Research Communications.* 160:801–807.

Cornelius, F. 1990. Variable stoichiometry in reconstituted shark Na,K-ATPase engaged in uncoupled efflux. *Biochimica et Biophysica Acta.* 1026:147–152.

Cornelius, F., and J. C. Skou. 1984. Reconstitution of $(Na^+ + K^+)$-ATPase into phospholipid vesicles with full recovery of its specific activity. *Biochimica et Biophysica Acta.* 772:357–373.

Cornelius, F., and J. C. Skou. 1985. Na^+-Na^+ exchange mediated by $(Na^+ + K^+)$-ATPase reconstituted into liposomes. Evaluation of pump stoichiometry and response to ATP and ADP. *Biochimica et Biophysica Acta.* 818:211–221.

Cornelius, F., and J. C. Skou. 1988. The sided action of Na^+ on reconstituted shark Na^+/K^+-ATP-ase engaged in Na^+-Na^+ exchange accompanied by ATP hydrolysis. II. Transmembrane allosteric effects on Na^+ affinity. *Biochimica et Biophysica Acta.* 944:223–232.

Dissing, S., and J. F. Hoffmann. 1983. Anion-coupled Na efflux mediated by the Na/K pump in human red blood cells. *Current Topics in Membranes and Transport.* 19:693–695.

Dissing, S., and J. F. Hoffmann. 1990. Anion-coupled Na efflux mediated by the human red blood cell Na/K pump. *Journal of General Physiology.* 96:167–193.

Forgac, M., and G. Chin. 1982. Na^+ transport by the (Na^+)-stimulated adenosine triphosphatase. *Journal of Biological Chemistry.* 257:5652–5655.

Garrahan, P. J., and I. M. Glynn. 1967. The sensitivity of the sodium pump to external Na. *Journal of Physiology.* 192:175–188.

Glynn, I. M., and S. J. D. Karlish. 1976. ATP hydrolysis associated with an uncoupled sodium flux through the sodium pump: evidence for allosteric effects of intracellular ATP and extracellular sodium. *Journal of Physiology.* 256:465–496.

Goldshleger, R., Y. Shahak, and S. J. D. Karlish. 1990. Electrogenic and electroneutral transport modes of renal Na/K ATPase reconstituted into proteoliposomes. *Journal of Membrane Biology.* 113:139–154.

Grinius, L. L., A. A. Jasaitis, Y. U. P. Kadziaukas, E. A. Liberman, V. P. Skulachev, V. P. Topali, L. M. Tsofina, and M. A. Vladimirova. 1970. Conversion of biomembrane-produced energy into electric form. *Biochimica et Biophysica Acta.* 216:1–12.

Hara, Y., and M. Nakao. 1986. ATP-dependent proton uptake by proteoliposomes reconstituted with purified Na^+,K^+-ATPase. *Journal of Biological Chemistry.* 27:12655–12658.

Karlish, S. J. D., R. Goldshleger, Y. Shahak, and A. Rephaeli. 1988. Charge transfer by the Na/K pump. *In* The Na^+,K^+-Pump, Part A: Molecular Aspects. J. C. Skou, J. G. Nørby, A. B. Maunsbach, and M. Esmann, editors. Alan R. Liss, Inc., New York. 519–524.

Karlish, S. J. D., and W. D. Stein. 1985. Cation activation of the pig kidney sodium pump: transmembrane allosteric effects of sodium. *Journal of Physiology.* 359:119–149.

Karlish, S. J. D., D. W. Yates, and L. M. Glynn. 1978. Conformational transitions between Na^+-bound and K^+-bound forms of (Na^+ + K^+)-ATPase, studied with formycin nucleotides. *Biochimica et Biophysica Acta.* 525:252–264.

Polvani, C., and R. Blostein. 1988. Protons as substitutes for sodium and potassium in the sodium pump reaction. *Journal of Biological Chemistry.* 32:16757–16763.

Polvani, C., and R. Blostein. 1989. Effects of cytoplasmic sodium concentration on the electrogenicity of the sodium pump. *Journal of Biological Chemistry.* 26:15182–15185.

Yoda, S., and A. Yoda. 1986. ADP- and K^+-sensitive phosphorylated intermediate of Na,K-ATPase. *Journal of Biological Chemistry.* 261:1147–1152.

Chapter 20

The Red Cell Na/K Pump Reaction Cycle: Ligand Signal Sites and Na Coupled Anion Transport

Joseph F. Hoffman, Reinaldo Marín, Harm H. Bodemann, Thomas J. Callahan, and Mark Milanick

Department of Cellular and Molecular Physiology, Yale University School of Medicine, New Haven, Connecticut 06510

This article summarizes work concerned with identifying types of internal (i) and external (o) Na and K sites on the red cell Na/K pump that not only function to modulate the pump's activity but also define the kind of transport reaction that the pump will carry out. These latter types of sites are referred to as signal sites. Thus modulatory (modifying) sites act qualitatively to affect the pump's activity, whereas signal sites serve to categorize various modes of the pump's operation. As described below, we attempt to sort out Na and K modulatory and signal sites and, where possible, to distinguish these types of sites from transport sites on the pump. Clearly the study of sided preparations is key to this type of analysis because otherwise the separate effects of alterations in the composition of the inside and outside phases would not be possible nor could they be properly evaluated. The various studies have been carried out with red cells as either intact cells, resealed ghosts, or inside-outside vesicles (IOVs).

It is well known that the pump's exchange rate of Na_i for K_o can be modulated by the competitive interactions of Na and K on the same side of the membrane without any appreciable effects on the pump's V_{max}. Thus Na_o can act to lower the apparent affinity of the Na/K pump for K_o (Post et al., 1960), while K_i can act to lower the pump's apparent affinity for Na_i (Hoffman, 1962; Garay and Garrahan, 1973; Knight and Welt, 1974). These types of interactions between Na and K on the same side of the membrane presumably occur on the same form of the phosphointermediate of the pump, recognizing that the form of the intermediate with which Na_i reacts is different from the form associated with K_o. On the other hand, it is not always clear whether they interact competitively at the same sites or allosterically at separate sites.

A variation on the above theme of interactions between Na and K is also seen (Sachs, 1986) during Na_i/Na_o and K_i/K_o types of exchange that represent partial modes of pump operation (see Glynn, 1985). Again the interactions are evidently competitive and it is also likely, but not proven, that the intermediates and sites that underlie the interaction of K_i or Na_o during Na_i/K_o exchange are the same as when the pump is operating in either of these partial modes.

In addition to Na and K interactions on the same side of the membrane, there are also cross-membrane interactions that can involve either Na or K on opposite sides of the membrane. One such instance is that the sigmoidicity of the K_o activation of K influx through the pump as influenced by Na_o (Post et al., 1960) may be dependent on the presence of K_i (Karlish and Stein, 1985). Another is that K_o can alter the pump's apparent affinity for inside ATP and, in a reciprocal manner, the pump's apparent affinity for K_o can be altered by ATP (Eisner and Richards, 1981). Two other types of cross-membrane interactions that modulate pump activity will be considered below. The first concerns the effects of Na_i and K_i in controlling the rate of ouabain binding to the outside of the red cell; the second considers the influence that Na_o and K_i have on the activity of the Na_i/K_o pump as altered by changes in the ratio, ATP/ADP.

Ouabain Binding

Here the properties of ouabain binding are used as a means for determining modulatory sites for Na_i and K_i on the red cell Na/K pump without delving into the mechanism(s) that underlie ouabain binding per se. The rate that ouabain binds to the external surface of the pump (the alpha subunit) is known to be markedly accelerated by internal ATP (Bodemann and Hoffman, 1976). In addition, the ouabain binding rate can also be affected by varying concentrations of Na_i and K_i but only when K_o is also present. (This may be a reflection of the fact that K_o antagonizes the binding of ouabain to the membrane [Glynn, 1957; Hoffman, 1966] since the ouabain binding rate in the absence of K_o is already maximal [Bodemann and Hoffman, 1976].) Current interest in the determinants of ouabain binding rate arose from the report of discrepant results that depended on the particular preset value of K_i. Thus it was found that when Na_i was increased, the rate of ouabain binding decreased at low values of K_i (Bodemann and Hoffman, 1976) but increased at high values of K_i (Joiner and Lauf, 1978). Since the pumped exchange of Na_i for K_o is activated in both instances by increasing values of Na_i, the results implicated the pump rate as a determining variable for ouabain binding. On the other hand, the results also implied that a value of K_i could be found that made the ouabain binding rate insensitive to changes in Na_i (cf. Bodemann et al., 1983). This value turned out to be $\sim$40 mmol K_i/liter ghosts. In addition, an analogous set of effects on the ouabain binding rate could also be observed when Na_i was maintained constant and K_i was changed; in this instance, the ouabain binding rate was found to be insensitive to changes in K_i when the value of Na_i was held at $\sim$30 mmol/liter ghosts. An explanation for these results evidently requires that at least two types of sites exist on the pump—one for Na_i and the other for K_i—that modulate the conformations of the pump associated with ouabain binding and are not necessarily associated with their pumped exchange. Whether or not these sites for Na_i and K_i are the same as the sites with which Na_i and K_i interact competitively, as referred to before, is not known, but they can be presumed to be the same until proven otherwise.

ATP/ADP Ratio

Here we use the results of manipulations in the ratio of ATP/ADP to expose the involvement of K_i and Na_o as modulators of Na/K pump rate. These studies were carried out on resealed ghosts made to contain varying concentrations of Na_i and K_i and a regenerating system in which ADP could be varied at set levels of ATP (Kennedy et al., 1986). It is important to note that the ouabain-sensitive efflux Na was always evaluated under circumstances where the pump was saturated with K_o (15–50 mM). This is to say that the ouabain-sensitive influx of Na_o was not only minimal, it was not changed by increasing ADP, eliminating the possibility that Na_i/Na_o exchange could be stimulated in this situation. On the other hand, the pumped exchange of Na_i for K_o was found to be inhibited by increasing the concentrations of ADP at a fixed concentration of ATP. Important for the present discussion is the fact that this inhibition is under the control of Na_o as shown in Fig. 1. Thus, where the pump is inhibited by high ADP relative to ATP at high Na_o (i.e., 150 mM) this inhibition is relieved by the removal of Na_o. But the situation is still more complicated because this effect of Na_o does not occur unless the concentration of K_i is high. Therefore, when the pump is inhibited by high Na_o, the inhibition can be reversed by lowering K_i (instead of Na_o). These effects of K_i and Na_o, by the way, do not appear to be dependent on their respective membrane gradients; for when systematic variations in the concentrations of Na_o and Na_i or K_i and K_o were made, the effects were sided, being dependent only on the relative concentrations of Na_o or

K_i. Since neither Na_o nor K_i are transported in this situation, these results again emphasize the presence of allosteric modulatory sites for Na_o and K_i on the pump. It should also be appreciated that the efficacy of these sites is dependent on the presence of a high concentration of ADP, although the identity of the underlying pump intermediates associated with the action of these sites is, at present, unknown.

Uncoupled Na Efflux

Having surveyed several examples of modulatory sites, we turn now to consider the properties of signal sites with regard to the actions that Na_o and K_o have on the form of the pump's transport reaction cycle. As first described by Garrahan and Glynn (1967), the red cell Na/K pump is known to continue to extrude Na when both Na

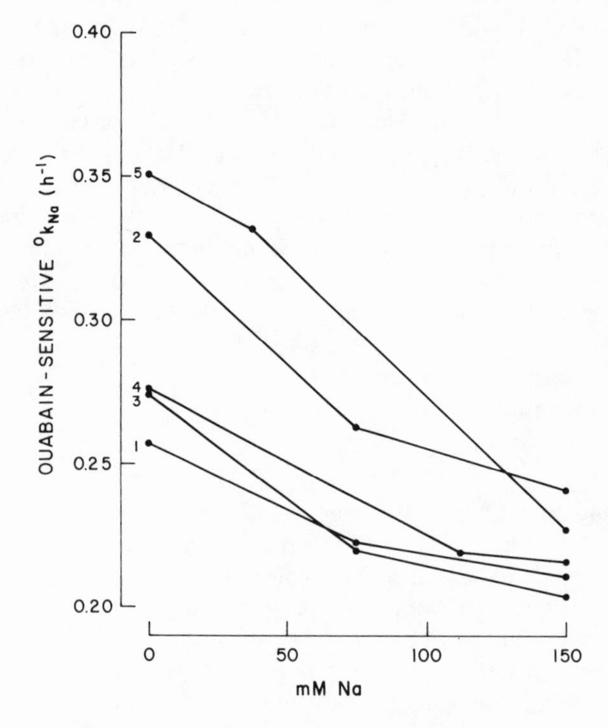

Figure 1. Inhibition of ouabain-sensitive Na efflux by elevation of Na_o at a controlled preset ratio of ATP/ADP. The medium contained the indicated concentrations of NaCl (substituted with choline) together with 15 mM KCl and 10 mM HEPES (pH 7.4). The mean ATP and ADP concentrations were, respectively, ~500 and 1,000 μmol/liter ghosts. (Figure taken from Kennedy et al., 1986.)

and K are removed from the external medium (see Kaplan, 1989). Because this ouabain-sensitive flux occurs in the absence of an exchangeable cation, it is referred to as uncoupled Na efflux. This flux is also known to be inhibited by 5 mM Na_o, but to a lesser extent than that inhibited by ouabain. Uncoupled Na efflux via the Na/K pump can therefore be divided into Na_o-sensitive and Na_o-insensitive components. We used DIDS-treated, SO_4-equilibrated human red blood cells suspended in HEPES-buffered (pH_o 7.4) $MgSO_4$ or $(Tris)_2SO_4$, in which we measured ^{22}Na efflux, $^{35}SO_4$ efflux, and changes in the membrane potential with the fluorescent dye, diS-C_3(5), as described by Dissing and Hoffman (1990). A principal finding is that uncoupled Na efflux occurs electroneutrally, in contrast to the pump's normal electrogenic operation, when exchanging Na_i for K_o. This electroneutral, uncoupled efflux of Na was found to be balanced by an efflux of cellular anions. The Na_o-sensitive efflux of Na_i was found to be twice the Na_o-sensitive efflux of $(SO_4)_i$, indicating that the stoichiometry of this cotransport is two Na^+ per $SO_4^=$, accounting for 60–80% of the electroneutral Na efflux. The remaining portion (that is, the ouabain-sensitive, Na_o-insensitive component) has been identified as PO_4-coupled Na transport and is discussed separately below. It is of interest to note that the SO_4 efflux (as well as the PO_4 efflux) that is coupled to Na efflux is inhibited when pH_o is lowered to below 6.8, evidently converting uncoupled Na efflux to Na_i/$proton_o$ exchange as described by Polvani and Blostein (1988). That uncoupled Na efflux

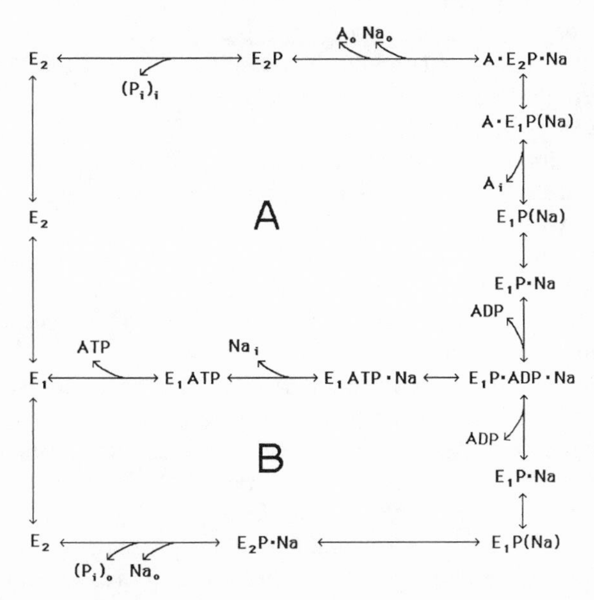

Figure 2. Reaction scheme of the Na,K-ATPase depicting two forms (cycles *A* and *B*) of uncoupled Na efflux. The symbols E_1, E_2, E_1P, and E_2P refer, respectively, to different conformations of the unphosphorylated and phosphorylated forms of the Na,K-ATPase (E). The subscripts i and o refer to ions inside and outside the membrane, respectively. The occluded form of Na is indicated by parentheses (see Glynn and Karlish, 1990). P_i represents orthophosphate, whereas A_i and A_o represent cytoplasmic anions such as SO_4 or chloride. Cycle *A*, representing the SO_4 component of uncoupled Na efflux, runs counterclockwise around the middle and upper part of the box, while cycle *B*, representing the PO_4 component of uncoupled Na efflux, runs clockwise around the middle and lower part of the box. As explained in the text, the idea of separating the two cycles is to emphasize the two sources of anions that are coupled to the transport of Na. In cycle *A*, cytoplasmic anions (A_i) are transported together with Na_i to the outside (A_o and Na_o) under circumstances where the liberated P_i remains inside as $(P_i)_i$. Note that since it is not known to which (if any) intermediate A_i first associates, it is arbitrarily assigned to $E_1P(Na)$. In cycle *B*, the phosphoenzyme ($E_2P{\cdot}Na$) transfers its phosphate directly to the outside as $(P_i)_o$ in concert with Na_o.

occurs as a cotransport with anions is supported by the result, obtained with resealed ghosts, that when internal and external SO_4 were substituted by the impermeant anion, $tartrate_{i,o}$, the efflux of Na was inhibited 60–80%. This inhibition could be relieved by the inclusion before DIDS treatment of 5 mM $Cl_{i,o}$. Addition of 10 mM K_o to tartrate ghosts, with or without $Cl_{i,o}$, resulted in full activation of Na/K exchange and the pump's electrogenicity. Thus it can be concluded that Na efflux in the uncoupled mode occurs by means of a cotransport with cellular anions. Fig. 2 shows a model scheme (cycle *A*) that depicts the cotransport of cytoplasmic anions, such as SO_4, with that of Na.

Although the above experiments were carried out on cells or ghosts where SO_4 had been substituted for Cl, uncoupled Na efflux clearly operates in the presence of Cl, with a preference for Cl over SO_4. This is shown by the fact that uncoupled Na efflux is unaffected by the systematic replacement of Cl for SO_4 but the efflux of SO_4 is gradually lost as the Cl/SO_4 ratio increases. The reason we used SO_4 rather than Cl in the experiments described above is that the uncoupled efflux of Cl, compared with the background flux (conductance) of Cl, is too low for accurate measurement.

To simplify the distinction between the cytoplasmic efflux of SO_4 from the substrate efflux of PO_4, we will refer to these two parts of ouabain-sensitive uncoupled Na efflux as the SO_4 and PO_4 components, respectively.

The characterization of the PO_4 component of uncoupled Na efflux has been carried out by R. Marín and J. F. Hoffman (unpublished results). To study the PO_4 component we loaded resealed ghosts to contain [γ-^{32}P]ATP together with an ATP regenerating system to keep the gamma-PO_4 of ATP at constant specific activity (Glynn and Karlish, 1976). We found that, in concert with uncoupled Na and SO_4 efflux, there was an accompanying efflux of $^{32}PO_4$ that, as pointed out before, was Na_o insensitive. PO_4 efflux was shown to depend on Na_i and increased to saturation as uncoupled Na efflux became maximal. Trapping up to 5 mM orthophosphate together with the [γ-^{32}P]ATP system before treatment with DIDS has no influence on the efflux of $^{32}PO_4$. When the trapped orthophosphate was labeled with ^{32}P together with the same concentration of nonradioactive ATP, there was no detectable ouabain-sensitive $^{32}PO_4$ efflux. These results indicate the reality of this group transfer of PO_4 as a component part of uncoupled Na efflux, presumably representing the direct transfer of PO_4 through the phosphoenzyme, ^{32}P-E(Na), as illustrated in Fig. 2 (cycle *B*). The residual Na efflux that takes place in the $tartrate_{i,o}$-loaded ghosts, referred to above, is fully accounted for by $^{32}PO_4$ efflux from [γ-^{32}P]ATP trapped inside as described before. K_o inhibits both the SO_4 and PO_4 effluxes with the same K_I ($\sim$ 0.25 mM K_o). But it was a surprise to find that the K_I for inhibiting the PO_4 efflux was independent of the Na_o concentration up to 150 mM Na_o. The surprise comes because this effect conflicts with the known Na_o and K_o competition of the Na/K pump (Post et al., 1960).

Therefore, the SO_4 and PO_4 components can be distinguished on the basis of the origin of the transported anion (cytoplasmic vs. substrate), the sensitivity to Na_o, and the action of K_o on the PO_4 efflux. Another difference between the two systems is that the PO_4 component, unlike the SO_4 component, cannot be converted to carry out Na_i/Na_o (with or without ADP) because of its insensitivity to Na_o and because this flux does not recognize Na_o as a surrogate K (cf. Lee and Blostein, 1980). An important implication of these results is that there are two types of Na/K pumps present in human red cells that only become evident when uncoupled Na efflux is studied. This appears curious in light of the fact that the only isoforms of the Na/K pump that are known to be in erythroid tissues are $\alpha 1$ and $\beta 1$. If there are two distinct types of pumps, these may not be detected by the use of antibodies that interact with common epitopes, but we nevertheless are currently probing the membranes with various isoform-specific monoclonal and polyclonal antibodies.

It should also be mentioned that the stoichiometry of Na_i to PO_4 transport, measured in the presence of 150 mM Na_o to inhibit the SO_4 component, is $\sim$2 Na/PO_4. Nevertheless, a major problem remains in being able to rationalize the stoichiometry of Na to the hydrolysis of ATP (i.e., 3Na/PO_4) with the two stoichiometries already noted for the Na/SO_4 and Na/PO_4 components of uncoupled Na efflux. It should also be noted that the ratio of SO_4 to PO_4 flux in uncoupled Na efflux is approximately 4 to 1, indicating that only a portion of the PO_4 hydrolyzed is, in fact, effluxed.

One reason for the above, somewhat detailed, analysis of uncoupled Na efflux is to emphasize that the pump complex can be converted from performing an electrogenic exchange of Na_i for K_o to an electroneutral cotransport system where anions are transported together with Na. The second reason, which is more to the point in the present context, is to emphasize the key roles played by Na_o and K_o in this conversion. Na_o acts at low concentration to inhibit the efflux of SO_4 and its coupled Na component, while at higher concentrations it converts, without the involvement of anions, the uncoupled Na efflux to Na_i/Na_o exchange. In this sense, Na_o in interacting with the outside of the pump acts as a signal for the pump to alter its mode of operation. In a similar sense, K_o also acts to signal the pump, converting uncoupled Na efflux into Na_i/K_o exchange with concomitant inhibition of all anion transport, PO_4 as well as SO_4. It is not clear what mechanisms underlie these types of signals that change the mode of transport, the internal charge structure of the pump, and perhaps subunit interactions. In addition, it is not known whether the sites involved are transport related or act allosterically. On the other hand, high affinity sites for Na_o have been defined (Cavieres and Ellory, 1975; Hobbs and Dunham, 1978; Sachs, 1987), and although their participation in uncoupled transport is not known, there is no reason to suspect that the sites are different.

Acknowledgments

This work was supported in part by National Institutes of Health grant HL-09906.

References

Bodemann, H. H., and J. F. Hoffman. 1976. Side-dependent effects of internal versus external Na and K on ouabain binding to reconstituted human red blood cell ghosts. *Journal of General Physiology* 67:497–525.

Bodemann, H. H., H. Reichmann, T. J. Callahan, and J. F. Hoffman, 1983. Side-dependent effects on the rate of ouabain binding to reconstituted human red cell ghosts. *Current Topics in Membranes and Transport.* 19:229–233.

Cavieres, J. D., and J. C. Ellory. 1975. Allosteric inhibition of the sodium pump by external sodium. *Nature.* 255:338–340.

Dissing, S., and J. F. Hoffman. 1990. Anion-coupled Na efflux mediated by the human red blood cell Na/K pump. *Journal of General Physiology.* 96:167–193.

Eisner, D. A., and D. E. Richards. 1981. The interaction of potassium ions and ATP on the sodium pump of resealed red cell ghosts. *Journal of Physiology.* 319:403–418.

Garay, R. P., and P. J. Garrahan. 1973. The interaction of sodium and potassium with the sodium pump in red cells. *Journal of Physiology.* 231:297–325.

Garrahan, P. J., and I. M. Glynn. 1967. The sensitivity of the sodium pump to external Na. *Journal of Physiology.* 192:175–188.

Glynn, I. M. 1957. The action of cardiac glycosides on sodium and potassium movements in human red cells. *Journal of Physiology.* 136:148–173.

Glynn, I. M. 1985. The Na,K-transporting adenosine triphosphatase. *In* The Enzymes of Biological Membranes. 2nd ed. A. Martinosi, editor. Plenum Publishing Corp., New York. 35–114.

Glynn, I. M., and S. J. D. Karlish. 1976. ATP hydrolysis associated with an uncoupled sodium flux through the sodium pump: evidence for allosteric effects of intracellular ATP and extracellular sodium. *Journal of Physiology.* 256:465–496.

Glynn, I. M., and S. J. D. Karlish. 1990. Occluded cations in active transport. *Annual Review of Biochemistry.* 59:171–205.

Hobbs, A. S., and P. B. Dunham. 1978. Interaction of external alkali metal ions with the Na-K pump of human erythrocytes. *Journal of General Physiology.* 72:381–402.

Hoffman, J. F. 1962. The active transport of sodium by ghosts of human red blood cells. *Journal of General Physiology.* 45:837–859.

Hoffman, J. F. 1966. The red cell membrane and the transport of sodium and potassium. *American Journal of Medicine.* 41:666–680.

Joiner, C. H., and P. K. Lauf. 1978. Modulation of ouabain binding and potassium pump fluxes by cellular sodium and potassium in human and sheep erythrocytes. *Journal of Physiology.* 283:177–196.

Kaplan, J. H. 1989. Active transport of sodium and potassium. *In* Red Blood Cell Membranes. P. Agre and J. C. Parker, editors. Marcel Dekker, Inc., New York. 455–480.

Karlish, S. J. D., and W. D. Stein. 1985. Cation activation of the pig kidney sodium pump: Transmembrane allosteric effects of sodium. *Journal of Physiology.* 359:119–149.

Kennedy, B. G., G. Lunn, and J. F. Hoffman. 1986. Effects of altering the ATP/ADP ratio on pump-mediated Na/K and Na/Na exchanges in resealed human red blood cell ghosts. *Journal of General Physiology.* 87:47–72.

Knight, A. B., and L. G. Welt. 1974. Intracellular potassium. A determinant of the sodium-potassium pump rate. *Journal of General Physiology.* 63:351–373.

Lee, K. H., and R. Blostein. 1980. Red cell sodium fluxes catalysed by the sodium pump in the absence of K^+ and ADP. *Nature.* 285:338–339.

Polvani, C., and R. Blostein. 1988. Protons as substitutes for sodium and potassium in the sodium pump reaction. *Journal of Biological Chemistry.* 263:16757–16763.

Post, R. L., C. R. Merritt, C. R. Kinsolving, and C. D. Albright. 1960. Membrane adenosine triphosphatase as a participant in the active transport of sodium and potassium in the human erythrocyte. *Journal of Biological Chemistry.* 235:1796–1802.

Sachs, J. R. 1986. Potassium-potassium exchange as part of the overall reaction mechanism of the sodium pump of the human red blood cell. *Journal of Physiology.* 374:221–244.

Sachs, J. R. 1987. Inhibition of the Na,K pump by vanadate in high-Na solutions. *Journal of General Physiology.* 90:291–320.

Chapter 21

Proton Transport, Charge Transfer, and Variable Stoichiometry of the Na,K-ATPase

Rhoda Blostein and Carlomaria Polvani

Departments of Medicine and Biochemistry, McGill University, Montreal, Quebec, Canada H3G-1A4

A striking decrease in the stoichiometry of Na:K(Rb) coupling when cytoplasmic Na^+ concentration is reduced to very low levels was observed in earlier studies carried out with inside-out membrane vesicles derived from human red cells (Blostein, 1983). This observation prompted us to investigate whether the reduced stoichiometry could be accounted for by the substitution of protons for sodium ions and/or whether the changed stoichiometry could be reflected by a change in the net charge transferred by the pump. That protons might act as surrogate Na^+ ions suggests a role for protons distinct from that due to ionization of groups accessible at the cytoplasmic face of the enzyme (Breitwieser et al., 1987) or to titrations that result in a conformational shift of the enzyme from the E_1-Na^+ form to the E_2-K^+ form (Skou and Esmann, 1980).

The first results in support of protons acting as surrogate Na^+ ions were obtained in experiments in which we showed that, in the absence of Na^+ and at low pH, the sodium pump catalyzes proton-activated Rb^+ (and presumably K^+) efflux from inside-out membrane vesicles (IOV) derived from human red cells (Blostein, 1985). Hara and co-workers provided further evidence for protons having a Na^+-like effect on the Na^+ sites of Na,K-ATPase (Hara et al., 1986) and for H^+ uptake into K^+-loaded Na,K-ATPase proteoliposomes (Hara and Nakao, 1986), although in the latter study they did not assess the ability of protons to mediate K^+ efflux. We subsequently showed that protons have not only cytoplasmic Na^+-like effects, but also extracellular K^+-like effects on the ATP hydrolytic reaction and phosphoenzyme intermediate of the Na,K-ATPase (Polvani and Blostein, 1988).

In this paper we review experiments that show directly pump-mediated transport of protons and that address the question of variable charge transfer associated with the change in stoichiometry of Na:K(Rb) coupling of the Na,K-ATPase.

Pump-mediated Proton Translocation

To obtain direct evidence for protons acting as surrogate Na^+ and/or K^+ ions, we monitored changes in intravesicular pH by following changes in the fluorescence of FITC-dextran–loaded IOV which had been prepared from red cells pretreated with the anion transport inhibitor 4′,4-diisothiocyano-2,2′-disulfonic stilbene (DIDS) to make them relatively impermeable to anions. With K^+-loaded vesicles equilibrated at pH 6.2, the condition under which proton-mediated Rb^+ efflux is optimal (Blostein, 1985), addition of ATP causes a rapid decrease in fluorescence, corresponding to a decrease in intravesicular pH, presumably reflecting pump-mediated H/K exchange (Fig. 1); as shown, this change is strophanthidin sensitive and is not apparent when the pH is increased to 6.8. Addition of Na^+ not only reverses the pH change, but results in a small increase in intravesicular pH, which, as discussed further below, is dissipated upon addition of relatively high concentrations of lipophilic cations such as tetra-*n*-butyl ammonium (TBA).

Although the experiments with FITC-dextran–loaded IOV were originally aimed at examining the role of protons as surrogate Na^+ ions, it was obvious that intravesicular FITC-dextran is an equally good probe of intravesicular alkalinization, which might occur under conditions in which alkali cation-free vesicles carry out pump-mediated Na^+ transport from cytoplasm to an extracellular milieu. Thus, although the flux known as "uncoupled Na efflux" is an anion-coupled electroneutral Na flux at physiological pH in red cells (Dissing and Hoffman, 1990), the question that remains is whether, at sufficiently high proton concentration, protons act as surrogate (extracellular) K^+ ions, in which case the sodium flux is coupled to the counter-transport of protons. In support of such a Na/H exchange is indirect evidence for a K^+_{ext} role of protons as shown by the following observations reported elsewhere (Polvani and Blostein, 1988): (*a*) With alkali cation-free IOV suspended

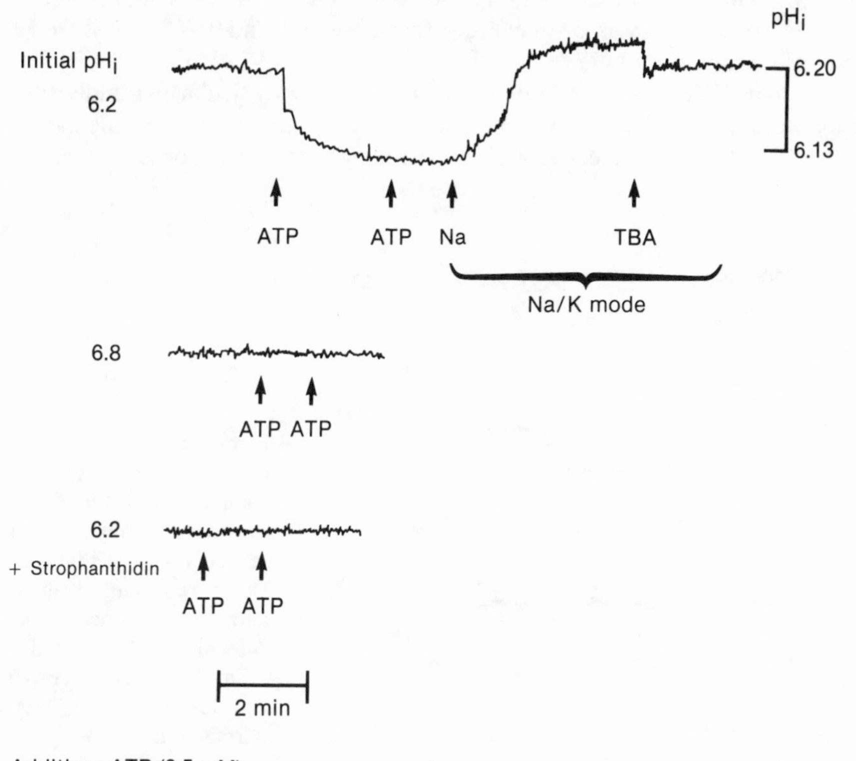

Figure 1. K^+- and ATP-dependent intravesicular pH changes: H/K mode. FITC-dextran–filled vesicles derived from DIDS-treated red cells were diluted 20-fold in 50 mM mannitol, 1 mM $MgSO_4$, 0.8 mM EDTA, 0.2 mM EGTA, 0.5 mM K_2SO_4, and 5 mM choline bitartrate titrated with Tris to either pH 6.2 or 6.8 as indicated, and then concentrated by centrifugation and equilibrated for 30 min at 37°C. 20 μl was then added to a fluorescence cuvette containing 2.7 ml of the same solution, but with K_2SO_4 omitted. 0.5 mM ATP (Na^+-free, titrated with Tris of pH 6.0) and (2 mM) Na_2SO_4 were added as indicated. Vesicles were preincubated with strophanthidin (0.2 mM) during the last 5 min of preincubation as indicated. Fluorescence measurements were carried out at 37°C. From Polvani and Blostein (1988) with permission.

in medium containing Na^+, Na^+ influx (normal efflux) is increased to a small but significant extent as pH is reduced below 7.0; this behavior is in contrast to normal Na/K exchange, which is reduced as pH is reduced. (*b*) As mentioned above, the Na-ATPase activity is increased and the steady-state level of phosphoenzyme is decreased as pH is decreased, effects that are expected of protons having K^+_{ext}-like effects at the extracellular surface.

The experiment shown in Fig. 2 was designed to test whether protons can be transported as surrogate K^+ ions. The results indicate that with IOV free of alkali cations and incubated at pH 6.2, strophanthidin-sensitive intravesicular alkalinization was observed when both ATP and Na^+ were added; neither ligand alone was effective (Fig. 2). This alkalinization is consistent with pump-mediated Na/H exchange, at least at relatively low pH, since its presence could not be detected at higher pH. The much smaller alkalinization observed with K^+ present inside (Na^+/K^+ exchange conditions) was distinct from that observed under either 0/K and Na/0 conditions as discussed below.

The conclusion that the foregoing changes in intravesicular pH under 0/K and Na/0 conditions with IOV equilibrated initially at pH 6.2 reflect pump-mediated H/K and Na/K exchanges, respectively, is supported by the following observations:

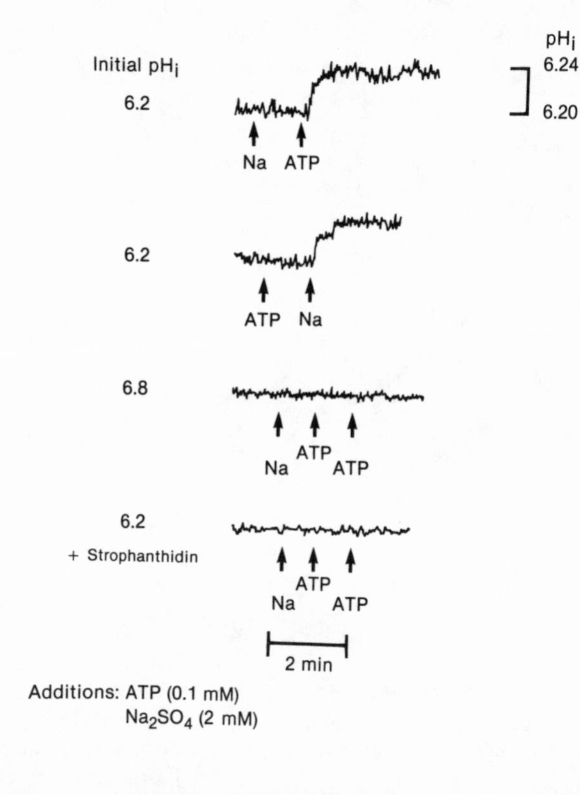

Figure 2. Na^+- and ATP-dependent intravesicular pH changes: Na^+/H^+ mode. Assays were carried out as described in Fig. 1. except that the vesicles were free of alkali cations and the final concentrations of ATP and Na_2SO_4 were 0.2 and 2.5 mM, respectively. From Polvani and Blostein (1988) with permission.

(*a*) As shown in Fig. 3, *A* and *B*, the strophanthidin-sensitive pH changes are not secondary to the redistribution of protons or hydroxyl ions in response to a membrane potential ($\Delta\psi$) since the fluorescent changes observed under 0/K and Na/0 conditions were unaffected by addition of a relatively high concentration (5 mM) of the lipophilic cation TBA; in contrast, the small ΔpH observed under Na/K exchange conditions is dissipated by TBA. The latter observation is consistent with the generation of a membrane potential, cytoplasmic (extravesicular) side negative, due to pump-mediated $3Na^+$ for $2K^+$ exchange; it also attests to the feasibility of using a lipophilic cation such as TBA, at least at relatively high concentration, to dissipate the membrane potential.

(*b*) Estimates of the apparent affinities for ATP, although based on the half-maximal change in fluorescence observed at steady-state, are in agreement with the apparent $K_{1/2}$ of ATP for proton-activated Rb^+ transport from IOV (Blostein, 1985) or pump-mediated acidification of K^+-filled Na,K-ATPase proteoliposomes (Hara and Nakao, 1986), and, in general, with the existence of a low affinity ATP site involved in K^+ translocation (Post et al., 1972).

(*c*) At low pH a net uphill exit of Rb^+ from IOV consistent with H/K exchange via the forward operation of the complete pump cycle was observed in earlier studies (Blostein, 1985).

(*d*) Proton-activated Rb efflux from IOV is not supported by either the nonhydrolyzable ATP analogue β,γ-methylene ATP or inorganic orthophosphate (Blostein, 1985).

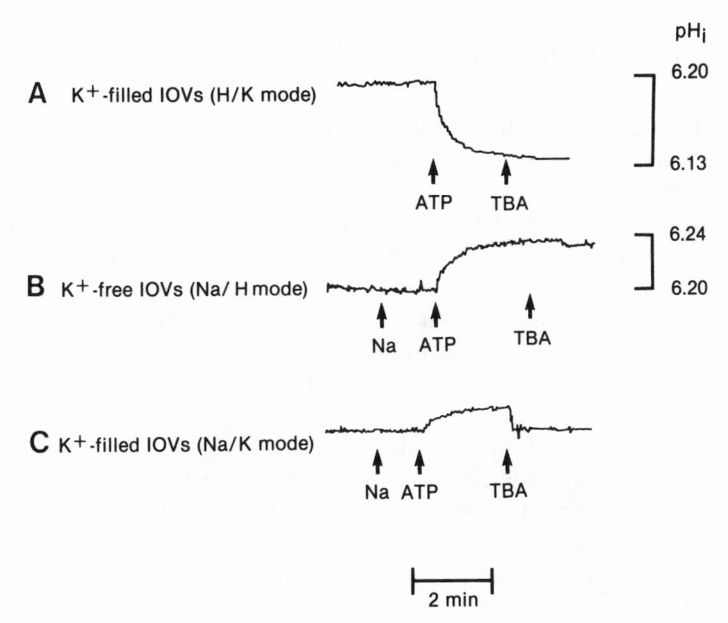

Figure 3. Tetrabutylammonium effects on pump-mediated intravesicular pH changes. Assays were carried out as described in Figs. 1 and 2. 5 mM TBA was added as indicated. From Polvani and Blostein (1988) with permission.

(*e*) Estimates of the initial rates of net proton fluxes calculated from (*i*) the buffering capacity of the IOV assessed by an NH_4Cl titration method (Roos and Boron, 1981) and (*ii*) the initial rates of acidification/alkalinization were generally similar to the rates of Rb^+ efflux under 0/Rb conditions and Na influx under Na/0 conditions (Polvani and Blostein, 1988).

Protons Replacing Sodium

In this paper we show that protons can replace Na^+ as the primary cation of the Na,K-ATPase. Relevant to our study and to the question of alternate uses of protons and other cations by transport enzymes in general is our recent evidence that Na^+ ions act as surrogate protons in the gastric H,K-ATPase reaction (Polvani et al., 1989). Although this result may not be surprising in view of the structural similarities of these two "P-type" cation translocating ATPases, it is particularly interesting to note the studies of Laubinger and Dimroth on the Na^+ translocating ATPase of the strictly anaerobic bacteria *Propionigenium modestum,* which uses a sodium cycle for energy conservation (Hilpert et al., 1984). Recently, the enzyme has been purified and shown to be a member of the F-type family (Laubinger and Dimroth, 1988). This particular F_0F_1 ATPase uses a Na^+ gradient to synthesize ATP (Laubinger and Dimroth, 1989). It is a primary Na^+ pump which can pump protons at sufficiently low pH. Moreover, this proton pump activity was operative only at very low Na^+ concentrations and was decreased as the Na^+ concentration increased. Thus, the alternate use of Na^+ and H^+ is not limited to P-type cation translocating ATPase but occurs in enzymes of the F-type as well. Whether this observation implies that the mechanism for cation/proton discrimination/translocation might be similar in P-type and F-type ATPases remains to be determined.

Protons Replacing Potassium

It is also shown that H^+ can replace K^+ ions. This is perhaps not surprising if one considers the selectivity for alkali cations of the extracellular K^+ sites compared with the selectivity of the cytoplasmic Na^+ sites. Only Li^+ ions were shown to replace Na^+, whereas a variety of cations can replace K^+ at the extracellular sites. In this context, protons resemble Li^+ more than K^+ or Rb^+ with respect to the ease with which they appear to be deoccluded during the reaction cycle. This is evidenced by the high apparent affinity for ATP of H^+ efflux (normal influx) under Na/H exchange conditions, in contrast to the low affinity for H^+ influx (normal efflux) under either H/K exchange or normal Na/K exchange conditions.

An unresolved question concerns the stoichiometry of electroneutral Na/H and K/H coupling. For H/K exchange it is difficult to reconcile the exchange of $3H^+$ for $3K^+$ per molecule ATP hydrolyzed with the well-documented exchange of $3Na^+$ for only $2K^+$ under Na/K exchange conditions, with evidence for electrogenic exchange of, presumably, $3Na^+$ for $2Na^+$ under conditions of Na^+ acting as surrogate K^+ ions (Lee and Blostein, 1980; Goldschleger et al., 1990), and with experiments demonstrating that only $2K^+$ ions are occluded per phosphorylation site (for examples, see Beaugé and Glynn, 1979; Glynn and Richards, 1982; Forbush, 1987; Shani et al., 1987). From these considerations, taken together with the observation that fewer than $3Na^+$ ions can be exchanged for $2K^+$ ions, it is likely that $2H^+$ ions are exchanged for $2K^+$ ions, at least at pH $\sim$6.0. The possibility that the pump can accommodate three protons resulting in electrogenic $3H^+/2K^+$ can only be tested with a preparation that remains active and cation impermeable under more acidic conditions.

Relevant to the question of the stoichiometry and molecular basis for electroneutral Na/H exchange, is the observation that in red cells at pH 7.4 removal of extracellular Na^+ and K^+ converts electrogenic Na/K exchange to electroneutral Na^+-anion cotransport (Dissing and Hoffman, 1990). These investigators suggest that a change in the internal charge structure of the pump is signaled by removal of extracellular Na^+ and K^+ from regulatory "signal" sites. Thus, in normal electrogenic Na/K exchange, three cytoplasmic Na^+ ions are thought to interact with two negatively charged and one neutral site, resulting in electrogenic transfer of one positive charge during the sodium translocation limb of the cycle; two extracellular K^+ ions then interact with the two vacated negative charges, resulting in no net mobile charge during the K^+ translocation limb of the cycle. Dissing and Hoffman

(1990) suggest that the change in the internal charge structure of the pump effected by removal of alkali cations is such that all three internal sites remain neutral while occupied by Na^+ and accompanying anions.

In order that electroneutral transport of $3Na^+$ accompanied by three anions is converted to electroneutral Na/H exchange as pH is reduced, an alteration in the internal charge structure must be signaled not only by removal of extracellular alkali cations but also by protonation of the enzyme. It is plausible that protonation reduces the number of Na^+ ions bound to internal sites, resulting in electroneutral $2Na^+/2H^+$ exchange. Alternatively, protonation may affect the charge structure at the external surface such that at acidic pH (pH $\leq$ 6.2) the transport of $3Na^+$ becomes coupled to the countertransport of $3H^+$. One way to gain insight into the question of Na/H stoichiometry would be to compare, at the same (acidic) pH, the Na:ATP coupling ratio under conditions of Na/H and Na/K exchange; i.e., to test whether removal of extracellular K^+ alters the Na^+:ATP stoichiometry.

Cation Binding Sites

A fundamental and unresolved aspect of the P-type enzymes is the structural basis for cation selectivity; namely, how these transport enzymes discriminate among different cations and how they determine the number of cations transported per reaction cycle. In trying to understand the molecular basis for cation selectivity, as well as in designing molecular models for cation transport, one must take into consideration the strong structural homologies encountered between P-type enzymes. Thus, only models for transport that can explain how small structural alterations lead to changes in ion selectivity for very different physico-chemical species of cations (e.g., from sodium to calcium) can be considered as valid.

Boyer proposed an attractive mechanism for proton translocation which accommodates the observed alternate use of protons and alkali cations (Boyer, 1988). Coordinating oxygen and nitrogen atoms from amino acid residues would complex, with only small changes in conformation, either the hydrated proton (hydronium ion) or (mainly dehydrated) alkali cations. In this regard, the transport sites may be similar to cyclic polyethers. The possibility that enzymes transport cations and protons through a similar mechanism is attractive from an evolutionary point of view, since it is rather difficult to envisage how adaptation to transport H^+, Na^+, or even Ca^{2+}, for example, would have occurred without major structural changes in the protein if a mechanism such as proton wire (Nagle and Morowitz, 1978) were used in the transport of protons. (For a more detailed discussion of cation binding and occlusion, see the recent review by Glynn and Karlish, 1990.)

Decreased Na/Rb(K) Coupling Associated with Altered Charge Transfer

Electrogenic Behavior of Na/K Exchange at pH 6.6

The decreased Na/Rb(K) coupling observed at low cytoplasmic Na^+ under conditions of physiological pH (Blostein, 1983) cannot be accounted for by replacement of sodium ions by protons since proton-dependent $Rb^+(K^+)$ efflux is markedly decreased as the pH is raised above pH 6.2, becoming imperceptible at neutral pH (Blostein, 1985). Accordingly, a series of experiments was carried out to determine whether, in fact, the electrogenicity of the pump is altered as the cytoplasmic Na^+ concentration is reduced.

As discussed in the preceding section, we showed that it is possible to monitor pH gradients across the red cell membrane by first treating the intact red cells with DIDS to reduce anion permeability and then incorporating FITC-dextran inside during the vesiculation process (Polvani and Blostein, 1988). To distinguish pH gradients generated by pumped protons from those reflecting the formation of a membrane potential, lipophilic cations such as TBA or tetraphenylphosphonium (TPP) at relatively high concentration have been used to dissipate ΔpH reflecting Δψ. This has already been illustrated by the experiments shown in Figs. 1 and 3.

Although strophanthidin-inhibitable, lipophilic cation–sensitive ΔpH may be used to assess the electrogenic behavior of the sodium pump in IOV (in particular to address the question of whether, in a relative manner, the electrogenicity of the sodium pump changes when the Na:K(Rb) coupling decreases as cytoplasmic Na is reduced [Blostein, 1983]), it was important to carry out experiments with vesicles at a pH high enough (pH 6.6) to reduce proton transport via H/K(Rb) exchange, yet, as discussed below, low enough to permit detection of a ΔpH. A main advantage of using FITC-dextran as a pH probe is that the potentials formed in either positive or negative directions should be equally well detected. Another advantage is that there is no evidence of interaction of the probe with either the membrane or intracellular constituents.

Fig. 4 shows experiments carried out with K^+-filled IOV equilibrated at pH 6.6. With ATP present, addition of Na^+ causes small but significant TPP-sensitive changes in intravesicular pH: either alkalinization in response to an inside positive membrane potential, or acidification due to an inside negative potential. As shown, these changes are strophanthidin sensitive. With varying amounts of Na^+ added, a maximal inside-positive potential is observed at cytoplasmic Na^+ concentrations >2 mM and a maximal inside-negative potential at cytoplasmic sodium concentrations <0.4 mM. With 0.8–1.0 mM Na^+, little fluorescence change is detected.

As shown in Fig. 5, the ATP-dependent voltage changes apparent at low and high cytoplasmic Na^+ concentrations are of similar magnitude (~ 1 mV, calculated from the ΔpH) but of opposite sign. Maximal changes are apparent at ~ 0.3 mM ATP, although a more quantitative evaluation of the apparent affinity for ATP at low and high cytoplasmic Na concentrations has not been technically feasible because of the relatively weak fluorescence signals obtained.

Distinct Exchanges of Potassium with Sodium and/or Protons at pH 6.2 and 6.6

The foregoing results of experiments carried out at pH 6.2 are consistent with the existence of pump-mediated exchange of $3Na^+$ for $2K^+$ at relatively high cytoplasmic Na (≥ 2 mM) resulting in the generation of a membrane potential, cytoplasmic side negative, and an exchange of $1Na^+$ for $2K^+$ at low cytoplasmic Na^+ (≤ 0.4 mM) resulting in the generation of a membrane potential, extracellular side negative. Direct measurements of Na:Rb coupling ratios at pH 6.6 indicate that the coupling ratios are indeed close to 1.5 and 0.5 at 2 and 0.2 mM Na^+, respectively (see Table I).

These results are in contrast to those obtained at pH 6.2, where changes in fluorescence were not dissipated by lipophilic cations. The most straightforward explanation for the behavior at pH 6.2 is that in the presence of small amounts of cytoplasmic Na^+, protons are not only transported in exchange for K^+ (presumably electroneutral $2H^+$ for $2K^+$ exchange, as discussed earlier), but are also cotransported with Na^+ in an electroneutral exchange of $1Na^+$ and $1H^+$ for $2K^+$. These two modes are likely to be active concurrently. In support of this notion are the combined results of experiments aimed at examining (*a*) the effect of adding 0.2 mM Na^+ on H^+ uptake and on Rb^+ efflux, and (*b*) the ratio of total Na influx to Rb efflux in IOV. As shown in Table I, the results are consistent with concomitant $2H/2K^+$ and $1Na + 1H/2K^+$ exchanges in the proportion ~60 and 40%, respectively. The different modes of exchange of $K^+(Rb^+)$ with Na^+ and/or H^+ are illustrated in Fig. 6.

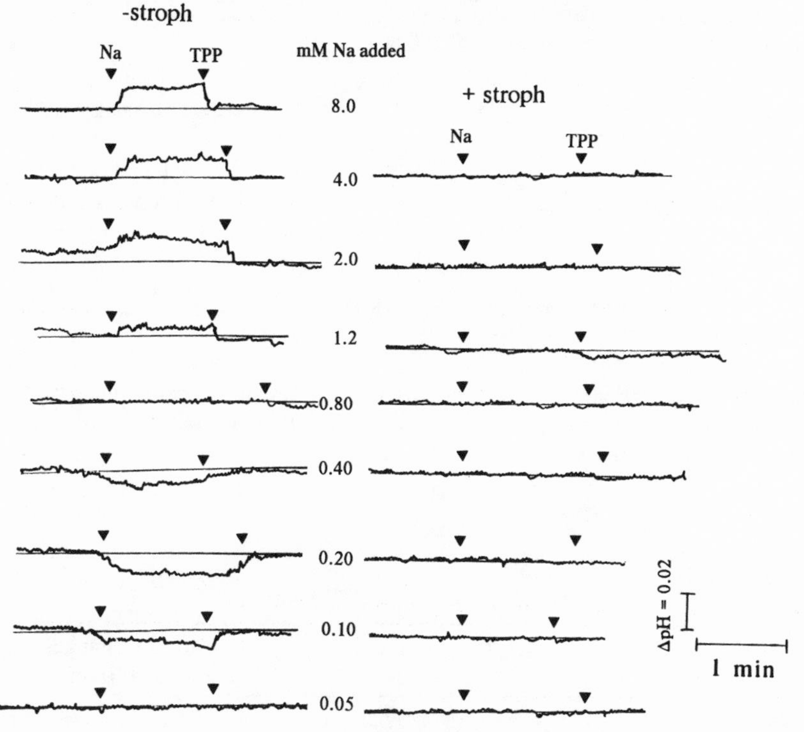

Figure 4. Effect of cytoplasmic Na concentration on electrogenic potentials. Experiments were carried out as described in Fig. 1, except that the final pH was adjusted to pH 6.6 with Tris. 0.5 mM ATP was present in the reaction medium and Na^+_{cyt} at various concentrations was added (left-hand arrow) as indicated. Electrogenic potentials are shown as TPP-sensitive ΔpH's (c.f. Fig. 3). From Polvani and Blostein (1989) with permission.

In the foregoing experiments with IOV, we have observed that even though the pump rate increases as pH increases, ΔpH actually decreases, becoming imperceptible at pH 7.0. Moreover, in experiments described elsewhere (Polvani and Blostein, 1989), we showed that the fluorescence changes observed after addition of valinomycin to generate transmembrane electrical gradients resulted in transient fluorescence changes that were of a magnitude and direction directly proportional to the calculated K^+ equilibrium potentials, a least up to ±10 mV. As in the case of the pump-generated signals, the fluorescence changes diminished as pH increased from 6.2 to 6.6, becoming undetectable at pH 7.0. The transient nature of the Δfluores-

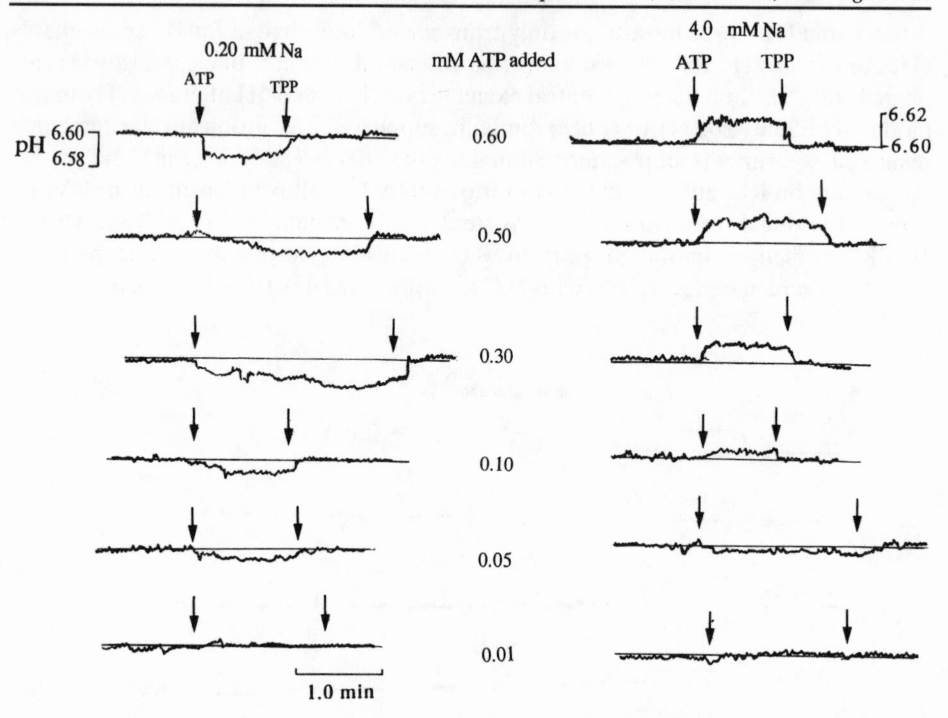

Figure 5. Effect of ATP concentration on electrogenic potentials. Experiments were carried out as described in Fig. 4, with the ATP concentration varied as indicated. From Polvani and Blostein (1989) with permission.

TABLE I
Exchanges of $Rb^+(K^+)$ with Na^+ and/or H^+ at pH 6.2 and pH 6.6

Measured fluxes					Cation exchanges (% of total Rb efflux)		
pH 6.2							
$[Na_{cyt}]$ nM	Rb^+ efflux*	Na^+ influx	H^+ influx*	Na^+ influx / Rb^+ efflux	3Na/2Rb	2H/2Rb	$1Na^+1H/2Rb$
0.0	1.00	—	1.00	—	—	100	—
0.2	~0.70	0.13	~0.50	0.19	0	64	36
2.0	2.00	3.00	0.00	~1.50	100	0	0

pH ≥6.6			
$[Na_{cyt}]$ mM	Na^+ influx / Rb^+ efflux	3Na/2Rb	1Na/2Rb
0.2	~0.50	0	100
2.0	1.50	100	0

*Rb^+ efflux and H^+ influx values shown were obtained from measurements of the rate constant for Rb^+ efflux and the decrease in percent fluorescence due to H^+ influx, respectively; both were normalized to a value of 1.0 for $[Na_{cyt}] = 0$. Values for Na^+ influx were also normalized to these values.

cence (ΔpH) changes and their tendency to plateau are probably due to the higher activity of the anion transporter (band 3) at the higher pH's and to greater dissipation of the gradients by higher concentrations of permeant, notably hydroxyl, anions. This behavior suggests that the failure to detect TPP-insensitive components of ΔpH at pH ≥ 6.6 does not necessarily indicate that H/K and Na/H exchanges are completely absent under these conditions. In fact, as shown previously (Blostein, 1985), Rb^+ efflux from IOV in the absence of cytoplasmic Na^+ is reduced but not absent as pH is increased to pH 6.6 (Blostein, 1985). By the same argument, the failure to detect lipophilic cation-sensitive ΔpH's as pH is reduced to 6.2, under which condition Δfluorescence due to $\Delta\psi$ becomes larger, argues strongly against the existence of electrogenic 1Na/2K exchange at low cytoplasmic Na^+ at this pH.

The possibility of altered stoichiometry has been considered by other investigators as well. Using Sartorius muscle from *Rana catesbeiana,* Marunaka has studied the effects of the Na^+ and K^+ concentrations on the ouabain-sensitive membrane

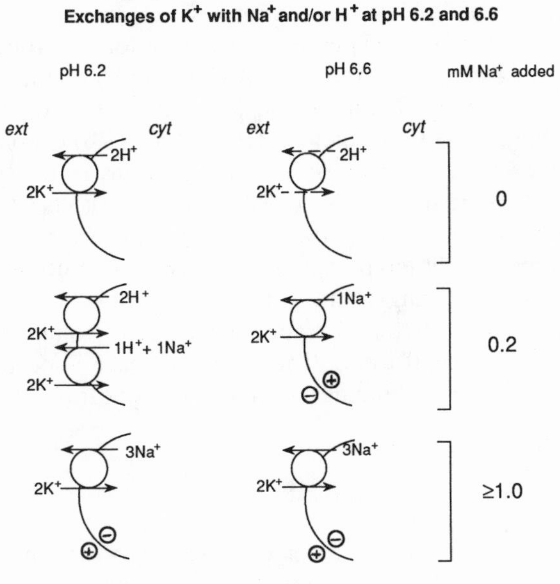

Figure 6. Predominant exchanges of potassium with sodium and/or protons at pH 6.2 and 6.6.

potential change (Marunaka, 1988). He found that when the Na^+ concentration was increased from 6 to 20 mmol/kg muscle the membrane potential increased eightfold while the sodium flux increased only fivefold. The possibility that an increase in membrane potential was due to an increase in membrane resistance at high Na^+ was excluded since the resistance was shown to be the same at both high and low Na^+ concentrations.

Karlish and co-workers have studied the formation of membrane potentials in pig kidney Na,K-ATPase reconstituted into phospholipid vesicles using the membrane potential–sensitive probe oxonol VI (Goldshleger et al., 1990). They showed that although the rate of Na/K exchange at 0.8 mM Na^+ should have sustained a detectable oxonol signal, assuming a Na:K coupling ratio of 1.5, the signal disappeared, becoming imperceptible below 0.8 mM Na^+. This suggested that the Na:K coupling becomes reduced below the normal 3:2 stoichiometry.

Evidence for the ability of the sodium pump to increase its electrogenicity comes from studies of the membrane potentials generated by the "uncoupled" Na^+ flux, and was also obtained by Goldshleger et al. (1990). They showed that at pH 7.0 the potential resulting from the "uncoupled" Na^+ flux (Na/0 conditions) was zero, but as pH was raised to 8.5 a membrane potential (inside positive) of 55 mV was observed. These results indicate that whereas the "uncoupled" Na^+ flux is electroneutral when the proton concentration is high, it becomes electrogenic as the proton concentration decreases. They suggest that at pH ≤ 7.0 the "uncoupled" Na^+ flux is really an electroneutral Na/H exchange, but as pH increases fewer protons are transported per Na^+ ion; i.e., $3Na^+$ for $2H^+$, or even $3Na^+$ for $1H^+$ (Goldshleger et al., 1990).

Conclusion

Inside-out vesicles from human red cells pretreated with DIDS are sufficiently anion impermeable to be used for measurements of pump-mediated proton transport. It is also evident that the residual proton and/or hydroxyl conductance of this preparation is sufficient to support the formation of a pH gradient in response to the generation of a membrane potential. Our experiments provide direct evidence for electroneutral H/K(Rb) and Na/H exchanges catalyzed by the Na,K-ATPase under conditions of acidic pH. Moreover, a combination of $2H^+$ for $2K^+$ exchange and $1H^+$ and $1Na^+$ for $2K^+$ exchange accounts for the low (~ 0.2) stoichiometry of Na:K(Rb) coupling observed at reduced cytoplasmic Na^+ concentration under conditions of low pH (pH ~ 6.0). At higher pH (pH ≥ 6.6) the reduced Na:K(Rb) coupling reflects a true change in stoichiometry as evidenced in a change in the net charge transferred by the Na,K-ATPase. Thus, at pH 6.6 the membrane potential, cytoplasmic side negative, associated with a ~ 1.5 Na:K(Rb) coupling changes to a membrane potential, cytoplasmic side positive, associated with a ~ 0.5 coupling as cytoplasmic Na^+ concentration is reduced from ≥ 2.0 to ≤ 0.4 mM. These findings have important implications regarding the mechanism of ion binding and charge translocation by the Na,K-ATPase.

References

Beaugé, L. A., and I. M. Glynn. 1979. Occlusion of K ions in the unphosphorylated sodium pump. *Nature.* 280L:510–512.

Blostein, R. 1983. The influence of cytoplasmic sodium concentration of the stoichiometry of the sodium pump. *Journal of Biological Chemistry.* 258:12228–12232.

Blostein, R. 1985. Proton-activated rubidium transport catalyzed by the sodium pump. *Journal of Biological Chemistry.* 260:829–833.

Boyer, P. D. 1988. Bioenergetic coupling to protonmotive force: should we be considering hydronium ion coordination and not group protonation? *Trends in Biochemical Sciences.* 12:5–7.

Breitwieser, G. E., A. A. Altamirano, and J. M. Russell. 1987. Effects of pH changes on sodium pump fluxes in squid giant axon. *American Journal of Physiology.* 253:C547–C554.

Dissing, S., and J. F. Hoffman. 1990. Anion-coupled Na efflux mediated by the human red blood cell Na/K pump. *Journal of General Physiology.* 96:167–193.

Forbush, B., III. 1987. Rapid release of ^{42}K and ^{86}Rb from an occluded state of the Na,K-pump in the presence of ATP and ADP. *Journal of Biological Chemistry.* 262:11104–11115.

Glynn, I. M., and S. J. D. Karlish. 1990. Occluded cations in active transport. *Annual Review of Biochemistry*. 59:171–201.

Glynn, I. M., and D. E. Richards. 1982. Occlusion of rubidium ions by the sodium pump: its implication for the mechanism of potassium transport. *Journal of Physiology*. 330:17–43.

Goldshleger, R., Y. Shahak, and S. J. D. Karlish. 1990. Electrogenic and electroneutral transport modes of renal Na/K ATPase reconstituted into proteoliposomes. *Journal of Membrane Biology*. 113:139–154.

Hara, H., and M. Nakao. 1986. ATP-dependent proton uptake by proteoliposomes reconstituted with purified Na,K-ATPase. *Journal of Biological Chemistry*. 261:12655–12658.

Hara, Y., J. Yamada, and M. Nakao. 1986. Proton transport catalyzed by the sodium pump: ouabain-sensitive ATPase activity and the phosphorylation of Na,K-ATPase in the absence of sodium ions. *Journal of Biochemistry*. 99:531–539.

Hilpert, W., B. Schink, and P. Dimroth. 1984. Life by a new decarboxylation-dependent energy conservation mechanism with Na^+ as coupling ion. *EMBO Journal*. 3:1665–1670.

Laubinger, W., and P. Dimroth. 1988. Characterization of the ATP synthase of *Proprionigenium modestum* as a primary sodium pump. *Biochemistry*. 27:7531–7537.

Laubinger, W., and P. Dimroth. 1989. The sodium ion translocating adenosinetriphosphatase of *Proprionigenium modestum* pumps protons at low sodium ion concentrations. *Biochemistry*. 28:7194–7198.

Lee, K. H., and R. Blostein. 1980. Red cell sodium fluxes catalyzed by the sodium pump in the absence of K^+ and ADP. *Nature*. 285:338–339.

Marunaka, Y. 1988. Effects of internal Na and external K concentration of Na-K coupling of Na,K-pump in frog skeletal muscle. *Journal of Membrane Biology*. 101:19–31.

Nagle, J. F., and H. J. Morowitz. 1978. Molecular mechanisms for proton transport in membranes. *Proceedings of the National Academy of Sciences, USA*. 75:298–302.

Polvani, C., and R. Blostein. 1988. Protons as substitutes for sodium and potassium in the sodium pump reaction. *Journal of Biological Chemistry*. 263:16757–16763.

Polvani, C., and R. Blostein. 1989. The effects of cytoplasmic sodium concentration on the electrogenicity of the sodium pump. *Journal of Biological Chemistry*. 264:15182–15185.

Polvani, C., G. Sachs, and R. Blostein. 1989. Sodium ions as substitutes for protons in the gastric H,K-ATPase. *Journal of Biological Chemistry*. 264:17854–17859.

Post, R. L., C. Hegyvary, and S. Kume. 1972. Activation by adenosine triphosphate in the phosphorylation of sodium and potassium ion transport adenosine triphosphatase. *Journal of Biological Chemistry*. 247:6530–6540.

Roos, A., and W. F. Boron. 1981. Intracellular pH. *Physiological Reviews*. 61:296–434.

Shani, M., R. Goldshleger, and S. J. D. Karlish. 1987. Rb occlusion in renal (Na,K)ATPase characterized with a simple manual assay. *Biochimica et Biophysica Acta*. 904:13–21.

Skou, J. C., and M. Esmann. 1980. Effects of ATP and protons on the Na:K selectivity of the $(Na^+ + K^+)$-ATPase studied by ligand effects on intrinsic and extrinsic fluorescence. *Biochimica et Biophysica Acta*. 601:386–402.

Electrogenic Aspects

Chapter 22

Kinetic Basis of Voltage Dependence of the Na,K-Pump

P. Läuger

Department of Biology, University of Konstanz, D-7750 Konstanz, Germany

The Na,K-pump, under physiological conditions, moves three Na^+ ions outward and two K^+ ions inward and thus translocates net charge across the membrane. The electrogenic nature of the pump becomes manifest in at least two ways: the pump acts as a generator of electric current, and its turnover rate depends on transmembrane voltage.

The overall electrogenic nature of the pumping reaction raises the problem of identifying the step (or steps) of the cycle in which charge actually moves inside the membrane. This question may be answered by correlating charge-translocating events with known reaction steps of the pump, such as conformational transitions or ion-binding and release steps.

Different experimental methods are available by which charge translocation in a pump protein can be studied. One method involves measuring the voltage dependence of pump current I_p in the stationary state (Gadsby and Nakao, 1989, and references cited therein). From the shape of the $I_p(V)$ curve studied under different ionic conditions, information on microscopic parameters of the pump can be obtained (Läuger and Apell, 1986; De Weer, 1990).

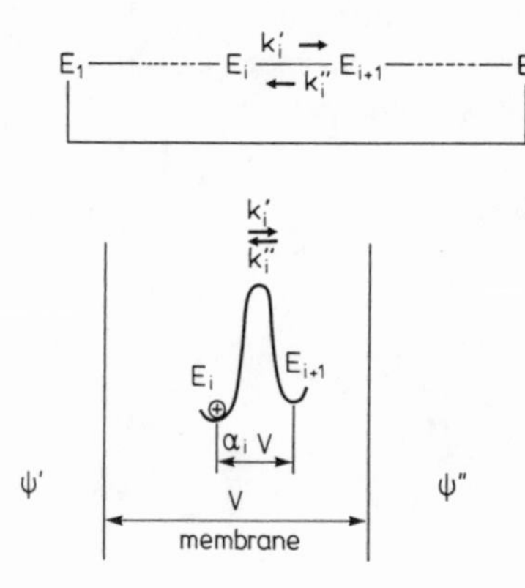

Figure 1. Reaction cycle of an ion pump involving transitions between states E_1, E_2, . . ., E_n of the pump molecule. The transition $E_i \rightarrow E_{i+1}$ is assumed to be associated with translocation of a univalent cation between two (quasi-)equilibrium positions. $\alpha_i V$ is the fraction of total voltage $V \equiv \psi' - \psi''$, which drops between the two energy minima corresponding to states E_i and E_{i+1}. k'_i and k''_i are the voltage-dependent rate constants of the transition $E_i \rightleftarrows E_{i+1}$.

A further method involves the measurement of transient currents after a fast perturbation of the system. If the transmembrane voltage is suddenly changed, a transient current is observed that decays with a characteristic time behavior to a steady-state current (Nakao and Gadsby, 1986). Alternatively, a step change of the concentration of a substrate may be carried out; such concentration-jump experiments are possible using "caged" compounds, e.g., caged ATP or caged Mg^{2+} (Fendler et al., 1985; Borlinghaus et al., 1987).

Dielectric Coefficients and Voltage Dependence of Rate Constants

The analysis of current–voltage measurements and transient–current experiments is based on the assumption that the overall transport process can be subdivided into transitions between discrete molecular states of the pump molecule (Fig. 1). Each transition $E_i \rightarrow E_{i+1}$ is described by the rate constants k'_i and k''_i in the forward and backward directions, respectively. The electrogenic behavior of the pump is completely determined by the voltage dependence of these rate constants.

We consider a transition $E_i \rightarrow E_{i+1}$ consisting in the (outward-directed) translocation of a Na^+ ion over a certain distance in the protein. If a voltage $V \approx \psi' - \psi''$ is applied across the membrane, a certain fraction of V, $\alpha_i V$, drops across the distance over which the ion is translocated. According to the theory of absolute reaction rates, the voltage dependence of the rate constants k_i' and k_i'' is described by a simple exponential relationship:

$$k_i' = \tilde{k}_i' \exp(\alpha_i u/2) \quad (1)$$

$$k_i'' = \tilde{k}_i''(-\alpha_i u/2) \quad (2)$$

where u is the transmembrane voltage, expressed in units of $RT/F \approx 25$ mV:

$$u \equiv \frac{\psi' - \psi''}{RT/F} = \frac{V}{RT/F} \approx \frac{V}{25 \text{ mV}} \quad (3)$$

$\tilde{k}_i'$ and $\tilde{k}_i''$ are the values of k_i' and k_i'' at zero voltage, and ψ' and ψ'' are the electric potentials on the cytoplasmic and extracellular sides, respectively. The parameter α_i is usually referred to as the dielectric coefficient of the transition $E_i \rightleftarrows E_{i+1}$.

The stationary pump current I_p is a unique function of the rate constants k_i' and k_i'' of the reaction cycle. This means that the $I_p(V)$ characteristic of the pump is completely determined by the set of the dielectric coefficients α_i, together with the zero-voltage values $\tilde{k}_i$ of the rate constants. The ultimate goal in the analysis of current–voltage curves and transient currents is to evaluate the set of the α_i and to interpret the α_i in terms of a microscopic model of the pump.

The dielectric coefficients thus play a central role in the microscopic description of electrogenic properties of ion pumps. If the membrane is considered as a homogeneous dielectric film of thickness d and if, in the transition $E_i \rightarrow E_{i+1}$, ν_i elementary charges are translocated over a distance a_i, the dielectric coefficient α_i is given by the product of ν_i times the fractional distance a_i/d:

$$\alpha_i \approx \nu_i a_i/d \quad (4)$$

In reality, a lipid membrane with embedded proteins is an inhomogeneous dielectric medium. This means that the dielectric coefficients can no longer be interpreted by a simple geometric picture; accordingly, the relation $\alpha_i = \nu_i a_i/d$ must be considered an approximation.

It is important to note that with the same set of dielectric coefficients that describes the steady-state current–voltage characteristic of the pump, one can also describe the transient currents observed in a voltage-jump of concentration-jump experiment. Consider the situation where all pump molecules are initially in state E_i. By a sudden voltage change or by another perturbation, transitions from E_i to E_{i+1} are induced. If the reaction $E_i \rightarrow E_{i+1}$ is associated with the translocation of a charge of magnitude $\nu_i e_o$, a compensatory charge $\alpha_i e_o$ flows in the external measuring circuit (Fig. 2), where α_i is the dielectric coefficient of the transition $E_i \rightleftarrows E_{i+1}$, which is proportional to ν_i (Eq. 4), and e_o is the elementary charge. The total time-dependent current $I_p(t)$ can be represented by the sum over all individual transition rates $\phi_i(t)$, multiplied by the corresponding dielectric coefficients (Läuger and Apell, 1986):

$$I_i(t) = e_o N \Sigma\, \alpha_i \Phi_i(t) \quad (5)$$

where N is the number of pump molecules in the membrane.

This means that the same microscopic information is obtained whether we determine the voltage dependence of a rate constant or measure the charge that is moved in the same transition. In practice, the methods are complementary.

Nature of Charge-translocating Reaction Steps

What voltage-dependent reaction steps may one expect in the overall transport process? Two different types of electrogenic steps may be distinguished: (*a*) conformational transitions associated with charge movement, and (*b*) ion-binding and -release reactions involving migration in access channels.

In a conformational transition, bound ions may move together with the liganding groups over a certain distance within the membrane dielectric. If, for example, three Na^+ ions are bound and the charge of the ligand system is $z_L = -2$, the change of activation energy at a given transmembrane voltage u is proportional to $(3 + z_L)\beta u$, where β is the relative dielectric distance over which the charges move. In addition to the bound ions and charged ligand groups, other charges in the protein may move during the conformational transition, and this may be accounted for by an

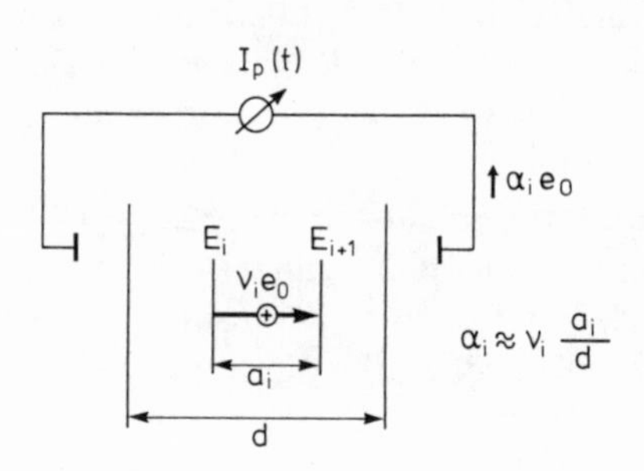

Figure 2. Analysis of transient pump currents $I_p(t)$. By an external perturbation, transitions between states E_i and E_{i+1} of the pump are induced. The transition $E_i \rightarrow E_{i+1}$ is associated with translocation of a charge of magnitude $\nu_i e_o$; accordingly, a compensatory charge $\alpha_i e_o$ flows in the external measuring circuit. α_i is the dielectric coefficient of the transition $E_i \rightleftarrows E_{i+1}$, which is proportional to ν_i, and e_o is the elementary charge.

additional term η. The voltage dependence of the rate constant k' of the conformational transition is thus given by

$$k' = \tilde{k}' \exp\{-[(3 + z_L)\beta + \eta]u/2\} \quad (6)$$

Little is known so far about movements of intrinsic protein charges other than bound ions or charged ligand groups, and for this reason the parameter η is usually neglected.

Access Channels

In a conformational transition ions may move inside the protein, but it is unlikely that the ions are translocated across the entire membrane dielectric in a single step. It is possible to circumvent this difficulty with the assumption that the pump protein contains access channels that connect the binding sites to the adjacent aqueous medium.

Regarding the structure of an access channel, two limiting cases are possible (Fig. 3). The access channel may consist of a wide opening (or vestibule) into which water and all kinds of ions may easily enter. Since such a wide channel has a high conductance, only a small fraction of an externally applied voltage drops across the length of the channel. Accordingly, the channel may be referred to as a low-field access channel.

In the other limiting case, the access channel is narrow and ion specific, admitting only the transported ions. Such a channel, which has a low conductance, is called a high-field access channel, or ion well.

In an ion well, part of the transmembrane voltage drops between the ion-binding site and the aqueous medium. As a consequence, the apparent equilibrium dissociation constant K of the ion becomes voltage dependent:

$$K'' = \tilde{K}'' \exp(\alpha'' u) \tag{7}$$

If the ion well is located at the extracellular side (as assumed here), an inside-positive voltage ($u > 0$) enhances dissociation of cations from the binding site.

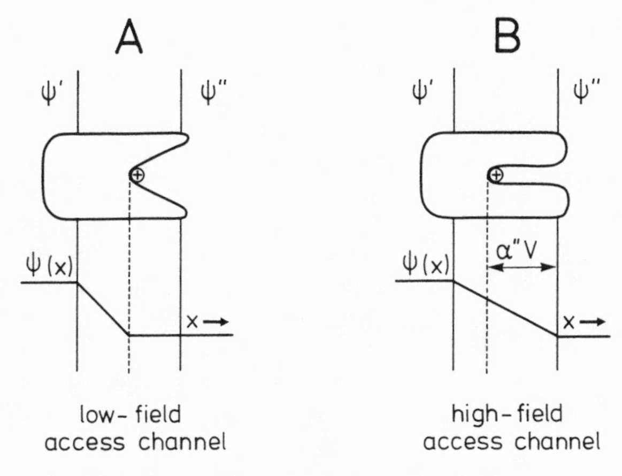

Figure 3. Two limiting cases for the structure of the access channel connecting an ion-binding site to the external medium. (*A*) Low-field access channel, consisting of a wide opening (or vestibule) into which water and all kinds of ions may freely enter. An externally applied voltage $V \equiv \psi' - \psi''$ does not lead to an appreciable potential drop across the channel. (*B*) High-field access channel (ion well) consisting of a narrow and ion-selective pore into which only the transported ions may enter. A fraction $\alpha''V$ of the externally applied voltage drops across the access channel.

Shape of the Current–Voltage Curve

Depending on the values of the rate constants and the dielectric coefficients of the individual reaction steps, the current–voltage characteristic of the pump can assume a variety of shapes. This may be illustrated by considering the simple four-state reaction cycle represented in Fig. 4.

In case A (Fig. 4) it is assumed that the cycle contains a single voltage-dependent reaction step which, in addition, is rate limiting. Increasing the voltage accelerates the rate-limiting step in the forward direction. This leads to an exponential dependence of pump current I_p on voltage V in the whole voltage range.

In case B, the rate-limiting step is voltage independent, but is preceded by a fast voltage-dependent reaction. In this case, although the rate constants of the slow step are voltage independent, the pump current nevertheless depends on voltage. This results from the fact that a voltage change shifts the equilibrium of the reaction preceding the rate-limiting step. A positive voltage increases the concentration of the reactant (state E_2) entering the rate-limiting step, and this increases the overall reaction rate. At large positive voltage, however, the equilibrium is completely shifted toward E_2, so that the current saturates.

In case C it is assumed that the pumping cycle contains two voltage-dependent steps of opposite polarity. For instance, release of Na^+ from an ion well at the extracellular face of the Na,K-pump is enhanced by an inside-positive voltage, but the following reaction, uptake of K^+ into an ion well, is inhibited. At large positive voltage, the reaction step with reverse voltage dependence is slowed down and eventually becomes rate limiting. This leads to a current–voltage curve with a region of negative slope.

It is remarkable that even a simple four-state reaction cycle can lead to very diverse shapes of the *I-V* curve. Of course, the operation of a real pump involves transitions between many states and is described by many rate constants and

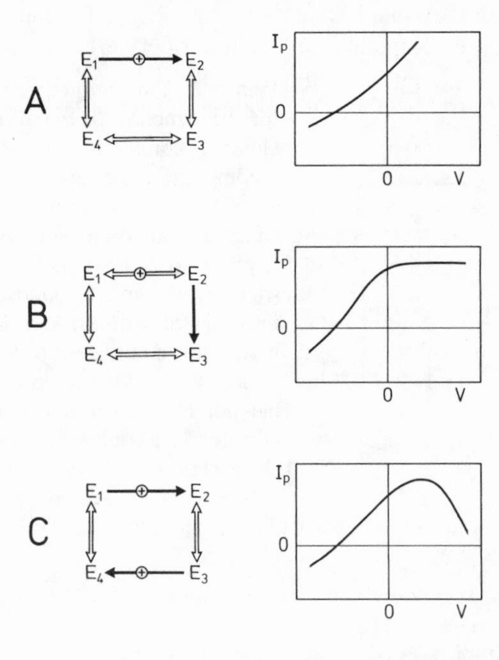

Figure 4. Possible shapes of current–voltage curves in a four-state pumping cycle. Rate-limiting reaction steps are indicated by a single-headed arrow and fast equilibrium reactions by a double-headed arrow. (*A*) Single charge-translocation step which is rate limiting. The $I_p(V)$ curve increases exponentially in the whole voltage range. (*B*) Single charge-translocating reaction preceding the rate-limiting step of the cycle. The $I_p(V)$ curve saturates at large voltage. (*C*) Two charge-translocating reaction steps of opposite voltage dependence. The $I_p(V)$ curve exhibits a region with negative slope.

dielectric coefficients. It may therefore seem difficult to extract useful microscopic information from current–voltage studies of ion pumps. But this is certainly not the case. The shape of the *I-V* curve depends on ion and substrate concentrations. Studying the *I-V* characteristic in a wide concentration range of ions, nucleotides, and inorganic phosphate provides a detailed data base from which microscopic parameters of the transport cycle can be evaluated.

Electrogenic Reaction Steps in the Post-Albers Cycle

Correlation between Charge-translocating Steps and Conformational Transitions

To study the question of which steps in the reaction cycle involve charge translocation, it is advantageous to choose conditions in which the pump can move through only part of the reaction cycle. Such experiments have been carried out with membrane fragments containing pump molecules in high density, which are bound to a planar lipid bilayer. After release of ATP from "caged" ATP in the presence of Na^+ and the absence of K^+, time-dependent currents are observed, reflecting transient charge movements in the pump protein (Fendler et al., 1985; Borlinghaus et al., 1987). In this condition the system is restricted to move through the sodium limb of the pumping cycle (Fig. 5). This means that charge translocation could occur either in the step in which the protein is phosphorylated and Na^+ is occluded ($Na_3 \cdot E_1 \cdot ATP \rightarrow (Na_3)E_1$-P), or in the subsequent deocclusion and release of Na^+ to the extracellular side ($(Na_3)E_1\text{-P} \rightarrow \ldots \rightarrow \text{P-}E_2 + 3Na^+_{ext}$).

To distinguish between these possibilities, two different experimental approaches are feasible. In the first set of experiments, transient charge movements are compared with the time course of conformational transitions. For this purpose, the pump protein is covalently labeled with 5-iodoacetamidofluorescein (IAF). When the pump protein undergoes a conformational transition from E_1 to E_2, the dye responds with a fluorescence decrease (Kapakos and Steinberg, 1986). The time course of the fluorescence signal, $\Delta F/F_o$, can be compared with the time course of the movement of charge Q (Stürmer et al., 1989). $Q(t)$ is obtained by integration of

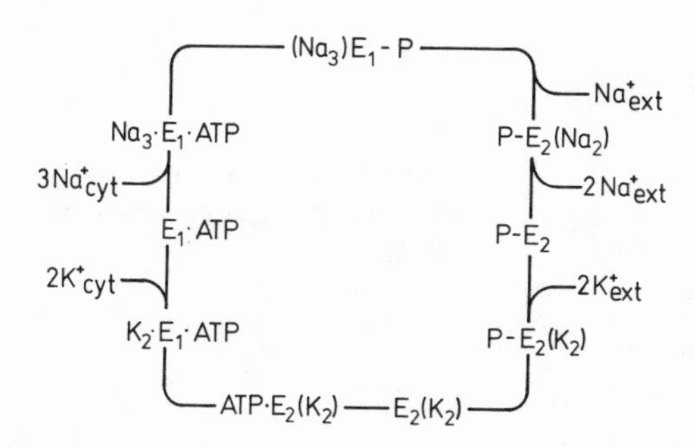

Figure 5. Post-Albers scheme of the pumping cycle, adapted from Forbush (1988) and Jørgensen and Andersen (1988). In the occluded states $(Na_3)E_1$-P, P-$E_2(Na_2)$, P-$E_2(K_2)$, and $E_2(K_2)$, the bound ions are unable to exchange with the aqueous phase. Dashes indicate covalent bonds and dots indicate noncovalent bonds. (From Stürmer et al., 1991*a*).

the pump current $I_p(t)$. The finding that $\Delta F(t)/F_o$ and $Q(t)$ closely agree indicates that charge translocation in the Na^+ transport route occurs after formation of the phosphorylated state $(Na_3)E_1$-P with occluded Na^+.

A second experiment involves blocking the cycle after the phosphorylation step. This can be done by chymotryptic cleavage of a bond in the α-chain. Under this condition phosphorylation and occlusion of Na^+ still occur, but the subsequent deocclusion of Na^+ is inhibited. After chymotrypsin modification of the protein, both the transient current and the IAF signal are abolished (Borlinghaus et al., 1987; Stürmer et al., 1989). This again shows that the charge-translocating event must occur after occlusion of Na^+.

In the K^+ limb of the cycle, a totally different behavior is observed. In the presence of K^+ and the absence of Na^+ the pump is initially in state $E_2(K_2)$ with occluded K^+. In this condition, ATP release from caged ATP may be expected to shift the equilibrium back toward E_1. This is indeed found in the IAF experiment in which ATP release induces a large positive fluorescence change. Under the same conditions, however, no transient current is observed (Stürmer et al., 1989). The finding that the reaction sequence from $E_2(K_2)$ to E_1 is electrically silent agrees with electrophysiological studies by Bahinski et al. (1988) and with studies of reconstituted Na,K-ATPase by Goldshleger et al. (1990).

Release of Na^+ at the Extracellular Side

The studies of transient currents discussed above have shown that deocclusion, with release of Na^+ to the extracellular side, is electrogenic. A possible explanation of this finding is that charge translocation results from movement of Na^+ in a narrow access channel, or ion well. This hypothesis can be tested in experiments using electrochromic dyes. When ions are released from a binding site that is buried in the membrane dielectric, the electric potential and the electric field strength change in the vicinity of the binding site. Such changes of field strength may be sensed by electrochromic dyes incorporated in the membrane (Klodos and Forbush, 1988; Bühler et al., 1991; Stürmer et al., 1991*a*, *b*). Electrochromic styryl dyes such as RH 421 (Fig. 6) have a delocalized positive charge and a localized negative charge and are thought to insert into the lipid bilayer with the axis of the molecule parallel to the hydrocarbon chains of the lipids (Grinvald et al., 1982).

When membrane fragments are labeled with RH 421, phosphorylation in the presence of Na^+ and absence of K^+ leads to a large increase of fluorescence, corresponding to the creation of a negative potential in the interior of the membrane.

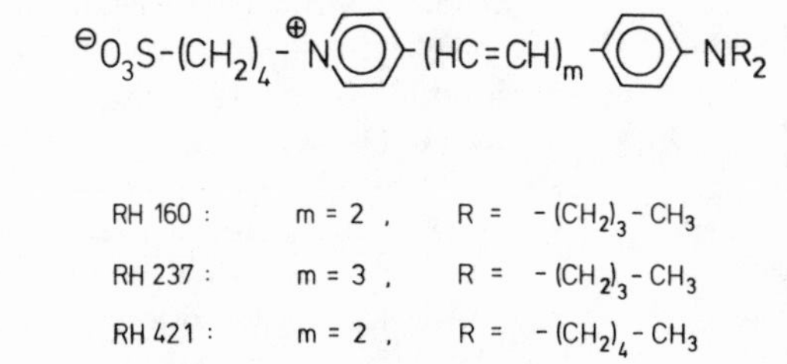

$$^{\ominus}O_3S{-}(CH_2)_4{-}\overset{\oplus}{N}C_5H_4{-}(HC{=}CH)_m{-}C_6H_4{-}NR_2$$

RH 160 : $m = 2$, $R = -(CH_2)_3-CH_3$

RH 237 : $m = 3$, $R = -(CH_2)_3-CH_3$

RH 421 : $m = 2$, $R = -(CH_2)_4-CH_3$

Figure 6. Structure of electrochromic styryl dyes, after Grinvald et al. (1982).

The fluorescence signal increases with Na^+ concentration, with half-maximal activation at a Na^+ concentration similar to the apparent Na^+ affinity at the cytoplasmic side. An obvious interpretation of this experiment is that after phosphorylation Na^+ is released at the extracellular face of the pump, and the associated change of electric field strength is sensed by the electrochromic dye (Fig. 7).

If this interpretation is correct, the signal amplitude should decrease at high Na^+ concentrations at which Na^+ release at the extracellular side is suppressed. This is indeed the case, as shown in Fig. 8 (curve labeled "0 TPP^+/TPB^-"). At high Na^+ concentration, the signal amplitude declines, reflecting the low Na^+ affinity at the extracellular release site. In these experiments, the ionic strength was held constant by addition of choline chloride, so that surface charge effects can be excluded.

Electrostatic Potentials Created by Adsorption of Lipophilic Ions: Effect on Na^+ Release

The notion that sodium is released from an ion well at the extracellular side can be further tested in experiments in the presence of lipophilic ions such as tetraphenylphosphonium (TPP^+) or tetraphenylborate (TPB^-). These ions are known to partition between water and a lipid membrane and to adsorb to a plane located

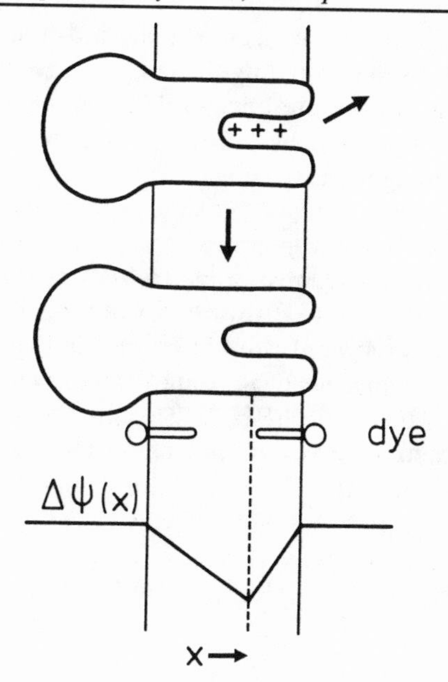

Figure 7. Release of Na^+ from a narrow access channel (ion well) at the extracellular side leads to a change of electric field strength in the membrane, which is sensed by the electrochromic styryl dye. The curve labeled "$\Delta\psi(x)$" indicates the change of electric potential ψ upon release of Na^+.

within the membrane dielectric a few tenths of a nanometer away from the membrane–solution interface (Andersen et al., 1978). When a lipophilic cation is bound to the membrane, a positive electrostatic potential is created inside the membrane, which should enhance release of Na^+ ions (Fig. 9). This expectation is borne out by fluorescence experiments in the presence of TPP^+, in which only a weak decline of signal amplitude is observed (Fig. 8, curve labeled "300 μM TPP^+;" Stürmer et al., 1991*a*, *b*). On the other hand, in the presence of TPB^- the signal

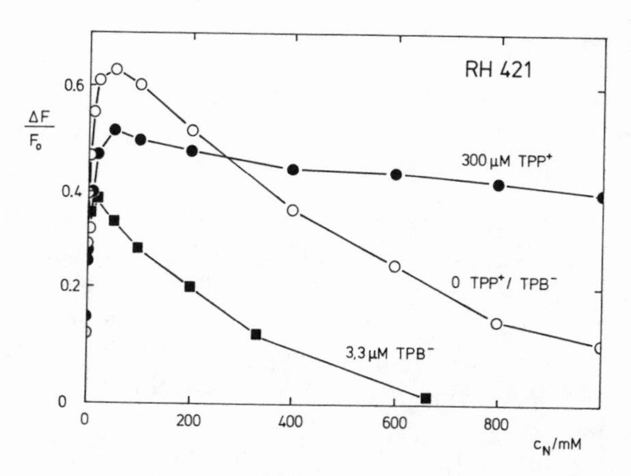

Figure 8. Change of fluorescence of RH 421 upon ATP-induced phosphorylation of Na,K-ATPase membranes, as a function of Na^+ concentration c_N in the presence and absence of tetraphenylphosphonium (TPP^+) or tetraphenylborate (TPB^-) (Stürmer et al., 1991*a*). The fluorescence change ΔF is referred to the fluorescence F_o before addition of ATP. ΔF was evaluated from time-resolved fluorescence experiments (ΔF is the stationary amplitude that is reached at long times). The ionic strength was held constant by addition of choline chloride. A suspension of membrane fragments (30 mg protein/ml) was added to a solution of 0.7 μM RH 421, 20 μM "caged" ATP, 10 mM $MgCl_2$, 1 M (NaCl + choline chloride), 1 mM EDTA, 30 mM imidazole chloride, pH 7.2, and 0 or 300 μM TPP^+ or 3.3 μM TPB^-. About 5 μM ATP was released by a 308-nm flash of 10 ns duration. The temperature was 20°C.

amplitude decreases already at much lower Na^+ concentrations, indicating that TPB^- increases the Na^+ affinity of the extracellular binding site (Fig. 8, curve labeled "3.3 μM TPB^-"). The fluorescence signals in the presence of TPP^+ or TPB^- were found to be insensitive to a variation in ionic strength by addition of choline chloride. This means that surface charge effects are negligible.

Binding of K^+ at the Extracellular Side

If release of Na^+ at the extracellular side leads to a decrease of electrostatic potential inside the membrane, then an opposite effect may be expected when K^+ is bound from the extracellular side. This prediction may be tested in experiments in which phosphorylation in the presence of Na^+ and absence of K^+ is induced by addition of ATP, leading to a large increase of the fluorescence of RH 421 (as described above). When K^+ is subsequently added to the medium, the fluorescence returns to almost

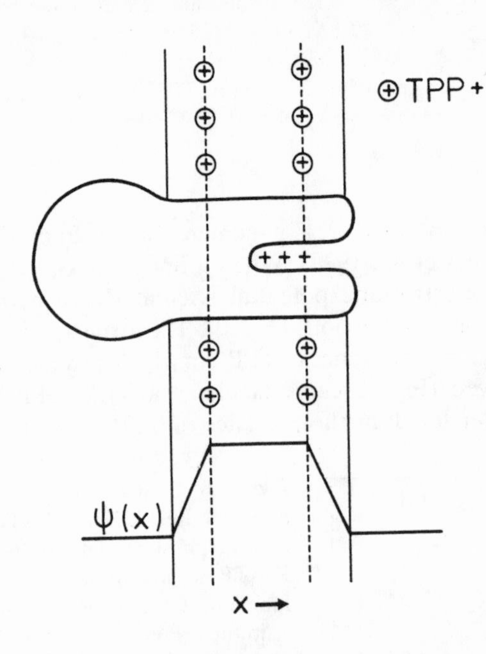

Figure 9. Creation of a positive electrostatic potential inside the membrane by adsorption of lipophilic cations (TPP^+).

the initial value (Klodos and Forbush, 1988; Stürmer et al., 1991*a*, *b*). Both the fluorescence increase in the presence of Na^+ and the fluorescence decrease upon subsequent addition of K^+ are nearly independent of ionic strength. A likely explanation for the K^+-induced fluorescence change is that K^+ binds to sites that are located inside the membrane dielectric. These sites may be identical to two of the sites from which Na^+ is released.

The notion that K^+ binds to a buried site inside the protein is further supported by experiments in which lipophilic cations are allowed to adsorb to the membrane. In the presence of TPP^+, higher K^+ concentrations are required for the fluorescence signal to reach saturation (Fig. 10). The observed dependence of the half-saturation concentration of K^+ on the concentration of TPP^+ is consistent with the expectation that the lipophilic cation electrostatically inhibits K^+ binding to a site located in the interior of the protein.

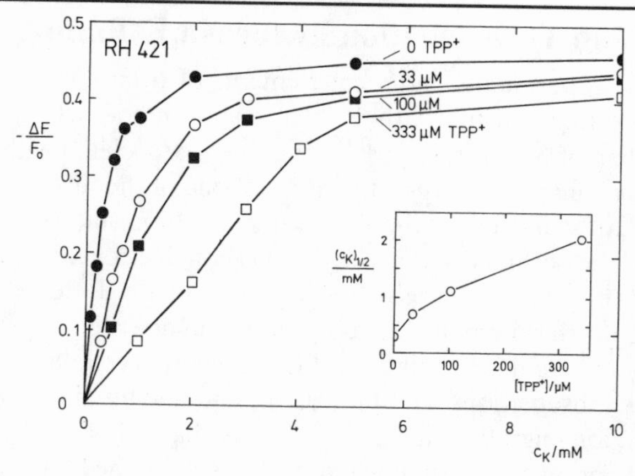

Figure 10. Change of fluorescence of RH 421 upon addition of K^+ to the phosphoenzyme, as a function of K^+ concentration c_K. The fluorescence change ΔF is referred to the fluorescence F_o before addition of ATP. ΔF was evaluated from time-resolved fluorescence experiments and taken as the difference of the stationary signal amplitudes at long times after release of ATP in the presence and in the absence of K^+. A suspension of membrane fragments (30 µg protein/ml) was added to a solution of 0.7 µM RH 421, 20 µM "caged" ATP, 10 mM $MgCl_2$, 50 mM NaCl, 1 M choline chloride, 1 mM EDTA, 30 mM imidazole chloride, pH 7.2, and various concentrations of KCl and tetraphenylphosphonium (TPP^+). In the absence of choline chloride the results were virtually identical. About 5 µM ATP was released by a 308-nm flash of 10 ns duration. The temperature was 20°C.

Ion Binding at the Cytoplasmic Side

When the pump is in state E_1 and Na^+ ions are bound to the cytoplasmic sites in the absence of ATP, the change of RH 421 fluorescence is small, indicating that the binding sites are located not far from the dielectric interface (Klodos and Forbush, 1988; Stürmer et al., 1991*a*, *b*). Only a very small fluorescence change is observed upon addition of K^+ to the medium. This indicates that binding of K^+ in state E_1 and transition to the occluded state ($E_2(K_2)$) are electrically silent processes.

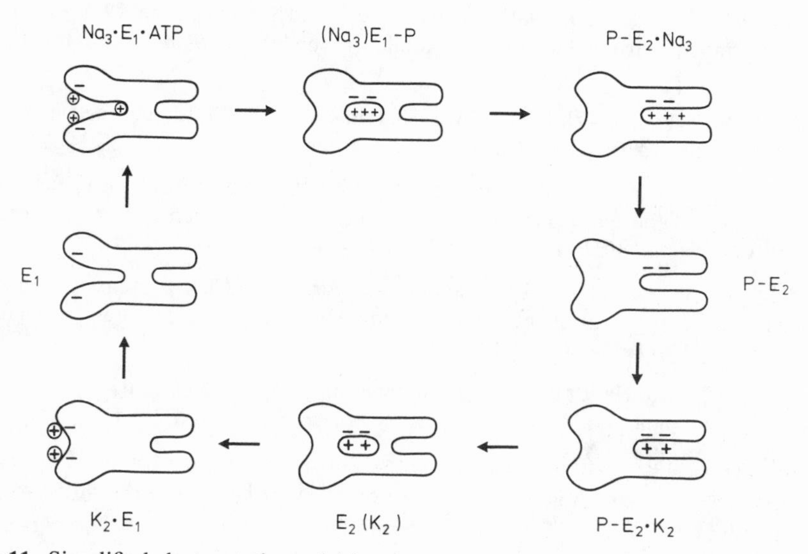

Figure 11. Simplified electrostatic model for charge translocation by the Na,K-pump. See text for explanations.

Electrostatic Model for Ion Translocation by the Na,K-Pump

The results of the fluorescence experiments and measurements of transient pump currents with the kidney enzyme are consistent with the electrostatic model schematically depicted in Fig. 11. Binding of Na^+ at the cytoplasmic side is associated with a minor charge translocation which may result from migration of one of the three Na^+ ions to a buried binding site. As indicated by the experiments with chymotrypsin-modified enzyme, formation of the occluded state $(Na_3)E_1$-P is nonelectrogenic. The process of Na^+ occlusion may thus involve merely the closing of a gate. Release of Na^+ to the extracellular side, on the other hand, represents a major electrogenic event. Another major electrogenic event is the binding of K^+ at the extracellular side. An obvious explanation of these observations is that release of Na^+ and binding of K^+ at the extracellular side involve ion migration in a narrow access channel.

It seems reasonable to assume, as has been repeatedly discussed in the literature (e.g., De Weer et al., 1988), that the liganding groups in the ion-binding sites bear two negative charges. This would explain the finding that the conformational rearrangements involved in occlusion of K^+ and the transition to state $K_2{\cdot}E_1$ with inward-facing ion-binding sites are electroneutral processes.

Acknowledgments

I want to thank H.-J. Apell, R. Bühler, W. Stürmer, and I. Wuddel for interesting discussions.

This work was supported by Deutsche Forschungsgemeinschaft (Sonderforschungsbereich 156).

References

Andersen, O. S., S. Feldberg, H. Nakadomary, S. Levy, and S. McLaughlin. 1978. Electrostatic interactions among hydrophobic ions in lipid bilayer membranes. *Biophysical Journal.* 21:35–70.

Bahinski, A., M. Nakao, and D. C. Gadsby. 1988. Potassium translocation by the Na^+/K^+ pump is voltage insensitive. *Proceedings of the National Academy of Sciences, USA.* 85:3412–3416.

Borlinghaus, R., H.-J. Apell, and P. Läuger. 1987. Fast charge translocations associated with partial reactions of the Na,K-pump. I. Current and voltage transients after photochemical release of ATP. *Journal of Membrane Biology.* 97:161–178.

Bühler, R., W. Stürmer, H.-J. Apell, and P. Läuger. 1991. Charge translocation by the Na,K-pump. I. Kinetics of local field changes studied by time-resolved fluorescence measurements. *Journal of Membrane Biology.* 121:141–161.

De Weer, P. 1990. The Na/K pump: a current-generating enzyme. *In* Regulation of Potassium Transport across Biological Membranes. L. Reuss, G. Szabo, and J. M. Russel, editors. University of Texas Press, Austin. 1–20.

De Weer, P., D. C. Gadsby, and R. F. Rakowski. 1988. Voltage dependence of the Na,K-pump. *Annual Review of Physiology.* 50:225–241.

Fendler, K., E. Grell, M. Haubs, and E. Bamberg. 1985. Pump currents generated by the purified Na^+K^+-ATPase from kidney on black lipid membranes. *EMBO Journal* 4:3079–3085.

Forbush, B., III. 1988. Occluded ions and Na,K-ATPase, *In* The Na^+,K^+-Pump. Part A: Molecular Aspects. J. C. Skou, J. B. Nørby, A. B. Maunsbach, and M. Esman, editors. Alan R. Liss, Inc., New York. 229–248.

Gadsby, D. C., and M. Nakao. 1989. Steady-state current-voltage relationship of the Na/K pump in guinea pig ventricular myocytes. *Journal of General Physiology.* 94:511–537.

Goldshleger, R., Y. Shahak, and S. J. D. Karlish. 1990. Electrogenic and electroneutral transport modes of renal Na/K ATPase reconstituted into proteoliposomes. *Journal of Membrane Biology.* 113:139–154.

Grinvald, A., R. Hildesheim, I. C. Farber, and L. Anglister. 1982. Improved fluorescent probes for the measurement of rapid changes in membrane potential. *Biophysical Journal.* 39:301–308.

Jørgensen, P.L., and J. P. Andersen. 1988. Structural basis for E_1-E_2 conformational transitions in Na,K-pump and Ca-pump proteins. *Journal of Membrane Biology.* 103:95–120.

Kapakos, J. G., and M. Steinberg. 1986. Ligand binding to [Na,K]-ATPase labeled with 5-iodoacetamidofluorescein. *Journal of Biological Chemistry.* 261:2084–2089.

Klodos, I., and B. Forbush III. 1988. Rapid conformatoinal changes of the Na/K pump revealed by a fluorescent dye, RH 160. *Journal of General Physiology.* 92:46*a*. (Abstr.)

Läuger, P., and H.-J. Apell. 1986. A microscopic model for the current-voltage behaviour of the Na,K-pump. *European Biophysical Journal.* 13:309–321.

Nakao, M., and D. C. Gadsby. 1986. Voltage dependence of Na translocation by the Na/K pump. *Nature.* 323:628–630.

Stürmer, W., H.-J. Apell, I. Wuddel, and P. Läuger. 1989. Conformational transitions and charge translocation by the Na,K-pump: comparison of optical and electrical transients elicited by ATP-concentration jumps. *Journal of Membrane Biology.* 110:67–86.

Stürmer, W., R. Bühler, H.-J. Apell, and P. Läuger. 1991*a*. Charge translocation by the Na,K-pump. II. Ion binding and release at the extracellular face. *Journal of Membrane Biology.* 121:163–176.

Stürmer, W., R. Bühler, H.-J. Apell, and P. Läuger. 1991*b*. Charge translocation by the sodium pump: ion binding and release studied by time-resolved fluorescence measurements. *In* The Sodium Pump: Recent Developments. J. H. Kaplan and P. De Weer, editors. The Rockefeller University Press, New York. In press.

Chapter 23

Voltage Dependence of the Na,K-ATPase Incorporated into Planar Lipid Membranes

Andreas Eisenrauch, Ernst Grell, and Ernst Bamberg

Max-Planck-Institut für Biophysik, D-6000 Frankfurt am Main 70, Germany

Introduction

The Na,K-ATPase (EC 3.6.1.37) is an electrogenic ion pump that exchanges three Na^+ ions and two K^+ ions across the cell membrane (Sen and Post, 1964; Eisner and Lederer, 1980; Lederer and Nelson, 1984; Avison et al., 1987). Therefore, the enzyme performs the translocation of one positive net charge out of the cell. Its electrogenicity has been demonstrated several times by electrophysiologic measurement of ouabain-sensitive membrane currents in various muscle and nerve cell types with high plasmalemmal ATPase densities (Thomas, 1969; Isenberg and Trautwein, 1974; Eisner and Lederer, 1980; Glitsch et al., 1982; Lederer and Nelson, 1984). Measurements of the voltage dependence, however, revealed two types of results. Several groups (Gadsby et al., 1985; De Weer, 1986; Glitsch and Krahn, 1986) observed monotonically increasing *I-V* relations which could be explained by electrogenic charge translocations in only one direction. Together with other independent results obtained with whole cells (Nakao and Gadsby, 1986; Bahinski et al., 1988) and in reconstituted systems (Borlinghaus et al., 1987; Fendler et al., 1987; Goldshleger et al., 1987; Nagel et al., 1987), these electrogenic events were correlated with the sodium translocation steps. In contrast, Lafaire and Schwarz (1986) as well as Schwarz and Gu (1988) reported bell-shaped *I-V* curves in *Xenopus* oocytes, implying electrogenic transport steps in both directions. In addition, recent data by Rakowski et al. (1990) showed that a negative slope of the *I-V* relationship could be attributed to voltage-dependent potassium binding before the potassium translocation steps.

The aim of the work presented here was to investigate the electrogenic properties and the current–voltage behavior of a highly purified preparation of pig kidney Na,K-ATPase reconstituted into artificial planar bilayer membranes.

Membrane fragments containing highly purified Na,K-ATPase from pig kidney were incorporated into planar lipid membranes by a modified Montal-Müller method. The ATPase was activated by an ATP concentration jump after photolysis of caged ATP. Pump currents were observed that were dependent on the presence of Na^+ ions and ATP, and sensitive to ouabain and vanadate ions. The voltage dependence of the pump currents was investigated under different conditions and is discussed on the basis of the Albers-Post scheme of the reaction cycle of the Na,K-ATPase. The shape of all *I-V* curves is monotonic, increasing with positive potentials of the ATP-containing compartment, demonstrating that only charge-carrying steps in the direction of the sodium translocation are present.

Materials and Methods

Materials

The Na,K-ATPase-containing membrane sheets were prepared from pig kidney red outer medulla according to the method of Jørgensen (1974). Caged ATP was synthesized following the modified protocol of Kaplan et al. (1978) as described previously (Fendler et al., 1985). The chemicals used were usually of analytical grade; in the experiments where only a single alkali cation species was used all salts were of suprapur grade (E. Merck, Darmstadt, Germany). Lipids were purchased from Avanti Polar Lipids, Inc. (Birmingham, AL). Ouabain was from Serva (Heidelberg, Germany).

Formation of Artificial Planar Bilayer Membranes

Artificial planar bilayer membranes were formed according to the method of Montal and Müller (1972), with modifications described earlier (Bamberg et al., 1981; Eisenrauch and Bamberg, 1990). Between the two halves of a Teflon cuvette there was a Teflon septum containing a hole ~150–200 μm in diameter. Both chambers were filled with the appropriate electrolyte solutions; the surface of the liquid was kept just below the hole. A droplet of 7.5 μl 2% glycerolmonooleate (GMO) in hexadecane was spread on the surface. After a few minutes the suspension containing the membrane fragments was added (~20 μg protein per chamber). Bilayers were formed by raising the electrolyte levels beyond the hole. Bilayer formation was monitored electrically by observing the membrane capacitance using sawtooth voltage patterns. Membrane conductance was monitored by application of voltage jumps.

Activation of the Na,K-ATPase

The Na,K-ATPase was activated via a concentration jump of ATP from a photolabile "caged ATP" (Kaplan et al., 1978). Since protein was added on both sides of the septum, the orientation of the ATPases in the membrane was random. Caged ATP was added to one compartment of the cuvette in order to activate only the pumps of one orientation. ATP was released from caged ATP by 125-ms UV light flashes from a mercury lamp (light intensity 2.0 W/cm^2). The flashes were applied at 10-min intervals to allow the reactants and the reaction products to redistribute in the cuvette. The concentrations of caged ATP were adjusted to generate defined ATP concentrations near the membrane. The amount of released ATP was measured independently by the luciferin/luciferase assay procedure (Boehringer Mannheim, GmbH, Mannheim, Germany). Pump currents were amplified and converted by a current–voltage converter and monitored on a digital storage oscilloscope that was connected to an XY-plotter. In all experiments Ag/AgCl electrodes were used. To avoid light artifacts, the electrodes were installed away from the light beam and electrically connected to the cuvette by agar-filled salt bridges.

Measurement of *I-V* Curves

Voltage-dependent pump currents were obtained by applying various voltages via a constant voltage source to one compartment of the cuvette, whereas the other was clamped to virtual ground. ATP release was initiated after application of the voltage step. Since the amplitudes of the pump currents varied between different experiments and decreased slowly during an experiment, *I-V* curves were expressed as stationary pump current amplitudes (always taken 10 s after the light flash) versus applied voltage, relative to current amplitudes at zero voltage (I/I_0). In practice, two nonzero voltage measurements were taken between two events in the absence of applied voltage and related to the interpolated zero voltage values at the time of the experiments in the presence of a voltage.

The theoretical reversal potentials were calculated using the known concentrations of ATP, ADP, and P_i. For the calculation of the conditions of the first flash it was assumed that all the free ATP that contaminates the caged ATP (~1%) was hydrolyzed at the time of the first flash (10 min after the addition of the cuvette). Due to the accumulation of the reaction products of the ATPase activity the reversal potential decreased by 100 mV after the seventh UV flash. For this reason not more than seven flashes per experiment were applied; the stability of the membranes, however, would normally have allowed more.

Results

Incorporation of the Na,K-ATPase into Planar Bilayers

Membrane fragments containing purified Na,K-ATPase were added symmetrically to both compartments of the cuvette which already contained preformed monolayers of 2% GMO in hexadecane on the electrolyte surface. Artificial planar bilayer membranes were formed from the ATPase containing lipid layers. Measurement of

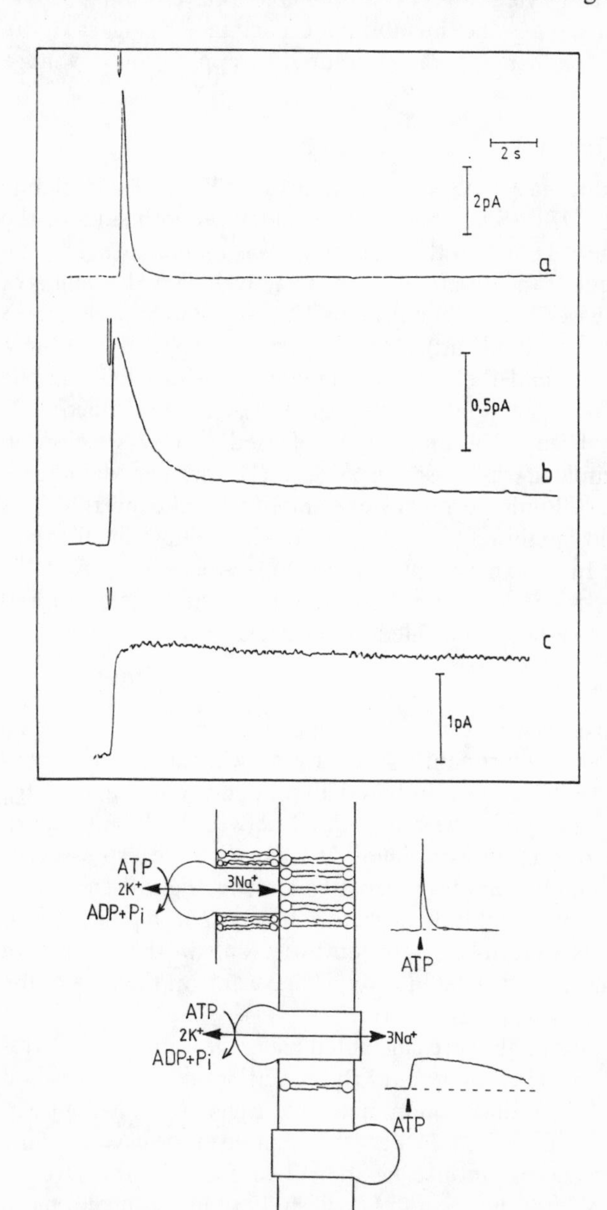

Figure 1. Pump current patterns of the Na,K-ATPase. Electrolyte: 130 mM NaCl, 20 mM KCl, 3 mM $MgCl_2$, 25 mM imidazole/HCl, pH 7.4, 1 mM DTT (on both sides), 15 μM free ATP (*cis* side only). (*Top*) Trace *b*, commonly observed combination of a peak current followed by a stationary current component. Trace *a*, occasionally no incorporation occurred and only capacitive peak currents were observed. Trace *c*, sometimes the number of incorporated pumps exceeded the attached ones and only the stationary currents component was visible. (*Bottom*) Schematic diagram of attached membrane fragments producing only capacitive currents (*upper part*) and incorporated pump molecules producing stationary currents (*lower part*). The normal current pattern is therefore believed to be a combined response.

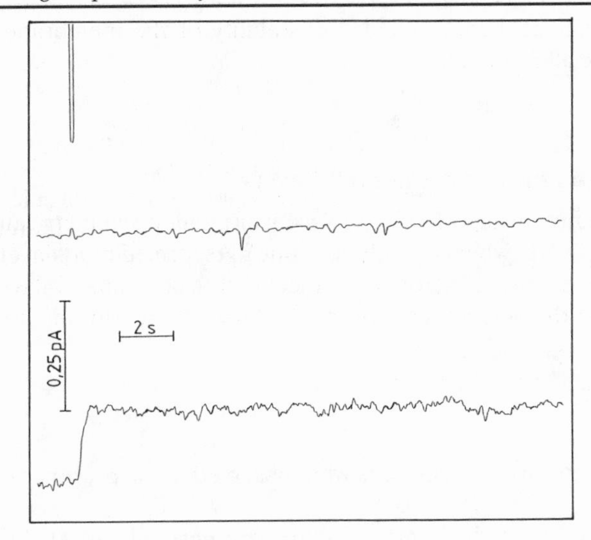

Figure 2. ATP dependence of the pump currents. Electrolyte as in Fig. 1, with the exception that caged ATP was replaced by caged ADP. *Upper trace,* caged ADP alone does not induce pump currents after its photolysis. *Lower trace,* addition of creatine phosphate (4 mM) and creatine kinase (200 U) led to pump currents after illumination.

the membrane capacitance as well as the occurrence of gramicidin A channels (not shown) demonstrated that true bilayer membranes were formed.

Upon release of 15 μM ATP from photolabile "caged ATP," stationary pump currents were observed in the absence of any ionophores, indicating that the ion pump is incorporated into the planar bilayer. The occurrence of capacitive current peaks in addition to the stationary current component shows that there are also many ATPases attached to the bilayer membrane (Fig. 1). This is confirmed by the observation that upon the addition of ionophores, e.g., 10 μM Monensin (an electroneutral cation exchanger) and 0.6 μM 1799 (an electrogenic protonophore), the stationary current amplitude increases to the level of the peak currents (Fendler et al., 1985). The release of ADP from "caged ADP" did not stimulate pump currents. Only in the presence of an ATP-generating system like creatine kinase/phosphocreatine did the release of ADP lead to pump currents (Fig. 2). In addition,

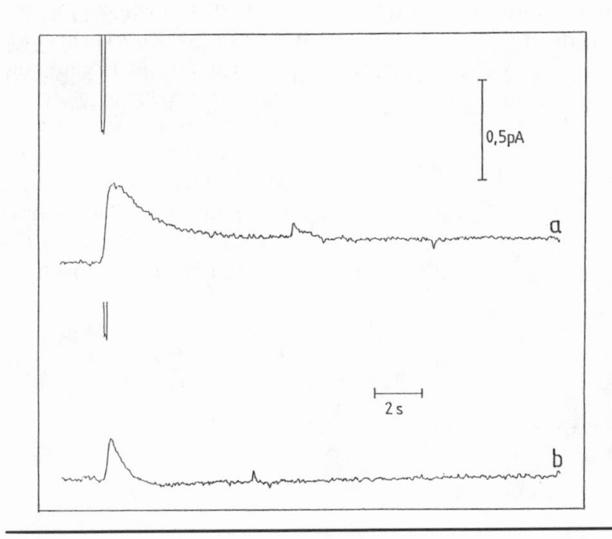

Figure 3. Action of ouabain on the Na,K-ATPase. Electrolyte: 130 mM NaCl, 20 mM KCl, 3 mM $MgCl_2$, 25 mM imidazole/HCl, pH 7.4, 1 mM DTT, 15 μM free ATP/flash. *Upper trace,* current response before ouabain application. *Lower trace,* disappearance of the stationary current component after addition of 200 μM ouabain (final concentration) to the "extracellular" compartment.

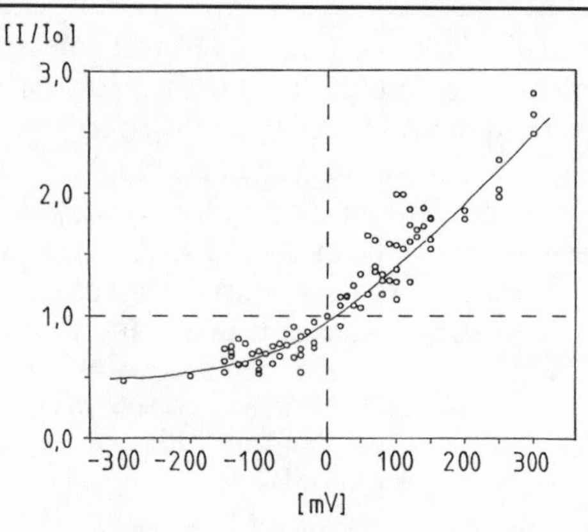

Figure 4. *I-V* curve without ion gradients. Electrolyte: 130 mM NaCl, 20 mM KCl, 3 mM $MgCl_2$, 25 mM imidazole/HCl, pH 7.4, 1 mM DTT (on both sides), 15 μM free ATP/flash (*cis* chamber).

without membrane fragments no current responses were observed in the presence of all other components.

The pump currents depend on the "intracellular" presence of 1–100 mM sodium and at least micromolar magnesium concentrations. The stationary current amplitudes were usually between 0.1 and 0.4 pA (10 s after the activating UV light flash); the largest stationary pump current was ~1 pA; the largest peaks had a size of 5–7 pA. The pump currents could be inhibited by addition of 200 μM ouabain to the "extracellular" compartment (Fig. 3) or by addition of 200 μM Na_3VO_4 to the ATP-containing chamber. In accordance with previous reports (Skou, 1957) the pH optimum of the currents (peak and stationary components) is at about pH 7.2–7.4 (not shown).

I-V Relationships of the Reconstituted Na,K-ATPase

I-V curves were obtained by applying various voltages before the photorelease of ATP from caged ATP, as described under Methods. Because of the varying amplitudes between experiments and a slight decrease of the amplitudes with time within each experiment, the data of the curves were expressed as relative stationary pump current amplitudes I/I_0 (see Methods).

A first *I-V* curve was obtained under symmetrical ionic conditions, except for the caged ATP which was present only in the *cis* compartment. In all experiments 15 μM ATP per UV flash was released in the vicinity of the membrane. It was possible to apply voltages up to ±300 mV. Over this range the stationary pump currents increase

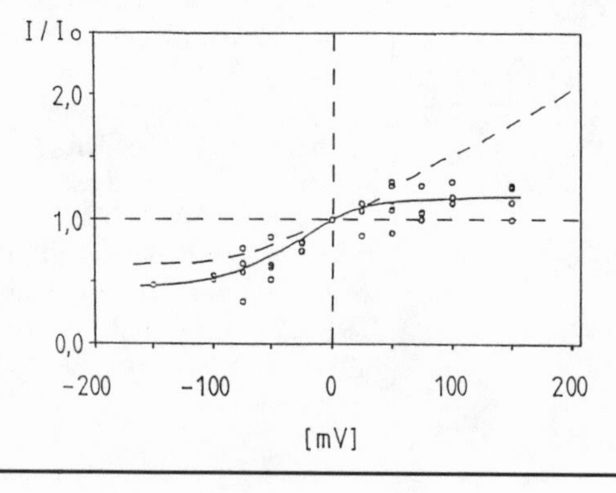

Figure 5. *I-V* curve with low sodium concentration. Electrolyte: 20 mM NaCl, 20 mM KCl, 3 mM $MgCl_2$, 25 mM imidazole/HCl, pH 7.4, 1 mM DTT (on both sides), 15 μM free ATP/flash (*cis* side only).

monotonically with depolarizing voltages (Fig. 4). The estimated reversal potential of the pump is in the range of several hundred millivolts. For comparison, the theoretical reversal potential calculated from the ATP, ADP, and P_i concentrations is about −750 mV, assuming one positive net charge transported per ATP.

A second *I-V* curve (Fig. 5) was taken with a sodium concentration reduced from 130 to 20 mM, which is near the $K_{M(Na)}$ in the presence of 20 mM KCl (Nagel et al., 1987). The reduction of the sodium concentration led to a marked decrease of the voltage dependence in the range of depolarizing voltages, whereas the estimated reversal potential remained unchanged.

Discussion

Reconstitution of the Na,K-ATPase into Planar Bilayer Membranes

In this work we tried to obtain the *I-V* relation of a highly purified pig kidney Na,K-ATPase. For this purpose we incorporated the ATPase into planar lipid membranes. It is assumed that an applied voltage drops over the integrated membrane proteins. In contrast, in experiments with purified ATPases attached to a black lipid membrane (Fendler et al., 1985, 1987; Nagel et al., 1987) it is not possible to decide which fraction of a membrane voltage drops across the pump-containing membranes.

The ATPase was provided as highly purified membrane fragments according to Jørgensen (1974). Incorporation was achieved by a modified Montal-Müller method as it was successfully applied in the case of bacteriorhodopsin (Bamberg et al., 1981) and sarcoplasmic reticulum Ca^{2+}-ATPase (Eisenrauch and Bamberg, 1990).

Pump currents could be observed after photo-release of ATP. The appearance of stationary currents in the absence of ionophores indicates that the pump is really integrated into the planar bilayer membrane. Its action, therefore, leads to a direct translocation of the cations across the membrane separating the two compartments. In addition to the incorporated pump molecules, a large number are still attached to the bilayer, thereby producing capacitive peak currents (Fig. 1). This could be demonstrated by adding the ionophores Monensin and 1799. Both ionophores increase the permeability of the membrane in a way that allows a flow of charges from the attached pumps across the bilayer. In this case the transient current disappears and the entire current response is an increased stationary current.

Na ions and ATP (from caged ATP) are indispensable. Contaminating potassium and magnesium levels are sufficient to stimulate considerable pump currents. The currents are sensitive to vanadate ions applied from the ATP side of the membrane and to ouabain acting from the opposite compartment. These findings together demonstrate that the currents stimulated by the release of ATP are pump currents derived from the Na,K-ATPase incorporated into planar bilayers.

Voltage Dependence of the Na,K-ATPase

I-V curves were measured as described under Methods and expressed as relative amplitudes I/I_0 versus applied voltage. The two curves obtained so far showed monotonically increasing pump current amplitudes with hyperpolarizing voltages. According to the theory (De Weer, 1984, 1986; Läuger, 1984, 1988), this indicates electrogenic steps in one transport direction only. With regard to the direction of the observed currents, it should be the translocation of the sodium ions that is electrogenic. This finding agrees with previous results from other experimental approaches (Nakao and Gadsby, 1986; Rephaeli et al., 1986; Borlinghaus et al., 1987; Fendler et al., 1987; Nagel et al., 1987). The potassium translocation step(s) should therefore be electroneutral as already claimed by others (Rephaeli et al., 1986; Goldshleger et al., 1987; Bahinski et al., 1988).

In every curve the estimated reversal potential is apparently in the range of several hundred millivolts. This agrees very well with the theoretical reversal potential (EMF) calculated from the ATP, ADP, and P_i concentrations, which is about −750 mV (in absence of ion gradients) assuming $n = 1$ transported net charges (see Eq. 1):

$$\mathrm{EMF} = (\Delta G_{\mathrm{ATP}} - \mathrm{A}_{\mathrm{osm}})/n \text{ with } \Delta G_{\mathrm{ATP}} = \Delta G^0 \ln [\mathrm{ATP}]/[\mathrm{ADP}][\mathrm{P_i}] \qquad (1)$$

Thus, our results are in agreement with the commonly accepted 3Na/2K stoichiometry of the Na,K-ATPase.

A pronounced voltage dependence should occur mainly in the case of a rate-limiting electrogenic step (De Weer, 1986). This is in contrast to several findings that some steps in the potassium branch of the Na,K-ATPase cycle should be slow (Karlish and Yates, 1978; Glynn et al., 1987), particularly in the presence of low ATP concentrations as in our experiments. However, Bahinski et al. (1988) showed that voltage dependence can also occur if a fast electrogenic step modulates the equilibrium concentrations of subsequent slow nonelectrogenic steps, thereby having effects on the overall reaction cycle rate. Apparently this model also explains our results.

A considerable decrease of the voltage dependence of the Na,K-ATPase was observed if the sodium concentration was reduced to 20 mM on both sides of the membrane. This finding must be interpreted as the effect of the lowered intracellular sodium concentration. Sodium-dependent steps (probably the binding of the sodium ions to their transport sites) now become slow, and therefore another rate-limiting step is created before the electrogenic step in the reaction cycle. This new electroneutral rate-limiting step is much less affected by the voltage-dependent fast step.

Acknowledgments

We thank Mr. W. Kümmel for the construction of several electrical devices and Mrs. A. Ifftner, Mr. E. Lewitzky, and Mr. G. Schimmack for excellent technical assistance.

This work was partially supported by the Deutsche Forschungsgemeinschaft (SFB 169).

References

Avison, M. J., S. R. Gullans, T. Ogino, G. Giebisch, and R. G. Shulman. 1987. Measurement of Na^+-K^+ coupling ratio of Na^+-K^+-ATPase in rabbit proximal tubules. *American Journal of Physiology.* 253:C126–C136.

Bahinski, A., M. Nakao, and D. C. Gadsby. 1988. Potassium translocation by the Na^+/K^+ pump is voltage insensitive. *Proceedings of the National Academy of Sciences, USA.* 85:3412–3416.

Bamberg, E., N. A. Dencher, A. Fahr, and M. P. Heyn. 1981. Transmembraneous incorporation of photoelectrically active bacteriorhodopsin in planar lipid bilayers. *Proceedings of the National Academy of Sciences, USA.* 78:7502–7506.

Borlinghaus, R., H. J. Apell, and P. Läuger. 1987. Fast charge translocations associated with partial reactions of the Na,K-pump. I. Current and voltage transients after photochemical release of ATP. *Journal of Membrane Biology.* 97:161–178.

De Weer, P. 1984. Electrogenic pumps: theoretical and practical considerations. *In* Electrogenic Transport: Fundamental Principles and Physiological Implications. M. P. Blaustein and M. Lieberman, editors. Raven Press, New York. 1–15.

De Weer, P. 1986. The electrogenic sodium pump: thermodynamics and kinetics. *In* Fortschritte der Zoologie: Membrane Control of Cellular Activity. H. C. Lüttgau, editor. Gustav Fischer Verlag, Stuttgart, New York. 387–399.

Eisenrauch, A., and E. Bamberg. 1990. Voltage-dependent pump currents of the sarcoplasmic reticulum Ca^{2+}-ATPase in planar lipid membranes. *FEBS Letters.* 268:152–156.

Eisner, D. A., and W. J. Lederer. 1980. Characterization of the electrogenic sodium pump in cardiac purkinje fibers. *Journal of Physiology.* 303:441–474.

Fendler, K., E. Grell, and E. Bamberg. 1987. Kinetics of pump currents generated by the Na^+,K^+-ATPase. *FEBS Letters.* 224:83–88.

Fendler, K., E. Grell, M. Haubs, and E. Bamberg. 1985. Pump currents generated by purified Na^+K^+-ATPase from kidney on black lipid membranes. *EMBO Journal.* 4:3079–3085.

Gadsby, D. C., J. Kimura, and A. Noma. 1985. Voltage dependence of Na/K pump current in isolated heart cells. *Nature.* 315:63–65.

Glitsch, H. G., and T. Krahn. 1986. The cardiac electrogenic Na pump. *In* Fortschritte der Zoologie: Membrane Control of Cellular Activity. H.C. Lüttgau, editor. Gustav Fischer Verlag, Stuttgart, New York. 401–417.

Glitsch, H. G., H. Pusch, T. Schumacher, and F. Verdonck. 1982. An identification of the K activated Na pump current in sheep purkinje fibres. *Pflügers Archiv.* 394:256–263.

Glynn, I. M., Y. Hara, D. E. Richards, and M. Steinberg. 1987. Comparison of rates of cation release and of conformational change in dog kidney Na,K-ATPase. *Journal of Physiology.* 383:477–485.

Goldshleger, R., S. J. D. Karlish, A. Rephaeli, and W. D. Stein. 1987. The effect of membrane potential on the mammalian sodium-potassium pump reconstituted into phospholipid vesicles. *Journal of Physiology.* 387:331–355.

Isenberg, G., and W. Trautwein. 1974. The effect of dihydro-ouabain and lithium-ions on the outward current in cardiac Purkinje fibers. *Pflügers Archiv.* 350:41–54.

Jørgensen, P. L. 1974. Isolation of (Na^+ +K^+)-ATPase. *Methods in Enzymology.* 32:277–291.

Kaplan, J. H., B. Forbush III, and J. F. Hoffman. 1978. Rapid photolytic release of adenosine 5-triphosphate from a protected analogue: utilization by the Na:K pump of human red blood cell ghosts. *Biochemistry.* 17:1929–1935.

Karlish, S. J. D., and D. W. Yates. 1978. Tryptophan fluorescence of (Na^+ + K^+)-ATPase as a tool for study of the enzyme mechanism. *Biochimica et Biophysica Acta.* 527:115–130.

Lafaire, A. V., and W. Schwarz. 1986. Voltage dependence of the rheogenic Na^+/K^+ ATPase in the membrane of oocytes of *Xenopus laevis. Journal of Membrane Biology.* 91:43–51.

Läuger, P. 1984. Thermodynamic and kinetic properties of electrogenic ion pumps. *Biochimica et Biophysica Acta.* 779:307–341.

Läuger, P. 1988. Electrogenic properties of the Na^+/K^+-pump. *In* The Ion Pumps: Structure, Function, Regulation. W. D. Stein, editor. Alan R. Liss, Inc., New York. 217–224.

Lederer, W. J., and M. T. Nelson. 1984. Sodium pump stoichiometry determined by simultaneous measurements of sodium efflux and membrane current in barnacle. *Journal of Physiology*. 348:665–677.

Montal, M., and P. Müller. 1972. Formation of bimolecular membranes from lipid monolayers and a study of their electrical properties. *Proceedings of the National Academy of Sciences, USA*. 69:3561–3566.

Nagel, G., K. Fendler, E. Grell, and E. Bamberg. 1987. Na^+ currents generated by the purified ($Na^+ + K^+$)-ATPase on planar lipid membranes. *Biochimica et Biophysica Acta*. 901:239–249.

Nakao, M., and D. C. Gadsby. 1986. Voltage dependence of the Na translocation by the Na/K pump. *Nature*. 323:628–630.

Rakowski, R. F., and C. L. Paxson. 1988. Voltage dependence of Na/K pump current in *Xenopus* oocytes. *Journal of Membrane Biology*. 106:173–182.

Rakowski, R. F., L. A. Vasilets, and W. Schwarz. 1990. Conditions for a negative slope in the current-voltage relationship of the Na/K pump in *Xenopus* oocytes. *Biophysical Journal*. 57:182*a*. (Abstr.)

Rephaeli, A., D. E. Richards, and S. J. D. Karlish. 1986. Electrical potential accelerates the E1P(Na)-E2P conformational transition of (Na,K)-ATPase in reconstituted vesicles. *Journal of Biological Chemistry*. 261:12437–12440.

Schwarz, W., and Q. Gu. 1988. Characteristics of the Na^+/K^+-ATPase from *Torpedo californica* expressed in *Xenopus* oocytes: a combination of tracer flux measurements with electrophysiological measurements. *Biochimica et Biophysica Acta*. 945:167–174.

Sen, A. K., and R. L. Post. 1964. Stoichiometry and localization of adenosine triphosphate-dependent sodium and potassium transport in the erythrocyte. *Journal of Biological Chemistry*. 239:345–352.

Skou, J. C. 1957. The influence of some cations on an adenosine triphosphatase from peripheral nerves. *Biochimica et Biophysica Acta*. 23:394–401.

Thomas, R. C. 1969. Membrane current and intracellular sodium changes in a snail neurone during extrusion of injected sodium. *Journal of Physiology*. 201:495–514.

Chapter 24

Variations in Voltage-dependent Stimulation of the Na^+/K^+ Pump in *Xenopus* Oocytes by External Potassium

Wolfgang Schwarz and Larisa A. Vasilets

Max-Planck-Institut für Biophysik, W-6000 Frankfurt/Main, Germany

Introduction

Under physiological conditions, the Na/K-ATPase transports three Na^+ ions out of the cell and two K^+ ions into the cell per ATP molecule that is split (see Fig. 1 *A*). This mismatch in charges translocated in opposite directions results in a steady-state current that is a measure of transport activity. The pump-generated current is voltage dependent, and it is believed that the voltage dependence reflects voltage-dependent steps in the reaction cycle. In the oocytes of *Xenopus laevis,* the voltage dependence of the steady-state current of the pump indicates two voltage-dependent steps (Lafaire and Schwarz, 1985,1986; Schweigert et al., 1988) and strongly depends on the ionic composition of the extracellular medium (Rakowski et al., 1991; see also Rakowski, 1991). Below, we will briefly summarize our observations on the effects of variations in extracellular $[Na^+]$ and $[K^+]$.

Under physiological conditions, i.e., high $[Na^+]$ (90 mM) and low $[K^+]$ (1–3 mM) the current–voltage relationship passes through a maximum with a positive slope at negative membrane potentials and a negative slope at positive potentials

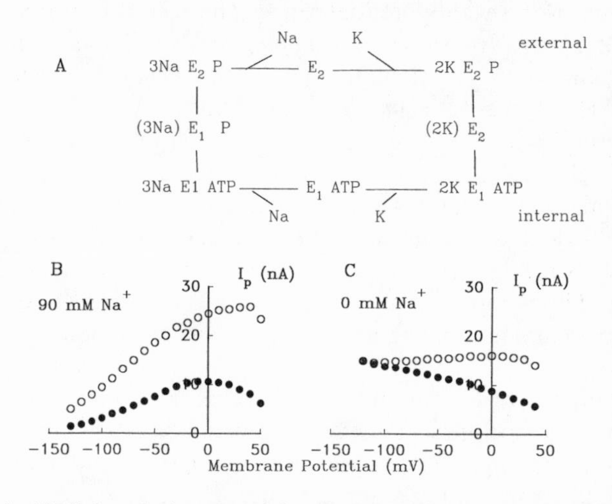

Figure 1. (*A*) Albers-Post reaction scheme for the $3Na^+/2K^+$ pump cycle. Ions in brackets represent occluded forms. (*B*) Voltage dependence of pump current at physiological extracellular $[Na^+]$ (90 mM) and (*C*) in Na^+-free solution where all NaCl was replaced by tetramethylammonium chloride. *Open circles,* pump current at 5 mM external $[K^+]$, *filled circles,* pump current at 1 mM (*A*) and 0.5 mM (*B*). Pump current was determined as steady-state K^+-sensitive current (Rakowski et al., 1991) in solution containing 5 mM $BaCl_2$ and 20 mM TEACl (see also Schweigert et al., 1988). Oocytes were preloaded with Na^+ by incubation for at least 1 h in nominally Ca^{2+}-K^+-free solution.

(Lafaire and Schwarz, 1985, 1986; see filled symbols in Fig. 1 *B*). It is now generally accepted that the positive slope originates from a voltage-dependent step involved in the Na^+-translocating half cycle of the reaction scheme for the Na^+/K^+ pump (compare Fig. 1 *A*) and is possibly related to Na^+ deocclusion and/or Na^+ release (Nakao and Gadsby, 1986; see also Läuger, 1991, and Gadsby, 1991). When we reported for the first time the existence of a negative slope (Lafaire and Schwarz, 1985), we suggested involvement of a voltage-dependent step during the K^+-translocating half cycle of the pump. However, the interpretation as well as the experimental protocol we applied was questioned (see, for example, DeWeer et al., 1988; Rakowski and Paxson, 1988). Usually pump current is determined as the difference between total membrane current with and without the pump operating, and it was claimed that the negative slope may not represent voltage dependence of pump current but rather modulation of parallel and independent ion pathways by pump inhibition. In most cells and tissues, pump current represents only a tiny fraction of total membrane current and the application of specific channel blockers is required to extract the pump current. In the past, subtraction of non–pump related current has led repeatedly to erroneous descriptions of current–voltage relationships for the Na^+/K^+ pump (Kostyuk et al., 1972; Isenberg and Trautwein, 1974; DeWeer and Rakowski, 1984). In the *Xenopus* oocytes, pump current is of about the same magnitude as non-pump current and hence much less susceptible to misinterpretation. Nevertheless, certain objections to our interpretation remained unanswered. The objections originated from the observation that (*a*) a negative slope could not be demonstrated in any other cell, and (*b*) analysis of the K^+-translocating steps in heart muscle (Bahinski et al., 1988) and in reconstituted systems (Fendler et al., 1985; Goldshleger et al., 1987; Stürmer et al., 1989) gave no indication of a second potential-dependent step. Recently we provided answers to the objections (Rakowski et al., 1991). We blocked all non–pump related K^+-sensitive currents with 5 mM Ba^{2+} and 20 mM TEA^+, and found that the remaining K^+-sensitive current was mediated by the Na^+/K^+ pump over a wide potential range. The results confirmed our previous observations of a negative slope in *Xenopus* oocytes (Lafaire and Schwarz, 1985, 1986) and allowed us to stipulate that in the *Xenopus* oocytes voltage-dependent binding of K^+ precedes the actual K^+ translocation.

Voltage-dependent K^+ binding was inferred from the observation that pump stimulation by external K^+ is voltage dependent. Elevation of the extracellular $[K^+]$ to 5 mM facilitates K^+ binding to the point where K^+ binding is no longer rate determining. Indeed, at 5 mM $[K^+]$ a current–voltage dependence of the Na^+/K^+ pump is obtained (open symbols in Fig. 1 *B*) that is very similar to the dependence reported for most other cells. The pump current increases with depolarization and shows saturation at positive potentials, and there is little if any indication of a negative slope.

Under conditions of facilitation of K^+ binding (high external $[K^+]$) and of Na^+ release (zero external $[Na^+]$), the pump current is nearly independent of membrane potential over a wide range (see open symbols in Fig. 1 *C* for 5 mM K^+ and 0 mM Na^+). This current–voltage dependence of the pump current in the *Xenopus* oocytes is very similar to the dependence seen in squid axon (Rakowski et al., 1989) and heart muscle (Nakao and Gadsby, 1989) in Na^+-free solution.

If in the absence of external Na^+, K^+ binding is made rate determining by lowering K^+ concentration, the negative slope in the current–voltage curve can be detected over the entire potential range.

We summarized the observations illustrated in Fig. 1, *B* and *C*, to emphasize that experiments performed in Na^+-free solution are particularly suited for detailed analysis of the voltage-dependent and K^+-sensitive step in the reaction cycle of the Na^+/K^+ pump. Under these conditions, the K^+-sensitive step seems to be the only voltage-dependent, rate-determining step in the transport cycle.

Results and Discussion

It is undoubted now that the activity of the Na^+/K^+ pump in the *Xenopus* oocytes is controlled by two voltage-dependent steps, but there still exists the question of whether this is also true for the Na^+/K^+ pump in other cells. For example, in heart muscle no indications were found for a second voltage-dependent step in the reaction cycle of the pump even at various extra- and intracellular Na^+ and K^+ concentrations (Nakao and Gadsby, 1988; Glitsch et al., 1989*a*, *b*). In earlier experiments, on the other hand, we had observed (Schwarz and Gu, 1988) that the Na^+/K^+ pump of *Torpedo* electroplax expressed in *Xenopus* oocytes exhibits a current–voltage relationship with both positive and negative slopes. Fig. 2 shows that a maximum can be observed for pump-generated current, as well as pump-mediated Na^+ efflux that can be measured in single oocytes under voltage clamp using microinjected $^{22}Na^+$ as a tracer (see below).

The stipulation of two voltage-dependent steps in the pump cycle, one of which is responsible for the occurrence of the negative slope at positive potentials, is based on the assumption that variations in membrane potential and external $[K^+]$ do not alter the $3Na^+/2K^+$ stoichiometry of the transport cycle. The stoichiometry of the Na^+/K^+ pump can be calculated from comparison of tracer flux measurements with current measurements. In oocytes with expressed pumps of *Torpedo* electroplax, $^{22}Na^+$ efflux is large enough to be detectable even at reduced extracellular $[K^+]$. The method for efflux measurements (Grygorczyk et al., 1989) is illustrated in Fig. 3.

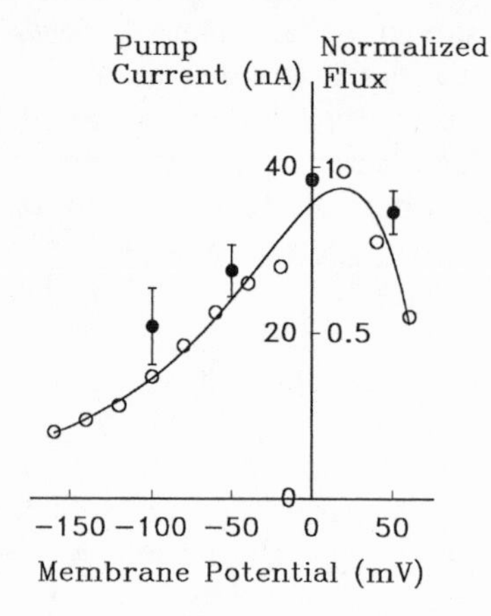

Figure 2. Voltage dependence of current (*open circles*) and $^{22}Na^+$ efflux (*filled circles*) generated by the Na^+/K^+ pump of electroplax of *Torpedo californica* (data were obtained from work by Schwarz and Gu, 1988). Oocytes were injected with 10 ng of cRNA for the α- and β-subunits of the Na^+/K^+ pump. For measurements of $^{22}Na^+$ efflux see Figs. 3 and 4.

Fig. 4 shows a typical efflux experiment performed under voltage clamp. The release of radioactivity was followed under different voltage clamp conditions for at least 20 min while the chamber was perfused with radioisotope-free solutions containing different K^+ concentrations. Both pump current and pump-generated efflux were determined as the differences of current or flux, respectively, in the presence and absence of a particular K^+ concentration. From the rate of pump-generated $^{22}Na^+$ efflux and intracellular Na^+ concentration, the number of Na^+ ions pumped per second can be estimated, and from pump-generated current determined in the same oocytes the number of charges transported per second can be estimated. Fig. 5 shows a comparison of these values for three different potentials. The data are compatible with the assumption that three Na^+ ions are translocated per net charge and that this ratio does not vary under the different experimental conditions chosen. In other words, the $3Na^+/2K^+$ stoichiometry is maintained when the medium is free of Na^+ and contains K^+ at concentrations (0.5 mM) significantly below the physiological level. In conclusion, the pump-generated current reflects the normal pump

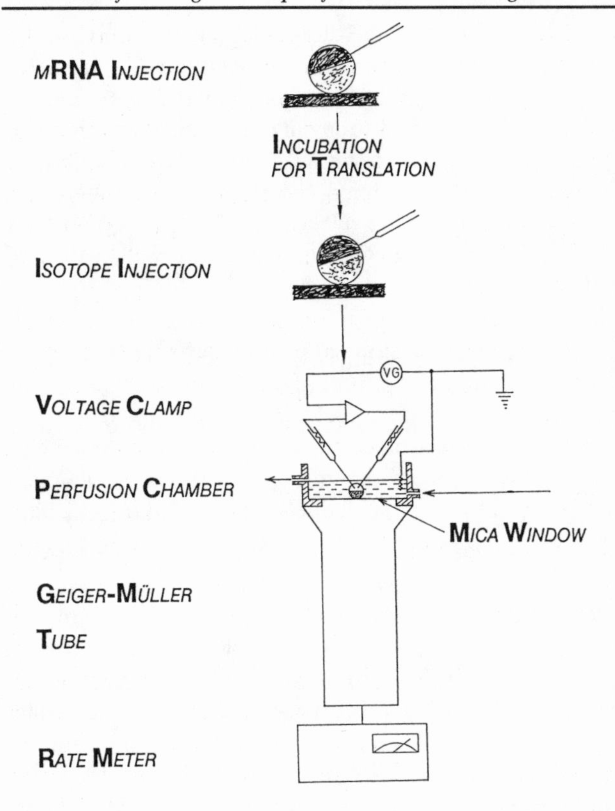

Figure 3. Experimental arrangement for measurements of $^{22}Na^+$ efflux under voltage clamp (see Grygorczyk et al., 1989) with oocytes that were injected with cRNA for the α- and β-subunits of the Na^+/K^+ pump of *Torpedo* electroplax 2 or 3 d before the experiment. Before the experiment an oocyte is injected with 20–50 nl $^{22}NaCl$ (~70 MBq/ml) and then placed into the perfusion chamber the bottom of which is formed by the mica window of a Geiger-Müller tube. Rate of efflux is determined from the exponential decline of radioactivity in the oocyte and can be followed under two-microelectrode voltage clamp.

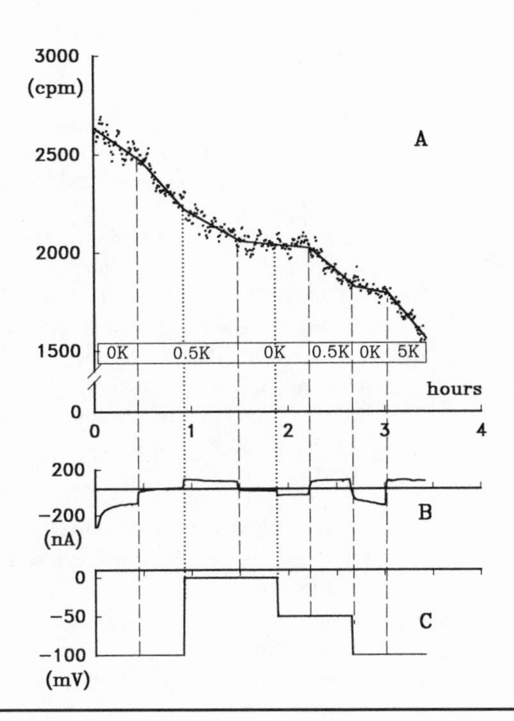

Figure 4. Efflux experiment performed under voltage clamp on an oocyte that was injected with 20 ng of cRNA for each subunit of the Na^+/K^+ pump of *Torpedo* electroplax. (*A*) Decline of radioactivity during perfusion with solutions containing different external $[K^+]$ as indicated, and at different membrane potentials as indicated in *C*. (*B*) Holding current necessary to clamp the membrane potential to the respective values.

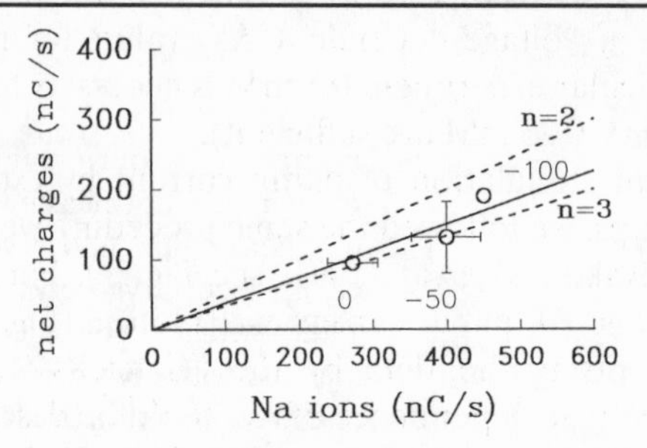

Figure 5. Comparison of the number of Na^+ ions with the number of charges translocated per second at three different membrane potentials (0, −50, and −100 mV). External $[K^+]$ was 0.5 mM. The data points represent averages (±SEM) from four experiments. The solid line is a linear fit to the data with a slope of 1/2.7.

activity with $3Na^+/2K^+$ stoichiometry under the various experimental conditions used here, and the voltage dependence for both the *Xenopus* and the *Torpedo* pump indicates the existence of a second voltage-dependent step in this normal pump cycle.

To investigate whether the negative slope of the *Torpedo* pump can also be attributed to voltage-dependent stimulation by external K^+, we performed experiments analogous to those we previously performed for the *Xenopus* pump (Rakowski et al., 1991). We measured current–voltage dependencies in Na^+-free solution at a range of different extracellular K^+ concentrations (Fig. 6*A*); pump current was determined as the difference current with and without extracellular K^+. After each change to another K^+ concentration, K^+-free solution was applied to allow linear drift corrections. In most experiments drift with time was negligible.

In experiments with *Torpedo* pump, just as obtained with *Xenopus* (and described above), reduction of extracellular $[K^+]$ from 5 to 0.5 mM results in the appearance of a negative slope in the current–voltage curve (Fig. 6*B*). The depen-

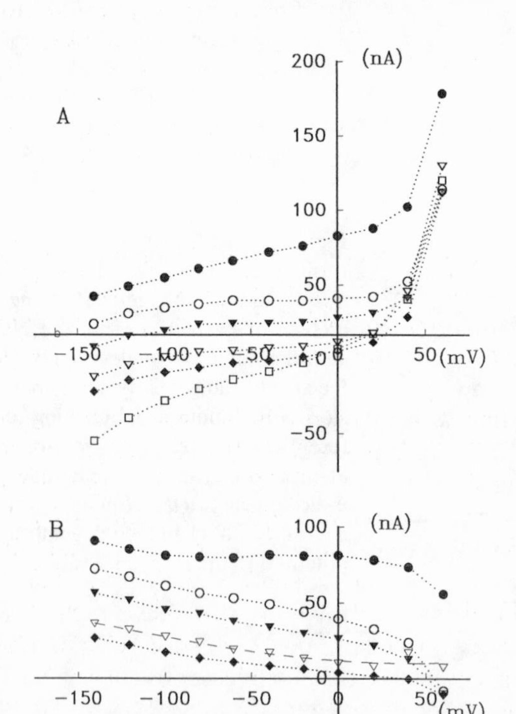

Figure 6. (*A*) Voltage dependence of total membrane current in oocytes that were injected with cRNAs for the *Torpedo* pump. Extracellular Na^+-free solutions contained different $[K^+]$, 5 mM $BaCl_2$, and 20 mM TEACl. The external $[K^+]$ was (in mM): 5 (*filled circles*), 0.5 (*open circles*), 0.25 (*filled triangles*), 0.1 (*open triangles*), 0.05 (*filled diamonds*), and 0 (*open squares*). (*B*) Voltage dependence of pump current at different extracellular $[K^+]$ (symbols as in A). Pump current was determined as the difference of total membrane current in presence and absence of extracellular $[K^+]$.

dence on $[K^+]$ may be described by a voltage-dependent $K_{1/2}$ value for pump stimulation by external K^+ (in this particular experiment 0.5 mM is necessary to give 50% stimulation at 0 mV, but at −100 mV 0.25 mM are sufficient).

To describe the voltage-dependent stimulation of pump current by external $[K^+]$ in oocytes containing *Torpedo* pumps, we followed the same procedure we used for studying *Xenopus* pump current (Rakowski et al., 1991; see Fig. 7). For each membrane potential, the $[K^+]$ dependence of pump current was plotted (Fig. 7*A*) and $K_{1/2}$ values were determined by fitting the data by the Michaelis-Menten equation. For the *Xenopus* pump, the voltage dependence of $K_{1/2}$ could be described by a single exponential. The steepness of the curve can be explained by assuming that movement in the electrical field of 0.37 of an elementary charge is associated with the K^+ binding (Fig. 7*B*, *inset;* Rakowski et al., 1991). In oocytes containing *Torpedo* pumps, the voltage dependence of $K_{1/2}$ is different.

For the whole potential range, the $K_{1/2}$ values are significantly larger (compare, for example, $K_{1/2}$ at $V = 0$ mV) and the potential-dependent variations of $K_{1/2}$ are much more pronounced. The voltage dependence shown in Fig. 7*B* does not directly allow quantitative estimates for the *Torpedo* pump alone, since the data were obtained from total pump current (i.e., the sum of *Torpedo* and *Xenopus* pump currents). The contributions of each of the two pumps were determined by the following procedure:

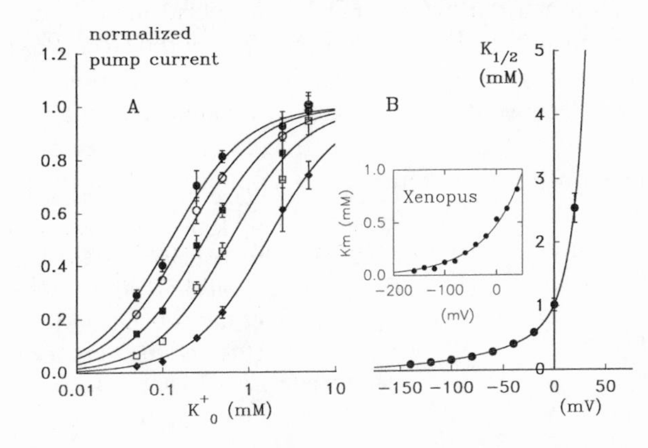

Figure 7. (*A*) Dependence of pump current in oocytes that were injected with cRNAs for the *Torpedo* pump on extracellular $[K^+]$ for different membrane potentials (in mV): −140 (*filled circles*), −100 (*open circles*), −60 (*filled squares*), −20 (*open squares*), and +20 (*filled diamonds*). The data points represent averages (±SEM) from five experiments. Solid lines represent fits of Michaelis-Menten kinetics to the data. (*B*) Voltage dependence of $K_{1/2}$ (i.e., $[K^+]$ that gives 50% stimulation) determined from data as shown in *A*. The inset shows $K_{1/2}$ values for stimulation of the *Xenopus* pump (taken from Rakowski et al., 1991).

Measurement of ouabain binding with radioactively labeled ouabain allows determination of the number of active pump molecules in the oocyte plasma membrane (Richter et al., 1984; Schmalzing et al., 1989). Comparison of the number of ouabain binding sites in control oocytes and oocytes injected with the cRNAs for the α- and β-subunits of the Na/K-ATPase of *Torpedo* electroplax suggests that ~30% of the pump molecules are endogenous pumps and 70% are *Torpedo* pumps. Though it was demonstrated (Schmalzing et al., 1991*b*) that the β-subunits of *Torpedo* (Tβ) can assemble with excess of endogenous *Xenopus* α-subunits (Xα) (Geering et al., 1989; Geering, 1991), it is unlikely that a significant fraction of Xα Tβ is formed as a consequence of the coinjection of cRNA for Tα with cRNA for Tβ. We therefore fitted the equation

$$I = I_{max}*\{0.7[K^+]/([K^+] + K_m^T) + 0.3[K^+]^{1.3}/([K^+]^{1.3} + K_m^{X1.3})\}$$

to data as shown in Fig. 7 *A*. The equation represents the sum of the two components in the 70:30 ratio of maximum pump current, I_{max}. Each component is described by a Michaelis-Menten-like expression with apparent K_m values. For the endogenous *Xenopus* pump, we used a Hill coefficient of $n = 1.3$ as in our previous work on the endogenous pump (Rakowski et al., 1991). The K_m values (K_m^X) for the various potentials were taken from the data shown in the inset of Fig. 7 *B*. For the *Torpedo* pump, a Hill coefficient of $n = 1$ gave the best fits when the K_m value (K_m^T) was used as the fit parameter. The voltage dependence of the K_m value for the *Torpedo* pump obtained by this fit procedure is shown in Fig. 8 in comparison with the date for the endogenous pump.

In contrast to the voltage dependence of the K_m^X value which can be described by a single exponential, in the *Torpedo* pump two exponentials are necessary with effective charges of 0.16 and 1 of an elementary charge. These results demonstrate that like the *Xenopus* pump cycle, the *Torpedo* pump cycle is influenced by a

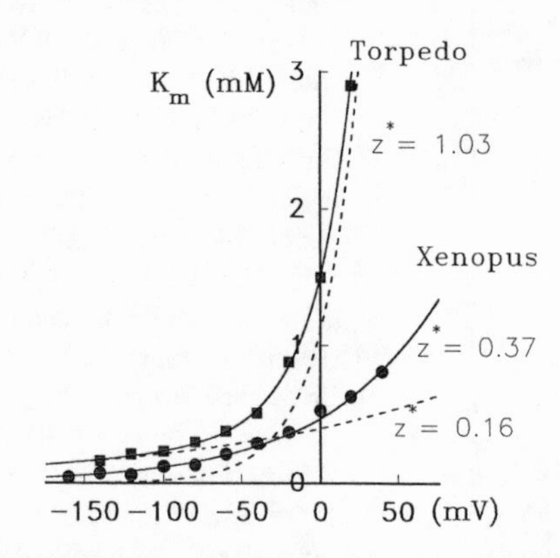

Figure 8. Voltage dependence of K_m values of pump stimulation by external K^+ for the *Xenopus* (*circles*) and the *Torpedo* (*squares*) pump. Solid lines represent fits of one (for *Xenopus*) or two (for *Torpedo*) exponentials (exp (z^*EF/RT) to the data.

voltage-dependent stimulation by external $[K^+]$. The voltage dependence is, however, more pronounced than that of the *Xenopus* pump, as indicated by one of effective charge of 1 for the *Torpedo* pump as compared with 0.37 for the *Xenopus* pump. A simple interpretation for the change in effective charge is that K^+ has to pass through an access channel to reach the binding site, but that the access channel in the *Torpedo* pump molecule differs from the channel in the *Xenopus* pump molecule. One could speculate that in the *Torpedo* pump the K^+ ions would have to pass in single file through a narrow access channel that is longer than in the *Xenopus* pump. Of course other interpretations are possible, including modification of the microscopic environment of the binding site.

Recently we have shown (Vasilets et al., 1990, 1991) that application of phorbol esters and analogues that are known to stimulate protein kinase C leads to a dramatic downregulation of the Na^+/K^+ pump. We demonstrated that this is primarily due to an endocytotic incorporation of pump molecules that are removed from the plasma membrane (see also Schmalzing et al., 1991*a*). Not all pump molecules, however, are removed from the cell surface. At least 20% remain in the surface membrane. We have speculated that the remaining pump molecules may be modified, and we wanted to investigate whether injection of, for example, diC_8, known to stimulate protein kinase C, leads to modulations in the voltage dependence of K_m for pump stimulation by external [K^+].

Fig. 9 summarizes the results of such experiments where we applied the same kind of analysis as described above. In oocytes containing only the endogenous Na^+/K^+ pump, diC_8 injection increases the K_m at all potentials. The voltage dependence can still be described by a single exponential but the effective charge has nearly doubled and increased to 0.5 of an elementary charge (Fig. 9*A*). The increase in effective charge suggests that stimulation of protein kinase C results in a modification of the access channel. If the size of the effective charge is interpreted as a measure of channel length, this could mean an increase in the length of the access channel. In oocytes containing the *Torpedo* pump, the voltage dependence of the K_m^T value is modified as well and can be described by an increase of the effective charge of one of the components by a factor of about 2 (Fig. 9*B*).

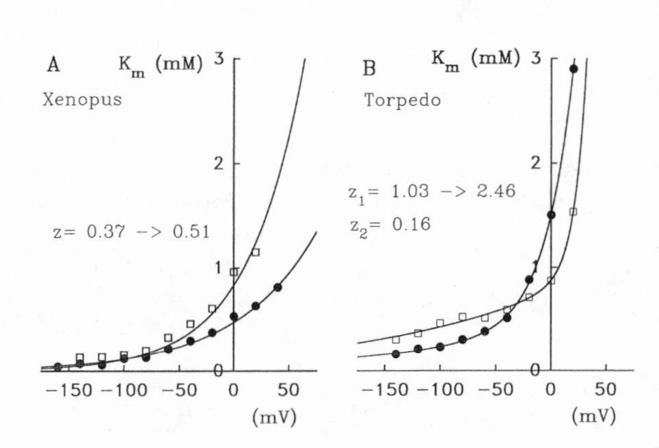

Figure 9. Variability of voltage dependence of apparent K_m values for pump stimulations by external K^+. Injection of diC_8 increases the effective charges derived from the exponential fits to the data as indicated. Data for control oocytes are represented by filled circles, and for diC_8-injected oocytes by open squares. (*A*) Effect of diC_8 injection on the *Xenopus* pump. Solid lines represent fits of single exponentials to the data. (*B*) Effect of diC_8 injection on the *Torpedo* pump. Solid lines represent fit of the sum of two exponentials to the data.

Conclusion

The voltage dependencies of the currents generated by the *Xenopus* pump as well as by the *Torpedo* pump demonstrate that in both species two voltage-dependent steps influence pump activity; stimulation of the pump by voltage-dependent K^+ binding can lead to a negative slope in the current–voltage relationship.

The voltage dependence of the K_m value for pump stimulation shows species differences. These differences can explain the discrepancies with respect to the occurrence of negative slopes. The *Torpedo* pump might represent one extreme with a long, narrow access channel ("high-field" access in the terms of Läuger, 1991) leading to pronounced voltage-dependent K^+ binding. The pump in heart muscle could represent the other extreme with a very short access channel ("low-field" access) and, consequently, nearly voltage-independent K^+ binding. The *Xenopus* pump would represent an intermediate.

The voltage-dependent stimulation of the pump by external [K^+] can be modulated by diacylglycerol analogues presumably via stimulation of protein kinase C.

Acknowledgments

We gratefully acknowledge the discussion and comments of Dr. Hermann Passow and Dr. Günther Schmalzing. We thank Dr. M. Kawamura for kindly providing us with the cDNAs for the α- and β-subunits of the Na/K-ATPase of electroplax of *Torpedo californica.*

L. A. Vasilets was a recipient of a Max-Planck stipend and supported by the Deutsche Forschungsgemeinschaft (DFG). Further support was provided by the DFG (SFB 169).

References

Bahinski, A., M. Nakao, and D. C. Gadsby. 1988. Potassium translocation by the Na^+/K^+ pump is voltage insensitive. *Proceedings of the National Academy of Sciences, USA.* 85:3412–3416.

De Weer, P., and R. F. Rakowski. 1984. Current generated by backward-running electrogenic Na pump in squid giant axons. *Nature.* 309:450–452.

Fendler, K., E. Grell, M. Haubs, and E. Bamberg. 1985. Pump currents generated by the purified Na^+,K^+-ATPase from kidney an black lipid membranes. *EMBO Journal.* 4:3079–3085.

Gadsby, D. C., M. Nakao, and A. Bahinski. 1991. Voltage-induced Na/K pump charge movements in dialyzed heart cells. *In* The Sodium Pump: Structure, Mechanism, and Regulation. J. H. Kaplan and P. De Weer, editors. The Rockefeller University Press, New York. 355–371.

Geering, K. 1991. Posttranslational modifications and intracellular transport of sodium pumps: importance of subunit assembly. *In* The Sodium Pump: Structure, Mechanism, and Regulation. J. H. Kaplan and P. De Weer, editors. The Rockefeller University Press, New York. 31–43.

Geering, K., I. Theulaz, F. Verry, M. T. Haeuptle, and C. B. Rossier. 1989. A role for the β-subunit in the expression of functional Na^+-K^+-ATPase in Xenopus oocytes. *American Journal of Physiology.* 257:C851–C858.

Glitsch, H. G., T. Krahn, and H. Pusch. 1989*a.* The dependence of sodium pump current on internal Na concentration and membrane potential in cardioballs from sheep Purkinje fibres. *Pflügers Archiv.* 414:52–58.

Glitsch, H. G., T. Krahn, and F. Verdonck. 1989*b.* Activation of the Na pump current by external K and Cs ions in cardioballs from sheep Purkinje fibres. *Pflügers Archiv.* 414:99–101.

Goldshleger, R., S. J. D. Karlish, A. Rephaeli, and W. D. Stein. 1987. The effect of membrane potential on the mammalian sodium-potassium pump reconstituted into phospholipid vesicles. *Journal of Physiology.* 387:331–355.

Grygorczyk, R., P. Hanke-Baier, W. Schwarz, and H. Passow. 1989. Measurement of erythroid band 3 protein-mediated anion transport in mRNA-injected oocytes of *Xenopus laevis. Methods in Enzymology.* 173:453–467.

Isenberg, G., and W. Trautwein. 1974. The effect of dihydroouabain and lithium ions on the outward current in cardiac Purkinje fibers. *Pflügers Archiv.* 350:41–54.

Lafaire, A. V., and W. Schwarz. 1985. Voltage-dependent ouabain-sensitive current in the membrane of oocytes of *Xenopus laevis. In* The Na/K-ATPase. I. Glynn and C. Ellory, editors. The Company of Biologists, Ltd., Cambridge, UK. 523–525.

Lafaire, A. V., and W. Schwarz. 1986. Voltage dependence of the reogenic Na^+/K^+ ATPase in the membrane of oocytes of *Xenopus laevis. Journal of Membrane Biology.* 91:43–51.

Läuger, P. 1991. Kinetic basis of voltage sensitivity. *In* The Sodium Pump: Structure, Mechanism, and Regulation. J. H. Kaplan and P. De Weer, editors. The Rockefeller University Press, New York. 303–315.

Nakao, M., and D. C. Gadsby. 1986. Voltage dependence of Na translocation by the Na/K pump. *Nature.* 323:628–630.

Nakao, M., and D. C. Gadsby. 1989. [Na] and [K] dependence of the Na/K pump current-voltage relationship in guinea ventricular myocytes. *Journal of General Physiology.* 94:539–565.

Rakowski, R. F. 1991. Stoichiometry and voltage dependence of the Na^+/K^+ pump in squid giant axons and *Xenopus* oocytes. *In* The Sodium Pump: Structure, Mechanism, and Regulation. J. H. Kaplan and P. De Weer, editors. The Rockefeller University Press, New York. 339–353.

Rakowski, R. F., D. C. Gadsby, and P. De Weer. 1989. Stoichiometry and voltage dependence of the sodium pump in voltage-clamped, internally dialyzed squid giant axon. *Journal of General Physiology.* 93:903–941.

Rakowski, R. F., and C. L. Paxson. 1988. Voltage dependence of Na/K pump current in *Xenopus* oocytes. *Journal of Membrane Biology.* 106:173–182.

Rakowski, R. F., L. A. Vasilets, J. LaTona, and W. Schwarz. 1991. A negative slope in the current-voltage relationship of the Na^+/K^+ pump in Xenupos oocytes produced by reduction of external $[K^+]$. *Journal of Membrane Biology.* 121:177–187.

Richter, H.-P., D. Jung, and H. Passow. 1984. Regulatory changes of membrane transport and ouabain binding during progesterone-induced maturation of Xenopus ooxytes. *Journal of Membrane Biology.* 79:203–210.

Schmalzing, G., W. Haase, and K. Geering. 1991*a*. Immunocytochemical evidence for protein kinase C-induced internalization of sodium pumps in *Xenopus laevis* oocytes. *In* The Sodium Pump: Recent Developments. J. H. Kaplan and P. De Weer, editors. The Rockefeller University Press, New York. In press.

Schmalzing, G., S. Kröner, and H. Passow. 1989. Evidence for intracellular sodium pumps in permeabilized *Xenopus laevis* oocytes. *Biochemical Journal.* 260:395–399.

Schmalzing, G., H. Omay, S. Kröner, H. Appelhans, and W. Schwarz. 1991*b*. Expression of exogenous β_1-subunits of Na/K pump in *Xenopus laevis* oocytes raises pump activity. *In* The Sodium Pump: Recent Developments. J. H. Kaplan and P. De Weer, editors. The Rockefeller University Press, New York. In press.

Schwarz, W., and Q. Gu. 1988. Characteristics of the Na^+/K^+-ATPase from *Torpedo californica* expressed in *Xenopus* oocytes: a combination of tracer flux measurements with electrophysiological measurements. *Biochimica et Biophysica Acta.* 945:167–174.

Schweigert, B., A. V. Lafaire, and W. Schwarz. 1988. Voltage-dependence of the Na-K ATPase: measurements of ouabain-dependent membrane current and ouabain binding in oocytes of *Xenopus laevis. Pflügers Archiv.* 412:579–588.

Stürmer, W., H. J. Apell, J. Wuddel, and P. Läuger. 1989. Conformational transitions and charge translocation by the Na,K pump: comparison of optical and electrical transients elicited by ATP concentration jumps. *Journal of Membrane Biology.* 110:67–86.

Vasilets, L. A., K. Mädefessel, G. Schmalzing, and W. Schwarz. 1991. Inhibition of the Na^+/K^+ pump in *Xenopus* oocytes by diacylglycerol analogues. *In* The Sodium Pump: Recent Developments. J. H. Kaplan and P. De Weer, editors. The Rockefeller University Press, New York. In press.

Vasilets, L. A., G. Schmalzing, K. Mädefessel, W. Haase, and W. Schwarz. 1990. Activation of protein kinase C by phorbol ester induces downregulation of the Na^+/K^+-ATPase in oocytes in Xenopus laevis. *Journal of Membrane Biology.* 118:131–142.

Chapter 25

Stoichiometry and Voltage Dependence of the Na^+/K^+ Pump in Squid Giant Axons and *Xenopus* Oocytes

R. F. Rakowski

Department of Physiology and Biophysics, University of Health Sciences/The Chicago Medical School, North Chicago, Illinois 60064

Introduction

Stoichiometry of the Na^+/K^+ Pump

It is well established that the stoichiometry of the Na^+/K^+ pump under physiological conditions is $3Na^+/2K^+$ (Post and Jolly, 1957; Garrahan and Glynn, 1967). However, as pointed out by Glynn (1984), the measurement of stoichiometry based on ouabain-sensitive fluxes of Na^+ and K^+ requires calculation of the ratio of two differences with a consequent summation of errors. The calculation of the ratio of unidirectional flux to net current provides an alternative method of measuring pump stoichiometry, if one makes the assumption that only Na^+ and K^+ are transported during the pump cycle. This method has the advantage that the measurements of flux and current can be carried out simultaneously on a single cell, thus allowing cancellation of errors in common factors in the computed stoichiometric ratio. The squid giant axon provides an almost ideal preparation for such studies since it can be internally dialyzed and voltage clamped (Brinley and Mullins, 1974; Rakowski, 1989). In principle, at least, the assumption that only Na^+ and K^+ are transported can be tested in separate experiments in which the ratio of ouabain-sensitive Na^+ efflux ($F\Delta\Phi_{Na}$) to ouabain-sensitive current (ΔI) is calculated (F is Faraday's number) and compared with the ratio of ouabain-sensitive K^+ influx to current ($F\Delta\Phi_K/\Delta I$). If only Na^+ and K^+ carry current and there is no significant Na^+/Na^+ or K^+/K^+ exchange, the following equation must be obeyed:

$$\frac{F\Delta\Phi_{Na}}{\Delta I} + \frac{F\Delta\Phi_K}{\Delta I} = 1 \tag{1}$$

Measurements under voltage clamp conditions also, in principle, permit a third approach to be used to measure pump stoichiometry. If one can measure the reversal potential of pump current or flux (V_{rev}), pump stoichiometry can be determined from the ratio of the change in reversal potential (ΔV_{rev}) to experimentally produced changes in the equilibrium potentials for Na^+ (ΔE_{Na}) or K^+ (ΔE_K) by the simultaneous equations

$$\frac{\Delta V_{rev}}{\Delta E_{Na}} = \frac{m}{(m - n)} \tag{2}$$

and

$$\frac{\Delta V_{rev}}{\Delta E_K} = \frac{-n}{(m - n)} \tag{3}$$

where m and n are the stoichiometric coefficients for Na^+ and K^+, respectively (De Weer et al., 1988). In practice, however, verification of Eq. 1 has only been attempted for the reverse mode of Na^+/K^+ pumping (De Weer and Rakowski, 1984) and no successful application of Eqs. 2 and 3 has been reported. Given the wide acceptance of $3Na^+/2K^+$ as the correct stoichiometry of the Na^+/K^+ pump, the point to be made is not simply to more precisely determine the stoichiometric ratio. Instead, interest centers on whether the voltage dependence of pump current arises from changes in the apparent stoichiometry of the pump, for example, as a result of enhancing electroneutral Na^+/Na^+ exchange by hyperpolarization, or whether hyperpolarization simply decreases pump rate.

Voltage Dependence of the Na^+/K^+ Pump

An obvious result of $3Na^+/2K^+$ pump stoichiometry is that the pump is electrogenic (Kernan, 1962; for reviews see Thomas, 1972; De Weer, 1975; Slayman, 1982). The current produced by the pump is outward and has a magnitude of $\sim 1\ \mu A\ cm^{-2}$ near 0 mV (Rakowski et al., 1989). For thermodynamic reasons at physiological levels of Na^+, K^+, ATP, and ADP, pump current reversal should occur near -240 mV (De Weer, 1984). This means that at least over some voltage range between -240 and 0 mV, pump current must be voltage dependent. The shape of that voltage dependence gives mechanistic information about the voltage-dependent steps in the pump cycle. Initially experiments addressed the questions of whether the Na^+/K^+ pump is, in fact, voltage dependent and what the precise shape of the pump *I-V* relationship under physiological conditions is. More recently, however, experiments have begun to address the broader aim of characterizing the pump *I-V* relationship under a variety of experimental conditions in order to extract mechanistic information about the Na^+/K^+ pump cycle. Thus far, only a very limited set of experimental conditions have been explored. This chapter will summarize the most recent information available on the stoichiometry and voltage dependence of the Na^+/K^+ pump in squid giant axons and *Xenopus* oocytes, and will reach some tentative conclusions about the charge translocating steps in the pump cycle.

Studies Conducted on Squid Giant Axons

Area Match Requirement

Fig. 1 is a diagram of the experimental apparatus for the measurement of changes in voltage clamp holding current and ^{22}Na efflux in internally dialyzed squid giant axons. The chamber is based on the design of Brinley and Mullins (1974). The axon is cannulated at both ends and the hardware required for voltage clamping and internal dialysis is inserted into the interior of the axon through the glass cannulas. Stopcock grease is used to separate the center portion of the axon from the cut ends, and this center pool is superfused with artificial seawater that can be collected and analyzed to determine ^{22}Na efflux across the axon membrane in the center pool. Although the stopcock grease provides physical isolation of the center pool, it does not completely electrically isolate the center pool from the ends. To ensure that the measured current derives from the same membrane area as the isotopic efflux, it is necessary to ensure that there is no current flow from the ends by means of two ancillary voltage clamp amplifiers that maintain the center and end pools at the same potential. A detailed description of the voltage clamp system has been published (Rakowski, 1989) and an extensive series of control experiments have been conducted that demonstrate the equality of unidirectional ^{22}Na efflux and current through TTX-sensitive Na^+ channels into Na^+-free seawater (Rakowski et al., 1989).

Simultaneous Measurements of ^{22}Na Efflux and Pump Current

The end-pool clamp system has made it possible to make accurate simultaneous measurements of Na^+/K^+ pump–mediated ^{22}Na efflux and current at various membrane potentials. A summary of the results is shown in Fig. 2 for both Na^+-free and 390 mM Na^+ seawater. The vertical scale in *A* is three times that in *B* when scaled by Faraday's number, so that points that are the same distance above the horizontal axis in *A* and *B* imply a movement of three Na^+ outward for each net charge transported outward during a pump cycle. Assuming that only Na^+ and K^+ are transported, the data imply a stoichiometry of $3Na^+/2K^+$ independent of membrane potential. The voltage dependence of pump current and ^{22}Na efflux is slight in Na^+-free seawater, and increases markedly in 390 mM Na^+ seawater.

Statistical Comparison of the Measured Value of $F\Delta\Phi_{Na}/\Delta I$ with the Stoichiometric Ratios Calculated for All Possible Values of *m* and *n*

If all of the data shown in Fig. 2 are combined, the overall mean value of the ratio $F\Delta\Phi_{Na}/\Delta I$ in these experiments is 2.87 ± 0.07 (n = 25), which is not significantly different from the ratio of 3.0 expected for a $3Na^+/2K^+$ stoichiometry. These

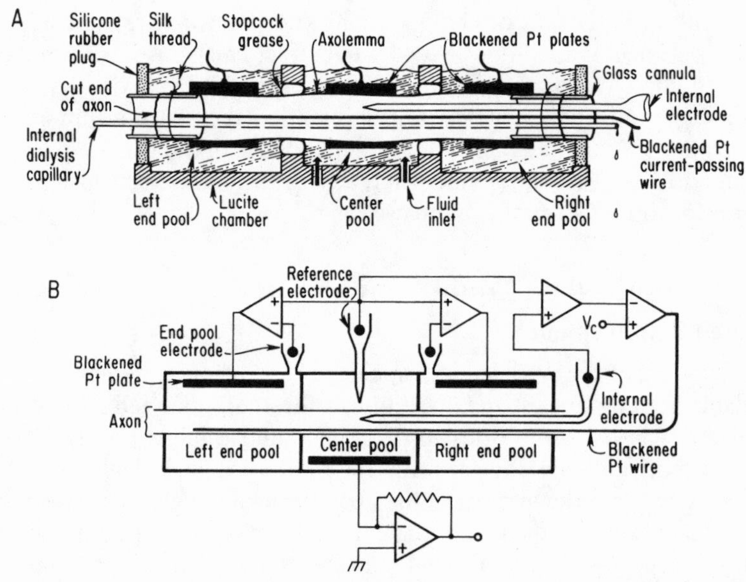

Figure 1. Diagram of experimental chamber and end pool voltage clamp apparatus. (*A*) Dialysis chamber. Stopcock grease is used to isolate the center experimental pools from the end pools. Fluid enters the center pool from below at a rate of 2 ml/min and is aspirated from above. (*B*) Main and end pool clamp circuit. Black circles indicate calomel half-cells. The end pool amplifiers null the potential difference between the center and end pools by passing current through a blackened platinum plate in each end pool. From Rakowski et al., 1989.

measurements are among the most precise determinations of Na^+/K^+ pump stoichiometry performed so far. The coefficient of variation (~12%) is not much greater than would be expected for the ratio of two differences, each subject to small (<5%) errors. The precision of the data makes it possible to statistically test the observed flux to current ratio against all possible integer values of the stoichiometric coefficients m and n. Calculated values of the ratio $m/(m - n)$ are tabulated in Table I for values of m and n between 1 and 11. The only positive values of the ratio $m/(m - n)$ that are not significantly different from the experimentally determined ratio are 3.0 for a $3Na^+/2K^+$ stoichiometry (or higher multiples such as $6Na^+/4K^+$) and 2.75 at the mechanistically improbable value of $11Na^+/7K^+$ (The next higher compatible ratio is $14Na^+/9K^+$). Also listed in the table is the value of the equilibrium potential for the Na^+/K^+ pump calculated from the values given in the legend. For all values of m greater than 10, the equilibrium potential is positive and therefore these values may be excluded based on the measurement of outward pump current at 0 mV. Similarly, the values at $6Na^+/4K^+$ and $9Na^+/6K^+$, and the value at $8Na^+/5K^+$ (which in any case is significantly different from the observed ratio at $P < 0.01$) can be excluded based on the predicted positive value of the equilibrium potential. We can conclude, therefore, that the only value of pump stoichiometry that is statistically compatible with the measured value of $F\Delta\Phi_{Na}/\Delta I$ and that has a negative value of pump equilibrium potential is $3Na^+/2K^+$.

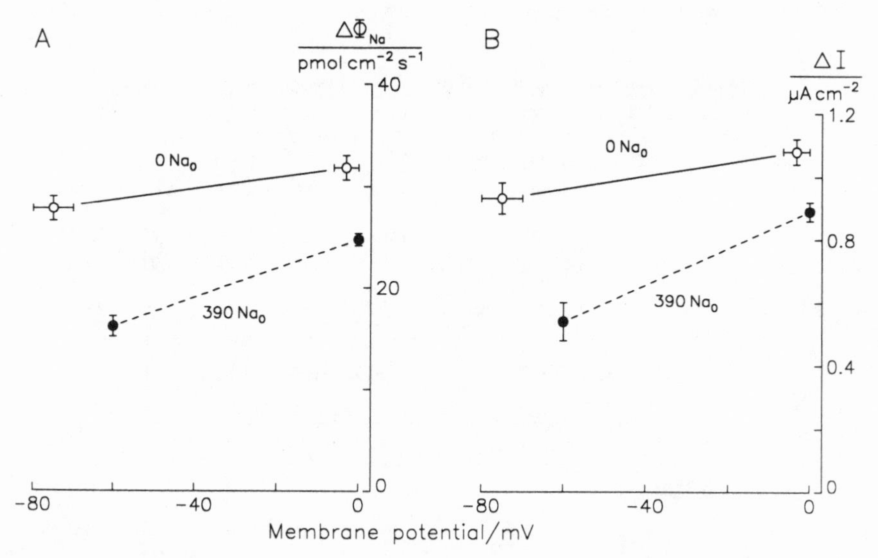

Figure 2. Na^+/K^+ pump currents and fluxes in either 390 or 0 mM Na seawater. (*A*) Mean values of H_2DTG-sensitive flux. (*B*) Mean values of H_2DTG-sensitive current. Error bars indicate ±SEM. From Rakowski et al., 1989.

Which Na^+ Translocation Steps Are Voltage Dependent?

In addition to establishing that the stoichiometry of the fully activated forward mode of operation of the pump is $3Na^+/2K^+$, independent of membrane potential, the data in Fig. 2 indicate that under these experimental conditions (high [ATP] and $[Na]_i$, low [ADP]) there is no substantial electroneutral Na^+/Na^+ exchange, since the presence of unidirectional Na^+ efflux that was unaccompanied by net charge movement would have increased the ratio of $F\Delta\Phi_{Na}/\Delta I$ beyond its theoretical value of 3.0. The data also directly demonstrate that at least one step in the process of Na^+ translocation is voltage dependent since the unidirectional Na^+ efflux is voltage dependent. There are four steps in the Post-Albers (Cantley et al., 1984) model of the Na^+/K^+ pump to be considered: (1) internal Na^+ binding, (2) Na^+ occlusion, (3) the E_1 to E_2 conformational change, and (4) external Na^+ unbinding and release. In

TABLE I
Calculated Values of $F\Delta\Phi_{Na}/\Delta I$ and $E_{Na/K}$ for Various mNa/nK Pump Stoichiometries

	mNa											
nK	0	1	2	3	4	5	6	7	8	9	10	11
0	X	1.00* −540	1.00* −240	1.00* −140	1.00* −90	1.00* −60	1.00* −40	1.00* −26	1.00* −15	1.00* −7	1.00* 0	1.00* +5
1	-	Neut.	2.00* −390	1.50* −165	1.33* −90	1.25* −53	1.20* −30	1.17* −15	1.14* −4	1.13* +4	1.11* +10	1.10* +15
2	-	-	Neut.	**3.00** −2.40	2.00* −90	1.67* −40	1.50* −15	1.40* 0	1.33* +10	1.29* +17	1.25* +23	1.22* +27
3	-	-	-	Neut.	4.00* −90	2.50* −15	2.00* +10	1.75* +23	1.60* +30	1.50* +35	1.43* +39	1.38* +41
4	-	-	-	-	Neut.	5.00* +60	**3.00** +60	2.33* +60	2.00* +60	1.80* +60	1.67* +60	1.57* +60
5	-	-	-	-	-	Neut.	6.00* +210	3.50* +135	2.67‡ +110	2.25* +98	2.00* +90	1.83* +85
6	-	-	-	-	-	-	Neut.	7.00* +360	4.00* +210	**3.00** +160	2.50* +135	2.20* +120
7	-	-	-	-	-	-	-	Neut.	8.00* +510	4.50* +285	3.33* +210	**2.75** +173
8	-	-	-	-	-	-	-	-	Neut.	9.00* +660	5.00* +360	3.67* +260
9	-	-	-	-	-	-	-	-	-	Neut.	10.00* +810	5.50* +435
10	-	-	-	-	-	-	-	-	-	-	Neut.	11.00* +960
11	-	-	-	-	-	-	-	-	-	-	-	Neut.

For each stoichiometry $F\Delta\Phi_{Na}/\Delta I$ is calculated as $m/(m-n)$. $E_{Na/K}$ (in millivolts) is calculated assuming 1 ATP per mNa/nK cycle, $E_{Na} = +60$ mV, $E_K = -90$ mV, and $E_{ATP} = -600$ mV. The value of $E_{Na/K}$ is given beneath the stoichiometric ratio. Negative values of $m/(m-n)$ and their corresponding reversal potentials are omitted since such values would imply a net inward pump current in the forward mode. Values in boldface are not significantly different from the experimental mean of 2.87 ± 0.07 ($n = 25$); $P > 0.05$, two-tailed Student's t test. All other values are significantly different from the experimental mean: * at $P < 0.001$, ‡ at $P < 0.01$, level of significance. X, not physically meaningful.

abbreviated form these four steps may be written as follows:

$$\text{binding (1)} \quad 3Na_i^+ + E_1 \cdot ATP \rightleftarrows Na_3 \cdot E_1 \cdot ATP$$

$$\text{occlusion (2)} \quad Na_3 \cdot E_1 \cdot ATP \rightleftarrows (Na_3)E_1P + ADP$$

$$\text{conformational change (3)} \quad (Na_3)E_1P \rightleftarrows Na_3 \cdot E_2P$$

$$\text{release (4)} \quad Na_3 \cdot E_2P \rightleftarrows E_2P + 3Na_o^+$$

To prevent electroneutral Na^+/Na^+ exchange, one of these steps must be irreversible. Step 2 is probably the irreversible step since conditions were chosen in which the [ATP] in the dialysate was high (5 mM) and [ATP] was backed up by two ADP rephosphorylating systems, pyruvate kinase (5 mM phosphoenol pyruvate was present in the dialysate) and arginine phosphokinase (5 mM arginine phosphate present). By maintaining a low ADP concentration these enzyme systems can be expected to make the Na^+ occlusion step essentially irreversible. Fig. 2 also shows that the steepness of the relationship between forward pump rate and voltage is sensitive to the presence of external Na^+. These facts, taken together, strongly suggest that either the conformational change (step 3) or Na^+ release (step 4), or both, is/are the voltage dependent step(s) in the process of Na^+ translocation. A similar conclusion was reached by Stürmer et al. (1989) from the similarity of the time course of fluorescence changes associated with the E_1 to E_2 conformational transition and with charge translocation. None of the above experimental findings, however, allows a clear choice between step 3 or 4 since the voltage-dependent step may either itself be rate limiting or may determine the level of enzyme intermediate that enters the rate-limiting step. Attention has been placed on the conformational change (step 3) as likely to be the charge-translocating and hence voltage-dependent step (Glynn, 1984; Nakao and Gadsby, 1989); however, this hypothesis requires re-evaluation in the light of recent evidence provided by studies of the voltage dependence of electroneutral Na^+/Na^+ exchange. The fact that both hyperpolarization and increasing external [Na] slow the rate of the forward-going Na^+/K^+ pump raises the possibility that rebinding or reocclusion of Na^+ at its external-facing sites are charge-translocating steps. The view that a reverse rate constant is rate determining is supported by the measurements of strophanthidin-sensitive pre–steady-state charge movement by Nakao and Gadsby (1986), which directly show that the reverse rate constant for charge translocation is increased by hyperpolarization and is more steeply voltage dependent than the forward rate constant.

Voltage Dependence of Electroneutral Na^+/Na^+ Exchange

The Na^+/K^+ pump can be essentially limited to steps 1–4 by performing experiments under internal and external K^+-free conditions. If both ATP and ADP are present, the primary mode of operation of the Na^+/K^+ pump will be electroneutral $3Na^+/3Na^+$ exchange catalyzed by the reversible operation of these four steps. As a control for the presence of "uncoupled" or other electrogenic Na^+ efflux modes, ouabain-sensitive current can be monitored in voltage clamp experiments. If either step 3 or 4, or both, is voltage dependent, then despite the fact that $3Na^+/3Na^+$ exchange is electroneutral, it should be voltage dependent. Glynn (1984) once remarked that it might seem "perverse" to look for voltage dependence of an electroneutral exchange process, but as he pointed out, there is nothing inconsistent in this as long as the forward and backward rates are equally affected. The data in Fig. 3 show the results of an experiment conducted to examine the voltage dependence of Na^+/Na^+ exchange (Gadsby, D. C., R. F. Rakowski, and P. De Weer, manuscript submitted for publication). Depolarization from 0 mV to +30 mV decreased ^{22}Na efflux from an internally dialyzed squid giant axon exposed to internal and external K^+-free media and hyperpolarization increased ^{22}Na efflux in a saturating manner. After application of a saturating concentration of the Na^+/K^+ pump–specific inhibitor dihydrodigitoxigenin (H_2DTG), the remaining background ^{22}Na efflux is insensitive to changes in membrane potential. The application of H_2DTG at −75 mV produced no measurable change in holding current, demonstrating that the measured change in ^{22}Na efflux was balanced by Na^+ influx and was, in fact, electroneutral. These data show that electroneutral Na^+/Na^+ exchange is voltage dependent. The voltage dependence is consistent with an enhancement of the reverse rate of Na^+ translocation by hyperpolarization up to a saturating level at about −45 mV.

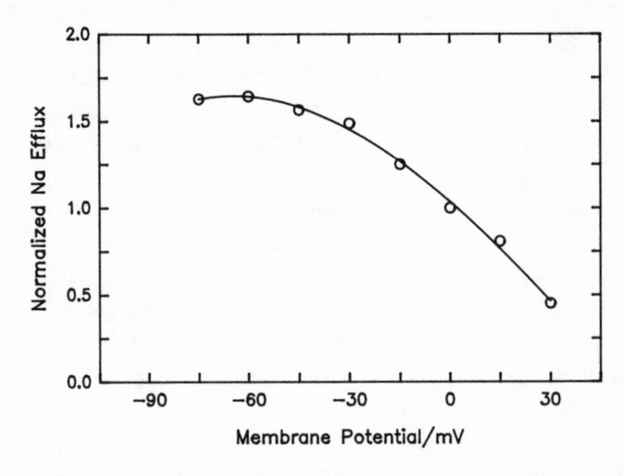

Figure 3. Voltage dependence of Na^+/Na^+ exchange. ^{22}Na efflux was measured from an axon in external and internal K^+-free solution in 400 mM Na^+ seawater, dialyzed with 5 mM internal ATP and ADP and 50 mM Na^+. Hyperpolarization increases Na^+ efflux under these Na^+/Na^+ exchange conditions. After the addition of 100 μM H_2DTG to stop the pump-mediated exchange, the remaining non–pump-mediated Na^+ efflux voltage was insensitive. Symbols represent the H_2DTG-sensitive difference measured at each voltage, normalized to the value at 0 mV (Gadsby, D. C., R. F. Rakowski, and P. De Weer, manuscript submitted for publication).

Effect of External [Na^+] on Na^+/Na^+ Exchange

Although it is gratifying that Na^+/Na^+ exchange is voltage dependent, as one would predict from the established voltage and [Na^+] dependence of forward pump rate, what more can we learn from an examination of the voltage dependence of Na^+/Na^+ exchange? One important fact is the external [Na^+] dependence of Na^+/Na^+ exchange at various voltages. Preliminary results at a holding potential of 0 mV indicated that reduction of external [Na] reduced ^{22}Na efflux in a linear fashion, consistent with a very low apparent affinity for activation of Na^+/Na^+ exchange by external [Na^+]. When the membrane potential was changed to −30 mV, however, ^{22}Na efflux was a hyperbolic function of $[Na]_o$ with an apparent K_m of ~200 mM. When the membrane potential was lowered further to −60 mV, the apparent K_m for activation of Na^+/Na^+ exchange was further reduced to ~50 mM. Thus, the apparent K_m for activation of Na^+/Na^+ exchange by external [Na^+] seems voltage dependent. The voltage dependence is in the direction expected for the behavior of an external ion well for Na^+ binding (Läuger and Apell, 1986): hyperpolarization enhances the rebinding of Na^+ at its external sites, resulting in an increased apparent affinity.

Studies Conducted on *Xenopus* Oocytes

Are K^+ Translocation Steps Voltage Dependent?

In contrast to the abundant evidence now available that Na^+ translocation is voltage dependent, it is generally thought that K^+ translocation is not. ATP-activated Rb^+/Rb^+ exchange is not affected by membrane potential (Goldshlegger et al., 1987). Strophanthidin-sensitive transient charge movement has not been observed under K^+/K^+ exchange conditions in response to voltage steps (Bahinski et al., 1988) or release of caged ATP (Fendler et al., 1985), and despite the presence of fluorescent signals thought to indicate conformational changes of the Na^+/K^+ pump, Stürmer et al. (1989) did not detect transient pump currents under K^+/K^+ exchange conditions. These observations provide strong evidence that the conformational transitions involved in K^+/K^+ exchange do not move charge through the membrane field and hence are voltage insensitive. However, these findings do not rule out the possibility that the binding of K^+ to its external high affinity transport sites might be voltage dependent. Lafaire and Schwarz (1986) found evidence for the existence of a region of negative slope in dihydroouabain (DHO)-sensitive difference currents measured in *Xenopus* oocytes. At least two voltage-dependent steps are required to account for regions of both positive and negative slope in the pump *I-V* relationship for a simple, unbranched pump cycle. These results were questioned, however (Rakowski and Paxson, 1988), since possible artifactual contributions to the DHO-sensitive current by ionic current had not been entirely eliminated. Since the findings of Lafaire and Schwarz (1986) were the sole evidence suggesting that a second voltage-dependent step might be present in the Na^+/K^+ pump cycle, it was important to re-examine this question under experimental conditions that were adequate to demonstrate the absence of contaminating sources of current. That re-examination has been done (Rakowski et al., 1991) and the reason for the previous apparent discrepancy is now clear. To observe effects of membrane potential on the activation of the pump by external K^+, it is necessary to perform the experiments at K^+ concentrations that are below saturation.

Experiments at Low $[K^+]$ in Na^+-free Solution

There are two principal sources of artifactual current that may change upon starting or stopping the Na^+/K^+ pump: (*a*) K^+ currents that result from changes of $[K^+]$ in restricted diffusion spaces, and (*b*) Na^+/Ca^{2+} exchange current that may result from changes of $[Na^+]$. The results of an experiment designed to eliminate these possible artifacts are shown in Fig. 4. The experiment was performed with 20 mM tetraethylammonium and 5 mM Ba^{2+} in the bathing solution, which was shown to be sufficient to block ionic current changes produced by adding or removing 5 mM K^+ in oocytes exposed to a saturating concentration of ouabain. The external solution was also Ca^{2+} free (replaced by Ni^{2+}), a condition that should prevent internal Na^+ activation of outward Na^+/Ca^{2+} exchange current. Fig. 4 *B* shows that the addition of 5 mM K^+ produces an approximately parallel upward shift of the steady-state *I-V* relationship measured in K^+-free solution. The K^+-sensitive difference current is shown as the filled circles in Fig. 4 *C*, and represents the nearly voltage-independent Na^+/K^+ pump

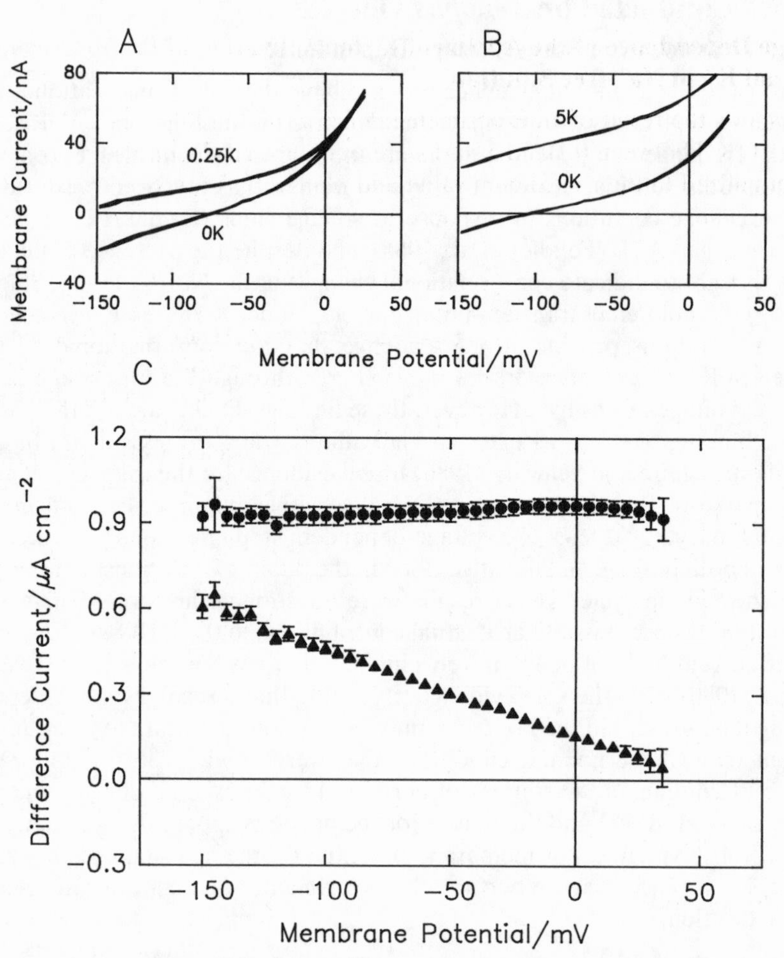

Figure 4. *I-V* measurements in *Xenopus* oocytes in Ca^{2+}- and Na^+-free solution in the presence of 2 mM Ni^{2+}. (*A*) Membrane currents measured using the 5-mV step, down-up-down staircase protocol described by Rakowski and Paxson (1988) in 0.25 mM K^+ and K^+-free solution. Two *I-V* determinations were done 5 min apart in each solution. The *I-V* curves in each condition are nearly superimposed and do not show significant hysteresis. (*B*) Similar *I-V* curves measured in the same oocyte in 5 mM K^+ and K^+-free solution 20 min after the measurements in *A*. (*C*) K^+-sensitive difference currents from *A* and *B* obtained by subtraction of the K^+-free control *I-V* data from the data obtained in 5 mM K^+ (*filled circles*) or 0.25 mM K^+ (*filled triangles*). The error bars represent ± SEM. Modified from Rakowski et al., 1991, by permission.

I-V relationship expected in Na^+-free solution. The current activated by 0.25 mM K^+ in Fig. 4*A* and shown as the K^+-sensitive difference current in Fig. 4*C* (filled triangles), however, clearly has a negative slope over the entire range of voltages examined.

Voltage Dependence of the Apparent K_m for Activation of Pump Current by External K^+ in Na^+-free Solution

Fig. 5 shows the result of an experiment similar to that in Fig. 4 except that several values of $[K^+]$ between 0.1 and 5 mM were examined. If the data at a given voltage are normalized to their maximum value and plotted as a function of external $[K^+]$,

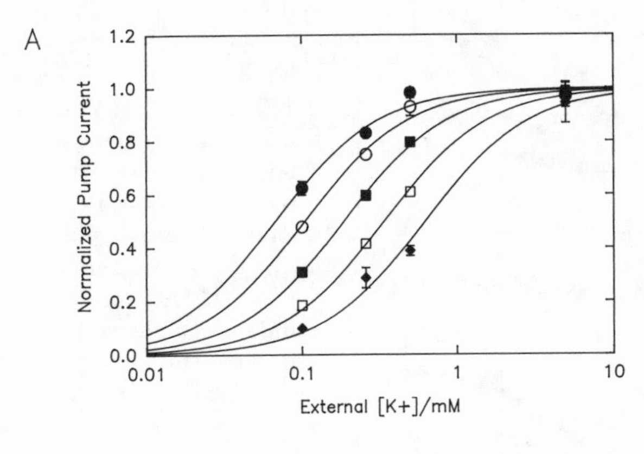

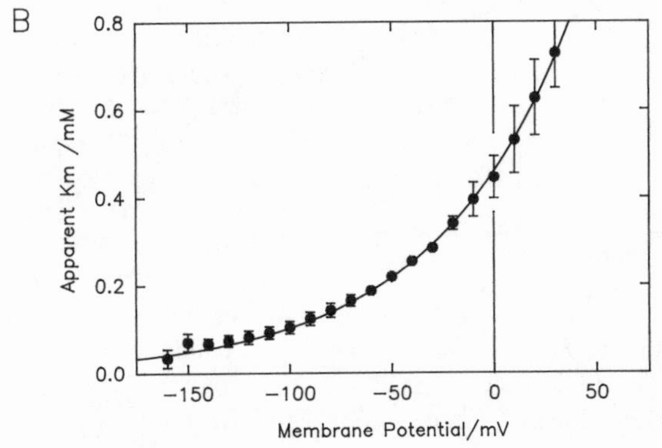

Figure 5. Voltage dependence of the apparent K_m for activation of Na^+/K^+ pump current by external $[K^+]$ in Na^+-free solution. The data are average values (± SEM) from eight oocytes. (*A*) External $[K^+]$ dependence of normalized Na^+/K^+ pump current at various voltages (*filled circles,* −140 mV; *open circles,* −100 mV; *filled squares,* −60 mV; *open squares,* −20 mV; *filled diamonds,* +20 mV). The curves represent the best fit of the Hill equation to the data at each voltage, assuming a Hill coefficient of 1.3. (*B*) Voltage dependence of the apparent K_m. Data are similar to those shown in *A* but are plotted here at 5-mV increments. From Rakowski et al., 1991, by permission.

the family of curves (here plotted at only 20-mV increments) in Fig. 5*A* are obtained. These data were fit by the Hill equation with a Hill coefficient of 1.3 to obtain estimates of the apparent K_m at each potential. The full data set (now plotted for 5-mV increments) is shown in Fig. 5*B*. The apparent K_m is voltage dependent and can be described by the simple exponential function $K_m = K_m(0) \exp(\alpha FV/RT)$, where $K_m(0)$ is the apparent K_m at 0 mV, V is the membrane potential, F, R, and T have their usual meanings, and α represents the dielectric coefficient defined by Läuger and Apell (1986). The values of $K_m(0)$ and α found by a least-squares fit to the data are 0.46 mM and 0.37, respectively. One simple interpretation of these data is that external K^+ ions must pass through a high-field access channel (ion well) in which ~ 18% of the membrane field is traversed by each of the $2K^+$ ions that bind. A possible alternative explanation is that membrane potential affects the distribution of enzyme intermediates so as to increase the apparent affinity for external K^+ upon hyperpolarization. These two alternative hypotheses cannot be distinguished by steady-state kinetic measurements.

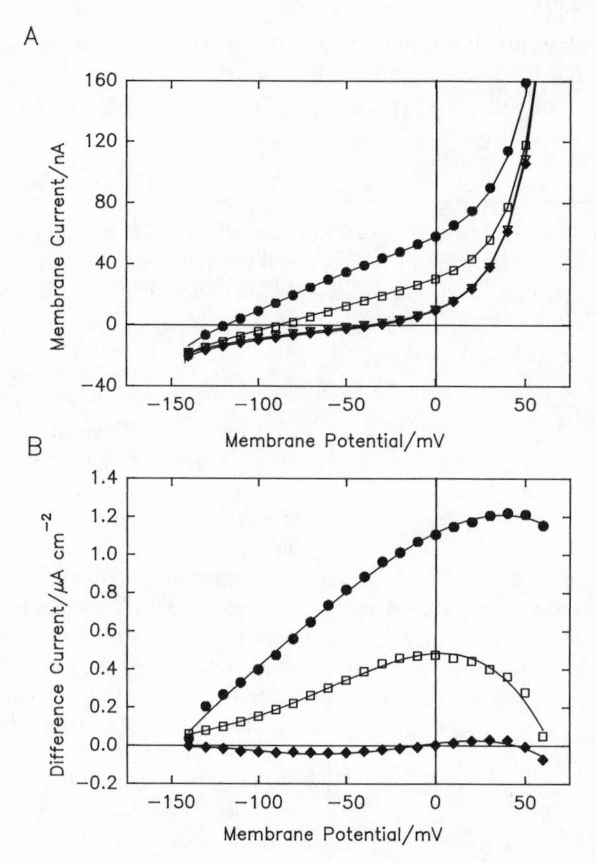

Figure 6. Na^+/K^+ pump *I-V* relationships in *Xenopus* oocytes in 90 mM Na^+ solution. (*A*) *I-V* relationships determined using the pulse protocol of Lafaire and Schwarz (1986) in various solutions. Two *I-V* measurements were made in each condition. The symbols represent the first determination. Solid lines connect the data points measured 5 min later in the same solution. Different symbols refer to different external $[K^+]$: *filled circles,* 5 mM; *open squares,* 1 mM; *filled diamonds,* K^+-free solution. The downward-directed open triangles that overlap the filled diamonds represent measurements made in K^+-free solution in the presence of 50 μM DHO. (*B*) Voltage dependence of Na^+/K^+ pump current at 5 mM (*filled circles*) and 1 mM (*open squares*) external $[K^+]$. Na^+/K^+ pump current is determined as the difference between the *I-V* measurements at the respective K^+ concentrations and the control measurements in K^+-free solution. The filled diamonds represent an additional control showing that there is only a small residual difference between the measurements in the presence of DHO and in K^+-free solution. From Rakowski et al., 1991, by permission.

A Negative Slope in the Pump *I-V* Relationship at Normal Extracellular $[Na^+]$

Since the pump *I-V* relationship has a negative slope in Na^+-free solution when external $[K^+]$ is reduced below saturation, we might also expect a negative slope at normal external $[Na^+]$ when $[K^+]_0$ is reduced. Fig. 6 shows that this is the case. The two lowest curves in Fig. 6*A* that are virtually superimposed represent the *I-V* relationships measured in K^+-free solution and in the presence of 50 μM DHO. Addition of 5 mM K^+ in the absence of DHO (filled circles) produces an additional outward current at all voltages. The difference between the current measured in 5 mM K^+ and K^+-free conditions is plotted in Fig. 6 *B* (filled circles). As previously reported (Rakowski and Paxson 1988), the Na^+/K^+ pump *I-V* relationship under these conditions of saturating external [K^+] is a sigmoid function of membrane potential with pump current increasing on depolarization and apparently saturating near +20 mV. When external [K^+] was lowered to 1 mM the *I-V* relationship shown by the open squares was obtained. The pump current magnitude is decreased at all voltages compared with that measured in 5 mM K^+, but in addition, at positive potentials there is a region of negative slope, presumably resulting from the decrease in apparent affinity for external K^+.

Conclusions

Na^+/K^+ Pump Stoichiometry

Simultaneous measurements of changes in ^{22}Na efflux and voltage clamp holding current in internally dialyzed squid giant axons have shown that under conditions of full forward activation of the pump, its stoichiometry remains $3Na^+$/1 net charge, independent of membrane potential. This implies that the decrease in pump current seen on hyperpolarization results from a decrease in pump rate. The voltage-dependent step or steps under these saturating external [K^+] conditions must be either the E_1 to E_2 conformational change or external Na^+ release steps, since Na^+ occlusion has been made essentially irreversible by lowering [ADP] as much as possible.

Voltage Dependence of Na^+/Na^+ Exchange

As predicted from the external [Na^+] and voltage dependence of pump current, pump-mediated Na^+/Na^+ exchange, although it is electroneutral, is voltage dependent. Consistent with the view that the voltage dependence lies in a reverse rate constant that is increased by hyperpolarization, negative voltages promote Na^+/Na^+ exchange, with saturation occurring at about −45 mV and half-activation at about 0 mV. The external [Na^+] dependence of Na^+/Na^+ exchange indicates that the apparent affinity for Na^+ activation of the exchange is increased by hyperpolarization. This is consistent with, but does not prove, the existence of an external ion well through which Na^+ must pass to be released from its external low affinity sites.

Voltage-dependent K^+ Binding Step

There is substantial evidence to support the view that the E_2 to E_1 conformational change that occurs during K^+ translocation does not involve charge movement through the membrane field and is not voltage dependent. On the other hand, experiments on *Xenopus* oocytes under conditions that are adequate to block passive K^+ channel current and Na^+/Ca^{2+} exchange have clearly shown that the pump *I-V* relationship has a negative slope in Na^+-free solution when $[K^+]_o$ is reduced below saturation. The voltage dependence of the apparent affinity for activation of pump current by K^+ under these conditions can be accounted for by a requirement that K^+ ions pass through ~18% of the membrane field to reach their external high affinity binding sites.

Negative Slope in the Pump *I-V* Relationship

When $[K^+]_o$ is reduced below saturation in *Xenopus* oocytes, the Na^+/K^+ pump *I-V* relationship has a negative slope over the entire measured voltage range (−150 to

+20 mV) in Na^+-free solution and a negative slope at positive voltages at normal $[Na^+]$ (90 mM). The existence of regions of both positive and negative slope in the *I-V* relationship requires that, for an unbranched reaction cycle, there must be at least two voltage-dependent steps in the Na^+/K^+ pump cycle. Based on the effects of changing external $[Na^+]$ and $[K^+]$, it appears that one of these steps is $[Na^+]_o$ dependent and the other $[K^+]_o$ dependent. A simple hypothesis that can account for the voltage dependence of the apparent affinities for external Na^+ and K^+ is that these ions must move through a high-field limited access channel (ion well) when moving between their external ion binding sites and the bulk extracellular fluid. It is not known whether these ions traverse the same or different ion wells. It is also possible that voltage affects the apparent affinity of these ions by a voltage-dependent redistribution of enzyme intermediates. The use of field-sensing styryl dyes (Stürmer et al., 1990) may provide a means of distinguishing between these two general classes of models. In any event, it is now clear that the apparent affinity for both Na^+ and K^+ is affected by membrane potential and that kinetic models of pump behavior should include appropriate voltage-dependent rate constants.

Acknowledgments

I thank my collaborators, P. De Weer, D. C. Gadsby, J. La Tona, C. Paxson, W. Schwarz, and L. A. Vasilets, for their help and advice.

Major funding for this work was provided by NIH grant NS-22979.

References

Bahinski, A., M. Nakao, and D. C. Gadsby. 1988. Potassium translocation by the Na^+/K^+ pump is voltage insensitive. *Proceedings of the National Academy of Sciences, USA.* 85:3412–3416.

Brinley, F. J., Jr., and L. J. Mullins. 1974. Effects of membrane potential on sodium and potassium fluxes in squid axons. *Annals of the New York Academy of Sciences.* 242:406–434.

Cantley, L. C., C. T. Carilli, R. L. Smith, and D. Perlman. 1984. Conformational changes of Na, K-ATPase necessary for transport. *Current Topics in Membranes and Transport.* 19:315–322.

De Weer, P. 1975. Aspects of the recovery processes in nerve. *In* MTP International Review of Science: Physiology Series One. Vol III: Neurophysiology. C. C. Hunt, editor. Butterworths & Co., Ltd., London. 231–278.

De Weer, P. 1984. Electrogenic pumps: theoretical and practical considerations. *In* Electrogenic Transport: Fundamental Principles and Physiological Implications. M. P. Blaustein and M. Lieberman, editors. Raven Press, New York. 1–15.

De Weer, P., D. C. Gadsby, and R. F. Rakowski. 1988. Voltage dependence of the Na-K pump. *Annual Review of Physiology.* 50:225–241.

De Weer, P., and R. F. Rakowski. 1984. Current generated by backward-running electrogenic Na pump in squid giant axons. *Nature.* 309:450–452.

Fendler, K., E. Grell, M. Haubs, and E. Bamberg. 1985. Pump currents generated by the purified Na^+, K^+-ATPase from kidney on black lipid membranes. *EMBO Journal.* 4:3079–3085.

Garrahan, P. J., and I. M. Glynn. 1967. Factors affecting the relative magnitudes of the sodium:potassium and sodium:sodium exchanges catalysed by the sodium pump. *Journal of Physiology.* 52:189–216.

Glynn, I. M. 1984. The electrogenic sodium pump. *In* Electrogenic Transport: Fundamental Principles and Physiological Implications. M. P. Blaustein and M. Lieberman, editors. Raven Press, New York. 33–48.

Goldshlegger, R., S. J. D. Karlish, A. Rephaeli, and W. D. Stein. 1987. The effect of membrane potential on the mammalian sodium-potassium pump reconstituted into phospholipid vesicles. *Journal of Physiology.* 387:331–355.

Kernan, R. P. 1962. Membrane potential changes during sodium transport in frog sartorius muscle. *Nature.* 193:986–987.

Lafaire, A. V., and W. Schwarz. 1986. The voltage dependence of the rheogenic Na^+/K^+ ATPase in the membrane of oocytes of *Xenopus laevis. Journal of Membrane Biology.* 91:43–51.

Läuger, P., and H.-J. Apell. 1986. A microscopic model for the current-voltage behavior of the Na,K-pump. *European Biophysical Journal.* 13:309–321.

Nakao, M., and D. C. Gadsby. 1986. Voltage dependence of Na translocation by the Na/K pump. *Nature.* 323:628–630.

Nakao, M., and D. C. Gadsby. 1989. [Na] and [K] dependence of the Na/K pump current-voltage relationship in guinea pig ventricular myocytes. *Journal of General Physiology.* 94:539–565.

Post, R. L., and D. C. Jolly. 1957. The linkage of sodium, potassium, and ammonium active transport across the human erythrocyte membrane. *Biochimica et Biophysica Acta.* 25:118–128.

Rakowski, R. F. 1989. Simultaneous measurement of changes in current and tracer flux in voltage-clamped squid giant axon. *Biophysical Journal.* 55:663–671.

Rakowski, R. F., D. C. Gadsby, and P. De Weer. 1989. Stoichiometry and voltage dependence of the sodium pump in voltage-clamped, internally dialyzed squid giant axon. *Journal of General Physiology.* 93:903–941.

Rakowski, R. F., and C. L. Paxson. 1988. Voltage dependence of Na/K pump current in *Xenopus* oocytes. *Journal of Membrane Biology.* 106:173–182.

Rakowski, R. F., L. A. Vasilets, J. La Tona and W. Schwarz. 1991. A negative slope in the current-voltage relationship of the Na^+/K^+ pump in *Xenopus* oocytes produced by reduction of external $[K^+]$. *Journal of Membrane Biology.* 121:177–187.

Slayman, C. L. 1982. Historical introduction to proceedings of symposium on electrogenic ion pumps, at Yale in 1981. *Current Topics in Membranes and Transport.* 16:xxxi–xxxvii.

Stürmer, W., H.-J. Apell, I. Wuddel, and P. Läuger. 1989. Conformational transitions and charge translocation by the Na, K pump: comparison of optical and electrical transients elicited by ATP-concentration jumps. *Journal of Membrane Biology.* 110:67–86.

Stürmer, W., R. Bühler, H.-J. Apell, and P. Läuger. 1990. Charge translocation by the Na/K pump. Kinetics of local field changes studied by time-resolved fluorescence measurements. *Journal of General Physiology.* 96:75*a*. (Abstr.)

Thomas, R. C. 1972. Electrogenic sodium pump in nerve and muscle cells. *Physiological Reviews.* 52:563–594.

Chapter 26

Voltage-induced Na/K Pump Charge Movements in Dialyzed Heart Cells

David C. Gadsby, Masakazu Nakao, and Anthony Bahinski

The Rockefeller University, New York 10021

Introduction

The Post-Albers scheme (Fig. 1; Albers et al., 1968; Post et al., 1969) for the kinetic cycle of the Na/K pump is supported by a good deal of evidence, both from biochemical studies of the distributions of enzyme intermediates and from kinetic analyses of pump-mediated, steady-state, unidirectional or net ionic fluxes, or currents (e.g., Glynn, 1985, 1988). According to that scheme, the pump assumes one of two major conformations, E_1 or E_2, either of which may be phosphorylated (E_1-P, E_2-P). Cytoplasmic Na ions promote phosphorylation of the E_1 conformation of the pump, occluding the Na to yield E_1-P (Na_3) which spontaneously relaxes to the E_2 conformation, E_2-P·3Na, deoccluding the Na ions and releasing them to the exterior. Binding of external K then promotes enzyme dephosphorylation and K occlusion in $E_2(K_2)$. Deocclusion of K accompanies reversion to the E_1 conformation, and the K ions are released to the cell interior. This normal, ATP-dependent export of three Na ions and import of two K ions is associated with the net extrusion of a single elementary charge per turnover (e.g., Thomas, 1972), and so at least one step in the transport cycle must involve charge movement through the membrane. Because the forward and backward transition rates of any step that moves charge within the membrane's electric field must vary with membrane potential, such steps are responsible for voltage-dependent properties of the transport cycle. Identification and characterization of the charge-translocating steps can provide important mechanistic information (Hansen et al., 1981; Chapman et al., 1983; De Weer, 1984; Reynolds et al., 1985; Läuger and Apell, 1986).

Measurements of steady-state transport rates at various membrane potentials have suggested that the Na translocation limb of the transport cycle incorporates the predominant electrogenic event (reviewed in De Weer et al., 1988*a*; Apell, 1989). However, more detailed kinetic information can be extracted from results of perturbation/relaxation experiments (e.g., Hansen et al., 1983; Läuger and Apell, 1988), and the effects of two kinds of perturbations have been examined. *Chemical* perturbations, in the form of jumps in the concentration of one of the pump's substrates, have been combined with measurements of the resulting changes in pumped radiotracer ion flux (e.g., Forbush, 1984, 1985; Karlish and Kaplan, 1985), in enzyme conformation (Rephaeli et al., 1986*a, b;* Stürmer et al., 1989), or in pump current (Fendler et al., 1985, 1987; Borlinghaus et al., 1987). Here we review our complementary studies of the effects of *electrical* perturbations, in the form of jumps of membrane potential, on Na/K pump currents in voltage-clamped, internally dialyzed cardiac myocytes. The advantages of using voltage jumps, rather than concentration jumps, to gain information about charge-translocating steps are that (*a*) they directly perturb *only* those steps, by instantaneously altering the voltage-sensitive forward and reverse transition rate constants, and (*b*) their influence can be examined over a range of membrane potentials which are monitored precisely. The main disadvantage of our experimental approach is that it uses intact cells (since, at present, pump current density is too low in sided, purified, reconstituted systems) whose surface membranes contain many current sources other than the Na/K pump. This means that, despite a fairly successful campaign to minimize other components of membrane current (Gadsby et al., 1985; Gadsby and Nakao, 1989), the pump-mediated current has to be determined as strophanthidin-sensitive current by subtraction of current records obtained after inhibiting the pump with a maximally

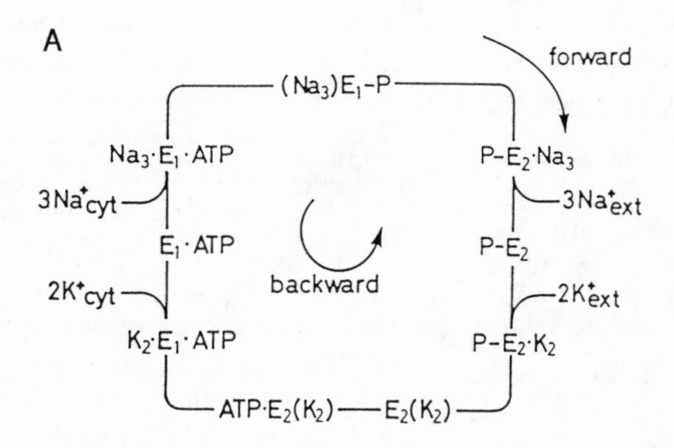

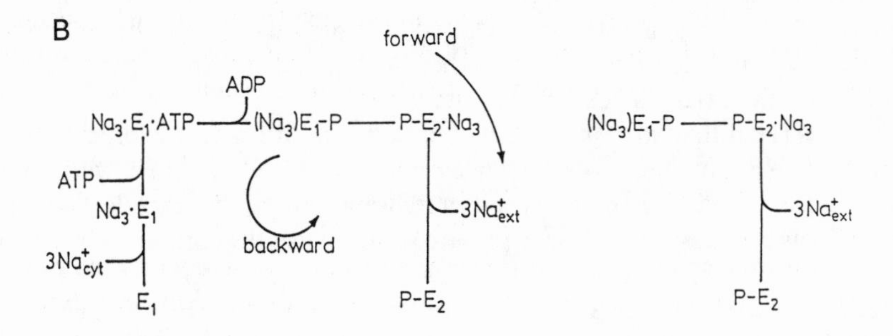

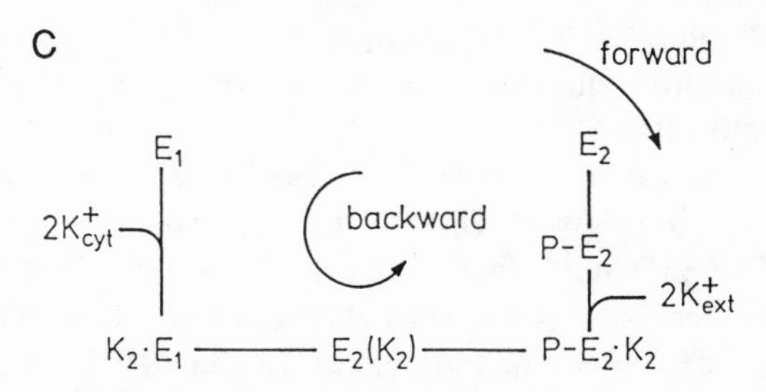

Figure 1. (*A*) Simplified Post–Albers kinetic scheme for the ion transport cycle of the Na/K pump. E_1 and E_2 are conformations of the enzyme with ion-binding sites exposed, respectively, to the cytoplasm or to the extracellular medium. In the states with "occluded" ions, $(Na_3)E_1$-P and $E_2(K_2)$, the bound ions are unable to exchange with either cytoplasmic or extracellular aqueous phase. Dashes represent covalent bonds and dots represent noncovalent bonds. The binding and dissociation reactions of ATP, ADP, and inorganic phosphate have been omitted for clarity. ATP bound at a low affinity site accelerates the deocclusion of K ions, $E_2(K_2) \rightarrow K_2{\cdot}E_1$. (*B*) *Left,* Reduced kinetic scheme for Na ion transport by the Na/K pump under Na/Na exchange conditions, in the absence of K ions. The binding and dissociation of ATP and ADP are incorporated. *Right,* Transport reactions of the Na/K pump remaining in the presence of ATP and Na, but absence of K and ADP. (*C*) Reaction sequence of the Na/K pump in the absence of Na ions under conditions of K/K exchange. Binding and dissociation of ATP and inorganic phosphate are not shown. (From Läuger and Apell, 1988.)

effective concentration of strophanthidin from currents elicited by identical voltage jumps just before the pump was inhibited.

Materials and Methods

The experimental details have been described previously (Nakao and Gadsby, 1986; Bahinski et al., 1988; Gadsby and Nakao, 1989). Briefly, cells were isolated from guinea pig ventricles after partial digestion with collagenase in nominally Ca-free Tyrode's solution as described by others (Isenberg and Klöckner, 1982; Matsuda et al., 1982). A high [K], Ca-free solution (Isenberg and Klöckner, 1982; Bendukidze et al., 1985) washed the collagenase from the partially digested heart, which was cut open and stored at 4°C in the same solution. Cells in that solution were introduced into the recording chamber on the movable stage of an inverted microscope and allowed to settle on the glass coverslip bottom for ~1 min before they were superfused with warmed Tyrode's solution containing (in mM): 145 NaCl, 5.4 KCl, 1.8 $CaCl_2$, 0.5 $MgCl_2$, 5.5 dextrose, and 5 HEPES/NaOH (pH 7.4). All external solutions passed through glass water-jacketed heating coils close to the chamber, and temperature was monitored with a micro thermistor bead and kept near 36°C (except when deliberately varied as described in the text).

GΩ seals (Hamill et al., 1981) were obtained with fire-polished pipettes with tip diameters of ~5 μm and tip resistances of ~1 MΩ when filled with Tyrode's solution. Before rupturing the enclosed membrane patch, the Tyrode's solution in the pipette was exchanged for K-free pipette solution usually containing (in mM): 50 NaOH, ~85 CsOH, 85 aspartic acid, 20 TEACl, 2 $MgCl_2$, 5.5 dextrose, 10 EGTA, 10 MgATP, 5 $Tris_2$-creatine phosphate, 5 pyruvic acid, and 10 HEPES (pH 7.4). The apparatus for exchanging the solution inside the pipette has been described previously (Soejima and Noma, 1984; Sato et al., 1985). When the membrane patch was ruptured, the wide pipette tip allowed rapid equilibration of the pipette solution with the cell interior (half-time $\ll$ 1 min), monitored as an inward shift of current at the −40-mV holding potential, due primarily to the loss of cellular K (Matsuda and Noma, 1984; Sato et al., 1985).

The voltage-clamped cell was then superfused with a modified K-free, Ca-free Tyrode's solution containing (in mM): 150 NaCl, 2.3 $MgCl_2$, 2 $BaCl_2$, 0.5 $CdCl_2$, 5.5 dextrose, and 5 HEPES/NaOH (pH 7.4). The Ba blocked K channels, the Cd prevented Ca channel currents (carried by Ba), and the lack of internal or external Ca prevented Na/Ca exchange current (Kimura et al., 1987). Variations of these intracellular and extracellular solutions, as specified in the text and legends, were used to sustain four modes of Na/K pump activity (e.g., Glynn, 1985), namely, Na/Na exchange, forward or backward Na/K exchange, or K/K exchange. Pipette [Na] was varied by substitution with Cs; external [Na] was lowered by equimolar replacement with *N*-methyl-D-glucamine (NMG); K was added to external solutions by substitution for Na (or NMG). All external and internal solutions had osmolalities close to 300 mosmol/kg.

Na/K pump current was determined as the difference between whole-cell current recorded in the absence and in the presence of 0.5–2 mM strophanthidin. Strophanthidin was added from a 0.5 M solution in DMSO; control measurements showed that up to 0.5% (by volume) DMSO had no effect on membrane currents (see, for example, Figs. 3 *A* (*b*), 3 *B* (*b*) and 6 *C*.

Current and voltage signals were low-pass filtered at 2 kHz, digitized (12-bit resolution) on-line at 8 kHz, and stored in a computer for analysis. Up to 55% of the series resistance between pipette interior and cell membrane was compensated by summing a fraction of the clamp output to the command potential. The voltage clamp amplifier was connected to the pipette and to the chamber via 3 M KCl-filled half cells to minimize voltage errors due to liquid junction potentials. For this reason, no corrections have been applied.

Results

Voltage-induced Na/K Pump Charge Movements Associated with Na Translocation Steps

Because the voltage dependence of steady-state Na/K pump current in frog blastomeres was found to be greatly attenuated after removal of extracellular Na (Béhé and Turin, 1984), a result subsequently confirmed with guinea pig cardiac myocytes (see Fig. 5; and Gadsby and Nakao, 1987), *Xenopus* oocytes (Rakowski and Paxson, 1988), and squid giant axons (Rakowski et al., 1989), Na translocation steps seemed an appropriate place to begin our search for voltage-induced pump charge movements. Fig. 2 shows superimposed traces of currents from a cell equilibrated with the 50 mM Na, 10 mM ATP pipette solution and the K-free, 150 mM Na external solution (conditions limiting pump activity to the sequence of steps in Fig. 1 *B*), first in the absence (*a*), and then in the presence (*b*), of strophanthidin. The strophanthidin-sensitive currents, shown in *c* and *d,* were obtained by subtracting records in *b* from those in *a.* The lack of steady strophanthidin-sensitive current at all potentials

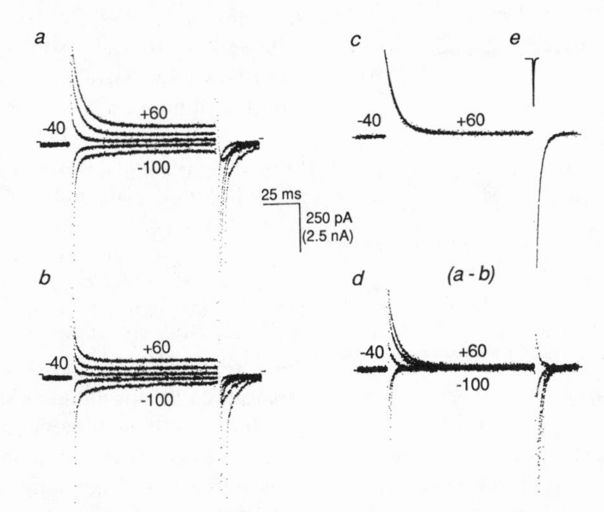

Figure 2. Voltage-induced transient pump currents recorded from a cardiac cell superfused with K-free, 150 mM Na solution and dialyzed with 50 mM Na, 10 mM ATP pipette solution. *a* and *b*, Superimposed traces of whole-cell currents elicited by 100-ms voltage pulses from the holding potential, −40 mV, to +60, +20, −20, −60, and −100 mV in the absence (*a*), and then in the presence (*b*), of 0.5 mM strophanthidin. *c* and *d*, Strophanthidin-sensitive (i.e., Na/K pump) currents, elicited by computer subtraction of records in *b* from counterparts in *a*: *c*, transient pump currents elicited by pulse to +60 mV with superimposed, least-squares, exponential fits (τ_{on} [+60 mV] = 8 ms; τ_{off} [−40 mV] = 4 ms); *d*, superimposed, transient pump currents for all pulses in *a* and *b*. *Inset* (*e*), Capacity current (decay time constant = 0.5 ms) during 10-mV hyperpolarization, without series resistance compensation; after compensation (as used for all other records), capacity currents decayed fully within 1 ms. Total cell capacitance, 156 pF. Current scale is 250 pA for all records except *e* (2.5 nA). (From Nakao and Gadsby, 1986.)

confirms the absence of any electrogenic transport cycle. Voltage steps nevertheless elicited exponentially decaying transient pump currents, which were directed outward following a positive voltage step, and inward following a negative voltage step. The exponential relaxation was slower at positive than at negative potentials.

That these transient pump currents are closely associated with Na translocation steps is suggested by the findings, illustrated in Fig. 3, that they depend on the presence of intracellular ATP, as well as both extracellular and intracellular Na, all known requirements for Na/Na exchange (Garrahan and Glynn, 1967*a;* Glynn and Hoffman, 1971). These requirements suggest that the charge movements involve Na-loaded, phosphorylated forms of the Na pump, and since pipette [ATP] and [Na] were 10 and 50 mM, respectively, and presumably saturating (cf. Nakao and Gadsby,

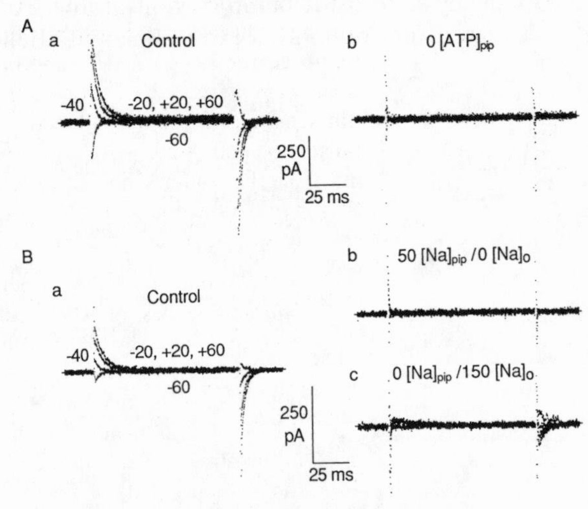

Figure 3. The transient pump currents require intracellular ATP and Na, as well as extracellular Na. All five sets of records show superimposed strophanthidin-sensitive currents (obtained as in Fig. 2) induced by 100-ms voltage steps to +60, +20, −20, and −60 mV from the holding potential, −40 mV. (*A*) *a*, Solutions as in Fig. 2; *b*, ~15 min after switching to pipette solution without ATP, creatine phosphate, or dextrose (all replaced by Cs-aspartate); cell capacitance, 194 pF. (*B*) *a*, Solutions as in Fig. 2; *b*, records obtained ~9 min before those in *a*, during exposure to Na-free external solution (Na replaced by NMG), but 50 mM Na pipette solution; *c*, records obtained ~25 min before those in *a*, during exposure to 150 mM Na external solution, but Na-free pipette solution (Na replaced by Cs); cell capacitance, 109 pF. The small transient current remaining in *c* presumably reflects the difficulty of maintaining zero intracellular [Na] in the face of the large inward electrochemical gradient and no means of net Na extrusion, as well as the high affinity of the intracellular pump sites for Na (Garay and Garrahan, 1973). (From Nakao and Gadsby, 1986.)

1989), they are consistent with a greatly reduced kinetic scheme such as that in Fig. 1 *C,* in which the deocclusion and/or release of Na to the outside are assumed to be voltage sensitive (e.g., Läuger and Apell, 1988). Further support for involvement of a voltage-dependent quasi-equilibrium like that in Fig. 1 *C* derives from the findings that there is no steady-state current, and that integration of the transient currents (e.g., those in Fig. 2 *c*) reveals that the same quantity of charge moves at the beginning and end of each voltage step, but in opposite directions. In addition, we have found that the transient currents are practically abolished by 2 μg/ml oligomycin B, which is known to inhibit Na/Na exchange (Garrahan and Glynn, 1967*b*) but not ATP/ADP exchange (Fahn et al., 1966), and hence is believed to prevent the E_1-P → E_2-P conformational change.

Voltage Dependence of Quantity of Charge and Rate of Relaxation

Fig. 4 provides information on voltage-dependent characteristics of the transient pump currents. In *A*, the quantity of charge moved during the "on" transients (i.e., the time integral of the current at the beginning of the step) is shown plotted against the step membrane potential. The amount of charge moved is a saturable function of voltage and is reasonably well approximated by a Boltzmann relation describing a two-state, voltage-dependent equilibrium governed by movement of a single charge through the entire membrane field. Fig. 4 *B* shows the rate constants for exponential

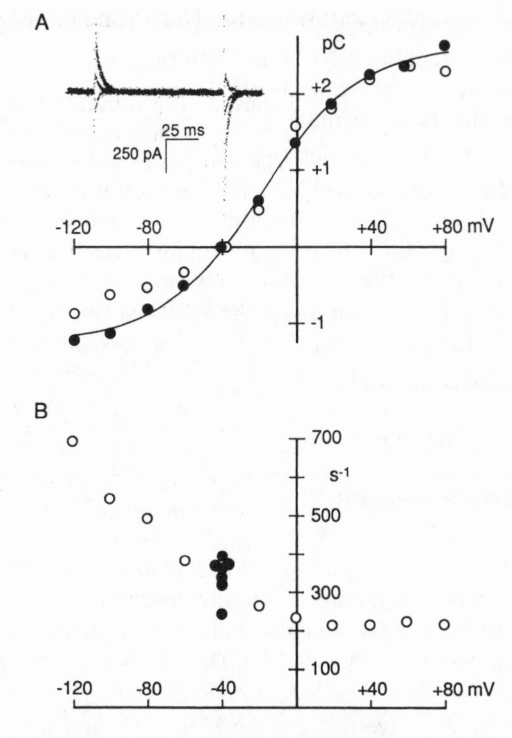

Figure 4. Voltage dependence of quantity of charge moved (*A*) and of exponential decay rate constant (*B*) during voltage-induced transient pump currents. (*A*) *Inset,* Superimposed sample records of strophanthidin-sensitive currents induced by pulses to +60, +20, −20, and −60 mV from the −40-mV holding potential; *graph,* time integral [$Q(V)$] of current transients elicited by each pulse, plotted against pulse potential (V). Open circles show charge measured directly from the records, and filled circles show charge estimated by extrapolation of the exponential current decay toward the start of the voltage step. Only "on" charge movement is plotted. The smooth curve, derived from the Boltzmann relation, shows $\Delta Q(V)/\Delta Q_{max} = 1/[1 + \exp(V' - V)/k]$, where $\Delta Q(V) = Q(V) - Q_{min}$, $\Delta Q_{max} = Q_{max} - Q_{min}$, V' is the potential at the midpoint, and k provides a measure of the "equivalent charge." The values used were $Q_{min} = -1.23$ pC, $Q_{max} = +2.63$ pC, $V' = -20$ mV, and $k = 26.5$ mV. (*B*) Rate constants of exponential fits to transient currents plotted against membrane potential: open circles from "on" transients; filled circles from "off" transients at −40 mV after repolarization from $-60 \text{ mV} < V < +80$ mV. Solutions as in Fig. 2; cell capacitance, 208 pF; temperature, 35°C. (From Nakao and Gadsby, 1986.)

relaxation of the transient currents as a function of the voltage during the relaxation. These equilibration rates increase as the membrane potential is made more negative, but seem to reach a minimum level at positive voltages. Since depolarization elicits outward currents, the forward process should reflect outward movement of positive charge (or inward movement of negative charge) and should be accelerated at positive membrane potentials, whereas the backward process should be accelerated at negative potentials. The equilibration rate constant after a voltage jump is given by the sum of the forward and backward, pseudo–first order transition rate constants at the new voltage, and so we can conclude from the results in Fig. 4 *B* that the backward rate constant is relatively strongly voltage dependent, whereas the forward rate constant is little, if at all, affected by membrane potential.

Influence of Temperature

Lowering the temperature slows the relaxation of the transient pump currents at all voltages by about the same factor (some threefold per 10°C) without affecting the quantity of charge moved (Nakao and Gadsby, 1986). Thus, at 0 mV the relaxation rate is $\sim 200\ s^{-1}$ at 36°C (Fig. 4 *B*), but is diminished to $\sim 40\ s^{-1}$ at 20°C.

Influence of Extracellular Na Concentration

Preliminary results indicate that the relaxation rates of the transient pump currents are also diminished by reduction of the extracellular Na concentration, $[Na]_o$, but are increased by raising $[Na]_o$ above 150 mM (Nagel, et al., 1991). However, the size of this effect varies with membrane potential, being virtually absent at positive voltages but becoming progressively more marked at increasingly negative potentials. The effect of lowering $[Na]_o$ is roughly equivalent to a leftward shift (i.e., toward more negative voltages) of the relaxation rate constant–voltage curve (e.g., Fig. 4 *B*). The lack of effects at positive voltages (where the forward voltage-sensitive rate constant should dominate the relaxation rate) implies that it is the backward, and not the forward, rate constant that is sensitive to $[Na]_o$. Moreover, the leftward shift of the rate–voltage curve on lowering $[Na]_o$ implies a qualitative equivalence between raising $[Na]_o$ and making the membrane potential more negative; such an equivalence is characteristic of "ion well" effects in which ions traverse some fraction of the electrical field of the membrane, via a high-field access channel (Mitchell, 1969; Läuger and Apell, 1986, 1988), before exerting their influence. Furthermore, it seems clear from the effective concentration range that this influence of external Na is mediated via a low affinity site.

Fig. 5 shows that, as already mentioned, external Na ions exert a similarly voltage-dependent influence, albeit inhibitory, on steady-state outward pump current recorded during forward Na/K exchange (Nakao and Gadsby, 1989). In that case, the apparent affinity with which external Na ions inhibited forward Na/K pumping was also very low (average $K_{0.5} \sim 360$ mM at -40 mV), and was increased as the membrane potential was made more negative ($K_{0.5} \sim 90$ mM at -120 mV). The curves fitted to the points in Fig. 5 show that this inhibitory influence of $[Na]_o$ can be mimicked by a simple two-state kinetic cycle (upper part of Fig. 5; Hansen et al., 1981; De Weer, 1984, 1990) with one pair of voltage-independent, pseudo–first order rate constants, and one pair of asymmetric, voltage-dependent, pseudo-first order rate constants in which almost all of the voltage dependence falls on the backward rate (Nakao and Gadsby, 1986, 1989). To fit the results in Fig. 5, it was necessary merely to allow the backward voltage-dependent rate constant, β, to vary with $[Na]_o$, continuously increasing as $[Na]_o$ was raised from 1.5 to 150 mM. As described above, we have now obtained more direct evidence for such activation by $[Na]_o$ of the backward voltage-dependent rate.

Lack of Transient Pump Currents under K/K Exchange Conditions

The voltage dependence of the quantity of charge moved under restricted Na/Na exchange conditions (Fig. 4 *A*) implies that, during Na translocation, a single elementary charge moves per pump. If correct, and if the same charge movement accompanies Na translocation during Na/K exchange, then one consequence of that inference is that no other step in the Na/K transport cycle (Fig. 1 *A*) is required to involve charge movement, because only a single net charge is extruded during each 3Na/2K exchange cycle (Thomas, 1972; Eisner et al., 1981, 1983; Glitsch et al., 1982; Rakowski et al., 1989). Furthermore, if any additional step were to be involved in charge movement, yet another step would be required to move precisely the same increment of charge, but in the opposite direction, to ensure that the net result is extrusion of a single positive charge per cycle. To examine the possibility of charge movement during K translocation, we looked for voltage-induced transient pump currents under K/K exchange conditions (Fig. 1 *C*).

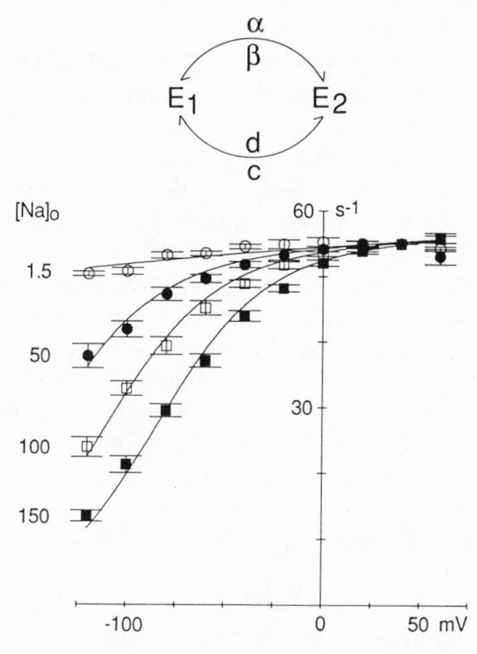

Figure 5. Graded influence of $[Na]_o$ on the shape of the steady-state current–voltage relationship for the Na/K pump at 50 mM $[Na]_{pip}$ and 5.4 mM $[K]_o$. Results from nine cells are summarized in the graph. Each Na/K pump current–voltage curve was normalized to its amplitude at +40 mV, and the resulting data at a given $[Na]_o$ were averaged. The mean normalized currents were then scaled by the average turnover rate, 55 s^{-1} (at +40 mV and 150 mM $[Na]_o$; Gadsby and Nakao, 1989) and the mean (±SEM) rates at the four $[Na]_o$ concentrations plotted against pulse voltage. *Open circles,* 1.5 mM $[Na]_o$; *filled circles,* 50 mM $[Na]_o$; *open squares,* 100 mM $[Na]_o$; *filled squares,* 150 mM $[Na]_o$. Overall, the amplitude of steady-state pump current at positive potentials showed no consistent variation with $[Na]_o$. The curves show least-squares fits to the pseudo two-state kinetic scheme (Hansen et al., 1981) illustrated above for movement of a single charge through the membrane field, and obey: rate $= (\alpha c - \beta d)/(\alpha + \beta + c + d)$, where c and d are voltage-independent rate constants, $\alpha = \alpha^0 \cdot \exp(\delta VF/RT)$, $\beta = \beta^0 \cdot \exp[-(1-\delta)VF/RT]$, α^0 and β^0 indicate values of the voltage-sensitive rate constants at 0 mV, and δ indicates the fraction of the membrane field that influences the forward voltage-sensitive rate constant, α. The curves demonstrate that the effect of lowering $[Na]_o$ can be mimicked by a progressive reduction of β, the backward voltage-sensitive rate constant. With β^0 and d fixed at 0 s^{-1}, and δ fixed at 0.1, best-fit values for α^0 and c were 490 and 61 s^{-1}, respectively, for the data at 1.5 mM $[Na]_o$. Then, holding all parameters but β^0 constant, curves were fitted to the data at 50, 100, and 150 mM $[Na]_o$, yielding values for β^0 of 3, 8, and 22 s^{-1}, respectively. (From Nakao and Gadsby, 1989.)

However, it was first necessary to demonstrate the adequacy of conditions for sustaining reverse K translocation by driving the entire Na/K pump reaction cycle backwards (Garrahan and Glynn, 1967*c*), generating inward current (De Weer and Rakowski, 1984). To accomplish that, the physiological transmembrane gradients of

[Na] (150 mM external [Na]; ~10 mM internal [Na]) and [K] (145 mM internal [K]; ~5 mM external [K]) were steepened by omitting Na from the pipette solution and K from the bath solution, and 5 mM ADP and 5 mM inorganic phosphate (P_i) were added to the pipette solution. Strophanthidin then caused an *outward* shift of steady-state current at the −40-mV holding potential (see Fig. 6 *A*), reflecting inhibition of *inward* pump current (De Weer and Rakowski, 1984; Bahinski et al., 1988; De Weer et al., 1988*b*). The strophanthidin-sensitive inward currents at various voltages (Fig. 6 *B, a − b*) were obtained by appropriate subtraction; their steady amplitudes increased as the membrane potential was made more negative, from being very small at +40 mV toward an apparent plateau level near −100 mV. Fig. 6 *B* shows that, during reverse Na/K exchange, voltage steps elicited pre–steady state transient pump currents before establishment of the new stationary levels of inward pump current, as also found for forward Na/K exchange (Bahinski et al., 1988).

The steady inward current confirms reverse Na/K pump cycling and so demonstrates that the 5 mM P_i and 145 mM K in the pipette solution were adequate to sustain reverse K translocation. We already knew that 5.4 mM $[K]_o$ and millimolar levels of pipette [ATP] were adequate to drive forward K translocation. So, to initiate pump-mediated K/K exchange, we suddenly removed all external Na and added 5.4 mM external K while retaining the high-K, Na-free pipette solution including 1 mM ATP and 5 mM P_i (Fig. 6 *A*). After that solution change neither transient nor steady-state strophanthidin-sensitive currents could be recorded (Fig. 6 *C, c − d*). This result suggests that K translocation by the Na/K pump does not involve net charge movement within the membrane field. Specifically, the K occlusion and deocclusion steps, $E_2\text{-P}\cdot K_2 \rightarrow E_2(K_2)$ and $E_2(K_2) \rightarrow E_1\cdot K_2$ (cf. Fig. 1), must be voltage insensitive. But this experiment does not rule out the possibility of charge movement associated with the K ion binding step, $E_2\text{-P} + 2K \rightarrow E_2\text{-P}\cdot K_2$, because in the absence of extracellular Na ions the affinity of the pump for external K is extremely high ($K_{0.5}$ ~0.2 mM for forward Na/K exchange in these cells; Nakao and Gadsby, 1989), so that at 5.4 mM $[K]_o$ the binding sites can be expected to remain saturated (cf. Läuger and Apell, 1988). However, recent preliminary tests have indicated that no strophanthidin-sensitive charge movement can be recorded even when the experiment is repeated at 0.2 mM $[K]_o$.

Discussion

Na Translocation Pathway

A large body of now overwhelming evidence, garnered using a variety of experimental techniques, indicates that Na translocation includes a charge-moving, and hence voltage-sensitive, step (reviewed in De Weer et al., 1988*a;* Apell, 1989). Thus, in partially purified pumps reconstituted into phospholipid vesicles, large positive membrane potentials set with ionophores accelerated both Na/K transport at high [ATP] and the conformational transition ($E_1\text{-P} \rightarrow E_2\text{-P}$) of Na-loaded, fluorescein-labeled enzyme, phosphorylated via acetylphosphate (Karlish et al., 1985, 1988; Rephaeli et al., 1986*b;* Goldshlegger et al., 1987). In addition, in the presence of Na but absence of K, photo-release of caged ATP elicited transient pump currents in Na,K-ATPase-containing membrane fragments adsorbed onto lipid bilayers (Fend-

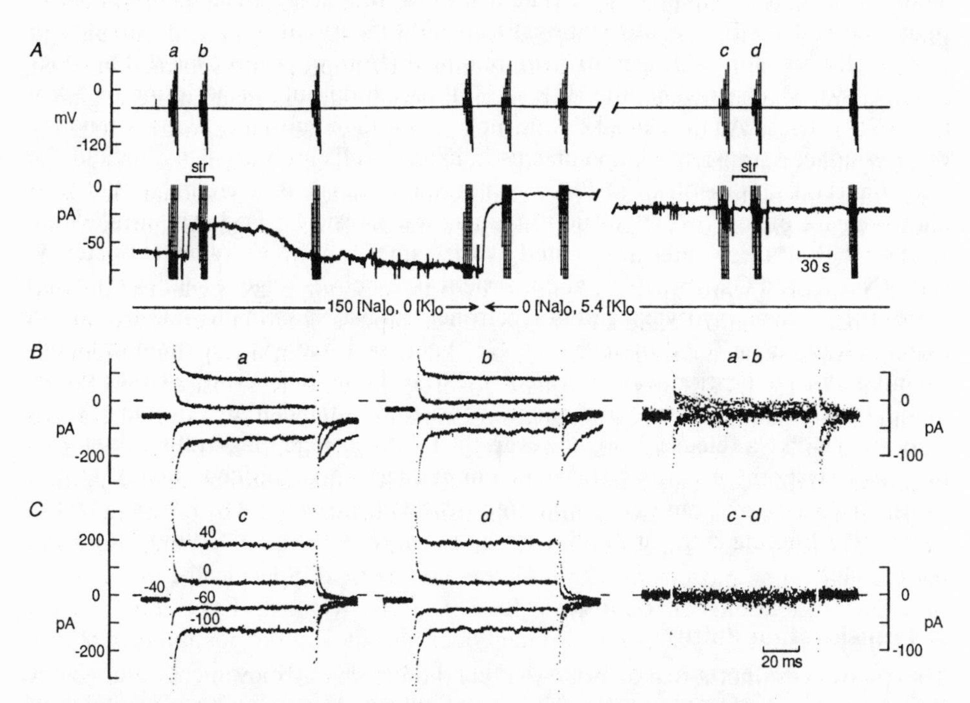

Figure 6. Transient currents and steady-state inward current generated by the pump during reverse Na/K exchange, and absence of strophanthidin-sensitive current under K/K exchange conditions. (*A*) Chart recording of membrane potential (*upper trace*) and whole-cell current (*lower trace*). The gap marks omission of 9 min of record; labels *a–d* denote acquisition of current–voltage data using 80-ms voltage pulses to potentials -140 mV $< V <$ $+60$ mV from the -40-mV holding potential; samples of the resulting currents are presented in *B* and *C*. The bars over the current recording mark exposures to 2 mM strophanthidin (*str*), and the arrows below signify the switch from K-free, 150 mM Na external solution to 5.4 mM K, Na-free (145 mM NMG) solution. As indicated at the top of the figure, the internal (pipette) solution was Na free, but included 145 mM K, 1 mM ATP, 5 mM ADP, and 5 mM P_i. (*B*) Superimposed traces of whole-cell currents (from *A*) elicited by voltage pulses to +40, 0, -60, or -100 mV, during backward pumping (*a*) or in the presence of strophanthidin (*b*). The high-gain, strophanthidin-sensitive currents to the right ($a - b$) were obtained by subtracting traces in *b* from counterparts in *a*. (*C*) Superimposed current records as in *B*, but recorded under K/K exchange conditions without (*c*) or with (*d*) strophanthidin. The high-gain, strophanthidin-sensitive currents ($c - d$) were obtained by appropriate subtraction. Cell capacitance, 177 pF. (From Bahinski et al., 1988.)

ler et al., 1985, 1987; Apell et al., 1987; Borlinghaus et al., 1987; Nagel et al., 1987), and voltage jumps elicited transient pump currents in cardiac myocytes (Nakao and Gadsby, 1986; Bahinski et al., 1988).

That charge translocation is associated with deocclusion and release of Na is now suggested by several findings. Thus, the ATP-initiated transient charge movements decay with the approximate time course of fluorescence changes believed to signal the E_1-P → E_2-P conformational transition (Stürmer et al., 1989), and both were abolished after mild chymotrypsin treatment (Borlinghaus et al., 1987; Stürmer et al., 1989), which prevents the E_1-P → E_2-P transition but not occlusion of Na in E_1- P(Na_3). Those ATP-induced charge movements in membrane fragments and the voltage-induced transient pump currents in cardiac cells both decay with about the same time constant (~ 40 s^{-1} at 20°C) as that of the transient ^{22}Na efflux initiated by photo-release of caged ATP inside tight membrane vesicles (Forbush, 1984, 1985), arguing that all three rates are limited by the same step. Measurements of steady-state (Nakao and Gadsby, 1989) and transient (see above; Nagel et al., 1991) pump currents in cardiac myocytes, and of electroneutral Na/Na exchange-mediated ^{22}Na efflux in squid axon (Gadsby, D. C., P. De Weer, and R. F. Rakowski, manuscript submitted for publication), all indicate that extracellular Na ions influence the Na/K pump in a voltage-dependent manner, consistent with charge movement being associated with Na release. If so, however, the relatively high temperature sensitivity of the decay both of cardiac transient pump currents (described above), and of transient ^{22}Na efflux from tight membrane vesicles (Forbush, 1984, 1985), mitigates against the limiting step for Na release being mere reversal of a diffusion-limited binding step.

K Translocation Pathway

The question of whether, and under what conditions, charge movement occurs in the K translocation limb of the cycle (Fig. 1 *C*) is presently controversial. In experiments with Na/K pumps reconstituted into phospholipid vesicles, large positive membrane potentials set with ionophores did not alter the rates of (ATP + P_i)-activated, vanadate-inhibited, Rb/Rb exchange, or of ATP-activated Na/K transport at low [ATP] (when K translocation should be rate limiting) (Goldshlegger et al., 1987), or of conformational changes of K-loaded, fluorescein-labeled enzyme (Rephaeli et al., 1986*a*). In bilayer experiments with Na/K pumps in adsorbed membrane fragments, release of caged ATP in Na-free solutions containing 1–10 mM K elicited virtually no transient current, despite the large fluorescence change observed with labeled enzyme in parallel experiments under identical conditions (Stürmer et al., 1989). As already described (Fig. 6), voltage jumps induced no transient strophanthidin-sensitive currents in cardiac myocytes exposed to Na-free, K-containing solutions under K/K exchange conditions (Bahinski et al., 1988).

While these results make it unlikely that the conformational changes associated with occlusion and deocclusion of K ions involve charge movement, in all of the experiments the external [K] was saturating so that voltage-dependent binding of K at extracellular sites could not be ruled out (Läuger and Apell, 1988). Indeed, the above suggestion that Na release from external sites could be voltage sensitive might be taken to imply that binding of external K could also depend on voltage. Although the failure, in preliminary experiments on cardiac myocytes (mentioned above), of voltage steps to elicit strophanthidin-sensitive charge movement under K/K exchange conditions, even at nonsaturating $[K]_o$ (0.2 mM), does not provide support for the idea that external K binding is voltage dependent, recent results from two other preparations are consistent with it. One indication comes from the $[Na]_o$ and $[K]_o$ dependence of fluorescence measurements (Stürmer et al., 1991) from electrochromic styryl dyes (Klodos and Forbush, 1988) incorporated into Na,K-ATPase membranes: the apparent affinities for Na release and K occlusion were altered when lipophilic ions changed the local electric field within the membrane, in a manner consistent with the existence of an ion well (Stürmer et al., 1991). The other indication comes from the finding of a similar voltage dependence of the apparent affinity for $[K]_o$ activation of Na/K pump current in *Xenopus* oocytes in Na-free solution (Rakowski, 1991; Rakowski et al., 1991). Under those conditions, at saturating $[K]_o$, the Na/K pump current of oocytes, like that of cardiac myocytes (Nakao and Gadsby, 1989), is practically voltage independent. But at low $[K]_o$ the Na/K pump current–voltage (*I-V*) relationship of oocytes has a pronounced negative slope over a wide voltage range (Rakowski et al., 1991), unlike results from cardiac cells (Nakao and Gadsby, 1989). Indeed, so far, cardiac cells have provided no consistent evidence for a convincing negative slope region in the Na/K pump *I-V* relationship between −140 and +60 mV over a wide range of intra- and extracellular [K] and [Na] (Nakao and Gadsby, 1989). The usual caveats (e.g., De Weer et al., 1988 *a, b;* Gadsby and Nakao, 1989) concerning contamination of steroid-sensitive currents with nonpump currents notwithstanding, the negative slope regions observed by Rakowski et al. (1991) at negative membrane potentials in oocytes bathed in Na-free solutions seem unlikely to be artifactual. The lack of corresponding marked negative slopes in Na/K pump *I-V* relationships of cardiac myocytes is somewhat puzzling in view of the clear influence of voltage on the apparent affinity for external Na inhibition of pump current in the same cells (Fig. 5), which is consistent with extracellular ions interacting with the pump via a high-field access channel (Nakao and Gadsby, 1989). Whether this seeming discrepancy reflects technical difficulties or real differences between pump isoforms and/or lipid environment remains to be established. Finally, it must be borne in mind that apparent affinities derived from transport rates are rarely good estimates of binding affinities, since they are generally distorted by the kinetic properties of other steps in the transport cycle.

Acknowledgments

We thank Dr. Paul F. Cranefield for constant encouragement, and Peter Hoff for technical assistance.

This work has been supported by the National Institutes of Health (HL-14899 and HL-36783), and the American Heart Association, New York City Affiliate, and the Japan Heart Foundation. D. C. Gadsby is a Career Scientist of the Irma T. Hirschl Trust.

References

Albers, R. W., G. J. Koval, and G. J. Siegel. 1968. Studies on the interaction of ouabain and other cardioactive steroids with Na, K-activated ATPase. *Molecular Pharmacology.* 4:324–336.

Apell, H.-J., 1989. Electrogenic properties of the Na,K pump. *Journal of Membrane Biology.* 110:103–114.

Apell, H.-J. R. Borlinghaus, and P. Läuger. 1987. Fast charge-translocations associated with partial reactions of the Na/K pump. II. Microscopic analysis of transient currents. *Journal of Membrane Biology.* 97:179–191.

Bahinski, A., M. Nakao, and D. C. Gadsby. 1988. Potassium translocation by the Na/K pump is voltage insensitive. *Proceedings of the National Academy of Sciences, USA.* 85:3412–2416.

Béhé, P., and L. Turin. 1984. Arrest and reversal of the electrogenic sodium pump under voltage clamp. *Proceedings of the 8th International Biophysical Congress.* 304. (Abstr.)

Bendukidze, Z., G. Isenberg, and U. Klöckner. 1985. Ca-tolerant guinea-pig ventricular myocytes as isolated by pronase in the presence of 250 μM free calcium. *Basic Research in Cardiology.* 80 (Suppl. 1): 13–18.

Borlinghaus, R., H.-J. Apell, and P. Läuger. 1987. Fast charge-translocations associated with partial reactions of the Na/K-pump. I. Current and voltage transients after photochemical release of ATP. *Journal of Membrane Biology.* 97:161–178.

Chapman, J. B., E. A. Johnson, and J. M. Kootsey. 1983. Electrical and biochemical properties of an enzyme model of the sodium pump. *Journal of Membrane Biology.* 74:139–153.

De Weer, P. 1984. Electrogenic pumps: theoretical and practical considerations. *In* Electrogenic Transport: Fundamental Principles and Physiological Implications. M. P. Blaustein and M. Lieberman, editors. Raven Press, New York. 1–15.

De Weer, P. 1990. The Na/K pump: a current-generating enzyme. *In* Regulation of Potassium Transport across Biological Membranes. L. Reuss, J. Russell, and G. Szabo, editors. University of Texas Press, Austin. 5–28.

De Weer, P., D. C. Gadsby, and R. F. Rakowski. 1988*a*. Voltage dependence of the Na/K pump. *Annual Review of Physiology.* 50:225–241.

De Weer, P., D. C. Gadsby, and R. F. Rakowski. 1988*b*. Stoichiometry and voltage dependence of the sodium pump. *In* The Na^+,K^+-Pump, Part A: Molecular Aspects. J. C. Skou, J. G. Nørby, A. B. Maunsbach, and M. Esmann, editors. Alan R. Liss, Inc., New York. 421–434.

De Weer, P., and R. F. Rakowski. 1984. Current generated by backward-running electrogenic Na pump in squid giant axons. *Nature.* 309:450–452.

Eisner, D. A., W. J. Lederer, and S.-S. Sheu. 1983. The role of intracellular sodium activity in the anti-arrhythmic action of local anaesthetics in sheep Purkinje fibres. *Journal of Physiology.* 340:239–257.

Eisner, D. A., W. J. Lederer, and R. D. Vaughan-Jones. 1981. The dependence of sodium pumping and tension on intracellular sodium activity in voltage-clamped sheep Purkinje fibres. *Journal of Physiology.* 317:163–187.

Fahn, S., G. J. Koval, and R. Albers. 1966. Sodium-potassium-activated adenosine triphosphatase of *electrophorus* electric organ. *Journal of Biological Chemistry.* 241:1882–1889.

Fendler, K., E. Grell, and E. Bamberg. 1987. Kinetics of pump currents generated by the Na^+K^+-ATPase. *FEBS Letters.* 224:83–88.

Fendler, K., E. Grell, M. Haubs, and E. Bamberg. 1985. Pump currents generated by the purified Na^+K^+-ATPase from kidney on black lipid membranes. *EMBO Journal.* 4:3079–3085.

Forbush, B. 1984. Na^+ movement in a single turnover of the Na pump. *Proceedings of the National Academy of Sciences, USA.* 81:5310–5314.

Forbush, B., III. 1985. Rapid ion movements in a single turnover of the Na^+ pump. *In* The Sodium Pump: 4th International Conference on Na,K-ATPase. I. M. Glynn and J. C. Ellory, editors. The Company of Biologists, Cambridge, UK. 599–611.

Gadsby, D. C., J. Kimura, and A. Noma. 1985. Voltage dependence of Na/K pump current in isolated heart cells. *Nature.* 315:63–65.

Gadsby, D. C., and M. Nakao. 1987. [Na] dependence of the Na/K pump current–voltage relationship in isolated cells from guinea-pig ventricle. *Journal of Physiology.* 382:106P. (Abstr.)

Gadsby, D., and M. Nakao. 1989. Steady-state current–voltage relationship of the Na/K pump in guinea-pig ventricular myocytes. *Journal of General Physiology.* 94:511–537.

Garay, R. P., and P. J. Garrahan. 1973. The interaction of sodium and potassium with the sodium pump in red cells. *Journal of Physiology.* 231:297–325.

Garrahan, P. J., and I. M. Glynn. 1967*a*. The behaviour of the sodium pump in red cells in the absence of external potassium. *Journal of Physiology.* 192:159–174.

Garrahan, P. J., and I. M. Glynn. 1967*b*. The stoichiometry of the sodium pump. *Journal of Physiology.* 192:217–235.

Garrahan, P. J., and I. M. Glynn. 1967*c*. The incorporation of inorganic phosphate into adenosine triphosphate by reversal of the sodium pump. *Journal of Physiology.* 192:237–256.

Glitsch, H. G., H. Pusch, Th. Schumacher, and F. Verdonck. 1982. An identification of the K activated Na pump current in sheep Purkinje fibres. *Pflügers Archiv.* 394:256–263.

Glynn, I. M. 1985. *In* The Enzymes of Biological Membranes. 2nd ed. Membrane Transport, Vol. 3. A. N. Martonosi, editor. Plenum Publishing Corp., New York. 35–114.

Glynn, I. M. 1988. Overview: The coupling of enzymatic steps to the translocation of sodium and potassium. *In* The Na^+,K^+-Pump, Part A: Molecular Aspects. J. C. Skou, J. G. Nørby, A. B. Maunsbach, and M. Esmann, editors. Alan R. Liss, Inc., New York. 435–460.

Glynn, I. M., and J. F. Hoffman. 1971. Nucleotide requirements for sodium: sodium exchange catalyzed by the sodium pump in human red cells. *Journal of Physiology.* 218:239–256.

Goldshlegger, R., S. J. D. Karlish, A. Rephaeli, and W. D. Stein. 1987. The effect of membrane potential on the mammalian sodium-potassium pump reconstituted into phospholipid vesicles. *Journal of Physiology.* 387:331–355.

Hamill, O. P., A. Marty, E. Neher, B. Sakmann, and F. J. Sigworth. 1981. Improved patch-clamp techniques for high resolution current recordings from cells and cell-free membrane patches. *Pflügers Archiv.* 391:85–100.

Hansen, U.-P., D. Gradmann, D. Sanders, and C. L. Slayman. 1981. Interpretation of current–voltage relationships for "active" ion transport systems: I. Steady-state reaction kinetic analysis of class-I mechanisms. *Journal of Membrane Biology.* 63:165–190.

Hansen, U.-P., J. Tittor, and D. Gradmann. 1983. Interpretation of current–voltage relationships for "active" ion transport systems: II. Nonsteady-state reaction kinetic analysis of class-I mechanisms with one slow time-constant. *Journal of Membrane Biology.* 75:141–169.

Isenberg, G., and U. Klöckner. 1982. Calcium tolerant ventricular myocytes prepared by preincubation in a "KB Medium." *Pflügers Archiv.* 395:6–18.

Karlish, S. J. D., R. Goldschleger, Y. Shahak, and A. Rephaeli. 1988. Charge transfer by the Na/K pump. 5th International Conference on Na,K-ATPase, Aarhus, Denmark. Alan R. Liss, Inc., New York. 519–524.

Karlish, S. J. D., and J. H. Kaplan. 1985. Pre-steady-state kinetics of Na^+ transport through the Na,K-pump. *In* The Sodium Pump: 4th International Conference on Na,K ATPase. I. M. Glynn and C. Ellory, editors. The Company of Biologists Ltd., Cambridge, UK. 501–506.

Karlish, S. J. D., A. Rephaeli, and W. D. Stein. 1985. Transmembrane modulation of cation transport by the Na,K-pump. *In* The Sodium Pump: 4th International Conference on Na,K ATPase. I. M. Glynn and C. Ellory, editors. The Company of Biologists Ltd., Cambridge, UK. 487–499.

Kimura, J., S. Miyamae, and A. Noma. 1987. Identification of sodium–calcium exchange current in single ventricular cells of guinea-pig. *Journal of Physiology.* 384:199–222.

Klodos, I., and B. Forbush III. 1988. Rapid conformational changes of the Na/K pump revealed by a fluorescent dye, RH-160. *Journal of General Physiology.* 92:46*a*. (Abstr.)

Läuger, P., and H.-J. Apell. 1986. A microscopic model for the current–voltage behaviour of the Na/K pump. *European Biophysical Journal.* 13:309–321.

Läuger, P., and H.-J. Apell. 1988. Transient behaviour of the Na/K pump: microscopic analysis of nonstationary ion-translocation. *Biochimica et Biophysica Acta.* 944:451–464.

Matsuda, H., and A. Noma. 1984. Isolation of calcium current and its sensitivity to monovalent cations in dialyzed ventricular cells of guinea pig. *Journal of Physiology.* 357:553–573.

Matsuda, H., A. Noma, Y. Kurachi, and H. Irisawa. 1982. Transient depolarization and spontaneous voltage fluctuations in isolated single cells from guinea pig ventricles. *Circulation Research.* 51:142–151.

Mitchell, P. 1969. Chemiosmotic coupling and energy transduction. *Theoretical and Experimental Biophysics.* 2:159–216.

Nagel, G., K. Fendler, E. Grell, and E. Bamberg. 1987. Na^+ currents generated by the purified (Na^+K^+)-ATPase on planar lipid membranes. *Biochimica et Biophysica Acta.* 901:239–249.

Nagel, G. A., M. S. Suenson, M. N. Nakao, and D. C. Gadsby. 1991. Transient and stationary Na/K pump currents in guinea pig ventricular myocytes at external sodium concentrations up to 250 mM. *Biophysical Journal.* 59:340*a*. (Abstr.)

Nakao, M., and D. C. Gadsby. 1986. Voltage dependence of Na translocation by the Na/K pump. *Nature.* 323:628–630.

Nakao, M., and D. C. Gadsby. 1989. [Na] and [K] dependence of the Na/K pump current–voltage relationship in guinea-pig ventricular myocytes. *Journal of General Physiology.* 94:539–565.

Post, R. L., S. Kume, T. Tobin, B. Orcutt, and A. K. Sen. 1969. Flexibility of an active center in sodium-plus-potassium adenosine triphosphatase. *Journal of General Physiology.* 55:306s–326s.

Rakowski, R. F. 1991. Stoichiometry and voltage dependence of the Na^+/K^+ pump in squid giant axons and *Xenopus* oocytes. *In* The Sodium Pump: Structure, Mechanism, and Regulation. J. H. Kaplan and P. De Weer, editors. The Rockefeller University Press, New York. 339–353.

Rakowski, R. F., D. C. Gadsby, and P. De Weer. 1989. Stoichiometry and voltage dependence of the sodium pump in voltage-clamped, internally dialyzed squid giant axon. *Journal of General Physiology.* 93:903–941.

Rakowski, R. F., and C. L. Paxson. 1988. Voltage dependence of the Na/K pump current in *Xenopus* oocytes. *Journal of Membrane Biology.* 106:173–182.

Rakowski, R. F., L. A. Vasilets, J. La Tona, and W. Schwarz. 1991. A negative slope in the current–voltage relationship of the Na^+/K^+ pump in *Xenopus* oocytes produced by reduction of external [K^+]. *Journal of Membrane Biology.* 121:177–187

Rephaeli, A., D. E. Richards, and S. J. D. Karlish. 1986*a*. Conformational transitions in fluorescein-labeled (Na,K) ATPase reconstituted into phospholipid vesicles. *Journal of Biological Chemistry.* 261:6248–6254.

Rephaeli, A., D. E. Richards, and S. J. D. Karlish. 1986*b*. Electrical potential accelerates the $E_1P(Na)$–E_2P conformational transition of (Na,K)-ATPase in reconstituted vesicles. *Journal of Biological Chemistry.* 261:12437–12440.

Reynolds, J. A., E. A. Johnson, and C. Tanford. 1985. Incorporation of membrane potential into theoretical analysis of electrogenic ion pumps. *Proceedings of the National Academy of Sciences, USA.* 82:6869–6873.

Sato, R., A. Noma, Y. Kurachi, and H. Irisawa. 1985. Effects of intracellular acidification on membrane currents in ventricular cells of the guinea pig. *Circulation Research.* 57:553–561.

Soejima, M., and A. Noma. 1984. Mode of regulation of the ACh-sensitive K-channel by the muscarinic receptor in rabbit atrial cells. *Pflügers Archiv.* 400:424–431.

Stürmer, W., H.-J. Apell, I. Wuddel, and P. Läuger. 1989. Conformational transitions and charge translocation by the Na,K pump: comparison of optical and electrical transients elicited by ATP-concentration jumps. *Journal of Membrane Biology.* 110:67–86.

Stürmer, W., R. Bühler, H.-J. Apell, and P. Läuger. 1991. Charge translocation by the Na,K-pump. II. Ion binding and release at the extracellular face. *Journal of Membrane Biology.* 121:163–176.

Thomas, R. C. 1972. Electrogenic sodium pump in nerve and muscle cells. *Physiological Reviews.* 52:563–594.

List of Contributors

Anthony Bahinski, The Rockefeller University, New York, New York

Ernst Bamberg, Max-Planck-Institut für Biophysik, Frankfurt am Main, Germany

Andrew Barnstein, Department of Physiology, Vanderbilt University School of Medicine, Nashville, Tennessee

Rhoda Blostein, Departments of Medicine and Biochemistry, McGill University, Montreal, Quebec, Canada

Harm H. Bodemann, Department of Cellular and Molecular Physiology, Yale University School of Medicine, New Haven, Connecticut

Thomas J. Callahan, Department of Cellular and Molecular Physiology, Yale University School of Medicine, New Haven, Connecticut

Victor A. Canfield, Department of Cell Biology, Yale University School of Medicine, New Haven, Connecticut

Elena Chertova, Shemyakin Institute of Bioorganic Chemistry, USSR Academy of Sciences, Moscow, USSR

Dar Chow, Department of Molecular and Cell Biology, University of California, Berkeley, California

Flemming Cornelius, Institute of Biophysics, University of Aarhus, Aarhus, Denmark

Julia Dorfman, Department of Biochemistry, McGill University, Montreal, Quebec, Canada

Roman Efremov, Shemyakin Institute of Bioorganic Chemistry, USSR Academy of Sciences, Moscow, USSR

Andreas Eisenrauch, Max-Planck-Institut für Biophysik, Frankfurt am Main, Germany

Douglas M. Fambrough, Department of Biology, The Johns Hopkins University, Baltimore, Maryland

Klaus Fendler, Max-Planck-Institut für Biophysik, Frankfurt am Main, Germany

Bliss Forbush III, Department of Cellular and Molecular Physiology, Yale University School of Medicine, New Haven, Connecticut

John G. Forte, Department of Molecular and Cell Biology, University of California, Berkeley, California

Jeffrey P. Froehlich, Laboratory of Cardiovascular Science, National Institute on Aging, National Institutes of Health, Baltimore, Maryland

David C. Gadsby, The Rockefeller University, New York, New York

K. Geering, Institut de Pharmacologie et Toxicologie de l'Université de Lausanne, Lausanne, Switzerland

R. Goldshleger, Department of Biochemistry, Weizmann Institute of Science, Rehovot, Israel

Ernst Grell, Max-Planck-Institut für Biophysik, Frankfurt am Main, Germany

Charles M. Grisham, Department of Chemistry, University of Virginia, Charlottesville, Virginia

Philippe Gros, Department of Biochemistry, McGill University, Montreal, Quebec, Canada

Rachel W. Hammerton, Department of Molecular and Cellular Physiology, Stanford University School of Medicine, Stanford, California

Maura Hamrick, Department of Biology, The Johns Hopkins University, Baltimore, Maryland

Hans Hebert, Structural Biochemistry Unit, CNT, Novum, Karolinska Institutet, Huddinge, Sweden

Joseph F. Hoffman, Department of Cellular and Molecular Physiology, Yale University School of Medicine, New Haven, Connecticut

Elisabeth M. Inman, Department of Biology, The Johns Hopkins University, Baltimore, Maryland

Jørgen Jensen, Institutes of Biophysics and Physiology, University of Aarhus, Aarhus, Denmark

Peter Leth Jørgensen, Biomembrane Research Center, August Krogh Institute, Copenhagen University, Copenhagen, Denmark

Jack H. Kaplan, Department of Physiology, University of Pennsylvania, Philadelphia, Pennsylvania

S. J. D. Karlish, Department of Biochemistry, Weizmann Institute of Science, Rehovot, Israel

Masaru Kawamura, Department of Biology, University of Occupational and Environmental Health, Kitakyushu, Japan

Cindy Klevickis, Department of Chemistry, University of Virginia, Charlottesville, Virginia

Irena Klodos, Institute of Biophysics, University of Aarhus, Aarhus, Denmark

Bruce C. Kone, Department of Biology, The Johns Hopkins University, Baltimore, Maryland

Theresa A. Kuntzweiler, Department of Chemistry, University of Virginia, Charlottesville, Virginia

P. Läuger, Department of Biology, University of Konstanz, Konstanz, Germany

Richard M. Lebovitz, Department of Biology, The Johns Hopkins University, Baltimore, Maryland

M. Victor Lemas, Department of Biology, The Johns Hopkins University, Baltimore, Maryland

Robert Levenson, Department of Cell Biology, Yale University School of Medicine, New Haven, Connecticut

Jerry B. Lingrel, Department of Molecular Genetics, Biochemistry and Microbiology, University of Cincinnati College of Medicine, Cincinnati, Ohio

Svetlana Lutsenko, Shemyakin Institute of Bioorganic Chemistry, USSR Academy of Sciences, Moscow, USSR

Reinaldo Marín, Department of Cellular and Molecular Physiology, Yale University School of Medicine, New Haven, Connecticut

Arvid B. Maunsbach, Department of Cell Biology at the Institute of Anatomy, and the Biomembrane Research Center, University of Aarhus, Aarhus, Denmark

Helen McNeill, Department of Molecular and Cellular Physiology, Stanford University School of Medicine, Stanford, California

Mark Milanick, Department of Cellular and Molecular Physiology, Yale University School of Medicine, New Haven, Connecticut

Nikolai Modyanov, Shemyakin Institute of Bioorganic Chemistry, USSR Academy of Sciences, Moscow, USSR

Masakazu Nakao, The Rockefeller University, New York, New York

W. James Nelson, Department of Molecular and Cellular Physiology, Stanford University School of Medicine, Stanford, California

Shunsuke Noguchi, Department of Biology, University of Occupational and Environmental Health, Kitakyushu, Japan

Jens G. Nørby, Institutes of Biophysics and Physiology, University of Aarhus, Aarhus, Denmark

Curtis T. Okamato, Department of Molecular and Cell Biology, University of California, Berkeley, California

John Orlowski, Department of Molecular Genetics, Biochemistry and Microbiology, University of Cincinnati College of Medicine, Cincinnati, Ohio

Bhavani G. Pathak, Department of Molecular Genetics, Biochemistry and Microbiology, University of Cincinnati College of Medicine, Cincinnati, Ohio

Carlomaria Polvani, Departments of Medicine and Biochemistry, McGill University, Montreal, Quebec, Canada

Robert L. Post, Department of Molecular Physiology and Biophysics, Vanderbilt University Medical School, Nashville, Tennessee

Elmer M. Price, Department of Molecular Genetics, Biochemistry and Microbiology, University of Cincinnati College of Medicine, Cincinnati, Ohio

R. F. Rakowski, Department of Physiology and Biophysics, University of Health Sciences/The Chicago Medical School, North Chicago, Illinois

Karen J. Renaud, Department of Biology, The Johns Hopkins University, Baltimore, Maryland

Jayson Rome, Department of Biology, The Johns Hopkins University, Baltimore, Maryland

John R. Sachs, Department of Medicine, State University of New York at Stony Brook, Stony Brook, New York

Wolfgang Schwarz, Max-Planck-Institut für Biophysik, Frankfurt/Main, Germany

Elisabeth Skriver, Department of Cell Biology at the Institute of Anatomy, and the Biomembrane Research Center, University of Aarhus, Aarhus, Denmark

Delores Somerville, Department of Biology, The Johns Hopkins University, Baltimore, Maryland

W. D. Stein, Institute of Life Science, Hebrew University, Jerusalem, Israel

John M. McD. Stewart, Department of Chemistry, University of Virginia, Charlottesville, Virginia

Kuniaki Suzuki, Department of Molecular Physiology and Biophysics, Vanderbilt University Medical School, Nashville, Tennessee

Kathleen J. Sweadner, Neurosurgical Research, Massachusetts General Hospital, Boston, Massachusetts; and Department of Cellular and Molecular Physiology, Harvard Medical School, Boston, Massachusetts

Kunio Takeyasu, Department of Physiology, The University of Virginia School of Medicine, Charlottesville, Virginia

D. M. Tal, Department of Biochemistry, Weizmann Institute of Science, Rehovot, Israel

Michael M. Tamkun, Department of Physiology, Vanderbilt University School of Medicine, Nashville, Tennessee

Joseph P. Taormino, Department of Biology, The Johns Hopkins University, Baltimore, Maryland

Larisa A. Vasilets, Max-Planck-Institut für Biophysik, Frankfurt/Main, Germany

Barry A. Wolitzky, Department of Molecular Genetics, Hoffmann-LaRoche Inc., Nutley, New Jersey